S0-AWV-405

Joachim K. Seydel and Michael Wiese

Drug-Membrane Interactions

Methods and Principles in Medicinal Chemistry

Edited by
R. Mannhold
H. Kubinyi
G. Folkers

Joachim K. Seydel and Michael Wiese

Drug-Membrane Interactions

Analysis, Drug Distribution, Modeling

Series Editors

Prof. Dr. Raimund Mannhold
Biomedical Research Center
Molecular Drug Research Group
Heinrich-Heine-Universität
Universitätsstraße 1
40225 Düsseldorf
Germany
raimund.mannhold@uni-duesseldorf.de

Prof. Dr. Hugo Kubinyi
BASF AG Ludwigshafen
c/o Donnersbergstraße 9
67256 Weisenheim am Sand
Germany
kubinyi@t-online.de

Prof. Dr. Gerd Folkers
Department of Applied Biosciences
ETH Zürich
Winterthurer Str. 190
8057 Zürich
Switzerland
folkers@pharma.ethz.ch

Authors

Prof. Dr. Joachim K. Seydel
Research Center Borstel
Center for Medicine and Biosciences
Parkallee 1-40
23845 Borstel
Germany
jseydel@fz-borstel.de

Prof. Dr. Michael Wiese
Institute of Pharmaceutical Chemistry
Rheinische Friedrich-Wilhelms-Universität
An der Immenburg 4
53121 Bonn
Germany
m.wiese@uni-bonn.de

Library of Congress Card No. applied for.

British Library Cataloguing-in-Publication Data:
A catalogue record for this book is available from the British Library.

Die Deutsche Bibliothek – CIP Cataloguing-in-Publication Data:
A catalogue record for this publication is available from Die Deutsche Bibliothek

Printed in the Federal Republic of Germany

Printed on acid-free paper

Typesetting TypoDesign Hecker GmbH, Leimen
Printing betz-druck GmbH, Darmstadt
Bookbinding Großbuchbinderei
J. Schäffer GmbH & Co. KG, Grünstadt

ISBN 3-527-30427-4

Contents

Preface

There is no life without cells and there are no cells without membranes. The cell membrane of animal cells is very important in that it protects the organelles and also keeps undesired particles from entering the cell body. Proteins span the membrane and touch both the inside and outside of the cell. Their function is to interact with molecules outside the cell which includes the ability to serve as protein receptors for hormones, to bind to other cells in wound healing and in the immune response, and to transport molecules into and out of the cell. What is outlined in these few sentences, however, is a very complex biological system which has been described only phenomenologically for a long time.

Structurization, organization, compartmentation within a living organism as well as its shape are functions of membranes. But structuring a living organism by compartmentation implicates communication between the different compartments. Hence membranes have to allow for or even enable communication. Membranes are bilayers made up of phospholipid molecules. Whether they cover organelles, as the endoplasmatic reticulum within a cell, or whether they form the cell wall itself, membranes have to provide communication and transport tools to serve the cell or its organelles with the "necessities of life", which are information and energy.

Proteins, often highly glycosylated and integrated into the membranes are responsible for most of the functions. But what we have learned from a more detailed inspection of these functional proteins is that they are by far not independent from the membrane itself. There is obvious cooperativity. Caveolae, small surface pits in the plasma membrane, already known since the 1950s, have become known to be initiated and formed by a protein named caveolin, which is tightly complexed to the phosphate bilayer by palmitoyl moieties sticking into the outer leaflet of the bilayer (M. Drab et al., Science 293, 2449 (2001)). Those caveolae are thought to be involved in endothelial transcytosis, lipid regulation and several signalling processes. Cholesterol, as an important stabilizing factor of membranes, controls the production of caveolin and hereby surmounts by far the "simply biophysical" function that it was thought to have for a long time.

Ion channels - regulated by membrane potential or ligands - provide and support complex signalling processes in nerve cells, surrounded by membranes that act as insulators all the way down the signal flow to the target or as communicators at the synapses. We have learned to interfere with these processes with modern therapeutics, mostly drugs targeted to membrane receptors and ion channels. Much less is known about the membrane itself as a target. If it regulates activity of proteins embedded in membranes and/or their function, then the membrane might be a target for therapeutics as well.

Calcium antagonistic dihydropyridines are thought to penetrate the membrane and travel in the bilayer until they hit their membrane integrated target protein. Thus, the membranes even act as an ingenious tool to reduce the dimensionality of navigation for drugs, bringing it down to a two-dimensional problem of travelling within the membrane instead of the three-dimensional navigation in the cytosol or the interstitium.

Having talked so far about human or animal cells, there are different examples from the other kingdoms of life. Most common antibiotics interfere with the bacterial cell wall, which is much different from the animal cell wall. Consisting of a mesh-type peptidoglycan polymer, this murein sacculus is wrapped around a "normal" phospholipid bilayer and forms a double cell wall for optimized protection of the bacteria from outside. In addition, it provides the organism with the necessary rigidity to withstand considerable osmotic pressure. According to the type of bacteria the cell wall has different architecture and even mediates important mechanisms of resistance. Hence bacterial cell walls are an excellent target for antimicrobial therapy and parts of this book show how traditional enzyme inhibitors interfere with bacterial membrane biosynthesis and exert a synergistic action.

Fungi use an even more sophisticated strategy to protect themselves by integrating chitin, the major constituent of the insect exoskeleton, or sometimes cellulose, into their cell wall. Chitin synthase, a protein residing in the phospholipid bilayer, is obvoiusly controlled by membrane tension, which again is a function of sterol concentration in the fungal membrane. If one hits ergosterol biosynthesis with lanosterol demethylase inhibitors or squalene epoxidase inhibitors, one changes the membrane stability of the fungal cell and hence stops chitin synthesis.

What, in contrast, is the relevance of artificial membranes, of phospholipid bilayers? Do such systems disclose enough or the right biophysical properties of real membranes that would enable us to design compounds interacting with such sophisticated targets like membranes? Those are the questions that the volume on *Drug-Membrane Interactions* by Joachim Karl Seydel and Michael Wiese is tracing. After a short introduction into membrane architecture and properties, biophysical analytics is the main topic of chapters 2 and 3. The authors go deep into the discussion of octanol/water modelling of partitioning of compounds into membranes. The fourth chapter is devoted to transport whereas target interaction and hence the pharmacodynamics is reflected in chapter 5. An outlook is given in chapter 6, where virtual membranes play the main role. Michael Wiese gives us the state of the art of modelling membranes and their interactions with ligands.

The editors are grateful to the authors that they have devoted their precious time to compile and structure the huge amount of information on that topic. Gudrun Walter and Frank Weinreich from Wiley-VCH did a very good job in producing this volume.

As we learn more and more about the great importance of membranes and their functions for understanding biological mechanisms, this book will have benefits not only for scientists interested in drug development, but to everyone involved in life science research.

January 2002

Raimund Mannhold, Düsseldorf
Hugo Kubinyi, Ludwigshafen
Gerd Folkers, Zürich

Foreword

In recent decades, it has become increasingly clear that knowledge of drug-membrane interactions is essential for the understanding of drug activity, selectivity, and toxicity. At the same time, there has been a large increase in the number of physicochemical analytical methods available for analyzing and quantifying various aspects of drug-membrane interactions.

In my laboratory we became interested in the role of drug-membrane interactions in multidrug resistance to antibacterial and antitumor therapy and in the explanation for toxic effects (lipidosis). We found that drug resistance could not be explained solely by changes in target proteins or, in the case of lipidosis, by the octanol-water partition coefficient, but was dependent also on the degree of interaction with membrane constituents. Later, the neuroleptic activity of flupirtin analogs was found to be better explained by their degree of membrane interaction than by their octanol-water partitioning.

Findings on the interaction of chemicals with biological membranes or model liposomes and methods for studying such interactions are mainly published in journals of biophysics or biochemistry. Only a few papers are published in journals of medicinal chemistry or pharmaceutical sciences. Even monographs on medicinal chemistry often lack some detailed information on this subject.

It has commonly been assumed that transfer processes can be modeled in terms of simple bulk-phase thermodynamics. However, in many circumstances this assumption seems to be incorrect. Bulk thermodynamics cannot be applied when the solutes (especially amphiphilic drugs) partition into amphiphilic aggregates such as bilayer membranes. It is important to remember that a bilayer consisting of phospholipids is a "solvent" with an interfacial phase and a high surface/volume ratio.

It is astonishing that it has taken such a long time to appreciate the importance of drug-phospholipid interactions in membranes for cell functioning and drug action. This despite the fact that, as Thudicum stated as long ago as 1884, "Phospholipids are the centre, life and chemical soul of all bioplasm whatsoever, that of plants as well as of animals."

Cell membranes, composed of lipids and proteins, function as a permeability barrier, maintain ion gradients across the membrane and steady state of fluxes, and possess recognition sites for communication and interaction with other cells. The lipid composition of membranes determines the organization of proteins in the membrane.

Disturbance of such a complex system by the uptake of exogenous compounds, drugs, or lipids, such as cholesterol, which differ in their structure and physicochemical properties can lead to changes in membrane fluidity and/or permeability, to phase separation, and to domain formation. Alterations in such properties can induce severe changes in the performance of the cell, including the function of transmembrane receptor proteins and proteins responsible for signal transduction. The degree of membrane alteration will depend on both the structure and physicochemical properties of the drug molecules and lipids involved.

The aims of this book are to highlight and summarize for medicinal and pharmaceutical chemists some important properties of phospholipid bilayers; to explain, using examples, analytical tools for determining thermotropic and dynamic membrane properties and the possible effects of drugs on such membrane properties; and, finally, to discuss examples of the importance of drug-membrane interactions for drug pharmacokinetics (absorption, distribution, accumulation) as well as drug efficacy, selectivity, and toxicity.

This is not a book written for particular specialists, as aspects of many different fields are considered. Rather, the intention was to highlight the importance for drug-membrane interaction of membrane composition and the dynamic molecular organization of membranes and to point out the effects of such interactions on membrane properties (Chapter 1). The short description of membrane properties and the possible changes which can arise from drug-membrane interactions may help readers to understand that membranes are not just material but a collection of chemicals. Membranes are asymmetric (the two layers can be compared to a bimetallic strip), and this results in a differential tension across the membrane. The tension can be reduced or increased by drug-membrane interactions. Such interactions are not always sufficiently described by the octanol-water partition coefficient, but can easily be followed in liposomal membrane models (Chapter 2). Several tools are available to analyze and quantify the various aspects of drug-membrane interactions. Such interactions not only alter the physicochemical properties of membranes, but also influence and determine drug localization, orientation, and conformation within the membrane (Chapter 3). In consequence, drug-membrane interaction can influence drug transport, absorption, distribution, selectivity, efficacy, and resistance. This is detailed in Chapters 4 and 5. Finally, tools and examples to model such interactions are outlined in Chapter 6 by M. Wiese. The aim of this book will be achieved if it inspires medicinal chemists to look more frequently into possible effects of drug-membrane interaction in drug research and development.

The wide variety of fields of research involved and the continuing flood of new papers and results made it necessary to select examples. Thus, I apologize if I have overlooked papers equally or even more important than those cited.

Finally, I would like to thank my colleagues, Prof. Dr. Ilza Pajeva of the Bulgarian Academy of Sciences and Dr. K.-J. Schaper of the Research Center Borstel, for their support and fruitful discussions, and Mrs. Bouchain for designing figures and graphs. Thanks are also due to my wife, Frauke, for proofreading, my daughter, Dr. Wiebke Seydel, for help in improving the English, and to both for their cheerful encouragement of my efforts to overcome so many difficulties throughout this work.

August 2001

Joachim K. Seydel
Research Center Borstel

List of Abbreviations

3-D	three-dimensional
ADR	adriamycin
ANN	artificial neural network
AS	*o*-acetylsalicylic acid
ATP	adenosine triphosphate
AZT	3'-azido-3'-deoxythymidine
BBPS	bovine brain phosphatidylserine
CAC	critical aggregate concentration
CF	carboxyfluorescein
CMC	critical micelle concentration
CNS	central nervous system
CSF	cerebrospinal fluid
CVFF	consistent valence force field
DAG	diacylglycerol
DEPE	dieladoylphosphatidylserine
DHEA	dehydroepiandrosterone
DHFR	dihydrofolate reductase
DLPC	dilauroylphosphatidylcholine
DMPC	dimyristoylphosphatidylcholine
DODCI	3,3'-diethyloxadicarboxyamine iodide
DOPC	dioleoylphosphatidylcholine
DOPE	dioleoylphosphatidylethanolamine
DOPS	1,2-dioleylphosphatidylserine
DPG	dipalmitoyl-glycerol
DPH	1,6-diphenyl-1,3,5-hexatriene
DPPA	dipalmitoylphosphatidic acid
DPPC	dipalmitoylphosphatidylcholine
DPPE	dipalmitoylphosphatidylethanolamine
DPPG	dipalmitoylphosphatidylglycerol
DPPS	dipalmitoylphosphatidylserine
DSC	differential scanning calorimetry
DSPC	distearoylphosphatidylcholine
ESR	electron spin resonance
FTIR	Fourier transform infrared

HBD	hydrogen bond donor
HMGR	HMG-CoA reductase
HPLC	high-performance liquid chromatography
5-HT	5-hydroxytryptamine
IAM	immobilized artificial membrane
ILC	immobilized liposome chromatography
ITC	isothermal titration calorimetry
log D	log P at physiological pH
LPS	lipopolysaccharide
LUV	large unilamellar vesicle
MAS	magic angle spinning
MC	Monte Carlo
MD	molecular dynamics
MES	maximal electro shock
MIC	minimum inhibitory concentration
MDR	multidrug resistance
MLR	multiple linear regression
NMR	nuclear magnetic resonance
NOE	nuclear Overhauser effect
NSAID	non-steroidal anti-inflammatory drug
OPLS	optimized parameters for liquid systems
PAMPA	parallel artificial membrane permeation assay
PC-HTS	physicochemical high-throughput screening system
PGDP	propylenglycoldipelargonate
PGE	prostaglandin E_1
PKC	protein kinase C
PL	palmitoyllecithin
PLA	phospholipase A_2
PLS	partial least squares
POPC	palmitoyloleoylphosphatidylcholine
POPS	palmitoyloleoylphosphatidylserine
PA	phosphatidic acid
PC	phosphatidylcholine
PE	phosphatidylethanolamine
PI	phosphatidylinositol
PS	phosphatidylserine
PSA	polar surface area
Py3Py	1,3-bis(1-pyrene)propane
QLS	quasielastic light scattering
QSAR	quantitative structure-activity relationship
RP-HPLC	reversed phase high performance liquid chromatography
SAR	structure-activity relationship
SPM	synaptosomal plasma membrane
SPR	surface plasmon resonance
THC	(–)-Δ^8-tetrahydrocannabinol

THF	tetrahydrofuran
TOE	transient Overhauser effect
TMC	N-trimethyl-chitosan
TPSA	total polar surface area

Introduction

Joachim K. Seydel

Interest in drug design has focussed mainly on the interaction of ligand molecules with proteins, in the form of specific receptors and enzymes. Most of the target proteins are embedded in membranes, and it is assumed that the biological activity of ligands arises as a result of binding to the membrane-embedded proteins. The lipid environment is considered to play a more passive role. There is, however, increasing evidence that the influence of ligand–membrane interaction on drug activity and selectivity has been underestimated. The so-called "non-specific" interaction of drugs with membrane constituents in fact involves an interaction with specific phospholipid structures. Although the lipid layer is a dynamic fluid, it is highly organized. Membranes do not consist of lipids only, but possess polarized phosphate groups and neutral or positively or negatively charged head groups, and they are highly structured and chiral. Interaction with such structures can have a decisive influence on drug partitioning, orientation, and conformation. It also influences the physicochemical properties and functioning of the membrane. Thus, drug–membrane interactions play an important role in drug transport, distribution, accumulation, efficacy, and resistance.

The perturbation of biological membranes by various classes of drugs can lead to changes in membrane curvature or to phase separation and thus to changes in protein conformation. Therefore, drug-membrane interactions are an important factor in drug action. At the macroscopic level, ligand–membrane interactions are manifested as changes in the physical and thermodynamic properties of "pure" membranes or bilayers. Depending on the composition of the membrane and the structure of the ligand molecules, the interaction can favor or prevent drug activity or toxicity.

Fortunately, most of the perturbations that can occur in complex biological membranes upon interaction with drug molecules can be studied and simulated *in vitro* and quantified by available physicochemical techniques, using as a model artificial membranes (bilayers, liposomes), which are readily created.

1
Function, Composition, and Organization of Membranes

Joachim K. Seydel

1.1
The Physiology of Cells and the Importance of Membranes for their Function

All living cells are surrounded by one or several membranes. The membrane defines the cell as a living unit and separates the cell from its surroundings; it separates intracellular from extracellular domains. Highly differentiated organisms are comparable to a federation of cells in which groups of cells are specialized in particular functions and are connected through complex communication networks. Any disorder in the communication of such complex systems influences the functioning of the organism. It reduces the readiness for reactions, decreases the ability to adapt to changes in the environmental conditions, and can, finally, lead to reduction in efficiency or to death.

Cells can communicate with each other in three ways:

1) By direct contact through a nexus or "gap junction" (which is involved in the transport of material from cell to cell and the transfer of electrical signals). This type of communication requires the cells to be in direct contact with each other.
2) Via "receptors", for example sugar molecules, positioned on the cell surface, which allow contact of cells and the initiation of reciprocal contact. A precondition for this type of communication is that at least one of the two cells is mobile and can approach the other cell.
3) By secretion of chemical compounds (cytokines, hormones, transmitters) that can be perceived as a signal by another cell at a certain distance.

The outer membrane, the plasmalemma, efficiently protects the cell from the environment while, at the same time, carrying out functions important for cell metabolism: the uptake of substrates and the elimination of toxic compounds. Substrate exchange with the environment is controlled by transport proteins embedded in the membrane (energy-requiring pumps such as Na^+,K^+-ATPase, or other transport units such as the Na^+/glucose cotransporter and sodium and calcium ion channels) [1].

It seems miraculous that a membrane about 10 nm thick can preserve extreme gradients of intra- and extracellular ions, amino acid and protein concentration (Figure 1.1). For example, the ratio of intracellular to extracellular ion concentration for Na^+ is 10:140 and for Ca^{2+} is 0.0001:2.5. Transmembrane concentration gradients of solvents, ions, pH, etc. are essential for cellular functions, for example the production of ATP, which cannot occur in the absence of a transmembrane gradient.

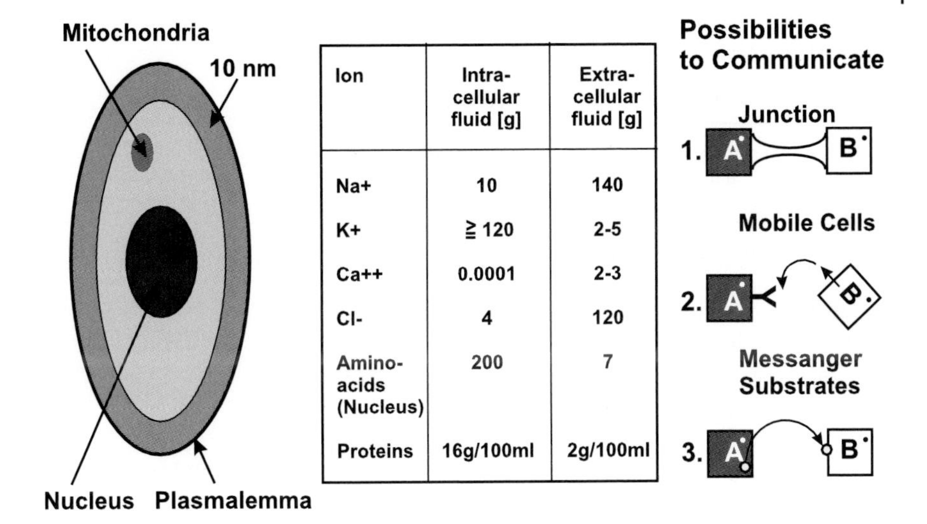

Ion	Intra-cellular fluid [g]	Extra-cellular fluid [g]
Na+	10	140
K+	≧ 120	2-5
Ca++	0.0001	2-3
Cl-	4	120
Amino-acids (Nucleus)	200	7
Proteins	16g/100ml	2g/100ml

Fig. 1.1 Schematic drawing of cell construction, concentration gradients for some ions and metabolites, and different methods of cell communication.

Another property of cell membranes in addition to compartmentalization is their ability to fuse. This is important for intracellular vesicle transport between intracellular organelles as well as, for example, for the fusion of enveloped viruses with target cell membranes.

1.2 Composition and Organization of Membranes

1.2.1 Mammalian Membranes

Membranes consist mainly of proteins and lipoids, these frequently having quite different functions. Proteins determine the functional properties of the membrane, lipoids the matrix, for example the construction. In principle, membranes consist of a phospholipid bilayer into which proteins are integrated. Phospholipid molecules consist of two long-chain fatty acids, each of which is esterified with one of the hydroxy groups of glycerol; the third hydroxy group of glycerol is connected to a phosphoric acid which is substituted by another substructure, for example choline (phosphatidylcholine), serine or sugar. This arrangement confers an amphiphilic character on the phospholipids. The apolar fatty acid chains are lipophilic, whereas the polar head groups and the polarized phosphate groups are hydrophilic. As a result of these properties, phospholipids associate readily in water to form the characteristic bilayers, with the polar head groups directed to the surrounding water and the fatty acid chains turned toward each other and directed to the inner part of the bilayer.

The stability of the bilayer depends on the segregation of the hydrocarbon residues from the watery phase and the polar interaction of the head groups with water, i.e. on

Tab. 1.1 Phospholipid charged groups and their electrostatic properties. (Adapted from Table 2.2 of ref. 2 with permission from Macmillan)

Phospholipid ionizable groups	Electrostatic properties
1. Primary phosphate (phosphatidic acid)	pK_1 3.9, pK_2 8.3
2. Secondary phosphate (phosphatidylinositol, cardiolipin)	$pK_a < 2.0$
3. Secondary phosphate + quaternary amine (sphingomyelin, phosphatidylcholine)	Isoelectric in pH range 3–10
4. Secondary phosphate + amine (phosphatidylethanolamine)	Net negative at pH 7.4
5. Secondary phosphate + amine+ carboxyl (phosphatidylserine)	Net negative at pH 7.4

the type and charge of the head groups. In addition, the phosphate groups bear a partial negative charge. Dissociation occurs at about pH 2–3 (see Table 1.1) [2].

Strong polar interactions also occur between carbohydrates attached to glycolipids and certain lipids, such as phosphatidylinositol. In membranes containing charged lipids, electrostatic repulsion occurs between similarly charged leaflets on either side of the bilayer. This prevents a decrease in the thickness of the structure. In addition, electrostatic repulsion supports lateral cohesion between the hydrocarbon chains and stabilizes the bilayer. The charged surface will attract oppositely charged mobile counter-ions from the aqueous phase. At equilibrium they will be distributed according to the electrostatic potential, to form the so-called electrical double layer.

Fatty acid composition, phospholipid composition, and cholesterol content can be modified in many different ways in intact mammalian cells. These changes alter membrane fluidity [3] and cell surface curvature and as a result can affect a number of cellular functions [4], including carrier-mediated transport, the activity and properties of some membrane-bound enzymes, phagocytosis, endocytosis, immunological and chemotherapeutic cytotoxicity, prostaglandin production, and cell growth. The effects of lipid modification on cellular function are very complex. Thus, it has not until now been possible to make any generalizations or to predict how a given system will respond to a particular type of lipid modification.

Cholesterol – an essential component of mammalian cells – is important for the fluidity of membranes. With a single hydroxy group, cholesterol is only weakly amphipathic. This can lead to its specific orientation within the phospholipid structure. Its influence on membrane fluidity has been studied most extensively in erythrocytes. It was found that increasing the cholesterol content restricts molecular motion in the hydrophobic portion of the membrane lipid bilayer. As the cholesterol content of membranes changes with age, this may affect drug transport and hence drug treatment. In lipid bilayers, there is an upper limit to the amount of cholesterol that can be taken up. The solubility limit has been determined by X-ray diffraction and is

66 mol% for phosphatidylcholine and 51 mol% for phosphatidylethanolamine bilayers. The decisive factor is the head group effect of phospholipids. The acyl chain has no effect on cholesterol solubility [5].

More recently, McMullen *et al.* [6] investigated the effects of cholesterol on the thermotropic organization and behavior of a series of linear saturated phosphatidylethanolamine bilayers (PE). They found, in contrast, that addition of cholesterol resulted in a progressive decrease in enthalpy and in the temperature of chain-melting transition up to a concentration of 20–30 mol%. Higher concentrations of cholesterol led to a dramatic increase in temperature and total enthalpy of chain-melting for these cholesterol–PE mixtures. It is thought that cholesterol induces chain-melting of the highly ordered crystalline phase of pure PEs. The reason for this could be the limited solubility of cholesterol in gel-state PE bilayers and its ability to facilitate the formation of cholesterol-free lamellar crystalline phase in such systems [6].

The effects of cholesterol and cholesterol-derived oxysterols on adipocyte ghost membrane fluidity has been studied. It has been found that cholesterol and oxysterols interact differently with rat adipocyte membranes. Cholesterol interacts more with phosphatidylcholine located at the outer lipid bilayer whereas, for example, cholestanone seems to interact more with phospholipids located at the inner layer [7].

Another important component of membranes is Ca^{2+} ions, which bridge negatively charged head group structures, thus stabilizing the membrane. Displacement of Ca^{2+} ions by cationic drug molecules will necessarily lead to significant changes in membrane organization and properties.

Eucaryotic cells are generally more complex than procaryotic cells and possess a variety of membrane-bound compartments called organelles. These intracellular membranes allow diverse and more specialized functions. For detailed information, the reader is recommended to consult specialized textbooks [2, 8, 9].

Fatty acyl residues (R) commonly found in membrane lipids are summarized in Table 1.2 [2]. Generally, four lipid structures are mainly found in eucaryotic cells: phospholipids, sphingolipids, glycolipids, and sterols [2]. The various organs differ in their phospholipid composition (Table 1.3). As an example, the composition of the liver cell membrane is given [2].There is also a considerable difference in the proportion of phospholipids in different cell types and in different species. Figure 1.2

Tab. 1.2 The fatty acid residues (R) commonly found in membrane lipids. (Reproduced from Table 1.3 of ref. 2, with permission from Macmillan)

C-atoms	Fatty acyl substituents (R)	Common name
12	CH_3-$(CH_2)_{10}$-COO-	Lauric
14	CH_3-$(CH_2)_{12}$-COO-	Myristic
16	CH_3-$(CH_2)_{14}$-COO-	Palmitic
16	CH_3-$(CH_2)_5$-CH=CH-$(CH_2)_7$-COO-	Palmitoleic
18	CH_3-$(CH_2)_{16}$-COO-	Stearic
18	CH_3-$(CH_2)_7$-CH=CH-$(CH_2)_7$-COO-	Oleic
18	CH_3-CH_2-CH=CH-CH_2-CH=CH-CH_2-CH=CH-$(CH_2)_7$-COO-	Linolenic
18	CH_3-$(CH_2)_4$-CH=CH-CH_2-CH=CH-$(CH_2)_7$COO-	Linoleic

Tab. 1.3 Phospholipid composition in mol% total lipids of some liver cell membranes

Phospholipid	Plasma membranes	Golgi membranes	Lysosomal membranes
Phosphatidylcholine	34.9	45.3	33.5
Phosphatidylethanolamine	18.5	17.0	17.9
Phosphatidylinositol	7.3	8.7	8.9
Phosphatidylserine	9.0	4.2	8.9
Phosphatidic acid	4.4	–	6.8
Sphingomyelin	17.7	12.3	32.9

shows the variation of sphingomyelin to phosphatidylcholine content in the total lipid fraction of erythrocytes from various species and in ox endoplasmic reticulum in various organs [2]. The lateral mobility of lipids in membranes is another essential aspect [10].

Equally important is the observation that phospholipids are asymmetrically distributed in the two leaflets of the plasma membrane of eucaryotic cells. It has been found that phosphatidylserine, phosphatidylethanolamine and phosphoinositides are principally located in the inner monolayer [11], whereas phosphatidylcholine, sphingomyelin and glycolipids are mainly located in the outer leaflet of the membrane. An example of the asymmetrical distribution of phospholipids in human red blood cells and in the membrane of influenza virus is shown in Table 1.4 [12, 13]. Asymmetry is maintained partly through the activity of enzymes responsible for lipid synthesis and partly by the activity of specific proteins called "phospholipid flippases", which catalyze the exchange of lipids between the two leaflets. The aminophospholipid translocase was first discovered in human erythrocytes and transports aminophospholipids from the outer to the inner leaflet of the plasma membranes [14, 15]. The same activity has been found in many other cell types. Proteins of the multidrug resistance (MDR) family are also flippases or floppases, which can transport not only lipids but also amphiphilic drugs from the inner to the outer

Tab. 1.4 Asymmetrical distribution of phospholipids (mol%) in membranes of influenza viruses [13] and in red blood cells [12]

Influenza virus (outside/inside)				
TPL	SH	PC	PE	PS
30/73	6/26	6/7	11/28	4/12
Red blood cells (outside/inside)				
51/51	20/5	22/9	6/25	0/9

TPL, total phospholipid; SH, sphingomyelin; PC, phosphatidylcholine; PE, phosphatidylethanolamine; PS, phosphatidylserine.

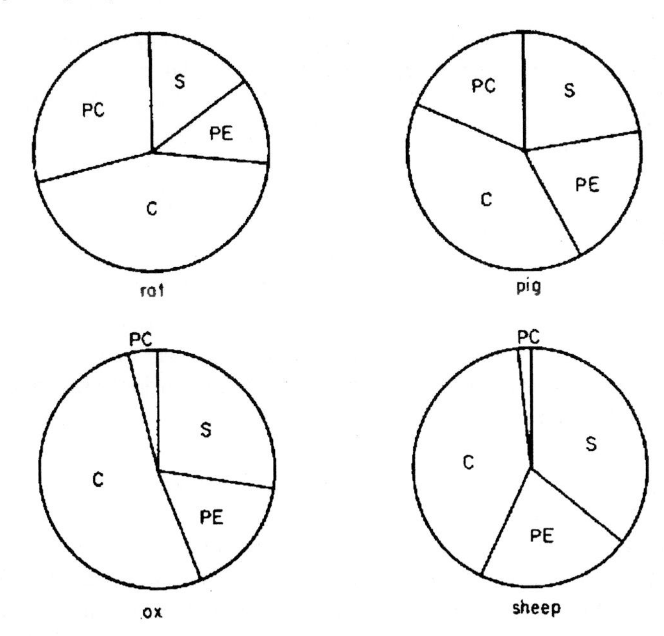

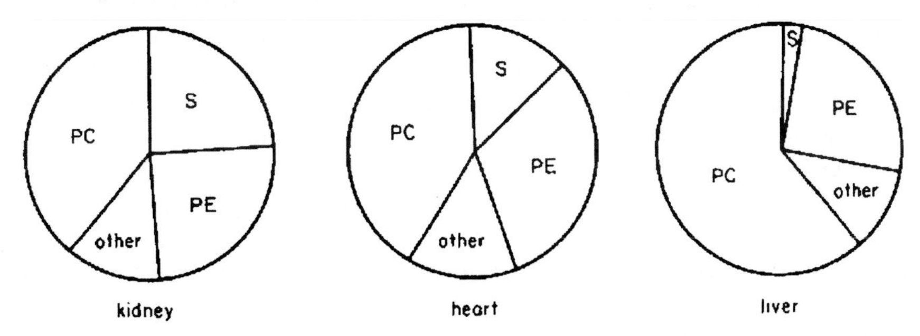

Fig. 1.2 (a) Variation in the proportion of sphingomyelin (S) to phosphatidylcholine (PC) and other constituents, such as cholesterol (C) and phosphatidylethanolamine (PE), in the total lipid fraction of erythrocytes from various species. (b) Sphingomyelin relative to total lipids present in the endoplasmic reticulum. (Reprinted from Fig. 1.14 of ref. 2 with permission from Macmillan.)

monolayer and vice versa [16, 17]. These membrane-located transport proteins – such as the MDR1 P-glycoprotein – play an essential role in multidrug resistance [16]. Reversal of multidrug resistance in cancer cells [18] and others is discussed in Section 5.2.

The biological function of lipid asymmetry and of proteins involved in the transmembrane traffic of lipids is multiple. Rapid reorientation of phospholipids in erythrocytes by flippases allows the erythrocyte cell membrane to remain in equilibrium while more lipids are being synthesized. On the other hand, ATP-dependent flippases generate asymmetrical membranes in accordance with the asymmetrical functions of the plasma membrane. The rapid reorientation of phospholipids in the plasma membrane of platelets upon cell stimulation involves a very active calcium-dependent phospholipid "scramblase" [19]. Lipid scrambling results in phosphatidylserine exposure on the outer leaflet of platelets, which in turn triggers the conversion of prothrombin into thrombin [20]. A review of the role of translocases in the generation of phosphatidylserine asymmetry has recently been published [21].

In the case of red blood cells, it is assumed that the progressive loss of lipid asymmetry, possibly associated with the entry of calcium, is a signal that the cell is aging. This signal, in turn, is recognized by macrophages and leads to cell destruction. Drugs which, for example, compete for calcium bound to phosphatidylserine could interfere with these processes and many other Ca^{2+}-dependent processes such as protein kinase C activation. The influence of asymmetry in membranes of different phospholipid composition on the fusion of liposomes has been studied and reported [22].

Another important aspect of phospholipid asymmetry in membranes with regard to drug permeation and drug distribution is the generation of surface tension or surface curvature. An increase in surface tension could change the membrane permeability either directly via a change in bilayer viscosity or indirectly via a change in protein conformation. The curvature of membranes – which is determined mainly by the volume, size, and charge of the phospholipid head groups and their distribution in the outer and inner leaflet – is an important factor for the functioning of embedded proteins. It is an indicator of the internal stress of the lipid layer. The internal stress is the tendency of the lipid system to adapt to a non-bilayer configuration as, for example, in the H_{II} phase. "The intrinsic radius of curvature is essentially a measure of the average mismatch between the minimum free energy projected areas of the hydrophilic and hydrophobic portions of the lipid molecules" [20]. In other words, "the spontaneous radius of curvature is a measure of the frustrated elastic curvature energy locked into bilayers" [23]. It is well known that many biomembranes contain phospholipid fractions that do not form bilayers under physiological conditions. Unsaturated phosphatidylethanolamines, for example, form non-lamellar phases such as the H_{II} phase [24, 25]. If such phospholipids are mixed with other phospholipids in a membrane, they change the properties of the bilayer. The mismatch between lateral tension in the polar and hydrocarbon zones of the lipid layer becomes large, corresponding to small values of the intrinsic radius of curvature. This is reflected in a change in the order of the hydrocarbon chains and can be observed by deuterium nuclear magnetic resonance (NMR) experiments. In the case of other lipids, such as phosphatidylcholine, the radius is large and the bilayer is therefore relaxed. The results of experiments that correlate the composition of bilayers with the operation or activity of certain intrinsic membrane proteins have shown that the range of the bilayer intrinsic radius of curvature for optimal function is limited. An example is the effect of lipid composition on the sarcoplasmic reticulum Ca^{2+}-

ATPase reconstituted in vesicles of different lipid composition [26]. Pumping efficiency (number of Ca^{2+} ions transported per molecule of ATP hydrolyzed) increased with the mole fraction of lipids that prefer H_{II}-phase formation. However, no such correlation was observed with dioleoylphosphatidylcholine or digalactosyldiglyceride, which do not form low-temperature H_{II} phases. It is important to note that it is not the chemical similarity but the phase preference of the phospholipid that is decisive.

Amphiphilic compounds are also known as potent modifiers of the bilayer intrinsic radius of curvature and utilize this property to act as a non-specific perturbator of membrane protein function [27]. Catamphiphilic drugs that can interact with the head groups or with the scramblases or flippases can change cell functioning.

Not only the composition but also the amount and type of integrated proteins differs in cell membranes. The ratio of protein to lipid in rat tissue membranes is given in Table 1.5 [2]. Membrane proteins can be divided into two classes, intrinsic and extrinsic [28]. Intrinsic membrane proteins are inserted to varying degrees into the hydrophobic core of the lipid bilayer and can be removed only by detergents or denaturants. Extrinsic membrane proteins are associated non-covalently with the surface of the membrane and can be removed from the bilayer by incubation in alkaline medium. Intrinsic membrane proteins can be classified into three groups: monotopic, bitopic, and polytopic (Figure 1.3) [29, 30]. Amino acids can change the degree of hydrogen binding within the bilayer, leading to changes similar to those obtained by the exchange of phosphatidylethanolamine for phosphatidylcholine.

The conformation of proteins integrated into a membrane depends not only on the type and sequence of the amino acids but also on the lipid environment, i.e. the composition of the membrane. The effect of lipid environments and external factors such

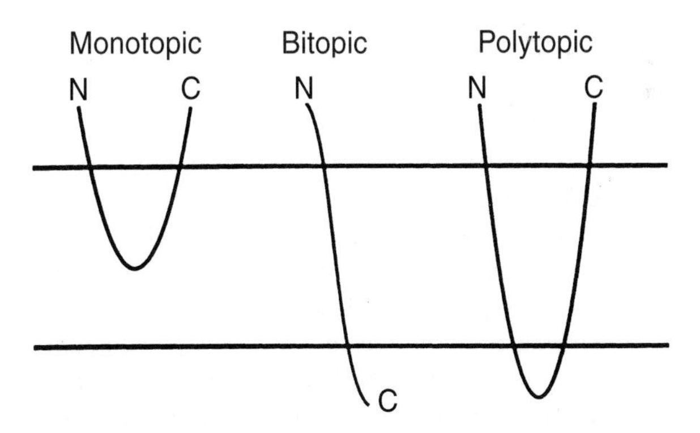

Fig. 1.3 Classes of intrinsic membrane proteins, as defined by Blobel [29]. The peptides chains end in amino (N) and carboxy (C) termini. The disposition of the N- and C-termini differs for each class of membrane protein. only two membrane spannings are indicated for the polytopic class; however, the number of membrane spannings is not limited for polytopic membrane proteins. (Reprinted from Fig. 1 of ref. 30 with permission from CRC Press.)

Tab. 1.5 Protein to lipid ratios of rat tissue membranes. (Reprinted from Table 1.5 of ref. 2, with permission from Macmillan)

Membrane	Protein/lipid	Cholesterol/ polar lipid
Plasma membranes		
Myelin	0.25[a]	0.95
Erythrocytes	1.1	1.0
Rat liver cells	1.5	0.5
Plasma membranes		
Nuclear membranes	2.0	0.11
Endoplasmic reticulum		
Rough	2.5	0.10
Smooth	2.1	0.11
Mitochondrial membranes		
Inner membrane	3.6	0.02
Outer membrane	1.2	0.04
Golgi membranes	2.4	–

a) myelin from central nervous system; higher ratios are found in peripheral nerve myelin.

as temperature, pH, and ionic strength on the three-dimensional (3-D) structure of proteins has been studied in reconstituted membranes [31]. The major driving forces for specific membrane associations of proteins are hydrophobic interactions and the peptide amphipathic character [32, 33].

Different amphipathic helical peptides can stabilize or disrupt membranes depending on the structure of the helix, which determines the nature of its association with the membrane. Features of peptides that are responsible for their specific properties and effects on membrane organization have been discussed [34, 35]. For example, peptides containing class A amphipathic helices can stabilize membranes by reducing negative monolayer curvature strain, whereas lytic amphipathic helical peptides can increase negative curvature strain and form pores composed of helical clusters [35]. The type of peptide involved is of importance for the functioning of the cells. It was found that a mixture of anionic amphiphilic peptides and charge-reversed cationic peptides induced a rapid and efficient fusion of egg phosphatidylcholine vesicles, but no fusion was observed with either peptide alone, probably because only the mixture of both peptides has an ordered helical structure [36].

The arrangement of the proteins within the membrane seems to depend to some extent on the electrostatic surface potential and interface permittivity. It is influenced by electrostatic interaction between the proteins, polar head groups of the phospholipid and ions within the aqueous medium of the membrane surface. This can be affected by exogenous molecules such as drugs. Phospholipid-induced conformational change in intestinal calcium-binding protein in the absence and presence of Ca^{2+} has been described [37]. There is, however, no doubt that hydrophobic interactions between peptides and membrane interfaces play an important role. A general frame-

Tab. 1.6 Whole-residue free energies of transfer, ΔG, from water to POPC interface (wif) and to *n*-octanol (woct). (Reprinted from Table 1 of ref. 38, with permission from Elsevier Science)

Amino acid	ΔG_{wif} (kcal/mol)[a]	ΔG w_{oct} (kcal/mol)[a]
Ala	0.17 ± 0.06	0.50 ± 0.12
Arg^+	0.81 ± 0.11	1.81 ± 0.13
Asn	0.42 ± 0.06	0.85 ± 0.12
Asp^-	1.23 ± 0.07	3.64 ± 0.17
Asp°	−0.07 ± 0.11	0.43 ± 0.13
Cys	−0.24 ± 0.06	−0.02 ± 0.13
Gln	0.58 ± 0.08	0.77 ± 0.12
Glu^-	2.02 ± 0.11	3.63 ± 0.18
Glu°	−0.01 ± 0.15	0.11 ± 0.12
Gly	0.01 ± 0.05	1.15 ± 0.11
His^+	0.96 ± 0.12	2.33 ± 0.11
His°	0.17 ± 0.06	0.11 ± 0.11
Ile	−0.31 ± 0.06	−1.12 ± 0.11
Leu	−0.56 ± 0.04	−1.25 ± 0.11
Lys^+	0.99 ± 0.11	2.80 ± 0.11
Met	−0.23 ± 0.06	−0.67 ± 0.11
Phe	−1.13 ± 0.05	−1.71 ± 0.11
Pro	0.45 ± 0.12	0.14 ± 0.11
Ser	0.13 ± 0.08	0.46 ± 0.11
Thr	0.14 ± 0.06	0.25 ± 0.11
Trp	−1.85 ± 0.06	−2.09 ± 0.11
Tyr	−0.94 ± 0.06	−0.71 ± 0.11
Val	0.07 ± 0.05	−0.46 ± 0.11

a) For both ΔG_{wif} and ΔG_{woct}, the signs have been reversed relative to those of the original publications to reflect the free energies of transfer from water phase.

work for the thermodynamics of membrane protein stability that is focussed on interfacial interactions has been reviewed and discussed by White and Wimley [38]. The authors experimentally determined a whole-residue interfacial hydrophobicity scale for the energy of transfer, ΔG, from water to palmitoyloleoylphosphatidylcholine (POPC) interface and to *n*-octanol. The difference in ΔG reflects the "important role of the peptide bond in partitioning and folding" (Table 1.6). Hydrophobic mismatch between proteins and lipid membranes has been discussed by Killian [39].

Perhaps the most important step in the development of our current understanding of biomembranes was the introduction of the fluid mosaic model [28] (Figure 1.4) [40]. This model describes the cell membrane as a fluid two-dimensional lipid bilayer matrix of about 50 Å thickness with its associated proteins. It allows for the lateral diffusion of both lipids and proteins in the plane of the membrane [41] but contains little structural detail. This model has been further developed and it has been assumed that the membrane consists of solid domains coexisting with areas of fluid-disordered membrane lipids that may also contain proteins [42]. This concept has

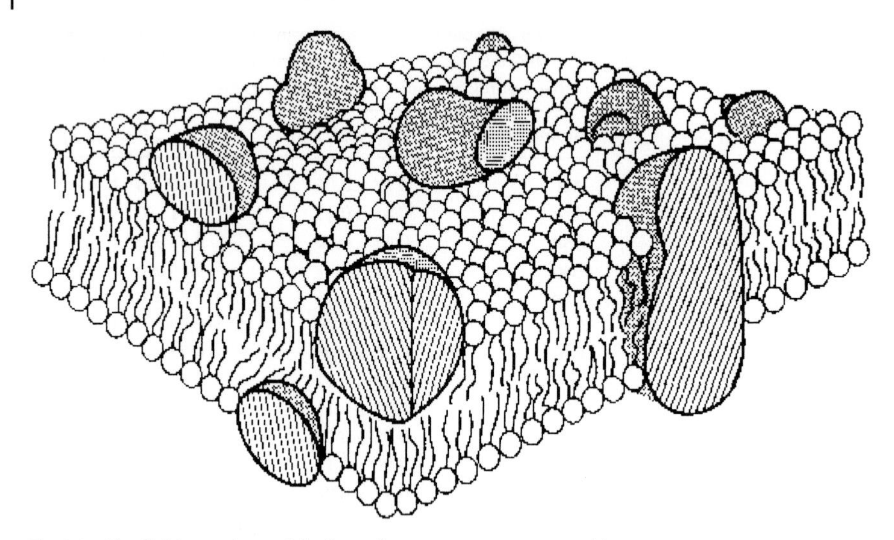

Fig. 1.4 The fluid mosaic model of membrane structure proposed by Singer and Nicolson [28]. (Reprinted from Fig. 1 of ref. 40 with permission from Wiley-VCH.)

been integrated into the fluid mosaic model, and the resulting model is called the plate model of biomembranes.

The concept of lamellar anisotropy was first proposed by Siekevitz [43]. Since then more data have become available, resulting in a refined model for the dynamic organization of biomembranes. The most important difference from the fluid mosaic model is that a high degree of spatiotemporal order also prevails in the liquid crystalline fluid membrane and membrane domains [44]. The interaction forces responsible for the ordering of membrane lipids and proteins are hydrophobicity, coulombic forces, van der Waals dispersion, hydrogen bonding, hydration forces, and steric elastic constraints. Specific lipid–lipid and lipid–protein interactions result in a precisely controlled and highly dynamic architecture of the membrane components. On the basis of recent experimental and theoretical progress in the understanding of the physical properties of lipid bilayer membranes, it is assumed that the lipid bilayer component of cell membranes is "an aqueous bimolecular aggregate characterized by a heterogeneous lateral organization of its molecular constituents." It is further assumed that the dynamically heterogeneous membrane states are important for membrane functions such as transport of matter across the membrane, and for enzymatic activity [45]. Evidence is accumulating that lipid bilayer structure and dynamics play an important role in membrane functionality. Several experimental examples have demonstrated that the physicochemical properties and phase behavior of lipid membranes are essential for the functioning of lipid-embedded and integrated enzymes.

The nature of the relation between lipid structure changes and enzyme activation has been investigated using, for example, fluorescence spectroscopy and calorimetry (see Section 3.1). Figure 1.5 shows the effect of temperature on the hydrolytic activi-

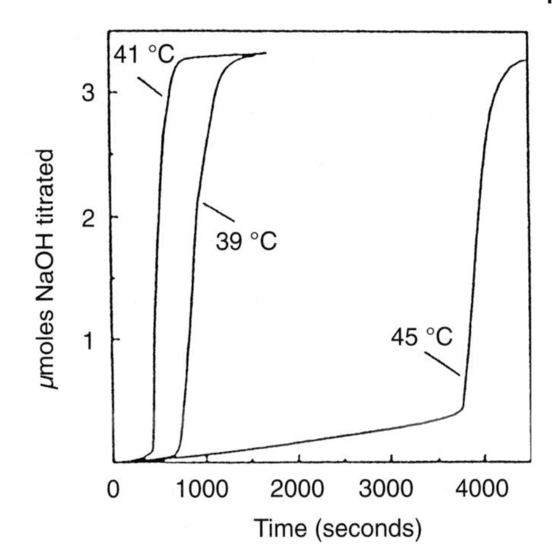

Fig. 1.5 Time courses of hydrolysis of DPPC LUVs by AppD49 PLA_2 at three different temperatures. (Reprinted from ref. 46, p. 3, with permission from Elsevier Science.)

ty of phospholipase A_2 (PLA_2) in large unilamellar vesicles (LUVs) of dipalmitoylphosphatidylcholine [46]. It was observed that the catalysis of phosphatidylcholine bilayers by PLA_2 starts only after a certain latency period to allow a sufficient formation of reaction products such as lysolecithin and fatty acids and their accumulation in the membrane. The influence of temperature near the main phase transition and of saturated long-chain diacylglycerols on the effectiveness of these reaction products was studied [47]. Fluorescence probes were used to evaluate the ability of the reaction products to perturb the bilayer and to promote enzyme binding to the membrane surface, and the effect of temperature and bilayer composition on this process [47]. The importance of the phase state of phospholipids – liquid crystalline or gel – for the proper function of enzymes has been established [48], and the role of membrane defects in the regulation of the activity of protein kinase C (PKC) has been studied [49]. It was found that the formation of defects alone is not sufficient to promote PKC activity. An important factor is the change in those membrane bilayer properties that are related to hexagonal phase formation. It was observed that PKC activity is strongly dependent on the ranking of phospholipids that are hexagonal phase formers, such as dioleoylphosphatidylethanolamine (DOPE) and dipalmitoylphosphatidylethanolamine (DPPE).

Most important for the regulation of the membrane architecture are membrane potential, intracellular Ca^{2+} concentration, pH, changes in lipid composition due to the action of phospholipases and cell–cell coupling as well as the coupling of the membrane to the cytoskeleton and the extracellular matrix. Membrane architecture is additionally modulated by ions, lipo- and amphiphilic hormones, metabolites, drugs, lipid-binding peptide hormones, and amphitropic proteins [44].

On the basis of the information presented, it is not surprising that membranes play an important role in drug action. The above-mentioned properties make lipid bilayers an important object of drug research in relation to drug targeting, permeation of membranes, and drug distribution. In addition, the reactions of liposome-encapsulated drugs with membrane-embedded receptors and drug delivery systems are being studied [50].

Clearly, the "construction" of membranes represents an important barrier to drug absorption, distribution, and efficacy depending on the structures of the membrane and the drug molecules. The hydrophobic inner part of the bilayer is an almost complete barrier to the diffusion of polar or charged drug molecules. In contrast, apolar drug molecules can easily penetrate the membrane. In this context it should, however, also be pointed out that the composition of mammalian membranes varies in different organs and species (Figure 1.2). This could be important in increasing selectivity in drug distribution, and accumulation and therefore also in improving drug efficacy.

1.2.2 Bacterial Membranes

Bacterial membranes have a much more complex construction than mammalian membranes. This enables bacteria to survive in the various environments of host organisms. Knowledge of the composition and functioning of bacterial membranes is therefore essential to the development of anti-infective drugs. In order to be effective, antibacterial agents not only have to have optimal pharmacokinetic properties such as uptake and distribution in the patient, but they must also be able to cross an additional barrier, the cell wall of the bacteria, so that they can reach the target site. This additional barrier is remarkable on account of its rigidity and permeability. The construction and structural uniqueness of this barrier is briefly described below.

With the exception of the cell wall-deficient *Mycoplasma*, all bacteria possess a more or less pronounced peptidoglycan layer that surrounds the protoplasm as a supporting layer of high rigidity and which can withstand the huge difference in osmotic pressure (up to 2000 kPa) between the cytoplasm and external milieu [51]. The murein network consists of long linear polysaccharide chains consisting of β-1,4-glycosidically connected *N*-acetylglucosamine and *N*-acetylmuraminic acid molecules, which are connected to each other by short peptide bridges. Bacteria are divided into two groups, Gram-positive and Gram-negative, depending on the nature of their cell wall.

In Gram-positive bacteria, the cell wall is 20–80 nm thick, and consists of 50% peptidoglycan, which is organized in up to 40 layers [52, 53]; in contrast, the bilayer membrane of a mammalian cell is only 10 nm thick. Other essential components are teichon and teichuronic acid, and acidic, water-soluble polymers that are connected by phosphodiester bridges. The large number of phosphate groups results in a strong negative charge on the cell surface which – under physiological conditions – is partially neutralized by alanyl residues and possibly by Mg^{2+} ions. A schematic drawing of the cell wall of a Gram-positive bacterium is shown in Figure 1.6. Recently, a completely new concept of the architecture of microbial murein was proposed by Dimitriev *et al.* [54]. In contrast to the classical concept, which assumes that cross-linked glycan chains are arranged in horizontal layers outside the plasma membrane, these authors postulate that the glycan strands in the bacterial cell wall run perpendicular to the plasma membrane. This could be of great importance for the understanding of the mechanism of action of antibiotics acting on cell wall function and synthesis.

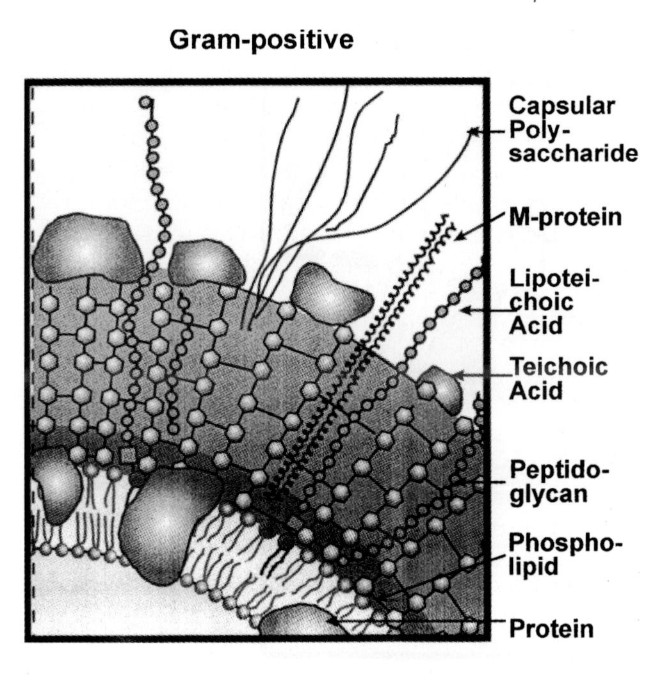

Fig. 1.6 Schematic drawing of the cell wall structure of Gram-positive bacteria.

Mycobacteria such as *Mycobacterium tuberculosis* hold a special position within the group of Gram-positive bacteria. Characteristic of species of this genus is their so-called acid fastness, which can be attributed to the large complex lipid portion, which consists of mycolic acids, that is 3-hydroxycarbonic acids substituted in the 2 and 3 position by long alkyl chains (up to C_{85}) (Figure 1.7) [55]. Recently, it was suggested that the impermeability of the mycobacterial cell wall is a result of the organization

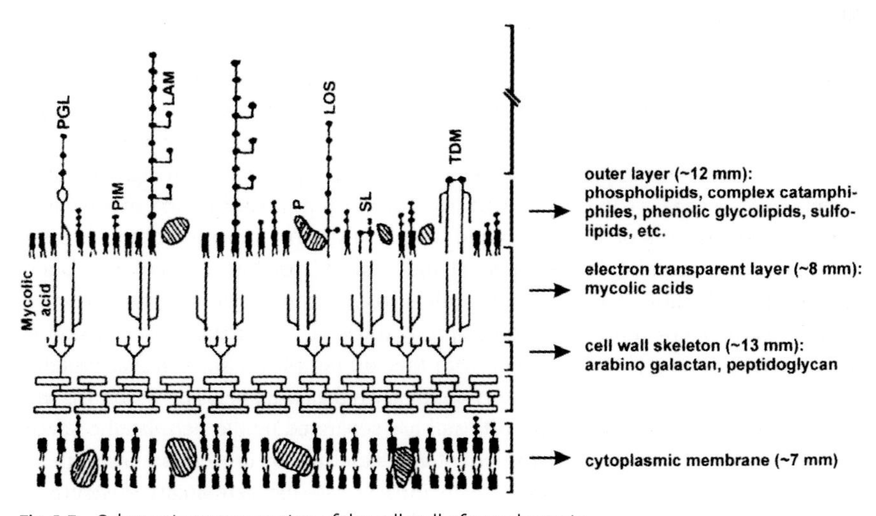

Fig. 1.7 Schematic representation of the cell wall of mycobacteria.

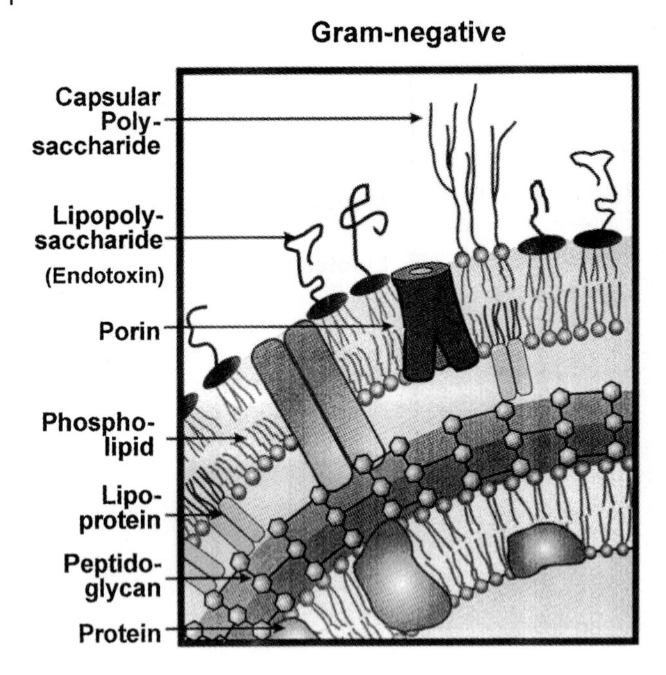

Fig. 1.8 Schematic drawing of the cell wall and membrane of Gram-negative bacteria.

of the cross-linked glycan strands, which, rather than forming horizontal layers, run perpendicular to the cytoplasmic membrane [56]. This is because these highly hydrated molecules "are under the strong stress of turgor pressure [56]".

In Gram-negative bacteria, the cell wall outside the cytoplasmic membrane has multiple layers and a complex constitution (Figure 1.8). It is a trilamellar structure composed of the cytoplasmic membrane, the murein sacculus, and the outer membrane. The murein sacculus and the outer membrane together form the true cell wall. The outer membrane protects the bacterium from aggressive physiological milieu, for example in the intestine of the host, and from host defense mechanisms. It is also known that the outer membrane of *Enterobacteria* and some other Gram-negative bacteria confers resistance to various antibiotics that are active against other bacterial species [57]. The cytosol is surrounded and protected by the inner membrane, which consists of a phospholipid double layer. The most important components of this membrane are phosphatidylserine, phosphatidylinositol, phosphatidylethanolamine, and phosphatidyl-glycerol [52]. In *Escherichia coli*, phosphatidylethanol-amine contributes up to 100% of the phospholipid portion of the inner membrane [58]. Associated with this are peripheral and integral proteins such as the enzymes of the respiratory chain as well as transport and structure proteins. The cytosol membrane and cell wall are separated by the periplasmic space. The cell wall is predominantly made up of and stabilized by murein. In Gram-negative bacteria – in contrast to Gram-positive bacteria – the peptidoglycan structure consists only of one layer and contributes only about 10% to the cell wall weight [51].

The layer typical of Gram-negative bacteria is the so-called outer membrane, which is connected to the peptidoglycan layer. This outer leaflet of the outer membrane is composed of an unusual lipid and lipopolysaccharide (LPS) rather than the glycerophospholipid normally found in biological membranes [59]. It is also a bilayer, but an asymmetrical one. The other layer, directed to the peptidoglycans, consists of phospholipids. The glucosaminedisaccharides of LPS are connected to long-chain fatty acids. This lipophilic component of LPS is called lipid A. LPS represents the endotoxic principle and is responsible for endotoxin-dependent septic shock [60, 61]. The hydrophilic part of the outer layer – consisting of polysaccharides covalently connected to glucosamine – is made up of the core region and a surface-specific side chain that is formed by identical repeating oligosaccharide units (repeating units). The polysaccharides of the LPS release no endotoxic activity. They do, however, possess, antigenic properties and are responsible for immunological reactions [62]. Another important component is proteins, for example lipoprotein OmpA, outer membrane protein A, and the pore formation proteins, for example OmpC, OmpF, and PhoE, or special transport proteins. These proteins contribute up to 50% to the mass of the outer membrane.

For chemotherapeutics to be effective against Gram-negative bacteria requires a balance between hydrophilic and hydrophobic properties. The reason for this is the characteristic construction of the bacterial cell wall, with an outer hydrophilic core, rich in polysaccharides, and a hydrophobic phospholipid bilayer. Only very small and hydrophilic drug molecules up to 600 Da can diffuse through membrane pores [63, 64].

Artificial asymmetric membranes composed of outer membranes of various species of Gram-negative bacteria and an inner leaflet of various phospholipids have been prepared using the Montal–Mueller technique [65]. Such planar bilayers have been used, for example, to study the molecular mechanism of polymyxin B–membrane interactions. A direct correlation between surface charge density and self-promoted transport has been found [66].

1.2.3 Fungal Membranes

The discussion of the cell envelope of fungi focusses on the example of *Candida albicans*. The fungal cell wall is a multilayer structure composed mostly of carbohydrates. As in bacteria, it determines cell shape, confers rigidity and strength, and prevents lysis from osmotic shock. Because of its limited porosity, the cell wall is a permeability barrier for large molecules [67]. In *Candida albicans* the cell wall consists primarily of glucans and mannoproteins [68]. Glucans are a mixture of branched β-1,3- and β-1,6-linked glucose polymers [69]. Mannoproteins, in contrast, are branched mannose polymers attached to a protein through a GlcNAc–GlcNAc group and an arginine residue. A minor but essential cell wall component is chitin, a linear β-1,4-linked GlcNAc homopolymer. It is associated with the septum (yeast form) and the apical region (hyphal form) [70].

The plasma membrane of fungi has the typical lipid bilayer structure, with phosphatidylcholine, phosphatidylethanolamine, and ergosterol as major lipid constituents. Its function is to protect the cytosol, and regulate the transport of molecules in and out of the cell. It is also the matrix for membrane-bound proteins, including those involved in cell wall synthesis. The composition of the plasma membrane has been discussed and reviewed [71]. It is important to note that – in contrast to mammalian cells – ergosterol, not cholesterol, is the most important component of fungal membranes. In this respect it is interesting to note that phospholipids and cholesterol are also a major component of the human immunodeficiency virus. Their molar ratio is close to 1, leading to an extremely rigid membrane. Ergosterol has at least two types of function in the membrane: a non-specific "bulking" function in the regulation of membrane fluidity and integrity and a specific "sparking" function in the regulation of growth and proliferation [72]. The inhibition of ergosterol synthesis is therefore one of the favored targets in antifungal therapy. It leads to a cessation of *Candida* growth, but not to cell wall lysis [73]. The effect of antifungal agents on membrane integrity in *Candida albicans* is summarized in Table 1.7 [73]. Assays for membrane integrity are based on the release of intracellular substances (potassium, aminoisobutyric acid) or the entry of substances that are normally excluded (methylene blue) [71].

Tab. 1.7 Effects of antifungal agents on membrane integrity of *Candida albicans*. (Reprinted from Table 2 of ref. 73, p. 107, with permission from Bertelsmann-Springer).

Antifungal agent	Concentration (μm)	Total release of: Potassium (%)	Total release of: Aminoisobutyric (%)	Percent of cells stained with methylene blue
None		< 5	< 5	0
Polyenes				
Amphotericin B	10	100	97	100
Nystatin	10	100	97	100
Azoles				
Clotrimazole	100	< 5	20	0
Econazole	100	47	98	0
Miconazole	100	66	94	60
Ketoconazole	100	< 5	20	0
Allylamines, thiocarbamates				
Naftifine	100	< 5	30	0
Tolnaftate	100	< 5	< 5	0

1.2.4 Artificial Membranes; Liposome Preparation, and Properties

During recent decades, the use of artificial phospholipid membranes as a model for biological membranes has become the subject of intensive research. As discussed above, biological membranes are composed of complex mixtures of lipids, sterols, and proteins. Defined artificial membranes may therefore serve as simple models of membranes that have many striking similarities with biological membranes. A comparison of some important physicochemical properties of biological and artificial membranes is given in Table 1.8 [2].

Tab. 1.8 The physical characteristics of artificial bilayers and biological membranes. (Reprinted from Table 2.3 of ref. 2, with permission from Macmillan)

Property	Phospholipid bilayer	Biological membranes
Thickness (nm)[a]	4.5–10	4–12
Interfacial tension (J/m^2)	2.0–60	0.3–30
Refractive index	1.56–1.66	approx. 1.6
Electrical resistance (Ω/m^2)	10^7–10^{13}	10^6–10^9
Capacitance (mF/m^2)[b]	3–13	5–13
Breakdown voltage (mV)	100–550	100
Resting potential difference (mV)	0.140	10–88

a) These measurements have been made by electron microscopy, X-ray diffraction and optical methods.
b) The capacitance was determined using an assumed dielectric constant.

Artificial membranes are used to study the influence of drug structure and of membrane composition on drug–membrane interactions. Artificial membranes that simulate mammalian membranes can easily be prepared because of the readiness of phospholipids to form lipid bilayers spontaneously. They have a strong tendency to self-associate in water. The macroscopic structure of dispersions of phospholipids depends on the type of lipids and on the water content. The structure and properties of self-assembled phospholipids in excess water have been described [74], and the mechanism of vesicle (synonym for liposome) formation has been reviewed [75]. While the individual components of membranes, proteins and lipids, are made up of atoms and covalent bonds, their association with each other to produce membrane structures is governed largely by hydrophobic effects. The hydrophobic effect is derived from the structure of water and the interaction of other components with the water structure. Because of their enormous hydrogen-bonding capacity, water molecules adopt a structure in both the liquid and solid state.

The structural dependence of phospholipid solutions on water content is called lysotropic polymorphism. At a water content of up to 30% dipalmitoylphosphatidylcholine (DPPC) forms lamellar phases consisting of superimposed bilayers. Increasing the water content results in heterogeneous dispersions formed by multilamellar structures, the so-called liposomes (see also Section 1.3.1).

Because of the complex composition of fatty acyl chains, a large variety of different molecules is present within each class of phospholipid, but they have in common their amphipathic nature. The packing of phospholipids depends on the normalized chain length difference between the *sn*-1 and *sn*-2 acyl chain. In addition to the bilayer structure of non-interdigitated acyl chains, mixed interdigitated and partially and totally interdigitated bilayer structures exist.

Liposome preparations have been shown to be suitable not only for studying special drug–membrane interaction effects *in vitro* but also for use as drug carriers. Various techniques have been developed and described to prepare homogeneous unilamellar or multilamellar liposome preparations of different sizes. The simple method of producing liposomes results in multilamellar vesicles of varying diameter in the range 0.2–0.5 μm [76]. Sonication of such preparations produces small unilamellar vesicles [77] of homogeneous size which have been extensively used as model membranes [78]. A method of preparing vesicles with high efficiency of encapsulation and a vesicle size distribution of the order of 0.4 μm has been described [79]. The conditions required to produce convenient and reproducible preparations of intermediate-size unilamellar liposomes using reverse-phase evaporation followed by extrusion through a polycarbonate membrane have been reported [80]. Large unilamellar vesicles have also been obtained using a stainless-steel extrusion device, allowing extrusion at elevated temperatures (above the phase transition temperature) [81–83]. Liposomes differing in phosphoinositide content were prepared using chromatography on immobilized neomycin, a procedure that could be useful in the preparation of asymmetric liposomes [84]. A new rapid solvent exchange (RSE) method has been developed which avoids the passage of lipid mixtures through an intermediate solvent-free state, thus preventing possible demixing of membrane components [85]. Two major mechanisms involved in liposome formation when starting from lipid-in-water suspensions have been discussed: fragmentation of bilayers with subsequent self-closure of bilayer fragments and budding off of daughter vesicles from mother liposomes [79].

When liposomes are loaded with drugs, phase transformation of a liposomal dispersion into a micellar solution can be induced depending on the phospholipid to drug ratio. Using small-angle scattering X-ray, a decrease in particle size (from 211 to 155.2 nm) and thickness (from 64 to 45.5 Å) was observed with increasing drug concentration. ^{32}P-NMR was then used to obtain information on drug localization in the membrane. Non-loaded liposomes showed only one signal in the spectrum, at 0.3 ppm, whereas three signals were observed in loaded liposome dispersions with phospholipid to drug ratios of 16:1 to 2:1. This shows that loading of liposomes with catamphiphilic drugs leads to large changes in physicochemical properties. Even at lipid to drug ratios of 1:1 a phase transformation from a liposomal dispersion to micellar solution occurred (particle size of 12.5 nm, loss of the ^{32}P resonance signal at 0.3 ppm), with only two remaining highfield shifted signals [86].

Liposomal preparations are of interest for studying not only drug–membrane interactions but also membrane permeability and drug transport.

The association and release of prostaglandin E_1 (PGE_1) from liposomes composed of different lipids and at different pH values has been studied using circular dichroism and differential scanning calorimetry. It was found that the association of protonated PGE_1 (pH 4.5) is highly dependent on the degree of saturation, the lamellarity, and the negative surface charge of the lipids used. More PGE_1 accumulated in the fluid-phase membrane than in the gel-phase membrane. The addition of cholesterol, which makes the membrane more rigid, led to a decrease in accumulation in both membrane phases. At pH 7.4 the release of PGE_1 from gel-phase membranes was significantly reduced compared with release from fluid-phase membrane [87].

An anomalous solubility behavior of ciprofloxacin in liposomes has been observed and determined by ^{1}H-NMR measurements. The drug was encapsulated in LUVs in response to a pH gradient to increase drug accumulation (up to 250 mmol/L). It could be shown that ciprofloxacin localized in the aqueous interior of the liposomes and self-associated in the form of small stacks. It did not, however, precipitate inside the liposomes despite its solubility in bulk aqueous solution of only 5 mmol/L under the same conditions of pH and ionic strength [88].

Factors that could influence uptake and retention of drugs containing amino groups have been investigated in LUVs, with a transmembrane pH gradient (ΔpH) induced by the addition of an acid to the interior of the LUVs [89]. It was found that factors that increase partitioning into the LUV bilayer increase retention properties. The presence of diammonium sulfate within the LUVs increased the uptake of doxirubicin and epirubicin to almost 100% (ΔpH 1.4 units). In comparison, loading of only 25% was achieved for ciprofloxacin and vincristine. No such improvement was achieved in the case of entrapped lipophilic alkylamines. The retention of ciprofloxacin and vincristine was improved when distearoylphosphatidylcholine was replaced by negatively charged distearoylphosphatidylglycerol.

The results are of general interest for the design of liposomes with better uptake and retention of drugs. It has been shown that prolonged retention *in vitro* of, for example, vincristine is accompanied by improved retention *in vivo* [90].

Liposomes can even be used as microreactors. Until recently, the utility of this technique was limited by the fact that they could be used for only a relatively short time because of depletion of the reaction mixture. It has now been shown that liposomes of 1-palmitoyl-2-oleoyl-*sn*-glycero-3-phosphocholine (POPC) can be used as a semipermeable microreactor after treatment with sodium cholate. It has been demonstrated that this allows a biochemical reaction to take place inside the liposomes but not in the external medium. Such cholate-induced POPC bilayers can also be used to insert enzymes [91].

1.3 Dynamic Molecular Organization of Membrane

1.3.1 Thermotropic and Lysotropic Mesomorphism of Phospholipids

The thermotropic phase transitions observed when solid-state phospholipids are heated above their melting point constitute an important physicochemical property of this class of compounds. This property can be exploited very effectively in the analysis of drug–membrane interactions, and is the basis for most physicochemical methods of studying such interactions, especially calorimetric experiments. The various motional modes present in lipid bilayers and methods of following and measuring them have been reviewed [92]. Phospholipids do not change directly from the crystalline to the isotropic fluid state, but at intermediate temperature pass through a liquid crystalline state. Both transition changes, from crystalline to liquid crystalline and from liquid crystalline to the isotropic fluid state, are associated with heat uptake, i.e. they are endothermic processes (Figure 1.9) [93]. When in the crystalline state, the fatty acid chains of phospholipids are in the extended conformation and in the *all trans* configuration. They form a regular, stiff crystal lattice stabilized by van der Waals forces. Electrostatic interactions between the head groups lead to addi-

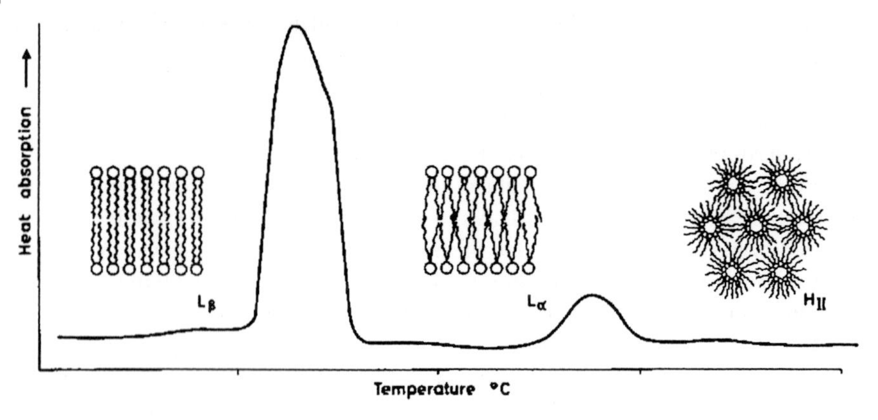

Fig. 1.9 Structures adopted by phospholipids in aqueous media. L_{β} and L_{α} (gel and liquid crystalline) states exist at low and intermediate temperatures, respectively, and the inverted cylinder (hexagonal) H_{II} state is found at elevated temperatures [93].

tional molecular interactions. During the first transition (crystalline to liquid crystalline), melting of the hydrocarbon chains occurs, as a result of which their mobility increases, the *gauche* configuration is adopted, and "kinks" in the chains are observed. Strong head group interactions prevent complete loss of the molecular connection. The second transition, from the liquid crystalline to the liquid phase, leads to movements of the phospholipid molecules in all directions. The inverted cylinder H_{II} or hexagonal phase can be observed at high temperatures. The dynamic molecular organization in vesicles and membranes has been discussed and reviewed [94, 95]. Thermodynamically, the gel-to-liquid crystalline phase transition occurs at a temperature at which the entropic reduction in free energy resulting from the chain melting counterbalances the decrease in bilayer cohesive energy. The latter arises from lateral expansion and from the energy needed to create *gauche* conformers in the hydrocarbon chain [96]. The gel-to-liquid crystalline phase transition temperature, T_t, of natural phospholipid membranes primarily depends on the length and chemical structure of the hydrocarbon chains and secondarily on the structure of the polar head groups.

Phase transition can be followed by various physical techniques, such as differential scanning calorimetry (DSC), ^{2}H-NMR, and electronic spin resonance (ESR), Fourier transform infrared (FTIR), and fluorescence spectroscopy. The various methods have been reviewed and their characteristics compared [97] (see also Chapter 3).

In addition to this thermotropic mesomorphism, a lysotropic mesomorphism is observed [98]. The phase transition temperature, T_t, for the transition from the crystalline to the liquid crystalline state decreases as a function of water content. The decrease in T_t is due to destabilization of the crystal lattice in the head group region by water molecules. This, in turn, decreases the interaction between the fatty acid chains. When the water content reaches a certain level, the phospholipids assume a thermodynamically optimal arrangements whereby the fatty acids are directed to the inner part and the hydrophilic head groups to the outer phase, which is in contact with the watery medium. By analogy with the 3-D crystal lattice of a salt, one can speak of a two-dimensional (2-D) "membrane crystal" which possesses a highly ordered state below the melting point and before phase transition [98].

The insertion of water molecules into the membrane head group region increases the space that can be occupied by the fatty acid chains and in consequence increases their mobility. The separation of any given head group, and therefore the decrease in T_t, is, however, limited because only limited amounts of water can be bound. The maximum amount of water that can be bound and the characteristic T_t at maximum hydration are a function of structure and charge of any given head group. Phospholipids which have taken up water molecules into the head group region below T_t are in the so-called gel phase. An increase in temperature will lead to phase transition from the gel to the liquid crystalline phase. This is paralleled by an increase in mobility of the fatty acid chains. The mobility of the methylene groups increases with increasing distance from the acyl groups. This goes hand in hand with the lateral diffusion of molecules within the membrane. However, because of the strong polar interaction in the head group region, the bilayer structure remains intact.

As already mentioned, the phase transition temperature depends on the phospholipid structure, including the number and conformation of double bonds, the length of the fatty acid chains, the charge and volume of the phospholipid head group, and the number of hydrogen bonds. Hydrogen bonds are important for the stabilization of the membrane and an increase in T_t. The influence of small structural changes on the phase transition of phospholipids is demonstrated in Table 1.9 using the example of alkyl branched diacylphosphatidylcholines [99]. A database of lipid phase tran-

Tab. 1.9 Thermodynamic characteristics of alkyl branched diacylphosphatidylcholines. (Adapted from Table 2 of ref. 99, with permission from the American Chemical Scociety)

	Lower transition		Main transition			
PC	T_c	ΔH_{cal}	T_t	ΔH_{cal}	CU	ΔS
DSPC	52.3	1.6	54.8	10.6	198	32
(1,2)-PC-4M			30.1	7.9	160	26
(1,2)-PC-6M			7.2	5.1	137	18
(1,2)-PC-16M	16.2	2.9	26.9	7.5	54	25
(2)-PC-4M			41.5	8.8	175	28
(2)-PC-6M			30.5	6.5	104	21
(2)-PC-8M			18.6	5.1	62	17
(2)-PC-10M			6.1	3.5	28	13
(2)-PC-12M			8.0	5.7	20	20
(2)-PC-16M			38.5	9.8	76	31
(1,2)-PC-4B			19.8	9.5	81	32

Units: T_c, °C; ΔH_{cal}, kcal/mol; ΔS, cal/mol K; CU, cooperativity unit.
DSPC, 1,2-distearoyl-sn-3-glycero-3-phosphatidylcholine numbers preceding PC indicate wether both or only the second chain is substituted

sition temperatures and enthalpy changes for synthetic and biologically derived lipids has been established. It also includes the measurement methods and experimental conditions used [100]. Recent studies on the principal transition behavior of phospholipid bilayers using combined approaches of high-resolution DSC and molecular mechanics simulation have been reviewed. An equation was derived allowing the prediction of T_t [95].

As stated, biological membranes are normally arranged as bilayers. It has, however, been observed that some lipid components of biological membranes spontaneously form non-lamellar phases, including the inverted hexagonal form (Figure 1.9) and cubic phases [101]. The tendency to form such non-lamellar phases is influenced by the type of phospholipid as well as by inserted proteins and peptides. An example of this is the formation of non-lamellar inverted phases by the polypeptide antibiotic Nisin in unsaturated phosphatidylethanolamines [102]. Non-lamellar inverted phase formation can affect the stability of membranes, pore formation, and fusion processes. So-called lipid polymorphism and protein–lipid interactions have been discussed in detail by Epand [103].

Pretransition

Upon heating, very pure uniform phospholipids show an additional small transition below the crystalline to liquid crystalline phase transition. According to Janiak *et al.* [104], this is due to a structural change in the head groups. The occurrence of pretransition is mainly determined by the structure of the head groups and their environment.

1.3.2 Phase Separation and Domain Formation

Phase separation is observed when mixtures of phospholipids are incorporated into a membrane and when the individual components exhibit certain differences in their structure. In this case, phase separation is indicated in the thermogram by two or more signals. Structural characteristics that could lead to phase separation are differences in chain length or different degrees of saturation. Mixtures of dipalmitoylphosphatidic acid (DPPA) and DPPC or of distearoyllecithin with dimyristoyllecithin, for example, show the phenomenon of phase separation. The reason for lateral lipid bilayer heterogeneity is the many-particle character of the lipid bilayer assembly. The existence of lateral domain organization in the nanometer range has been experimentally shown and can be predicted by computer simulation [105]. Important is the observation that in the area between two domains the structure of lipids is extremely disturbed.

Many ESR studies (see Section 3.3.6) using different systems have shown that phase separation in lipid layers may lead to a domain-like lateral structure. The area of domain formation can be extended over several hundred angstroms. In this context, charge-induced domain formation in biomembranes is of special interest for the medicinal chemist. In particular, the addition of Ca^{2+} to negatively charged lipids leads to domain formation [106]. Each lipid component is expected to have a characteristic spontaneous curvature. The Ca^{2+}-induced domains lead to protrusions in the

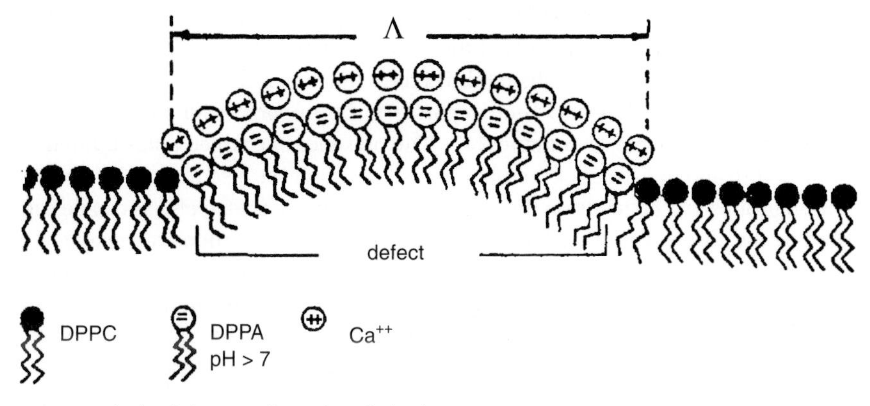

Fig. 1.10 The lipid domain of one phospholipid component (e.g. DPPA) forms a protrusion within the two-dimensional structure of the second component (e.g. DPPC). The protrusion is caused by differences in spontaneous curvature of the two phospholipids. (Reprinted from ref. 94 with permission from Wiley-VCH.)

membrane plane. Lateral variation in concentration in the plane of the membrane would then lead to a parallel variation in curvature (Figure 1.10) [94], which can be documented by electron microscopy.

In biological membranes differences not only in lipid composition and Ca^{2+} ions but also in membrane-integrated proteins can lead to phase separation as a result of local tightening of membrane regions [107]. The phase separation observed in the region of integrated membrane proteins is of importance for their functioning. According to Carruthers and Melchior [108], phase separation and domain formation create an area of rigid phospholipids around the proteins in an otherwise fluid surrounding. The resulting microheterogeneity is of the utmost importance for many membrane functions. At the border between rigid and fluid phospholipids the properties differ significantly from those of the pure phase. This can have a number of consequences [50]:

- Disarray of the structural molecular packing can enhance penetration by foreign molecules and their insertion (proteins, drugs) into the bilayer and will lead to an increase in permeability.
- Dynamic heterogeneity produces weakening of the mechanical properties, i.e. the compressibility and bending rigidity of the bilayer, because they are linked to the local membrane curvature.
- Lipid domains lead to compartmentalization of the lipid bilayer and affect the association and binding of proteins and enzymes and thereby their function.

The last point has been demonstrated in several contexts, most notably in protein kinase C [109, 110] and phospholipases [111–113], in coupling of G-proteins to their receptors [114], and in the activation of pancreatic lipase by colipase [115].

It has also been observed that sorting and transport of lipids and proteins can be mediated by the formation of small lipid-based domains, so-called rafts. The tendency to form such rafts has been studied using the example of sphingolipid–cholesterol. It was found that the integrity of the rafts is dependent on cholesterol, and that the association of proteins to rafts modifies their functions [116].

Domain formation has also been reported to be important for the budding of viruses from the plasma membrane [117]. The presence of charged lipids influences the structure and dynamics of membranes as a result of the electrostatic interaction of head groups and depends on the degree of dissociation. Strong effects can be triggered by the absorption of Ca^{2+}. Ca^{2+} ions and other divalent cations can change the phase transition behavior of mixtures of negatively charged and zwitterionic phospholipids and can lead to phase separation. They can also induce an increase in transition enthalpy. Phase transition temperature can be increased by several degrees centigrade. This effect is also of importance in Ca^{2+}-induced cell fusion.

Lipids exist in multiple phases. The simultaneous presence of several phases within a membrane can lead to lipid domain formation. One example would be the coexistence of gel and liquid crystalline phases, described in model membranes [118]. Most biomembranes are fluid. Therefore, it is important to understand the underlying mechanisms leading to domain formation [119]. Several reasons and examples are presently discussed, e.g. osmotically induced immiscibility [120], the phase segregation of dimyristoylphosphatidylcholine and distearoylphosphatidylcholine mixtures observed above main phase transition temperature [121], and non-ideal mixing of 16:0, 18:1 phosphatidylserine and di-12:1 phosphatidylcholine, as well as density fluctuation at the main phase transition [122]. In the beginning, hydrophobic mismatch was considered to be the mechanism by which integral membrane proteins attract lipids of matching hydrophobic thickness [123, 124]. Later, Lehtonen *et al.* [125] provided evidence that the formation of microdomains in liquid crystalline LUVs composed of phosphatidylcholines with homologous monounsaturated acyl chains of varying length can also be caused by hydrophobic mismatch of the constituent phospholipids. Domain formation leads to high rates of lateral diffusion. The results also offer an explanation for the chain length diversity observed in biological membranes [125].

The physical and functional properties of natural and artificial membranes which have been discussed in brief in this section have been the subject of extensive investigations. Several books and reviews have been published on these topics [8, 126]. The described features of membranes contribute to the physicochemical properties that make biological membranes highly structured fluids, in both space and time. They confer on the membranes particular structural, dynamic, and functional properties.

For the medicinal chemist it is of interest to note that such phenomena, i.e. phase separation and domain formation, can also be induced in artificial membranes by cationic amphiphilic drugs. An increase in the microheterogeneity of biological membranes and in consequence a decisive change in membrane function in a defined area must, therefore, be considered. It is mediated through indirect physicochemical interaction with amphiphilic drugs.

Depending on the relative affinities and concentrations, Ca^{2+} ions can displace amphiphilic drugs from binding sites in membranes and vice versa. This effect can easily be followed using labeled Ca^{2+} ions and phospholipid monolayers [127] (see Section 3.2). The interaction of larger polar molecules, such as aminoglycoside antibiotics [128], or of polyamines, such as spermine and spermidine [129], with charged phospholipids has been described.

Polypeptide antibiotics, such as gramicidin A and polymyxin B, are capable of increasing the permeability of bacterial membranes. As is to be expected, they change the phase transition, much like cholesterol [130]. These substances induce a tightening of fluid membranes and an increase in the fluidity of rigid membranes. It has been shown that polymyxin B produces phase separation and forms a Dimyristoylphosphatidylcholine (DMPG)-rich phase in DMPG/DMPC membranes [131].

Using cholesterol as the basis, a theoretical model to explain the observed changes in lipid domain interfacial area has been derived. The model shows that enhancement of lateral density fluctuation and lipid domain interfacial area caused by cholesterol is stronger at temperatures further from the transition temperature [132]. A decrease in the phase transition temperature of DPPC vesicles upon addition of antidepressants and phase separation at increased concentrations has also been reported [133].

1.4 Possible Effects of Drugs on Membranes and Effects of Membranes on Drug Molecules

Artificial membranes have become a subject of intensive research as a model for biological membranes. Experimental work on such artificial membranes or liposomes has demonstrated that their structural properties are strongly affected by membrane-associated molecules. Many drugs directly or indirectly influence cell membrane properties; for example, interactions between proteins and phospholipids or the formation of complexes between ligand molecules and phospholipids or sterols can lead to disruption of the membrane so that it becomes highly permeable. Also, some compounds affect the fluidity of membranes. This is particularly true for anesthetics, whose action has been studied and reported [134]. Methods of studying membrane fluidity have been described [135]. As discussed, changes in membrane fluidity can affect receptor and enzyme activity at both the static and dynamic level. Membrane fluidity also influences the ability of drugs to pass through the membrane, which can affect their efficacy.

Because of the unique structure of a lipoid matrix consisting of phospholipids and embedded proteins, the interaction of drug molecules with polar head groups, apolar hydrocarbons, or both, can induce several changes in the membrane. Consequently, the drug behavior is changed (diffusion, accumulation, and conformation) [136].

Events which can arise from drug–membrane interaction are summarized below.

(a) Possible effects of drugs on membrane properties:

- change in conformation of acyl groups (*trans–gauche*);
- increase in membrane surface (curvature);
- change in microheterogeneity (phase separation, domain formation)
- change in the thickness of the membrane;

- change in surface potential and hydration of head groups;
- change in phase transition temperature (fluidity, cooperativity);
- change in membrane fusion.

(b) Possible effects of the membrane on drug molecules:

- diffusion through membrane may become rate limiting;
- membrane may completely prevent the diffusion to the active site (resistance);
- membrane may bind drug (accumulation, selectivity, toxicity)
- desolvation of the drug in the membrane may lead to a conformational change of drug molecules or force them into a special orientation (efficacy).

Pioneering work which has shed new light on the importance of drug–membrane interaction in the understanding of drug action and drug development processes has been contributed by Herbette and coworkers [83, 137–142]. A general model has been discussed [83, 142]. This model is summarized in Figure 1.11 and takes account of the various possibilities which can become rate limiting for the interaction of drugs with membrane-bound receptors. The first part of the model describes how a lipophilic substance can trace specific or unspecific binding sites on membrane proteins. a) Binding can occur directly at those proteins sticking out of the membrane or after distribution into the lipid phase and lateral diffusion to the binding site of the membrane-integrated proteins. b) During the diffusion process, drug molecules move along a defined membrane bilayer axis and c) can adopt a well-defined orientation as well as conformation with respect to the bilayer axis. d) Alternatively, the highly structured environment of the bilayer can reduce diffusion and favor or hinder the orientation or optimal conformation of a drug molecule. For example, the preferred conformation or the equilibrium between conformers may change in the environment of the bilayer, which has a different polarity from water. The highly structured environment can also reduce the rate of diffusion, depending on the physicochemical properties of the drug molecules.

The second part of the model assumes that an influence on the functioning of the membrane-integrated proteins is possible even without a specific drug–protein interaction. This is understandable if one considers another parameter that characterizes cells or vesicles, namely their curvature. This is a measure of internal stress and depends on the tendency of bilayers to assume a non-bilayer conformation, for instance a hexagonal or cubic phase (see Section 1.1.3.1 and Figure 1.9). This transformation can, for example, be detected and measured by X-ray or neutron diffraction techniques.

Optimal functioning of some cell membranes and their integrated proteins affords a defined curvature. Changes in curvature induced by certain membrane-active molecules will lead to an indirect perturbation of the membrane, thus modifying or preventing agonist–receptor interaction and the biological response. Amphiphilic drugs have been shown to have such effects on curvature.

Because of the defined organization and composition of bilayers we cannot expect that drug transport, distribution, efficacy and the degree of resistance can always be sufficiently described by the octanol–water partition coefficient. Membranes do not consist only of the hydrocarbon lipids but possess polarized phosphate groups and neutral or positively or negatively charged head groups. They are highly structured

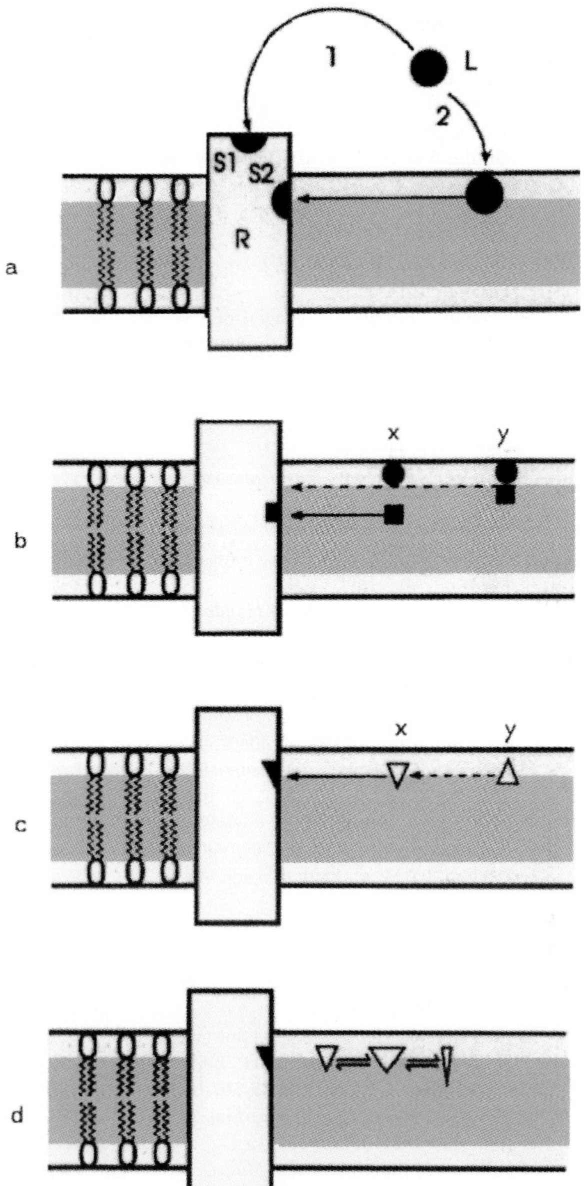

Fig. 1.11 Various possibilities for the rate-limiting steps in the approach to the receptor proteins embedded in the membrane. (Reprinted from ref. 83 with permission from the American Chemical Society.)

and can be asymmetric and chiral so that – besides overall hydrophobic forces – additional physicochemical properties of drug molecules are involved in drug–membrane interaction. This is especially true in the case of amphiphilic drugs. The "hydrophobic gravity center" of molecules, together with hydrogen binding and charge interaction, may become an important factor in the orientation within the membrane.

References

1 Lüllmann, H., Mohr, K., Ziegler, A., *Taschenatlas der Pharmakologie*, Thieme Verlag, Stuttgart **1990**.
2 Quinn, P.J., *The Molecular Biology of Cell Membranes*, Macmillan, London **1976**.
3 Epand, R.M., *Biochemistry* **1985**, *24*, 7092–7095.
4 Spector, A.A., Yorek, M.A., *J. Lipid Res.* **1985**, *26*, 1015–1035.
5 Huang, J., Bubolz, J.T., Feigenson, G.W., *Biochim. Biophys. Acta* **1999**, *1417*, 89–100.
6 McMullen, T.P.W., Lewis, R.N.A.H., McElhaney, R.N., *Biochim. Biophys. Acta* **1999**, *1416*, 119–134.
7 Lau, W.F., Das, N.P., *Experientia* **1995**, *51*, 731–737.
8 Yeagle, P., *The Structure of Biological Membranes*, CRC Press, Boca Raton **1992**.
9 Cevc, G. (Ed.), *Phospholipids Handbook*, Marcel Dekker, New York **1993**.
10 Lee, M.G., Jacobson, M., *Curr. Topics Membranes* **1994**, *40*, 111–142.
11 Redelmeier, T.E., Hope, M.J., Cullis, P.R., *Biochemistry* **1990**, *29*, 3046–3053.
12 Rothman, J.E., Lenard, J., *Science* **1977**, *195*, 743–753..
13 Verkleij, A.J., Zwall, R.F.A., Roelofsen, B., Comfurius, P., Kastelijn, D., Van Deenen, L.L.M., *Biochim. Biophys. Acta* **1973**, *323*, 178–195.
14 Devaux, P.F., Zachowski, A., *Chem. Phys. Lipids* **1994**, *73*, 107–120.
15 Williamson, P., Shlegel, R.A., *Mol. Membr. Biol.* **1994**, *11*, 199–216.
16 van Hellvoort, A., Smith, A.J., Sprong, H., Fritzsche, I., Schinkel, A.H., Borst, P., van Meer, G., *Cell* **1998**, *87*, 507–517.
17 Kamp, D., Haest, C.W.M., *Biochim. Biophys. Acta* **1996**, *1372*, 91–101.
18 Kellen, J.A. (Ed.), *Reversal of MultiDrug Resistance in Cancer*, CRC Press, Boca Raton **1994**.
19 Buton, X., *J. Biol. Chem.* **1996**, *271*, 6651–6658.
20 Schroit, A.J., Zwaal, R.F.A., *Biochim. Biophys. Acta* **1991**, *1071*, 313–329.
21 Diaz, C., Schroit, A.J., *J. Membr. Biol.* **1996**, *151*, 1–9.
22 Eastman, S.J., Hope, M.J., Wong, K.F., Cullis, P.R., *Biochemistry* **1992**, *31*, 4262–4268.
23 Gruner, S.M., Shyamsunder, E., *Ann. N.Y. Acad. Sci.* **1991**, *625*, 685–697.
24 Cullis, P.R., Hope, M.J., de Kruiff, B., Verklej, A.J., Tilcock, P.S., in *Phospholipids and Cellular Regulations*, J.F. Kuo (Ed.), CRC Press, Boca Raton **1985**, pp. 1–60.
25 Marsh, D., *Biochim. Biophys. Acta* **1996**, *1279*, 119–123.
26 Navarro, J., Tiovio-Kinnucan, M., Racker, E., *Biochemistry* **1984**, *23*, 130–135.
27 Herbette, L.G., *Dev. Cardiovasc. Med.* **1987**, *68*, 353–365.
28 Singer, S.J., Nicholson, G.L., *Science* **1972**, *175*, 720–731.
29 Blobel G., *Proc. Natl. Acad. Sci. U.S.A.* **1980**, *77*, 1496–1500.
30 McCarthy, M.P., in *The Surface of Biological Membranes*, P. Yeagle (Ed.), CRC Press, Boca Raton, **1991**, pp. 395–447.
31 Polovinikova, O.Y., Simonova, T.N., Tarahkovsky, Y.S., Nikonova, T.N., *Biol. Membr.* **1995**, *12*, 260–278.
32 Schwyzer, R., *Biochemistry* **1986**, *25*, 4281–4286.
33 Chow, M., Der, Ch. J., Buss, J.E., *Curr. Opin. Cell Biol.* **1992**, *4*, 629–636.
34 Epand, R.M., Yechiel, S., Segrest, J.P., Anantharamaiah, G.M., *Biopolymers* **1995**, *37*, 319–338.
35 Epand, R.M, Shai, Y., Segrest, J.P., *Biopolymers* **1995**, *37*, 339–348.
36 Murata, M., Kagiwada, S., Takahashi, S., Ohnishi, S., *J. Biol. Chem.* **1991**, *266*, 14353–14358.
37 Chiba, K., Mohri, T. *Biochemistry* **1989**, *28*, 2995–2999.
38 White St. H., Wimley, W.C., *Biochim. Biophys. Acta* **1998**, *1376*, 339–352.
39 Killian, J. A., *Biochim. Biophys. Acta* **1998**, *1376*, 401–416.
40 Winter, R., *Chem. unserer Zeit* **1990**, *24*, 71–81.
41 Edidin, M., *Annu. Rev. Biophys. Bioeng.* **1974**, *3*, 179–201.
42 Klausner, R.D., Wolf, D.E., *Biochemistry* **1980**, *19*, 6199–6203.

43 Siekewitz, P., *Annu. Rev. Physiol.* **1972**, *34*, 117–140.
44 Kinnunen, P.K.J., *Chem. Phys. Lipids* **1991**, *57*, 375–399.
45 Mouritsen, O.G., Jorgensen, K. *BioEssays* **1992**, *14*, 129–136.
46 Biltonen, R.L., *Thermodynamica Acta* **1990**, *172*, 1–10.
47 Bell, J.D., Burnside, M., Owen, J.A., Royall, M.L., Baker, M.L., *Biochemistry* **1996**, *35*, 4945–4955.
48 Lee, A.G., *Progr. Lipid Res.* **1991**, *30*, 323–348.
49 Senistra, G., Epand, R.M., *Arch. Biochim. Biophys.* **1993**, *300*, 378–383.
50 Mouritzen, O.G., Jørgensen, K., *Pharm. Res.* **1998**, *15*, 1507–1519.
51 Kayser, F.H., Bienz, K.A., Eckert, J., Lindemann, J. *Medizinische Mikrobiologie* 7. Aufl. Georg Thieme, Stuttgart **1989**.
52 Seltmann, G. *Die bakterielle Zellwand*, Gustav Fischer Verlag, Stuttgart **1982**.
53 Hammond, S.M., Ambert, P.A., Ycroft, A.N. *The Bacterial Cell Surface*, Croom Helm, London **1984**.
54 Dimitriev, B.A., Ehlers, St., Rietschel, E.Th., *Med. Microbiol. Immunol.* **1999**, *187*, 173–181.
55 Rastogi, N., *Res. Microbiol.* **1991**, *142*, 464–476.
56 Dimitriev, B.A., Ehlers, St., Rietschel, E.Th., Brennan, P.J., *Int. J. Med. Microbiol.* **2000**, *290*, 251–258.
57 Nikaido, H., Nakae, T., in *Advances in Microbial Physiolgy*, Vol. 20, A.H. Rose, J. Gareth (Eds.), Academic Press, London **1979**, pp. 163–250.
58 Korn, E.D. *Science* **1966**, *153*, 1491–1498.
59 Nikaido, H., *Science* **1994**, *264*, 382–388.
60 Westphal, O., in *Chemistry of Endotoxin*, Vol. 1, E.Th. Rietschel (Ed.), Elsevier Science, Amsterdam **1984**, pp. 17–23.
61 Jacobs, E.R., in *Pathophysiology of Endotoxin, Handbook of Endotoxin* Vol. 2, L.B. Hinshaw (Ed.), Elsevier Science, Amsterdam **1985**, pp. 1–15.
62 Jann, K., Jann, B., in *Chemistry of Endotoxin, Handbook of Endotoxin* Vol. 1, E.Th. Rietschel (Ed.), Elsevier Science, Amsterdam **1984**, pp. 138–186.
63 Nikaido, H., *Pharmacol. Ther.* **1985**, *27*, 197–231.
64 Seltmann, G., *Acta Biotechnol.* **1990**, *10*, 107–115.
65 Seydel, U., Schröder, G., Brandenburg, K., *J. Membr. Biol.* **1989**, *109*, 95–103.
66 Wiese A., Münstermann, T., Lindner, B., Kawahara, K., Zähringer, U., Seydel, U., *J. Membr. Biol.* **1998**, *162*, 127–138.
67 Zlotnik, H., Fernandez, M.P., Bowers, B., Cabib, E.J., *Bacteriology*, **1984**, *159*, 1018–1026.
68 Sullivan, P.A., Chiew, Y.Y., Molloy, C., Templeton, M.D., Shepherd, M.G., *J. Microbiol.* **1983**, *29*, 1514–1525.
69 Gopal, P.K., Shepherd, M.G., Sullivan, P.A., *J. Gen. Microbiol.* **1984**, *130*, 3295–3301.
70 Braun, P.C., Calderone, R.A., *J. Bacteriol.* **1978**, *135*, 1472–1477.
71 Georgopapadakou, N.H., Dix, B.A., Smith, S.A., Freudenberg, J., Funke, P.T., *Antimicrob. Agents Chemother.* **1987**, *31*, 46–51.
72 Rodriguez, R.J., Low, C., Bottema, C.D.K., Parks, L.W., *Biochim. Biophys. Acta* **1985**, *837*, 336–343.
73 Georgopapadakou, N.H., in *Perspectives in Antiinfective Therapy*, G.G. Jackson, H.D. Schlumberger, H.-J. Zeiler (Eds.), Springer-Verlag, Berlin **1989**, pp. 100–111.
74 Ching-hsien H., in *Phospholipid Binding Antibodies*, E.N. Harris, T. Exner, G.R.V. Hughes, R.A. Asherson (Eds.), CRC Press, Boca Raton **1991**, pp. 4–27.
75 Lasic, D.D., *J. Liposome Res.* **1995**, *5*, 431–441.
76 Szoka, F., Papahadjopoulos, D., *Annu. Rev. Biophys. Bioeng.* **1980**, *9*, 467–508.
77 Papahadjopoulos, D., Miller, N., *Biochim. Biophys. Acta* **1967**, *135*, 624–638.
78 Bangham, A.D., Hill, M.W., Miller, N.G.A., *Methods Membr. Biol.* **1974**, *1*, 1–68.
79 Szoka, F., Papahadjopoulos, D., *Proc. Natl. Acad. Sci. U.S.A.* **1978**. *75*, 4194–4198.
80 Szoka, F., Olso, F., Heath, T., Vail, W., Mayhew, E., Paphadjopoulos, D., *Biochim. Biophys. Acta* **1980**, *601*, 559–571.

81 Nayar, R., Hope, M.J., Cullis, P.R., *Biochim. Biophys. Acta* **1989**, *986*, 200–206.
82 Mac Donald, R.C., MacDonald, R.I., Menco, B.Ph.M., Takeshita, K., Subbarao, N.K., Lau-rong, H., *Biochim. Biophys. Acta* **1991**, *1061*, 297–303.
83 Mason, R.P., Rhodes, D.G., Herbette, L.G., *J. Med. Chem.* **1991**, *34*, 869–877.
84 Riaz, M., Weiner, N.D., Schacht, J., *J. Pharm. Sci*, **1989**, *78*, 172–175.
85 Buboltz, J.T., Feigenson, G.W., *Biochim. Biophys. Acta* **1999**, *1417*, 232–245.
86 Schütze, W., Müllert-Goymann, C.C., *Pharm. Res.* **1998**, *15*, 538–543.
87 Davidson, S.M.K., Cabral-Lilly, D., Maurio, F.P., Franklin, J.C., Minchey, S.R., Ahl, P.L., Janoff, A.S., *Biochim. Biophys. Acta* **1997**, *1327*, 97–106.
88 Maurer, N., Wong, K.F., Hope, M.J., Cullis, P.R., *Biochim. Biophys. Acta* **1998**, *1374*, 9–20.
89 Maurere-Spurej, E., Wong, K.F., Maurer, N., Fenske, D.B., Cullis, P.R., *Biochim. Biophys. Acta* **1999**, *1416*, 1–10.
90 Bonan, N.L., Mayer, L.D., Cullis, P.R., *Biochim. Biophys. Acta* **1993**, *1152*, 253–258.
91 Oberholzer, Th., Meyer, E., Amato, I., Lustig, A., Monnard, P.-A., *Biochim. Biophys. Acta* **1999**, *1416*, 57–68.
92 Blum, A., in *Phospholipid Handbook*, G. Cevc (Ed.), Marcel Dekker, New York **1993**, pp. 455–510.
93 Seydel, J.K. *Trends Pharmacol. Sci.* **1991**, *12*, 368–371.
94 Sackmann, E., *Ber. Bunsenges. Phys. Chem.* **1978**, *82*, 891–909.
95 Chung-hsien, H., Li, Sh., *Biochim. Biophys. Acta* **1999**, *1422*, 273–307.
96 Cevc, G., Marsh, D., *Phospholipid Bilayers, Physical Principles and Methods*, John Wiley, New York **1987**.
97 McElhaney, R.N., *Portland Press*, London, Proc. *7*, **1994**, pp. 31–48.
98 Chapman, D., in *Form and Function of Phospholipids*, G.B. Ansell, J.N. Hawthorne, R.M.C. Dawson (Eds.), BBA Library Vol. 3, Elsevier, Amsterdam, **1973**, pp. 117–142.
99 Menger, F.M., Wood, M.G., Zhou, Q.Z., Hopkins, H.P., Fumero, J., *J. Am. Chem. Soc.* **1988**, *110*, 6804–6810.
100 Caffrey, M., Moynikan, D., Hogan, J., *J. Chem. Inf. Comput. Sci.* **1991**, *31*, 275–284.
101 Luzzati, V., *Curr. Opin. Struct. Biol.* **1997**, *7*, 661–668.
102 Jastimi, R.E., Lafleur, M., *Biochim. Biophys. Acta* **1998**, *1418*, 97–105.
103 Epand, R.M., *Biochim. Biophys. Acta* **1998**, *1376*, 353–368.
104 Janiak, M.J., Small, D.M., Shipley, G.G., *Biochemistry* **1976**, *15*, 4575–4580.
105 Mouritzen, O.G., Jørgensen, K., *Curr. Opin. Struct. Biol.* **1997**, *7*, 518–527.
106 Silvius, J.R., *Biochemistry* **1990**, *29*, 2930–2938.
107 Horvath, L.L., Brophy, P.J., Marsh, D., *Biochemistry* **1990**, *29*, 2635–2638.
108 Carruthers, A., Melchior, D.L., *Trends Biochem. Sci.* **1986**, *11*, 23–34.
109 Hinderleiter, A.K., Dibble, A.R.G., Biltonen, R.L., Sando, J.J., *Biochemistry* **1997**, *36*, 6141–6148.
110 Yang, L., Glaser, M., *Biochemistry* **1995**, *34*, 1500–1506.
111 Kimmelberg, H.K., in *Dynamic Aspects of Cell Surface Organization*, G. Post, G.L. Nicolson (Eds.), North Holland, New York **1977**, pp. 205–293.
112 Basanez, G., Nieva, J.-L., Goni, F., Alonso, A., *Biochemistry* **1996**, *35*, 15183–15187.
113 Burack, W.R., Biltonen, R.L., *Chem. Phys. Lipids* **1994**, *73*, 209–222.
114 Neubig, R.R., *FASEB J.* **1994**, *8*, 939–946.
115 Momsen, W.E., Momsen, M.M., *Biochemistry* **1995**, *34*, 7271–7281.
116 Rietveld, A., Simons, K., *Biochim. Biophys. Acta* **1998**, *1376*, 467–479.
117 Luan, P., Glaser, M., *Biochemistry* **1994**, *33*, 4483–4489.
118 Vaz, W.C., *Mol. Membr. Biol.* **1995**, *12*, 39–43.
119 Mouritsen, O.G., Kinnunen, P.K.J., in *Membrane Structure and Dynamics*, K.M. Merz Jr., B. Roux (Eds.), Birkhäuser Publishing. **1996**, pp. 463–502.
120 Lehtonen, J.Y.A., Kinnunen, P.K.J., *Biophys. J.* **1995**, *68*, 525–535.
121 Melchior, D.L., *Science* **1986**, *234*, 1577–1580.

122 Mouritsen, O.G., Jørgensen, K., *Chem. Phys. Lipids* **1994**, *73*, 3–25.

123 Mouritsen, O.G., Bloom, M., *Biophys. J.* **1984**, *46*, 141–153.

124 Zhang, Y.-P., Lewis, R.N.A.H., Hodges, R.S., McElhaney, R.N., *Biochemistry* **1992**, *31*, 11579–11588.

125 Lehtonen, J.Y.A., Holopainen, J.M., Kinnunen, P.K.J., *Biophys. J.* **1996**, *70*, 1753–1760.

126 Lipowski, R., Sackmann, E. (Eds.), *Structure and Dynamics of Membranes, Handbook of Biological Physics, A & B*, Vol. 1 Elsevier, Amsterdam **1995**.

127 Lüllmann H., Plösch, H., Ziegler, A., *Biochem. Pharmacol.* **1980**, *29*, 2969–2974.

128 Lüllmann, H., Volmer, B., *Biochem. Pharmacol.* **1982**, *31*, 3169–3173.

129 Eklund, K.K., Kinnunen, P.K.J., *Chem. Phys. Lipids* **1986**, *39*, 109–117.

130 Chapman, D., Urbina, J., Keough, K.M.W., *J. Biol. Chem.* **1974**, *249*, 2512–2521.

131 Sixl, F., Watts, A., *Biochemistry* **1985**, *24*, 7906–7910.

132 Cruzeiro-Hanson, L., Ipsen, J.H., Mouritzen, O.G., *Biochim. Biophys. Acta* **1988**, *979*, 166–176.

133 Seydel, U., Brandenburg, K., Lindner, B., Moll, H., *Thermochim. Acta* **1981**, *49*, 35–48.

134 Aloia, R.C., Curtain, C.C., Gordon, L.M. (Eds.), *Drug and Anesthetic Effects on Membrane Structure and Function*, Wiley-Liss, NewYork **1991**.

135 Aloia, R.C., Curtain, C.C., Gordon, L.M. (Eds.), *Methods for Studying Membrane Fluidity* Wiley-Liss, New York, Vol. 1 Vol. 2 **1983**, Vol. 4 **1985**.

136 Seydel, J.K., Coats, E.A., Cordes, H.-P., Wiese, M., *Arch. Pharm. (Weinheim)* **1994**, *327*, 601–610.

137 Herbette, L.G., Chester, D.W., Rodes, D.G., *Biophys. J.* **1986**, *49*, 91–94.

138 Chester, D.W., Herbette, L.G., Mason, R.P., Joslyn, A.F., Triggle, D.J., Koppel, D.E., *Biophys. J.* **1987**, *52*, 1021–1030.

139 Herbette, L.G., Gruner, S.M., *Dev. Cardiovasc. Med.* **1987**, *68*, 353–365.

140 Mason, R.P., Chester, D.W., *Biophys. J.*, **1989**, *56*, 1193–1201.

141 Bae, S.-J., Kitamura, S., Herbette, L.G., Sturtevant, J.M., *Chem. Phys. Lipids* **1989**, *51*, 1–7.

142 Mason, R.P., Morling, J., Herbette, L.G., *Nucl. Med. Biol.* **1990**, *17*, 13–33.

143 Rhodes, D.G., Newton, R., Butler, R., Herbette, L.G., *Mol. Pharmacol.* **1992**, *42*, 596–602.

2
Octanol–Water Partitioning versus Partitioning into Membranes

Joachim K. Seydel

The oldest publication on structure–activity relationships (SARs) known to us from the literature is a paper by Cros from 1861. He compared the toxic effect of alcohols in various species after different routes of administration. He found an increase in toxic effect with decreasing water solubility, that is increasing lipophilicity. Cros was also the first to detect a maximum in activity followed by a decrease, i.e. a non-linear relationship, as a function of solubility of alcohols in water.

Forty years later Meyer [1], and at the same time Overton [2], observed a linear relationship between the activity of narcotics and their oil–water partition coefficient. An 40 years after that, a thermodynamic interpretation of this relationship was provided by Ferguson [3], which also explained "cut-off" of biological activity that is sometimes observed after a certain lipophilicity range has been passed.

The real breakthrough came in 1964 when Hansch and Fujita [4] described the logarithm of the biological activity within a series of compounds as a linear combination of different physicochemical properties of drug molecules. The lipophilicity was quantified as the partition coefficient, log P, in the system octanol–water. The authors also defined the contribution to the lipophilicity π, of substituents X:

$$\pi_x = \log P_x - \log P_H \tag{2.1}$$

where log P_x is the partition coefficient of the substituted derivative and log P_H of the parent structure. Since then, log $P_{oct.anol}$ ($P_{oct.}$) and the derived π values have been successfully used in thousands of examples and became the basis of quantitative SAR (QSAR) methodology. For details see refs. 5–9.

The reasons why octanol was chosen as the reference system are manifold:

- It is thought to have a membrane-like structure.
- It possesses hydrogen-bonding characteristics that do not differ too markedly between the two phases.
- The temperature dependence of log $P_{oct.}$ is lower than in hydrocarbon–water systems.
- It has a low solubility in water and a very low vapor pressure.
- It is transparent to UV radiation.
- Huge databases of experimentally determined log $P_{oct.}$ values, containing several thousand examples, are available.

In addition, various methods of calculating lipophilicity based on these data bases are available [10, 11].

Nevertheless, there is increasing evidence that the octanol–water partition coefficient may not be the most suitable partition coefficient in all cases. Meyer and Hemmi [12] concluded that oleyl alcohol would better describe the data of Overton [13] and Meyer [1] than olive oil. Diamond [14] found that a better correlation for the partitioning into *Nitella* membranes was obtained using olive oil than with isobutanol, and in 1979 [15] it was suggested that hydrophobic protein pockets are not well modeled by octanol. Rekker [16] discussed the possibility of using a range of solvent systems to describe various membrane systems, such as buccal membrane or the blood–brain barrier. On the basis of membrane partition coefficients, Herbette *et al.* [17] and Mason and Chester [18] have indeed shown that there can be an enormous discrepancy between the partitioning of drugs into the octanol–water systems and into biological membranes (see Chapter 4).

Numerous examples available from the application of the octanol–water system allow translation between different solvents or the modeling of a lipid membrane, provided the Collander relation [19] is valid.

$$\log P_{\mathrm{II}} = a \log P_{\mathrm{I}} + b \tag{2.2}$$

This means that partition coefficients obtained in one solvent system are related to those determined in another system. Seiler [20] derived the following equation under the assumption that a in Eq. 2.2 is 1 in the absence of hydrogen bonding. In such a way, I_{H} values (H-bond descriptors) have been derived for several functional groups and used in QSAR analysis.

$$\log P_{\mathrm{oct.}} = \log P_{\mathrm{cyclohexane}} + I_{\mathrm{H}} - 0.16 \tag{2.3}$$

The above equation is, however, invalid. In homologous series in which Eq. 2.2 holds, a is essentially the fraction value, $f_{\mathrm{CH2.\pi}}$. But it is also well established that f_{CH2} is a variable quantity in different solvent systems. Log P values in octanol and cyclohexane have been determined by Taft and coworkers [21, 22] for a series of non-amphitropic, monofunctional aliphatics. The partition coefficient obtained could be described and correlated by the following equations:

$$\log P_{\mathrm{oct.}} = 2.66V/100 - 0.96\pi^{*} - 3.38\beta + 0.24 \tag{2.4}$$

$n = 47 \qquad r = 0.991 \qquad s = 0.18$

$$\log P_{\mathrm{cyclohexane}} = 3.69V/100 - 1.15\pi^{*} - 5.65\beta - 0.05 \tag{2.5}$$

$n = 47 \qquad r = 0.998 \qquad s = 0.07$

where V is the molar volume and β and π^* are proton acceptor and polarity/polarizability parameters respectively. The volume term clearly indicates that it is easier for the non-polar moieties to partition into cyclohexane than into octanol. Therefore, the derivation of I_{H} in Eq. 2.3 is not sound. It is not a universal constant, but will vary with log P.

Using this line of argumentation, Leahy and coworkers [23, 24] came to the conclusion that Rekker's assumption [16], based on Eq. 2.3, that there is "no urgent need for differentiation between donor and acceptor solutes" should be treated with cau-

tion. According Leahy *et al.*, two conditions should be satisfied for the assumption to be valid:
1) Pure proton acceptors should retain the same proportionate strength from one solvent system to another.
2) The donor and acceptor abilities of amphitropic groups should remain in constant ratio.

The authors claim that neither condition is fulfilled and conclude that "no two-term equation will ever be found to relate two properly distinctive solvent systems, except in the trivial case that the substituents themselves are not discriminating." As a result, this disproves not only Eq. 2.2, but also other related two-term equations, such as those of Seiler [20] and Moriguchi [25].

Leahy and coworkers [23, 24] suggest four model systems for the description of "de novo real membranes:"
1) amphitropic, like water, for which octanol is the medium of choice;
2) inert, for which any alkane will be suitable;
3) a pure H-bond donor, for which chloroform will be the medium of choice; and
4) a pure H-bond acceptor.

In the last case the authors propose propylenglycoldipelargonate (PGDP) to complement octanol. An additional property of PGDP is its lower risk of extracting ion pairs compared to octanol.

Because PGDP exhibits greater resistance to phase separation and a much greater degree of UV opacity than other solvents, a high-performance liquid chromatography (HPLC) methodology similar to the one with octanol as the organic phase has been developed [23, 26]. Quite a large number of compounds have been studied. Their log P_{PGDP} values have been published and compared with their log $P_{oct.}$ values, and fragmental constants for selected solvent systems were derived (Table 2.1) [23, 27].

In a significant number of cases, large deviations from log $P_{oct.}$ in both directions were observed. In the following chapters further evidence is presented that log $P_{oct.}$ alone is unable to account for variations in biological selectivity. According to Leahy and coworkers [23, 24], this is the result of hydrogen-bonding substituents and not overall hydrophobicity alone. Mammalian membranes are amphitropic in nature as well as proton acceptors as a result of the polar and negatively charged surface of their phospholipids with varying head groups.

The concept of accessory binding sites [28] may implicate the role of different hydrogen-bonding characteristics. Thus, the supplementary use of PGDP or other proton acceptor systems may lead to a better understanding of drug–membrane interaction.

The first attempt to resolve this problem involved the use of Δlog P between two partition systems, namely octanol–water and hexane–water, as a model for the penetration of the blood–brain barrier [29]. This leads back to Seiler's I_H value [20] (for details see Section 4.2).

Partitioning of 121 solutes in five different solvent systems (octanol, heptane, chloroform, diethyl ether, and *n*-butylacetate) was carried out to determine the extent to which the hydrogen donor capacity of solutes affects the permeation of membranes,

Tab. 2.1 Fragment values for selected solvent systems. (Adapted from Table 2 of ref. 27)

Fragment*	Alkane	Octanol	$CHCl_3$	PGDP
C	0.20	0.19	0.20	0.19
H	0.22	0.17	0.21	0.16
CH_3	0.86	0.70	0.83	0.67
CH_2	0.64	0.53	0.62	0.51
H	0.42	0.36	0.41	0.35
C_6H_5	2.03	1.96	2.59	2.20
a F	0.44	0.31	0.26	0.30
a Cl	0.91	0.88	0.87	0.88
a Br	1.05	1.03	1.02	1.07
A CN	–2.37	–1.34	–0.82	–1.29
a CN	–0.99	–0.40	0.12	–0.54
a NO_2	–0.59	–0.11	0.34	–0.04
A NH_2	–3.87	–1.61	–2.47	–2.70
a NH_2	–2.03	–1.06	–1.33	–1.25
A NH A	(–3.57)	–1.96	(–2.64)	–2.95
A N (A) A	–3.68	–2.22		–2.75
A OH	–3.77	–1.66	(–2.21)	–2.48
a OH	–2.84	–0.50	–2.24	–1.03
A O A	(–2.39)	–1.55		–1.71
a O A	–0.80	–0.55	–0.30	–0.46
A CO A	–2.55	–1.75	(–1.42)	–1.81
a CO A	–1.77	–1.06	–0.63	–1.26
A CO_2 A	–1.82	–1.41		–1.34
a CO_2 A	–1.19	–0.55	–1.15	–0.54
A CO_2 H	–3.90	–1.08	(–2.43)	–2.12
a CO_2 H	–3.42	–0.09	–2.09	–1.05
A $CONH_2$	–5.53	–2.07	–2.24	–3.30
a $CONH_2$	–4.33	–1.32	–2.48	–2.56
A CONH A	(–5.55)	–2.45		–3.87
a CONH A	–4.65	–1.80	–2.46	–2.92
A CONH a	–4.59	–1.50	–2.54	–2.47
A CON (A) A		(–2.85)		–3.83
A NHCONH A		(–1.89)		–3.97
A NHCSNH A	(–5.86)	(–1.71)	(–2.40)	(–2.48)
A SO_2NH_2	–5.49	–1.95	–3.53	–2.60
a SO_2NH_2		–1.65	–2.83	–2.23
A SO A	–5.77	–2.69	–3.04	–4.10
A SO_2 A	–5.08	–2.52	–2.56	–2.80
a SO_2 A		–2.16	–1.55	–2.40

*) A, alkyl; a, aryl. Values in brackets are averaged values taken from literatur

especially the blood–brain barrier. The results suggest that the rate-limiting step in drug penetration of the blood–brain barrier "is the binding of drug to lipid-rich layers of the blood–brain barrier by the donation of hydrogen-bonds" [30].

El Tayar *et al.*, using micelle–water (mw) partition coefficients determined on a polyoxyethylene(23)–lauryl ether column [31], found a significant correlation between log P_{mw} and the activity of a series of antiplatelet agents, but no correlation when using log $P_{oct.}$ or calculated [10] log *P*-values (CLOGP) parameters.

Recently Rekker *et al.* [32] have developed an aliphatic hydrocarbon (ahc)–water fragmental system by stepwise interconnecting with the octanol–water partition values. It is based on the partition coefficients obtained for various series of structures, including hydrogen bond donors and acceptors, in the ahc–water system, where the aliphatic hydrocarbon can be cyclohexane, heptane, pentane, isooctane, nonane, dodecane, etc. In a first application of their f_{ahc} fragmental constants they were able to demonstrate the superiority of this parameter over the traditional $f_{oct.}$ fragmental constant in describing the equilibrium blood–brain concentration, log BB, of 24 compounds studied by Young *et al.* [29] and Abraham *et al.* [33]. (S = 95 % c. l.)

$$\log BB = 0.23\ (\pm 0.08)\ f_{oct} - 0.22\ (\pm 0.25) \tag{2.6}$$

$$n = 24 \qquad r = 0.778 \qquad s = 0.32 \qquad F = 34$$

$$\log BB = 0.18\ (\pm 0.035)\ f_{ahc} + 0.04\ (\pm 0.09) \tag{2.7}$$

$$n = 24 \qquad r = 0.935 \qquad s = 0.181 \qquad F = 153$$

HPLC should be an even more appropriate methodology for certain classes of compounds and substituents. Phospholipids or artificial membranes (liposomes) composed of various lipids would constitute the organic phase. Columns, including some with irreversibly bound phosphatidylcholine, are now commercially available. It has been found that, compared with log $P_{oct.}$, the antihemolytic activity of a series of phenothiazine neuroleptics was described significantly better by chromatographic indices using immobilized artificial membrane (IAM) columns ($r = 0.935$ and 0.812 respectively) [34]. Similarly, it has been shown for dihydropyridines that chromatographic indices determined on such IAM columns are useful. They are better descriptors of lipophilic and polar interactions with membranes, especially in the case of partially charged derivatives [35]. A comparison of membrane partition coefficients chromatographically measured using IAM columns (K'_{IAM}) or by partitioning into liposomes (K'_m) or in octanol–water ($K_{oct.}$) has been carried out for seven β-blockers, six imidazoline derivatives, and 10 imidazolidine derivatives. An excellent correlation was found between log K'_m and log K_{IAM} ($r = 0.907$, slope 0.994, intercept 0.496). Solute partitioning using octanol–water systems did not correlate with log K'_m ($r = 0.483$) or log K'_{IAM} ($r = 0.419$), except in the case of homologous series of compounds [36].

The most appropriate method of mimicking partition into biological membranes is by using artificial membranes composed of various phospholipids as model systems [37, 38]. The influence of phospholipid composition on drug partitioning can easily be studied (Table 2.2).

A standardized distribution model using artificial membranes has been developed by Wunderli-Allenspach and coworkers [39, 40]. The apparent partition coefficients of (*RS*)-propranolol and (*S*)-dihydroalprenolol were determined in the pH range 2–12 by

Tab. 2.2 Apparent partition coefficients of α-adrenoceptor agonists in liposome–buffer (log K'_m) systems as indicated and in *n*-octanol buffer (log P') at 37°C. (Adapted from Table 1 of ref. 37)

	Log K'_m [a]				
Compound	**(1)**	**(2)**	**(3)**	**(4)**	**log P'**
Oxymetazoline	1.94	2.50	2.96	1.16	–0.32
Xylometazoline	1.94	2.40	2.80	1.30	0.40
Cirazoline	1.72	2.23	2.65	1.15	0.53
Tramazoline	1.48	2.17	2.59	0.77	–0.62
Naphazoline	1.34	2.12	2.45	0.70	–0.52
Lofexidine	1.24	1.76	2.20	1.20	0.73
Clonidine	1.15	1.61	2.01	1.17	0.85
Tiamenidine	1.02	1.53	1.94	0.61	–0.17
Tetryzoline	0.95	1.43	1.80	0.55	–0.90

a) The numbers in brackets represent different liposome compositions: (1) DMPC; (2) DMPC/cholesterol/DCP (7:1:2 molar ratio); (3) DMPC/phosphatidylserine (3.5:1 molar ratio; the average molecular weight of brain phosphatidylserine was taken as 798 ± 4 on the basis of various fatty acid compositions; (4) DMPC/STA (3:1 molar ratio). The maximum standard deviation (SD) was ± 5%, although in most cases it was less than ± 2%. DCP, diacetyl phosphate; STA, stearylamine.

a two-chamber equilibrium dialysis using small unilamellar phosphatidylcholine liposomes. A universal buffer adjusted in ionic strength and osmolarity was used. The observed plateau (between pH 4.5 and 7.4) in the partitioning of the two drugs showed that ionized drugs also partition into the liposomal membrane. This is probably due to electrostatic attractive forces. Highly standardized experimental conditions also enabled the influence of the liposome preparation to be studied. To obtain small unilamellar vesicles in a homogeneous size distribution, liposomes were prepared by the controlled detergent dialysis method using sodium cholate as detergent. It is obvious that such a standardized phosphatidylcholine system resembles biological membranes much more closely than the classical octanol–buffer system. Thus, it is most likely more appropriate for describing the partitioning of drugs into membranes. Using this standardized equilibrium dialysis technique, Wunderli-Allenspach and coworkers measured the pH-dependent partitioning into PC liposomes of labeled propranolol, lidocaine, diazepam, warfarin, salicylic acid, and cyclosporin A. Partition coefficients for both the neutral and the ionized species were calculated [41] (Table 2.3). The development of the pH-metric titration technique was another decisive step. It was used to determine the liposomal membrane–water partition coefficients and pH-dependent lipophilicity profiles of eight ionizable drugs [42, 43]. The results are given in Table 2.4. Again, the data show that the membrane–water partition coefficients of ionized species are larger than log $P_{oct.}$, whereas the partition coefficients for neutral species are similar in both systems. Based on the electrostatic interaction of the ionized species, the „pH piston hypothesis“ was proposed to describe the observed difference in partitioning between the ionized and neutral forms. According to this hypoth-

Tab. 2.3 Analysis of partition data for acids, bases, and a neutral drug obtained with the phosphatidylcholine liposome–buffer system (fit parameters). (Adapted from Table 2 of ref. 41)

Drug	$P_{n/>pHm}$ [a]	$P_{n/<pHm}$	$P_{i/>pHm}$ [b]	$P_{i/<pHm}$	pK_a	pH_m [b]
(*RS*)-[^{1}H]Propranolol						
without cholate	1748 ± 38	–	577 ± 20	181 ± 146	9.22 ± 0.06	2.61 ± 0.43
with cholate[c]	2439	–	–	–	9.32	–
[^{14}C]Diazepam[d]	983 ± 9	–	–	–	–	–
[^{14}C]Lidocaine	115 ± 2	–	8.2 ± 2.3	2.5 ± 22	7.78 ± 0.07	2.34 ± 3.90
[^{14}C]Warfarin	2433 ± 7	1698 ± 31	25.0 ± 0.5	–	4.90 ± 0.01	2.13 ± 0.05
[^{14}C]Salicylic acid	317 ± 17	ND[e]	11.1 ± 3.0	–	3.00 ± 0.05	0.95 ± 0.19
[^{14}C]Cyclosporin A	8293 ± 114	5771 ± 511	–	–	–	2.74 ± 0.31

a) P_n and P_i, partition coefficient of neutral and ionized drug respectively.
b) pH_m, apparent pK_a of the organic phosphate group of the phosphatidylcholine molecule.
c) residual cholate to lipid ratio 1:700 (mol/mol).
d) unstable at pH < 5.
e) Few measured points. Partition data were analysed with the ProFit II non-linear curve fit program.

Tab. 2.4 Partition coefficients and ionization constants of the eight ionizable drugs measured in the systems octanol–water and DPOC–water using the pH-metric technique (literature values are printed in italics). (Adapted from Table 1 of ref. 43)

Drug	Octanol–water system			Liposomal membrane–water system	
	Log P^N_{oct}	Log P^I_{oct}	pK_a	Log $P^N_{membr.}$	Log $P^I_{membr..}$
Ibuprofen	3.97 ± 0.01	−0.05 ± 0.01	4.45 ± 0.04	3.80 ± 0.03	1.81 ± 0.05
Diclofenac	4.51 ± 0.01	0.69 ± 0.02	3.99 ± 0.01	4.45 ± 0.02	2.64 ± 0.04
5-Phenylvaleric acid	2.92 ± 0.02	−0.95 ± 0.02	4.59 ± 0.02	3.17 ± 0.02 *2.94*[a]	1.66 ± 0.03 *0.50*[a]
Warfarin	3.25 ± 0.01	−0.46 ± 0.02	4.90 ± 0.01	3.46 ± 0.01 *3.4*[b]	1.38 ± 0.03 *1.4*[b]
Propranolol	3.48 ± 0.01	0.78 ± 0.02	9.53 ± 0.01	3.45 ± 0.01 *3.51*[c] *3.28*[d]	2.01 ± 0.02 *2.56*[c] *2.76*[d]
Lidocaine	2.45 ± 0.02	−0.53 ± 0.36	7.96 ± 0.02	2.39 ± 0.02 *2.1*[b] *2.39*[c]	1.22± 0.04 *0. 9*[b] *1.49*[c]
Tetracaine	3.51 ± 0.01	0.22 ± 0.02	2.39 ± 0.02 8.49 ± 0.01	3.23 ± 0.02	2.1 1 f0.03
Procaine	2.14 ± 0.01	−0.8 I ± 0.03	2.29 ± 0.01 9.04 ± 0.01	2.38 ± 0.02	0.76 ± 0.09

a) Dimyristoylphosphatidylcholine liposome using the ultrafiltration method.
b) Egg phosphatidy1choline liposome using the dialysis method at 37 °C.
c) Egg phosphatidylcholine liposome using the pH-metric method.
d) Egg phosphatidylcholine liposome using the dialysis method at 37°C.

esis, the ionized species moves in the direction of the aqueous exterior. Consequently, the negatively charged phosphate groups will be the first to approach, and further

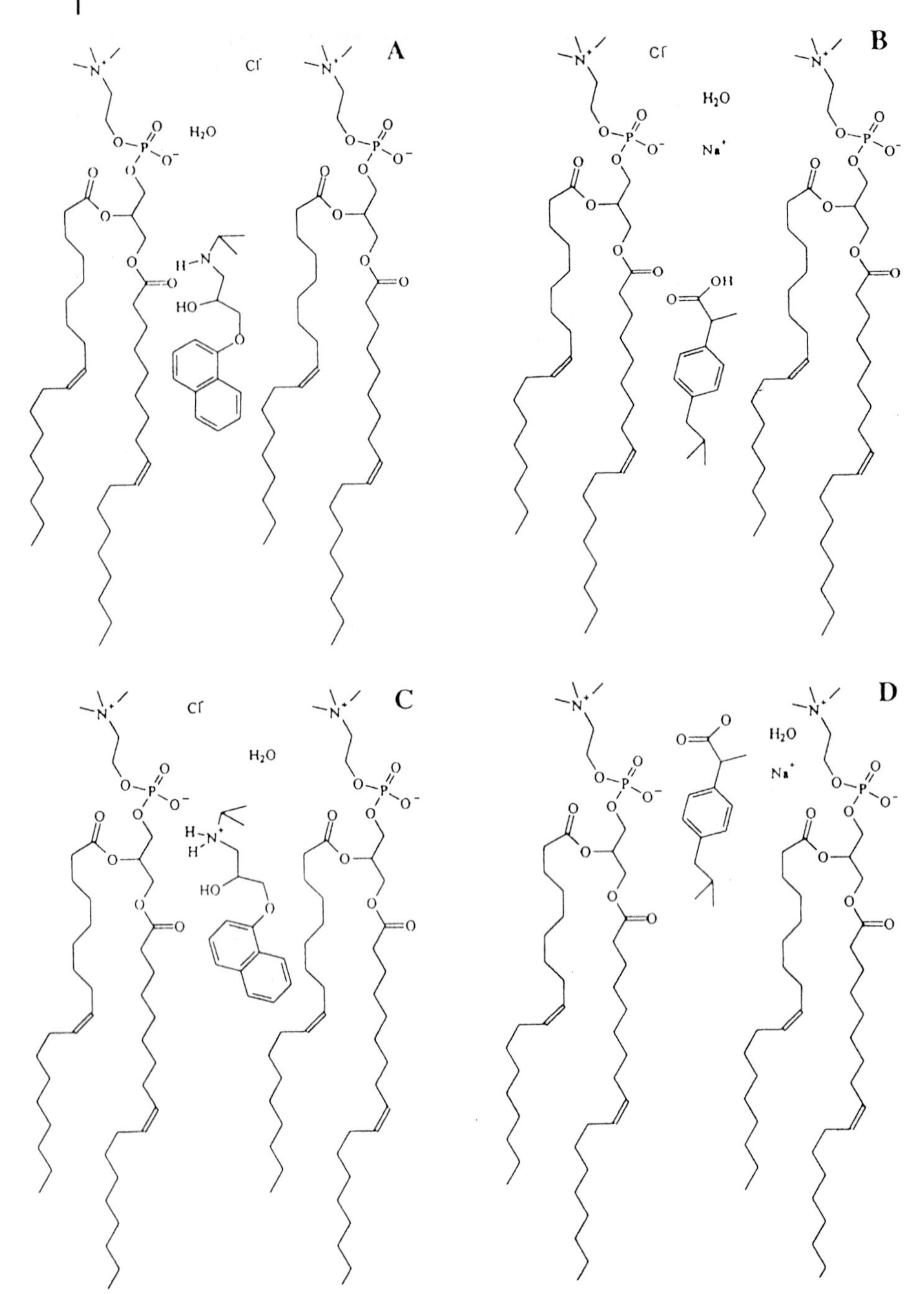

Fig. 2.1 Schematic drawing of the so-called piston theory showing the partitioning into DOPC liposomes of (A) propranolol in high-pH medium, (B) diclofenac in low-pH medium, (C) propranolol in low-pH medium, and (D) diclofenac in high-pH medium. (Reprinted from Fig. 3 of ref. 43 with permission from Kluwer Academic/Plenum Publishers.)

movement will bring the ionized species close to the positively charged trimethylammonium groups. This implies that weak acids will move farther than weak bases. The hydrophobic part of weak acids will therefore be more exposed to the aqueous exterior, and the membrane interaction will decrease. In order to keep charge neutrality, positive ions will move deeper into the membrane. According to the authors [43], the results show that ionized bases complement the charge/hydrogen bond structure of the phosphatidylcholine membrane better than ionizable acids (Figure 2.1).

A thorough investigation of the influence of various experimental conditions on the distribution of ionizable drugs into liposomes using the pH-metric log P titration method has recently been reported [44]. The base propranolol and the acid diclofenac were used as model compounds. The influence of temperature, equilibrium time, type of lipid, liposome preparation, lipid to drug ratio, and titration technique on the distribution was studied. It was concluded that small unilamellar vesicles are very suitable as a rapid partitioning system. They have the advantage that the testing can be done at high lipid concentrations, thus avoiding saturation phenomena. In addition, they can easily be prepared. Soybean phospholipid seems to be very suitable because its composition is more similar to biological membranes than, for example, DOPC. Also, it can be obtained in pure form and is relatively inexpensive. It is of interest that the authors found a significant shift in the pK_a of propranolol to lower basicity at higher temperatures. Thus, pK_a should be determined at the temperature of the partitioning experiment., For example, it has also been shown that the pH distribution behavior of ionized amines in octanol–water systems is not always the same as that obtained in model membrane systems. The amines studied were found to have an unexpectedly high membrane affinity [45] (Figure 2.2 and Table 2.5). Protonated amines generally exhibit high membrane partitioning, and this could be of interest in drug design. Results reported for the calcium channel antagonist amlodipine are a good example of this. At pH 7.4 log $D_{oct.}$ for amlodipine is 1.1, whereas log $P_{membr.}$ over all of the measured pH range is 3.75. This finding is of great importance because it had been assumed that amlodipine approaches the receptor through a

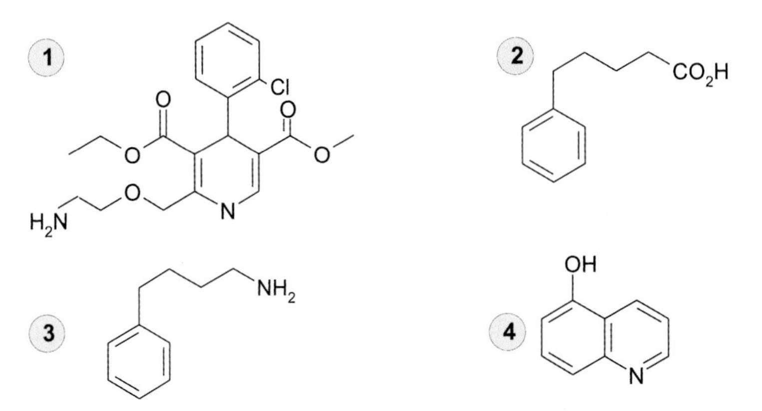

Fig. 2.2 Chemical structures of amlodipine (1), 5-phenylvaleric acid (2), 4-phenylbutylamine (3), and 5-hydroxyquinoline (4). (Reprinted from ref. 45 with permission from the American Pharmaceutical Association.)

Tab. 2.5 Parameter values derived from regression analysis of membrane distribution data. (Reprinted from ref. 45 with permission from the American Chemical Society)

Compound	Log $P_{membr.}$ [a]	Log $P^{+}_{membr.}$ [a]	Log $P^{-}_{membr.}$ [a]	pK_a (37 °C) from DMPC partitioning	pK_a
1	3.75[b]	3.75[b]			9.02 (24 °C)
2	2.94		0.50	5.10	4.88 (20 °C)
3	2.41	2.12		10.54	10.42 (not specified)
4	1.85[c]			5.13, 8.55	5.18, 8.56 (20 °C)

a) $P_{membr.}$, $P^{+}_{membr.}$, $P^{-}_{membr.}$, membrane distribution data of neutral, cationic, and anionic drug forms respectively (see Figure 2.2).
b) Mean of measurements at all pH values.
c) Maximum log $D_{membr.}$

membrane bilayer pathway [46, 47]; the assumption that the membrane partitioning of the charged form is due to ion pairing can be excluded. It can therefore be concluded that the ion pair enhanced partitioning observed in the octanol–water system is not important for the partitioning of ionized species into model phospholipid membranes and that the mechanism underlying this process must be a different one [45].

The partitioning of (*p*-methylbenzyl)-alkylamines in liposome–water and octanol–water systems has been investigated to reveal the complex mechanism governing the interaction between catamphiphiles and biomembranes [48]. Two experimental procedures were used:

1) potentiometric titration according to Avdeef [42, 49] to determine the partition coefficients of the neutral and cationic forms in the two solvent systems;
2) measurement of the NMR relaxation rates of the catamphiphiles in the presence of liposomes to validate the potentiometric results and to obtain additional insight into the mechanism of interaction between the solutes and liposomes (see also ref. 50).

The results are summarized in Tables 2.6 and 2.7. They clearly demonstrate that the relationship between neutral and cationic distribution in the lipid is more complex than the straightforward relation found for the partitioning of the two species in octanol. A similar conclusion can be derived by comparing log $D^{7.5}$ values for both systems (Figures 2.3 and 2.4). In contrast, a linear relation is seen when log D_{lip} is plotted against log NMR_{slope}. It can be concluded that for the *N*-alkyl homologs ($n >$ 4) intercalation in the membrane is dominated by electrostatic forces, i.e. the interaction of the cations with the liposomes involves the negatively charged phosphate groups. The effect should decrease with increasing bulk at the ammonium group. Hydrogen bonds and van der Waals forces determine the interaction for the neutral species and become stronger and stronger as the chain length increases. As reported in earlier papers on conformational analysis using nuclear Overhauser effect (NOE)-NMR experiments, the non-linear partitioning of benzylamines into the lipid phase in lecithin liposomes that occurs with increasing alkyl chain length is due to a conformational change from the extended to folded conformation [50, 51]. The negative

Tab. 2.6 CLOGP, partition coefficients in octanol–water[a], and related parameters as pK_a and log D of compounds 1–11. (Adapted from Table 1 of ref. 48)

CH_3–(C$_6$H$_4$)–CH_2–NH–$(CH_2)_n$–CH_3

Compound number (number of CH_2 groups)	CLOGP[b]	Virtual log P_{MLP}[c]	Log P^N[d]	Increment[e]	Log P^C[f]	Increment[g]	*diff* (log P^{N-C})[h]	pK_a	Log $D^{7.5}$[i]
1 (0)	2.01	1.95	1.96		–[j]		–	9.93	–0.43
				0.42		–			
2 (1)	2.54	2.75–2.81 (0.24)	2.38		-0.85		3.23	10.04	–0.11
				0.58		0.50			
3 (2)	3.07	3.08 –3.44 (0.36)	2.96		–0.35		3.31	9.98	0.54
				0.53		0.52			
4 (3)	3.60	3.39–3.97 (0.58)	3.49		0.17		3.32	9.98	1.07
				0.77		0.60			
5 (4)	4.13	3.83–4.50 (0.67)	4.26		0.77		3.49	10.08	1.73
				0.70		0.54			
6 (5)	4.65	4.22–5.09 (0.87)	4.96		1.31		3.65	10.17[j]	2.33
				0.16		0.25			
7 (6)	5.18	4.36–5.59 (1.23)	5.12		1.56		3.56	10.02[j]	2.52
8 (7)	5.71	4.91–6.24 (1.33)	–[k]	–	–[k]	–	–	9.47[j]	–
9 (8)	6.24	5.28–6.72 (1.44)	–[k]	–	–[k]	–	–	9.48[j]	–
10 (9)	6.77	5.73–7.25 (1.52)	–[k]	–	–[k]	–	–	9.48[j]	–
11 (10)	7.30	6.11–7.81 (1.70)	–[k]	–	–[k]	–	–	9.46[j]	–

a) Measured by potentiometry; the volume ratios of octanol and water were 0.067, 0.499, and 0.995.
b) Taken from the Pomona database.
c) Limits (range) of virtual log P_{MLP}, i.e. virtual log P of probable conformers obtained by molecular lipophilic potential (MLP).
d) log P of the neutral form: $n = 4$; SD < 0.03.
e) Increment in log P^N (neutral form) on addition of a CH_2 group.
f) Experimental log P of the cationic form: $n = 4$; SD < 0.03.
g) Increment in log P^C (cationic form) for the addition of a CH_2 group.
h) Experimental log P^N minus log P^C.
i) Calculated by the equation: $D = p^N \cdot \left(\frac{1}{1+10^{pK_a-pH}}\right) + p^c \cdot \left(\frac{10^{pK_a-pH}}{1+10^{pK_a-pH}}\right)$
j) The Yasuda–Shedlowsky approach (at least four points) was used.
k) Not measurable (solubility too low).

Tab. 2.7 Partition coefficient, log *P*, log *D*, and related parameters of compounds 1–7 in the liposome–water system.[a] (Adapted from Table 2 of ref. 48)

*	Log P^{N} [b]	Log $D^{7.5}$ [c]	Increment[d]	Log P^{C} [e]	Increment[f]	diff(log P^{N-C})[g]
1 (0)	3.09	2.54		2.54		0.55
			–0.03		–0.28	
2 (1)	3.06	2.27		2.26		0.80
			0.01		–0.15	
3 (2)	3.07	2.12		2.11		0.96
			–0.02		–0.57	
4 (3)	3.05	1.58		1.54		1.51
			0.45		0.30	
5 (4)	3.50	1.89		1.84		1.66
			0.70		0.59	
6 (5)	4.20	2.48		2.43		1.77
			0.20		0.28	
7 (6)	4.40	2.77		2.71		1.69

*) See Table 2.6 for the chemical structure of the investigated compounds.
a) Lipid–water ratios were 1.24. 4.65. 12.6 (in mg/mL); lipid/solutes ratios varied from 5 to 40.
b) Experimental log *P* of the neutral form: $n = 5$; SD < 0.05.
c) Calculated from the equation: $D = P^{N}[1/(1+10^{pK_a\text{-}pH})] + P^{C}[(10^{pK_a\text{-}pH})/(1+10^{pK_a\text{-}pH})]$
d) Increment in log P^{N} on addition of a CH_2 group.
e) Experimental log *P* of the cationic form: $n = 5$; SD < 0.05.
f) Increment in log P^{C} on addition of a CH_2 group.
g) log P^{N} minus log P^{C}.

slope up to compound 4 indicates a "negative contribution of hydrophobic interactions." In the liposome system, differences in the partition coefficients of neutral and protonated species are small compared with the differences observed in the octanol–water system. Benzylamines are a good example of this important observation. For a discussion of the correlation of partition behavior and the exerted biological effect of benzylamines, see Chapter 5.

Additional information, especially on the effects of solutes (drugs) on the bilayer, can be obtained by studying the partitioning of drugs into liposomes by DSC, NMR, and fluorescence techniques or other suitable spectroscopic techniques (see Chapter 3).

Taking all of the presented data into consideration, it is evident that the consequence of drug–membrane interaction for drug transport, distribution, and efficacy as well as drug resistance cannot be sufficiently described and explained by the partition coefficient of the drugs in organic bulk solvents. Octanol fails to mimic the pronounced interfacial character of the membrane phospholipid bilayer and the strong influence of the ionic interaction of charged drug molecules with the polar phospholipid head groups. This is because of the special properties of amphiphilic drugs and lipid membranes that affect drug–membrane interactions and drug partitioning. The partition behavior of amphiphiles is determined by their lipophilicity, acid–base equilibrium, hydrogen bonding potential, and conformational sensitivity to the polarity of the medium, i.e. the bilayer. Parameters affecting the partitioning of amphiphiles include the features of the lipid membrane, such as fatty acid chain length, saturation, branching,

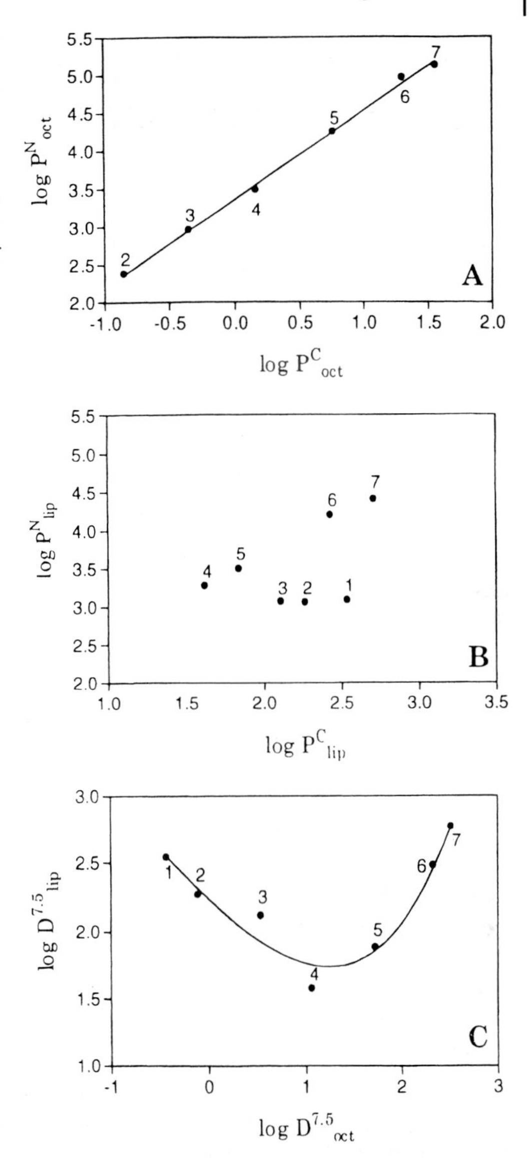

Fig. 2.3 Relative lipophilicities of compounds 1–7. (A) Relation between the partition coefficients of the neutral and cationic forms (log P^N and log P^C respectively) in octanol–water. (B) Relation between the partition coefficients of the neutral and cationic forms in the liposome-water system. (C) Bilinear relation between log $D^{7.5}$ in octanol-water and liposome-water. (Reprinted from Fig. 2 of ref. 48 with permission from Kluwer Academic/Plenum Publishers.)

head groups (dipole, pK_a), clustering, and chirality, as well as by membrane curvature, lateral pressure, surface potential, fluidity, domain formation, asymmetry (inner leaflet negatively charged, outer leaflet isoelectrical), and so forth.

Also, it is not the overall lipophilicity of a drug molecule determined in octanol–buffer that is important, but the 3-D distribution pattern of lipophilicity on the surface of the molecule, which may generate a hydrophobic-hydrophilic dipole. Such a dipole can determine the specific interaction and orientation of an amphiphilic drug in a highly structured biological membrane. This includes domain formation and accumulation, change in drug conformation, and so forth. All this is

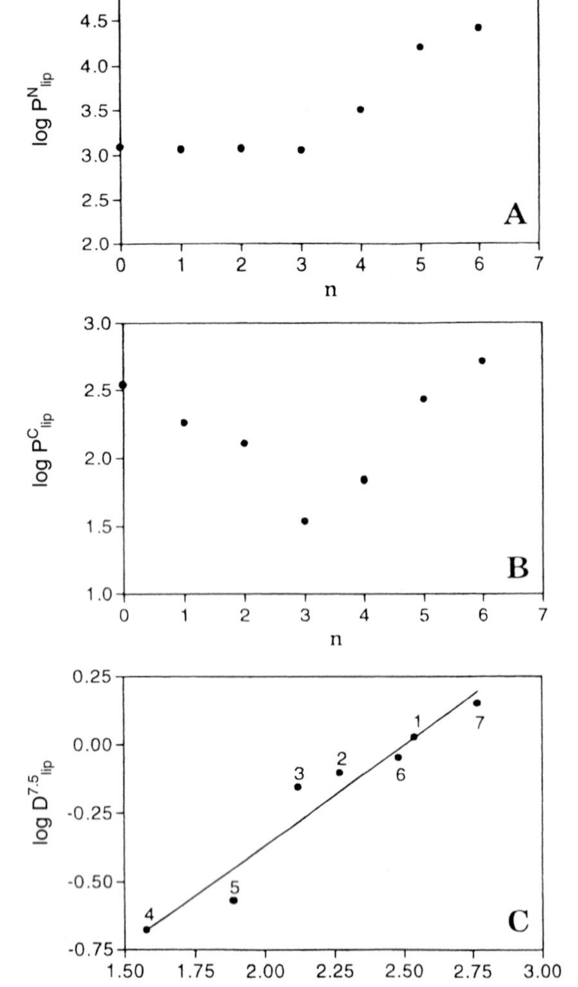

Fig. 2.4 Relation between the lipophilicity of compounds 1–7 and their other properties. (A) Variation of log $P^N_{lip.}$ with the number of methylene groups (n). (B) Variation of log $P^C_{lip.}$ with the number of methylene groups. (C) Linear regression between log NMR slope and lipophilicity parameters in the system liposome–water. (Reprinted from Fig. 3 of ref. 48 with permission from Kluwer Academic/Plenum Publishers.)

of special interest in case of catamphiphiles, which constitute a major proportion of the drugs presently used. To improve our understanding of drug selectivity, the octanol–water partition coefficient must be supplemented or replaced by information obtained with other partition systems. HPLC retention on IAM columns, partitioning into liposomes, and the use of the pH-metric technique seem to be more appropriate to model the anisotropic nature of biological and model membranes than simple log $P_{oct.}$ partition coefficients.

Nevertheless, log $P_{oct.}$ will remain an important tool because it has provided excellent service during the last three decades in QSAR studies, and large data bases are available. But we should remember also that membranes are not made of octanol and membranes are not uniformly made of one type of phospholipid.

References

1 Meyer, H., *Arch. Exp. Path. Pharm.* **1899**, *42*, 109–118.
2 Overton, E., *Studien über die Narkose zugleich ein Beitrag zur allgemeinen Pharmakologie*, G. Fischer, Jena **1901**.
3 Ferguson, J., *Proc. Roy. Soc.* **1939**, *127*, 387–404.
4 Hansch, C., Fujita, T., *J. Am. Chem. Soc.* **1964**, *86*, 1616–1626.
5 Seydel, J.K., Schaper, K.-J. *Chemische Struktur und Biologische Aktivität von Wirkstoffen, Methoden der Quantitativen Struktur-Wirkung-Analyse*, VCH, Weinheim **1979**.
6 Ramsden, Chr.A. (Ed.), *Comprehensive Medicinal Chemistry*, Vol. 4 *Quantitative Drug Design*, Pergamon Press, Oxford **1990**.
7 Kubinyi, H., *QSAR: Hansch Analysis and Related Approaches, Methods and Principles in Medicinal Chemistry*, R. Mannhold, P. Krogsgaard-Larsen, H. Timmerman (Eds.), VCH, Weinheim **1993**.
8 Hansch, C., Leo, A., *ACS Professional Reference Book*, American Chemical Society, Washington, DC **1995**.
9 Pliska, V., Testa, B., van de Waterbeemd, H., (Eds.), *Lipophilicity in Drug Action and Toxicology*, in: *Methods and Principles in Medicinal Chemistry*, R. Mannhold, H. Kubinyi, H. Timmerman (Series Eds.), VCH, Weinheim **1996**.
10 Leo, A., in *Comprehensive Medicinal Chemistry*, Vol. 4 *Quantitative Drug Design*, Chr.A. Ramsden (Ed.), Pergamon Press, Oxford **1990**, pp. 295–319.
11 Rekker, R.F., Mannhold, R., *Calculation of Drug Lipophilicity, The Hydrophobic Fragmental Constant Approach*, VCH, Weinheim **1992**.
12 Meyer, K.H., Hemmi H., *Biochem. Z.* **1935**, *277*, 39–71.
13 Overton, E.Z., *Phys. Chem.* **1897**, *22*, 189–209.
14 Diamond, J.M., Wright, E.M., *Annu. Rev. Physiol.* **1969**, *31*, 581–646.
15 Martin, Y.C., in *Drug Design*, Vol. VIII, E.J. Ariens (Ed.), Academic Press, London **1979**, pp. 2–68.
16 Rekker, R.F., *The Hydrophobic Fragment Constant*, Elsevier, Amsterdam, **1977**.
17 Herbette, L.G., Chester, D.W., Rhodes, D.G., *Biophys. J.*, **1986**, *49*, 91–94.
18 Mason, R.P., Chester, D.W., *Biophys. J.*, **1989**, *56*, 1193–1201.
19 Collander, R., *Acta Chem. Scand.* **1937**, *5*, 774–780.
20 Seiler, P., *Eur. J. Med. Chem.* **1974**, *9*, 473–479.
21 Kamlet, M.J., Abraham, M.H., Doherty, R.M., Taft, R.W., *J. Am. Chem. Soc.* **1984**, *106*, 464–466.
22 Taft, R.W., Abraham, M.H., Famini, G.R., Doherty, R.M., Abboud, J.L.M., Kamlet, M.J., *J.Pharm. Sci.* **1985**, *74*, 807–814.
23 Leahy, D.E., Taylor, P.J., Wait, A.R., *Quant. Struct.–Act. Relat.* **1989**, *8*, 17–31.
24 Leahy, D.E., Morris, J.J., Taylor, P.J., Wait, A.R., *J. Chem. Soc. Perkin Trans.* **1992**, *2*, 705–722, 723–731.
25 Moriguchi, I., *Chem. Pharm. Bull.* **1975**, *23*, 247–257.
26 Mirrlees, M.S., Moulton, S.J., Murphy, C.T., Taylor, P.J., *J. Med. Chem.* **1976**, *19*, 615–619.
27 Taylor, P.J., in *Comprehensive Medicinal Chemistry*, Vol. 4 *Quantitative Drug Design*, Chr.A. Ramsden (Ed.), Pergamon Press, Oxford **1990**, pp. 241–295.
28 Åriens, E.J., Simonis, A.M., in *Beta-Adrenoceptor Blocking Agents*, P.R. Saxena, R.P. Forsyth (Eds.), North-Holland, Amsterdam **1976**, pp. 3–27.
29 Young, C.R., Mitchell, R.C., Ganellin, C.R., Griffiths, R., Jones, M., Rana, K.K., Saunters, D., Smith, I.R., Sore, N.E., Wilks, T.J., *J. Med. Chem.* **1988**, *31*, 656–671.
30 El Tayar, N., Tsai, R.-S., Testa, B., Carrupt, P.-A., Leo, A., *J. Pharm. Sci.* **1991**, *80*, 590–598.
31 Tanaka, A., Nakamura, K., Nakanishi, I., Fujiwara, H., *J. Med. Chem.* **1994**, *37*, 4563–4566.
32 Rekker, R.F., Mannhold, R., Bijloo, G., de Vries, G., Dross, K., *Quant. Struct.–Act. Relat.* **1998**, *17*, 537–548.
33 Abraham, M.H., Chadha, H.S., Mitchell, R.C., *J. Pharm. Sci.* **1994**, *83*, 1257–1268.

34 Kaliszan, R., Nasal, A., Bucinski, A., *Eur. J. Med. Chem.* **1994**, *29*, 163–170.
35 Barbato, F., La Rotonda, M.I., Quaglia, F., *Eur. J. Med. Chem.* **1996**, *31*, 311–318.
36 Ong, S., Liu, H., Qui, X., Bhat, G., Pidgeon, C., *Anal. Chem.* **1995**, *67*, 755–762.
37 Choi, Y.W., Rogers, J.A., *Pharm. Res.* **1990**, *7*, 508–512.
38 Choi, Y.W., Rogers, J.A., *Pharm. Res.* **1989**, *6*, 399–403.
39 Pauletti, G.M., Wunderli-Allenspach, H., *Eur. J. Pharm. Sci.* **1994**, *1*, 273–282.
40 Krämer, St.D., Wunderli-Allenspach, H., *Pharm. Res.* **1996**, *13*, 1851–1855.
41 Ottiger, C., Wunderli-Allenspach, H., *Eur. J. Pharm. Sci.* **1997**, *5*, 223–231.
42 Avdeef, A., *Quant. Struct.–Act. Relat.* **1992**, *11*, 510–517.
43 Avdeef, A., Box, K.J., Comer, J.E.A., Hibbert, C., Tam, K.Y., *Pharm. Res.* **1998**, *15*, 209–215.
44 Balon, K., Riebesehl, B.U., Müller, B.W., *J. Pharm. Sci.* **1999**, *88*, 803–805.
45 Austin, R.P., Davis, A.M., Manners, C., *J. Pharm. Sci.* **1995**, *84*, 1180–1183.
46 Herbette, L.G., Vant Erve, Y.M.H., Rhodes, D.G., *J. Mol. Cell. Cardiol.* **1989**, *21*, 187–201.
47 Mason R.P., Campbell, S.E., Wang, S.-D., Herbette, L.G., *Mol. Pharmacol.* **1989**, *36*, 634–640.
48 Fruttero, R., Caron, G., Fornatto, E., Boschi, D., Ermonidi, G., Gasco, A., Carrupt, P.-A., Testa, B., *Pharm. Res.* **1998**, *15*, 1407–1413.
49 Avdeef, A. in *Lipophilicity in Drug Action and Toxicology*, V. Pliska, B. Testa, H. van de Waterbeemd (Eds.), VCH, Weinheim **1996**, pp. 109–139.
50 Seydel, J.K., Cordes, H.-P., Wiese, M., Chi, H., Croes, N., Hanpft, R., Lüllmann, H., Mohr, K., Patten, M., Padberg, Y., Lüllmann-Rauch, R., Vellguth, S., Meindl, W.R., Schönenberger, H., *Quant. Struct.–Act. Relat.* **1989**, *8*, 266–278.
51 Seydel, J.K., Albores-Velasco, M., Coats, E.A., Cordes, H.-P., Kunz, B., Wiese, M., *Quant. Struct.–Act. Relat.* **1992**, *11*, 205–210.

3
Analytical Tools for the Analysis and Quantification of Drug–Membrane Interactions

Joachim K. Seydel

Aqueous dispersions of lipids are attracting increasing interest for reasons other than their biological implications. Physicists are beginning to perceive them as systems which can successfully be investigated by physical methods despite their great complexity. The development of such methods for the study of drug–membrane interactions is also of great interest to medicinal chemists. The determination and quantification of the various effects of drugs on membranes and vice versa is a precondition for the understanding of drug action. Correlating drug effects on membranes of different composition with the structural properties of drugs under study can lead to new information. The effects can be quantified by the derivation of quantitative structure–interaction relationships. Physicochemical or spectral parameters describing the strength or type of interaction can also be used as "constraints" in the modeling of drug orientation and conformation in the environment of membranes. The most frequently used physical methods for the investigation of events or combinations of events leading to changes in membrane properties or drug conformation, orientation, and localization in the membrane environment are listed below. They will be introduced and discussed with regard to selected examples.

Methods for the study of drug–membrane interactions are:

- high-performance liquid chromatography (HPLC);
- displacement of $^{45}Ca^{2+}$ from phospholipid monolayers;
- differential scanning calorimetry (DSC);
- fluorescence techniques;
- Fourier transform infrared spectroscopy (FT-IR);
- electron spin resonance (ESR);
- X-ray and neutron diffraction;
- nuclear magnetic resonance spectroscopy (NMR);
- circular dichroism (CD);
- others;
- combined techniques.

3.1 High-performance Liquid Chromatography (HPLC)

3.1.1 Determination of the Retention Time on "Artificial Membrane" Columns

The use of octanol-covered reversed-phase columns for the determination of log $P_{oct.}$ or log $K'_{oct.}$ [1], instead of the more tedious shake flask method, has become common practice. If the chromatographic hydrophobicity measuring system is intended to model processes in the biophase, then the similarity between the measuring system and the biological system should be as close as possible. An HPLC system that can model drug–membrane interaction and transport of drugs through biological membranes should therefore consist of an aqueous phase and a defined phospholipid layer.

Such new RP-HPLC stationary-phase materials have been available for some years (Regis Chemical Company, Morton Grove, IL, USA). These so-called immobilized artificial membrane (IAM) columns consist of lipid molecules covalently bound to propylamine-silica. The unreacted propylamine moieties are end-capped with methylglycolate. The membrane lipid, phosphatidylcholine, possesses polar head groups and two non-polar hydrocarbon chains (C_{18}). One of the alkyl chains is linked to the propylamine-silica surface.

The log K' retention time determined for a series of drugs on such columns has successfully been used to describe the permeation of human skin by a series of alcohols and steroids, whereas the same biological data showed poor correlation with HPLC retention parameters determined on regular RP-HPLC columns [2, 3]. Kaliszan *et al.* [4] reported a weak correlation between retention data determined on an IAM column and reference hydrophobicity measurements, log P, for a series of psychotropic and antihistamine drugs. In another paper, this group reported the retention time on IAM columns of three series of drugs – β-adrenolytics, α-adrenomimetics, and phenothiazine neuroleptics – and compared it with log $D_{oct.}$ and log $K'_{oct.}$ After correction for different pH values in the experimental procedures, the log K'_{IAM} values correlated well with the reference log $D_{oct.}$ values and the log $K'_{oct.}$ for the β-adrenolytics ($r = 0.96$ and 0.94 respectively), but a poor correlation between these parameters was observed for the phenothiazines studied [5]. The authors also compared the ability of different descriptors of hydrophobicity to explain the variance in various pharmacokinetic properties of β-adrenolytics, determined by Hinderling *et al.* [6]. In all but two cases the statistics of regression describing pharmacokinetic parameters in terms of log K'_{IAM} were superior to those provided by log $K'_{SFoct.}$ (shake flask determination) and log $K'_{C18\,oct.}$ (HPLC C_{18}, covered with octanol) (see Tables 3.1).The capacity factors, K'_{IAM}, were calculated assuming that the dead volume of the column was the solvent disturbance signal given by methanol.

The antihemolytic activity of phenothiazines (log $1/C$) was also correlated with log K'_{IAM} (Tab. 3.2). Again, it was found that the corresponding equation with log $D_{oct.}$ is of significantly lower statistical value:

$$\log 1/C = 1.126\ (\pm 0.174)\ \log K'_{IAM} - 3.266\ (\pm 0.266) \tag{3.1}$$

$n = 8 \qquad s = 0.102 \qquad r = 0.935$

$$\log 1/C = 0.405\ (\pm 0.130)\ \log D_{oct} + 3.662\ (\pm 0.416) \tag{3.2}$$

$n = 7 \qquad s = 0.170 \qquad r = 0.812$

Tab. 3.1 Hydrophobicity parameters and antihaemolytic activity of a series of phenothiazine neuroleptics. (Reprinted from Tab. 5 of ref. 5, with permission from Elsevier Science)

No	Drug	pK_a [a]	Log P [b]	Log D [c]	CLOGP [d]	CLOGD [e]	log K'_{IAM} [f]	Log 1/C [g]
1	Chlorpromazine	9.30	5.35	3.05	5.20	2.90	1.440	4.721
2	Fluphenazine	8.10	4.36	3.23	5.90	4.77	1.401	5.000
3	Perphenazine	7.80	4.20	3.34	5.57	4.71	1.373	4.824
4	Prochlorperazine	8.10	–	–	6.15	5.02	1.782	5.222
5	Promazine	9.40	4.55	2.15	4.28	1.88	1.171	4.553
6	Promethazine	9.10	–	–	4.65	2.55	1.195	–
7	Propiomazine	9.10	–	–	5.00	2.90	1.265	–
8	Thioridazine	9.50	5.90	3.40	6.42	3.92	1.747	5.301
9	Trifluoperazine	8.10	5.03	3.90	6.48	5.35	1.754	5.222
10	Trifluopromazine	9.20	5.19	2.99	5.53	3.33	1.474	4.959

a) pK_a value.
b) Logarithm of *n*-octanol-water partition coefficient for the non-ionized form determined experimentally by the shake-flask method.
c) Value of experimental *n*-octanol-water hydrophobicity parameter corrected for ionization at pH 7.0.
d) Theoretical *n*-octanol-water partition coefficient for the non-ionized form of solute calculated by the fragmental method.
e) Value of theoretical *n*-octanol-water hydrophobicity parameter corrected for ionization at pH 7.0.
f) Logarithm of HPLC capacity factor determined on an immobilized artificial membrane column with acetonitrile-buffer pH 7.0 eluent.
g) Logarithm of reciprocal of antihaemolytic concentration.

Depending on the composition of the various membranes, it might be appropriate to use IAM columns with different phospholipids. As these columns are not commercially available, phosphatidylcholine (PC)- or phosphatidylserine (PS)-covered infusorial earth as the stationary phase and phosphate buffer as the mobile phase have been used to determine the retention time for a series of 2,3,6-triaminopyridines with anticonvulsant activity. Again, log K'_{PS} was superior to log $K'_{oct.}$ in describing the observed variance in biological activity [7].

The partition coefficients of triphenylalkylphosphonium homologs have been determined in gel bead-immobilized small or large unilamellar liposomes by chromatography [8]. It was claimed that the technique, immobilized liposome chromatography (ILC), is suitable for the determination of membrane partition coefficients of drugs.

Tab. 3.2 Statistical parameters (*n*, number of data points; *r*, correlation coefficient; *s*, standard error) of regression equations relating bioactivity data [6] to experimental hydrophobicity parameters. (Reprinted from Tab. 4 of ref. 5, with permission from Elsevier Science)

Pharma-cokinetic parameter	Hydrophobicity descriptor log K'_{IAM}			log $K'_{SF\ oct.}$			log K'_{C18}		
	n	*r*	*s*	*n*	*r*	*s*	*n*	*r*	*s*
Log r_τ	9	0.834	0.444	11	0.838	0.463	9	0.735	0.546
Log f_b	13	0.829	0.367	14	0.805	0.397	10	0.757	0.473
Log r_A	6	0.952	0.149	6	0.899	0.213	6	0.884	0.227
Log K_{BC}	10	0.896	0.180	11	0.863	0.219	9	0.765	0.260
Log V_{uss}	9	0.858	0.233	11	0.845	0.295	9	0.732	0.309
Log K_p	10	0.870	0.447	13	0.900	0.432	9	0.846	0.465
Log r	9	0.822	0.575	11	0.926	0.380	9	0.948	0.320
Log $B_{\%}$	8	0.800	0.274	7	0.740	0.324	7	0.711	0.338

r_τ, ratio of the fraction of drug bound and unbound to tissue.
f_b, fraction of drug bound to plasma.
r_A, ratio of the fraction of drug bound and unbound to albumin.
K_{BC}, true red cell partition coefficient.
V_{uss}, steady-state volume of distribution referenced to the unbound drug in plasma.
K_p, partition coefficient of drug between plasma protein and plasma water.
r, ratio of the fractions of the drug non renally and renally eliminated.
$B_{\%}$, percentage binding of drug to serum proteins.

3.2 Displacement of $^{45}Ca^{2+}$ from Phospholipid Head Groups

3.2.1 Studies of Drug–Membrane Interactions using Phospholipid Monolayers

A relatively simple method of studying and quantifying drug–membrane interactions is the use of planar monolayers. They provide an organized interfacial structure which is assumed to be similar to that found in biological membranes. This model can facilitate the study of the tendency of drugs to accumulate at an interface and aid understanding of the behavior of drugs at the surface of a cell membrane. Many anti-inflammatory processes are cell surface phenomena. The study of the effect of anti-inflammatory drugs on monolayers could be useful for an understanding of the underlying mechanism. Drug–phospholipid interaction can be characterized by changes in surface pressure [9], surface potential or binding, and displacement of calcium. The last mentioned is thought to be an appropriate technique to determine drug–phospholipid head group interaction under physiological conditions. As the drugs also change the surface pressure, care has to be taken that the replacement experiment is performed at constant surface pressure.

One of the first systematic studies using this technique measured the replacement of $^{45}Ca^{2+}$ ions from monolayers of three different phospholipids, namely PS, phosphatidylinositol (PI), and phosphatidylethanolamine (PE), at different pH values by 30 amphiphilic drugs [10]. As the capacity of a drug to replace $^{45}Ca^{2+}$ is a function of its concentration, the ID_{50} values found are valid only under the defined conditions. In this study, the $^{45}Ca^{2+}$ concentration was 1.2×10^{-5} M. The results are summarized in Table 3.3. The drug concentration required to displace $^{45}Ca^{2+}$ from the PS monolayers was highest for PS, as expected from the fact that binding of Ca^{2+} ions is strongest for PS. Among the local anesthetics, dibucaine showed the highest affinity.

Tab. 3.3 Displacement of Ca^{2+} (ID_{50}) from monolayers of phosphatidylserine (PS), phosphatidylinositol (PI) and phosphatidylethanolamine (PE) by various drugs. (Adapted from ref. 10)

Compound	ID_{50}		
	PS ($\times 10^5$ M)	PI ($\times 10^5$ M)	PE ($\times 10^5$ M)
1. Sodium chloride	3500	3700	–
2. 3-Amino pyridine	3000	–	–
3. 2-Amino pyridine	1200	–	–
4. Ammonia	1100	1000	–
5. Atenolol	450	250	200
6. Procaine	400	150	50
7. 4-Aminopyridine	300	–	–
8. I-Scopolamine	300	250	170
9. Atropine	200	70	50
10. Carticaine	120	65	50
11. Lidocaine	250	200	80
12. Metoprolol	90	50	45
13. Phentermine	70	40	20
14. Quinine	4.5	6	8
15. Quinidine	5	6	7
16. Verapamil	12	8	3
17. Dexetimide	12	20	10
18. Tetracaine	10	1.1	2
19. Chlorphentermine	8	5	3
20. Propranolol	3	3	1.2
21. Dibucaine	3.5	1.4	0.4
22. Imipramine	4	1.3	0.7
23. Amitriptyline	2.5	1.3	1
24. Amiodarone	1.6	1.7	1.8
25. Chloroquine	3	0.4	0.4
26. Perhexiline	1.5	0.7	0.6
27. 1-Chloramitriptyline	1	0.7	0.4
28. Iprindol	1	0.7	0.5
29. Chlorpromazine	0.7	0.7	0.4
30. Mepacrine	0.5	0.22	0.22

The pH dependence of Ca^{2+} displacement for selected compounds is shown in Figure 3.1. It can be concluded that the protonation of drugs is a precondition for binding to the polar head groups of phospholipids. The degree of binding is modified by the lipophilicity of the drug molecule.

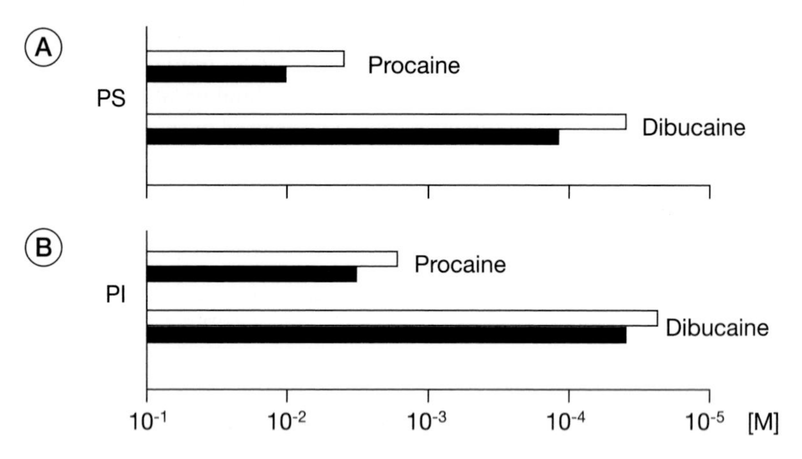

Fig. 3.1 Concentration of two local anesthetics, procaine and dibucaine, displacing, at two different pH values, 50% of the bound calcium from phosphatidylserine (PS) (A) and phosphatidylinositol (PI) (B). Open columns, pH 7.5; dark columns pH 9.5. (Reprinted from Fig. 4 of ref. 10 with permission from Elsevier Science.)

The effect of indomethacin and flurbiprofen, two non-steroidal anti-inflammatory drugs, on monolayer surface pressure has also been studied. An increase in the surface pressure of DPPC monolayers is observed as a function of drug concentration. The dependence of drug penetration on the initial pressure of the DPPC monolayer was also determined. It can be concluded that even at high initial pressure these drugs can penetrate the monolayer [9] (Figure 3.2). The results are in agreement with studies performed on bilayers using differential scanning calorimetry (DSC) [11].

Girke *et al.* [12] have reported on the relationship between Ca^{2+} displacement in PS monolayers (IC_{50}), the effect on phase transition, T_t, of dipalmitoylphosphatidic acid (DPPA), and the depression of cardiac functions for a series of β-blockers, antiarrhythmics, and other drugs. The relation with drug effect will be discussed in Section 5.1. The two parameters, ΔT_t, and IC_{50}, used to describe the displacement of $^{45}Ca^{2+}$ mirror different effects of the drug on the membrane ($r^2 = 0.48$). The ability to reduce T_t is probably related to the depth to which the hydrophobic part of the drug molecule penetrates the hydrocarbon chain. It has been shown that tetracaine penetrates most deeply (to carbon 6–8) and also induces a great decrease in T_t. The reduction in Ca^{2+} binding, however, is related to the uptake of the positively charged hydrophilic part into the head group region (see Table 3.8).

Recently, Klein *et al.* [13] studied the inhibition of Ca^{2+} binding to PS monolayers by cationic amphiphilic compounds with antiarrhythmic activity. The authors de-

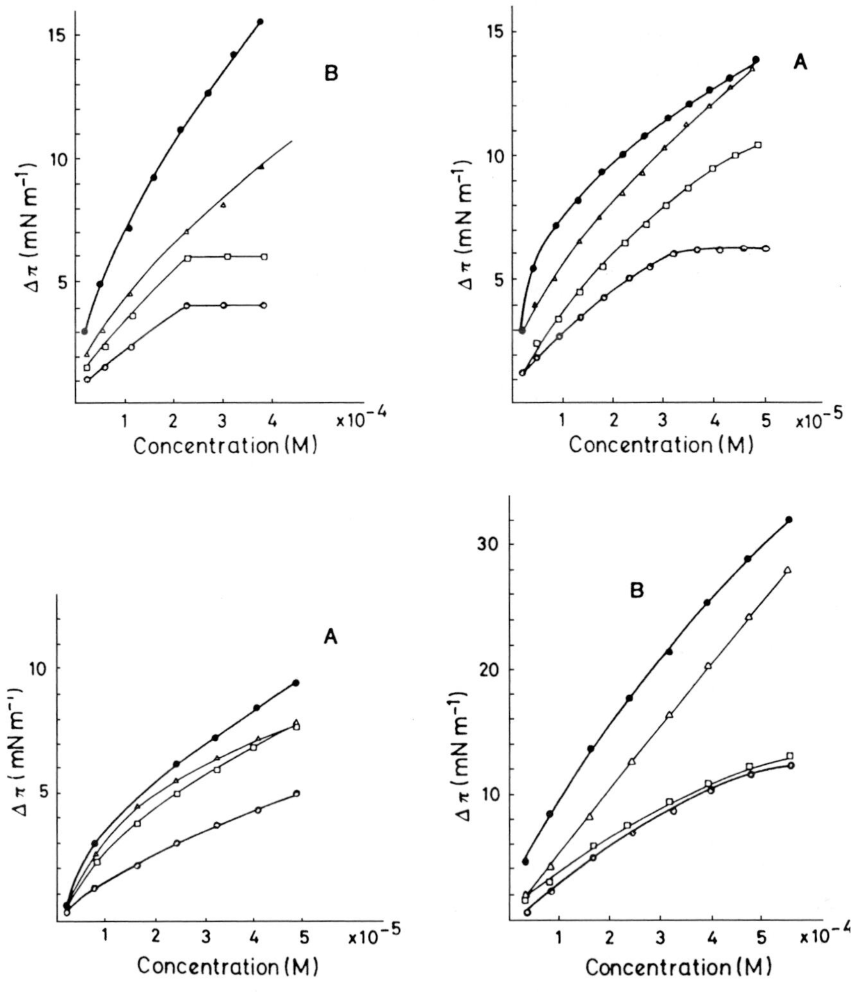

Fig. 3.2 (Top) Effect on the initial surface pressure of DPPC monolayer of indomethacin concentration in the subphase. $\pi_i = 10\,mN/m$ (●), 20 mN/m (△), 30 mN/m (□), 40 mN/m (O). (A) Acetic acetate, 0.1 M NaCl, pH 4.6. (B) Tris HCl, 0.1 M NaCl, pH 7.4. (Bottom) Effect on initial surface pressure of DPPC of flurbiprofen concentrations in the subphase. Conditions as given in (A) and (B). (Reprinted from Figs. 5 and 6 of ref. 9 with permission from Academic Press.)

rived a QSAR to characterize drug–membrane interaction. The following parameters were derived and used to describe the physicochemical properties of the studied compounds: total polar surface area (TPSA), dipole moment in the *Z*-direction (dipole-*Z*), and the relative position charge (RPCG). The RPCG is the charge of the most positive atom/total positive charge. The derived regression equation underlines the importance of the electrostatic properties of the catamphiphiles for Ca^{2+} displacement from the head group of phosphatidylserine.

$$-\log IC_{50} = -5.24 + 0.019\,\text{TPSA} + 0.111\,(\text{dipole-Z} - 1.34)^2 + 12.11\,\text{RPCG} \qquad (3.3)$$

$n = 12 \qquad r^2 = 0.98 \qquad s = 0.155 \qquad Q^2 = 0.79$

The opposite, the reversal of drug–membrane interaction by addition of Ca^{2+} ions, has been also studied by NMR spectroscopy using the example of multidrug resistance modifiers [14].

3.3 Differential Scanning Calorimetry (DSC)

3.3.1 Phase Transition and Domain Formation

The above-mentioned physicochemical properties of phospholipids lead to spontaneous formation of bilayers. Depending on the water–lipid ratio, on the type of phospholipids, and the temperature, the bilayer exists in different, defined mesomorphic physical organizations. These are the L_α high-temperature liquid crystalline form, the L_β gel form with restricted movement of the hydrocarbon chains, and an inverted hexagonal phase, H_{II} (see Sections 1.3.1 and 1.3.2).

The transition of a number of phospholipid chains from an ordered to a disordered state is of direct relevance to the state of the cell membrane – it is a characteristic parameter for each phospholipid. The dynamic molecular organization in vesicles and membranes has been described in detail [15] and is shown schematically in Figure 3.3. Using the thermodynamic DSC technique, changes in phase transition

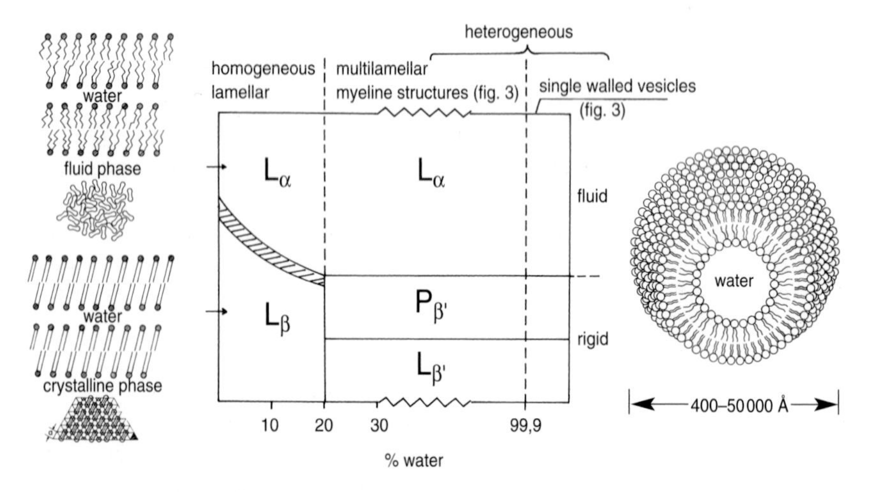

Fig. 3.3 Phase diagram of diacylphosphatidylcholines as suggested by X-ray, optical birefringence, and electron microscopy. L_d corresponds to the smectic A state and L_β, L_β', P_β' to the smectic B phase of thermotropic liquid crystals. The shaded area is a region of coexistence. (Reprinted from Fig. 2 of ref. 15 with permission from Wiley-VCH.)

properties can, for example, be measured as a function of the type of phospholipid, chain length [16], cholesterol content [17], and drug–lipid interaction. DSC is a non-perturbing, thermodynamic technique that measures the heat exchange associated with cooperative lipid phase transitions in model and biological membranes. Several modern DSC instruments allow an accurate, quick, and sensitive determination of lipid phase state. In addition, data analysis is straightforward and relies on equilibrium thermodynamic principles. DSC provides accurate information on temperature, enthalpy, entropy, and cooperativity during phase transition. Three parameters are particularly useful in describing drug–phospholipid interactions:

1) the change in main phase transition temperature, defined as the peak of the gel-to-liquid crystalline endotherm, T_t;
2) the calorimetric enthalpy, ΔH; and
3) the width at half-height, ΔT_t, of the phase transition profile, expressed in °C.

The first parameter is an expression of changes in membrane fluidity, i.e. the passage of a certain number of aliphatic chains from an ordered to a disordered state. The last is a measure of destabilization of the phospholipid assemblies, indicating a decrease in size of the cooperative unit, which is defined by the number of chains that change simultaneously.

The influence of substituent size, polarity, and location on the thermotropic properties of synthetic phosphatidylcholines has been studied by Menger *et al.* [18]. The effect of increasing membrane curvature on the phase transition has been investigated by DSC and FTIR [19]. In addition, a data bank, LIPIDAT, on lipid phase transition temperatures and enthalpy changes is available [20, 21].

As already mentioned, other factors influencing the thermotropic phase behavior are cholesterol and Ca^{2+} ions. Cholesterol is an essential component of biomembranes. It plays a major role in determining the fluidity of membranes. It makes the gel phase somewhat more fluid by disturbing the gel packing and makes the liquid crystalline phase more rigid by increasing the order of the bilayer. In the presence of Ca^{2+}, cholesterol alters the lateral phase behavior of particular phospholipids. It destabilizes the lamellar phase and promotes an inverted hexagonal II (H_{II}) phase [22]. In cellular processes involving Ca^{2+}, cholesterol modulates Ca^{2+}-induced cell fusion [23].

The effect of the adrenal steroid precursor dehydroepiandrosterone (DHEA) on PC and PS bilayers has been investigated by DSC in combination with the X-ray diffraction technique. It was found that when the molar fraction of the sterol is less than 0.2 it interacts with both types of liposomes by depressing the melting temperature and reducing the enthalpy of melting [24]. As the cholesterol content of mammalian membranes changes with age, this may be a factor worth considering in drug therapy. It may well influence drug permeation by altering the fluidity of the membranes in some organs. A short review on cholesterol–membrane interaction has been published [25].

Structure, phase behavior, and fusion of phospholipid membranes are affected by divalent cations, especially Ca^{2+}. Phosphatidylserine is the major acidic phospholipid in mammalian membranes. As a result of its anionic character, it binds strongly to cations. The importance of Ca^{2+}–PS interactions, for example in triggering membrane fusion [26], has stimulated many studies on model membranes using various techniques. Infrared spectroscopic studies have shown that Ca^{2+} binds to the phosphate ester group of PS in the presence of cholesterol up to 50%, as in the case of pure PS bilayers. In contrast, other IR data indicate that the presence of cholesterol induces disorder of the acyl chain packing, increases the degree of immobilization of the interfacial and polar regions, and increases the degree of dehydration of the Ca^{2+}–PS complexes [27]. The effect of protons or calcium on the phase behavior of PS–cholesterol mixtures has also been studied using DSC and X-ray diffraction [28].

The effects of small organic molecules such as *n*-alkanes, *n*-alkanols, fatty acids, and charged alkyl compounds on phase transition behavior of various phospholipids have been studied and reviewed by Lohner [29]. Although they are located differently in the membrane, all of these different molecules have been shown to induce similar effects on phospholipid phase transition.

In addition, the reversibility of phase transition in lipid–water systems has been studied [30]. It was observed that the relaxation times in the transition region and the lifetimes of the metastable phases are similar, and sometimes significantly longer than the times characteristic of the biomembrane processes. The question arises as to the physiological significance of the equilibration that occurs a long time after lipid phase transition.

Phospholipid bilayers can undergo morphological rearrangements to form other phases. The formation of one of these non-bilayer phases, the hexagonal H_{II} phase, is preceded by an increase in hydrophobicity at the bilayer surface and a destabilization of the bilayer structure. Some membrane additives promote, while others inhibit, the formation of the H_{II} phase. For example, it has been shown saturated PEs exhibit lamellar inverse hexagonal (H_{II}) phase transitions with increasing temperature, these depending on the length and chemical structure of the acyl chain. The formation of bicontinuous cubic phases in these phospholipid systems arises from an interplay between the packing of the chains and the hydration/hydrophilicity quality of

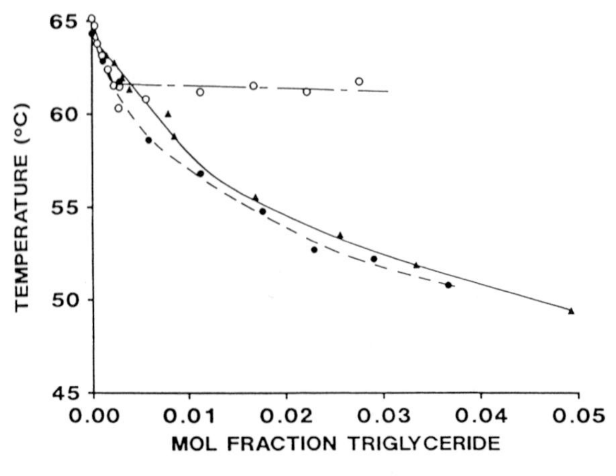

Fig. 3.4 Effect of triacylglycerols on the bilayer to H_{II} phase transition temperature of DEPE (▲, trilaurin; ●, triolein; O, tristearin). (Reprinted from Fig. 1 of ref. 32 with permission from Elsevier Science.)

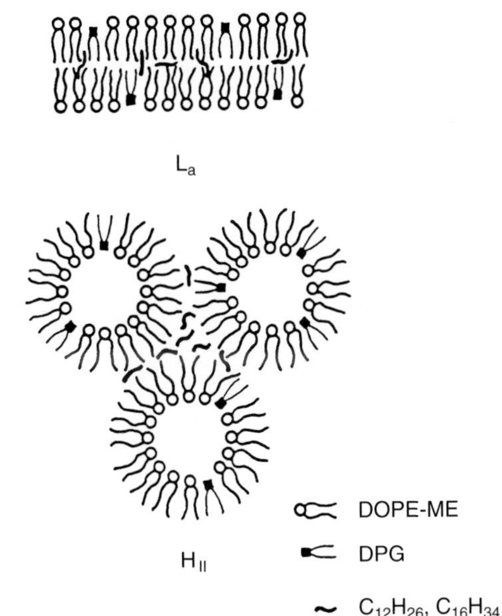

Fig. 3.5 Inferred positions of low levels of long-chain alkanes and diglycerides in the L_α (top) and H_{II} (bottom) phases. (Reprinted from Fig. 6 of ref. 34 with permission from the American Chemical Society.)

the head groups [31]. Such H_{II} phases can be promoted by low concentrations of triacylglycerol and alkanes [32] (Figure 3.4) [33]. The lamellar/inverted hexagonal phase transition (L_α/H_{II}) of phospholipid systems is very sensitive to the presence of hydrophobic "impurities" [34] (Figure 3.5). Addition of membrane stabilizers decreases the fusogenic behavior of membranes. Cholesterol sulfate is such a bilayer stabilizer. Several antiviral agents are also found to be bilayer stabilizers and have been shown to inhibit membrane fusion [35, 36].

The induction of aggregation and fusion by gramicidin has been shown for lamellar dioleoylphosphatidylcholine vesicles at peptide to lipid ratios exceeding 1:100.

The potency of a series of peptides in stabilizing the bilayer phase has been determined by DSC, and the following ranking was observed Z-Tyr-Leu-NH_2 = Z-Gly-Phe-NH_2 > Z-Ser-Leu-NH_2 > Z-Gly-Leu-NH_2 > Z-Gly-Gly-NH_2. The authors found a linear correlation between the HPLC retention time parameter, K', of peptides with varying activity against viral infections and the slope of the bilayer stabilization curve. The latter was determined using the DSC technique by plotting the bilayer to hexagonal phase transition temperature of dielaidoylphosphatidylethanolamine (DEPE) against the mole fraction of peptide present in the lipid [37].

Not only peptides or other biological compounds in membranes but also foreign compounds such as drug molecules influence phase transition. The usefulness of the DSC technique in measuring and quantifying drug–membrane interactions and in gaining insight into the mechanism of action has been shown in numerous experiments. Several examples shall be demonstrated.

The first is a systematic study of the ability of cationic amphiphiles to depress the transition temperature of DPPA liposomes [38]. Using 1-dimethylamino-3-x-propane

Tab. 3.4 Structure, physicochemical properties and activity by the indicated cationic compounds. (Reprinted from Tab. 1 of ref. 38, with permission from Elsevier Science)

R–	pK_a	Log *D*	ΔT_t (°C)	IC_{50} (µM)
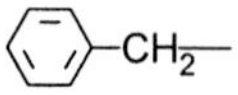 –CH_2–	11.0	1.5	29	2100
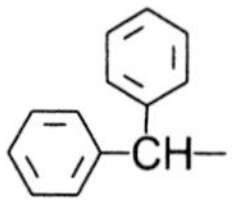 –CH–	10.0	2.8	28	40
 –CH_2–	9.9	3.0	53	9

Log *D*, log *P* at pH = 7.4; IC_{50}, concentration for Ca^{2+} displacement from dipalmitoylphosphatidic acid (DPPA) monolayers.

(Table 3.4) it could be shown that the depression of transition temperature depends on the spatial arrangement of the lipophilic moiety of the compounds. All three derivatives induced a new peak (control transition at 63.7 °C), indicating substance-containing domains in the bilayer. The peak induced at lower temperatures was almost identical for compounds **I** and **II** (ΔT = 28 °C and 29 °C respectively) and was independent of concentration. The transition signal induced by derivative **III**, however, was detected at a significantly lower temperature (11.1 ± 0.1 °C), ΔT amounted to 53 °C. Parallel inhibition studies of Ca^{2+} binding to DPPA monolayers showed a concentration-dependent displacement of Ca^{2+} ions by the three derivatives. The IC_{50} values are given in Table 3.4. They probably correspond to drug binding and depend on the lipophilicity of the three derivatives. The interesting result is that compounds **II** and **III** differed widely in their ability to shift the phase transition temperature of DPPA liposomes despite of being structural isomers with similar lipophilicity. In contrast, compounds **I** and **II**, which differ in their lipophilic properties, reduced T_t to the same extent. It follows that the lipophilicity of cationic compounds does not govern their perturbing activity on DPPA membranes, whereas the binding affinity to the phospholipid increases in parallel with the lipophilicity of the catamphiphiles. The change in transition temperature of DPPA by the three 1-dimethylamino-3x-propane derivatives again shows the importance of the position and orientation of the lipophilic moiety with respect to the hydrophilic area, the "hydrophilic–hydrophobic dipole".

Differential scanning calorimetry has also been used to study the interaction of DPPC liposomes with the neuromediators norepinephrine and 5-hydroxytryptamine and four antidepressant drugs (imipramine, indalpine, citalopram, and milnacipran) known to inhibit uptake of these neurotransmitters [39]. Changes in the thermograms, transition temperature maximum, T_t, and ΔT_t were determined as a function

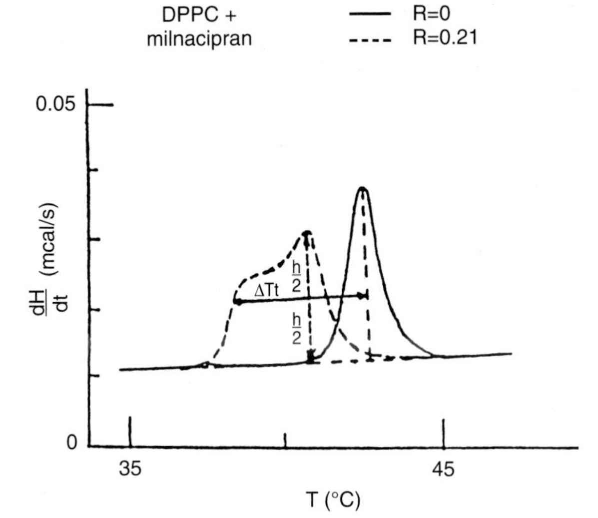

Fig. 3.6 Determination of ΔT_t for pure liposome transition (DPPC) and in the presence of a particular drug (milnacipran) at ratio, *R*. ΔT_t is measured as indicated at half the height of the main peak of liposome transition. (Reprinted from Fig. 10 of ref. 39 with permission of the American Chemical Society and the American Pharmaceutical Association)

of increasing drug concentration (Figure 3.6). The results are summarized in Table 3.5. Two types of interaction behavior are indicated. For 5-hydroxytryptamine (5-HT) and norepinephrine, only a small change in transition temperature and width of endothermal peak was observed. This indicates that both neuromediators are essentially localized in the aqueous phase. In consequence, the interaction with the DPPC liposomes is small. This is in agreement with the assumption that the recognition sites for carrier proteins of these drugs are directed toward the aqueous phase. It also explains why these two hydrophilic drugs cannot pass the blood–brain barrier. In the case of the other four drugs studied a new phase of decreased transition temperature occurred, indicating the location of the cationic molecules at the polar head groups of DPPC. Detailed analysis nevertheless shows differences between the four drugs in the produced ΔT_t. Lowering of the phase transition temperature and broadening of the peak indicate destabilization of the phospholipid assembly and thus a decrease in the cooperative unit. This directly correlates with the depth of penetration into the lipid bilayer. In the case of the neuromediators, milnacipran, indalpine, and to a lower extent citalopram, ΔT_t stabilizes with increasing drug concentration. In contrast, the most lipophilic substance, imipramine, exhibited a steady increase in ΔT_t with increasing drug concentration, indicating further penetration into the bilayer. Imipramine is known to block the uptake of 5-HT non-competitively. In contrast, indalpine and citalopram are competitive blockers of uptake. According to the authors drugs such as indalpine and citalopram, and especially milnacipran, are located in the hydrophilic region and therefore have a higher probability of competing directly with the neuromediators. The DSC results for imipramine suggest, however, that it localizes in both regions, i.e. it can at the same time interact with the polar head groups and the hydrocarbon chains. This is detected as a strong effect on ΔT_t. Accordingly, it can be hypothesized that the neurotransmitter uptake protein has recognition sites inside the bilayer, which may be occupied by imipramine. This could

Tab. 3.5 Temperatures (°C) of transition maxima (T_t) and of ΔT_t parameters characteristic of the peak width for various molar ratios (R) (see Figure 3-6). (Reproduced from Tab. 1 of ref. 39, with permission from the American Chemical Society)

Product	*R*	T_t	ΔT_t
5-Hydroxytryptamine	0.07	42.6±0.2	1.1±0.2
	0.17	42.3±0.3	1.2±0.2
	0.29	41.9±0.4	2.1±0.3
	0.44	40.3±0.2	3.1±0.2
	0.66	40.3±0.1	3.3±0.1
	0.80	40.4±0.1	3.2±0.1
Citalopram	0.02	43.1±0.2	0.1±0.1
	0.08	41.6±0.2	4.6±0.1
	0.15	38.8±0.2	5.5±0.3
	0.26	38.3±0.2	6.5±0.1
	0.35	37.8±0.3	6.9±0.1
	0.47	37.1±0.2	8.3±0.1
Norepinephine	0.08	42.6±0.5	1.3±0.3
	0.31	42.3±0.3	1.2±0.3
	0.46	42.2±0.5	1.6±0.5
	0.69	42.1±0.5	2.0±0.2
	0.82	41.1±0.4	2.8±0.2
Milnacipran	0.05	41.3±0.1	1.6±0.1
	0.21	41.0±0.1	3.9±0.1
	0.34	40.0±0.3	3.9±0.3
	0.44	39.1±0.1	4.0±0.1
	0.56	38.8±0.1	4.1±0.1
Indalpine	0.03	41.7±0.3	2.0±0.1
	0.14	38.2±0.1	6.2±0.1
	0.24	36.5±0.1	7.1±0.2
	0.39	36.0±0.1	7.7±0.1
Imipramine	0.05	41.5±0.2	1.5±0.5
	0.35	33.1±0.3	10.6±0.2
	0.58	29.7±0.2	14.1±0.2

lead to conformational change and block the uptake of neurotransmitters (non-competitive inhibition) [40]. The ability of imipramine to penetrate membrane structures may be the reason for its postsynaptic activity. This sedative effect is unique to imipramine and is not shared by the other three drugs.

The effect of halothane ($CF_3CHBrCl$) on the lateral surface conductance and membrane hydration has been studied by Yoshida and coworkers [41]. Below the pretransition temperature, the activation energy of the ion movement (H_3O^++OH^-) was 18.1 kJ/mol, which corresponds to that of the spin–lattice relaxation time of water (18 kJ/mol) above pretransition the activation energy increased to 51.3 kJ/mol. Halothane did not show any effect on the ion movement when the temperature was below the pretransition temperature. When the temperature exceeded the pretransition temperature, the authors observed at 0.35 mM halothane (equilibrium concen-

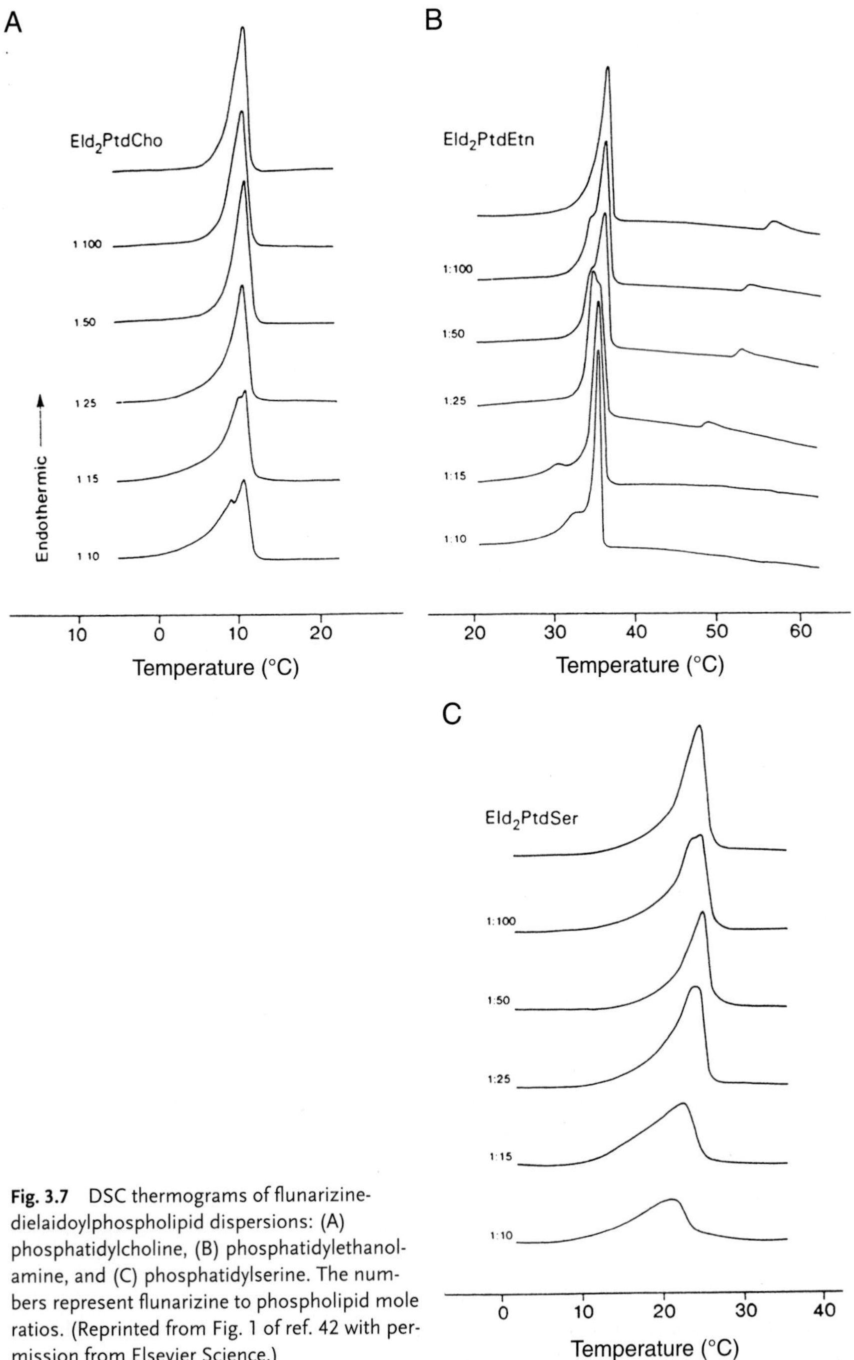

Fig. 3.7 DSC thermograms of flunarizine-dielaidoylphospholipid dispersions: (A) phosphatidylcholine, (B) phosphatidylethanolamine, and (C) phosphatidylserine. The numbers represent flunarizine to phospholipid mole ratios. (Reprinted from Fig. 1 of ref. 42 with permission from Elsevier Science.)

Tab. 3.6 Structures and some physical parameters of the investigated drugs. (Reprinted from Tab. 1 of ref. 43, with permission of the American Society for Biochemistry and Molecular Biology)

Drug	Structure	pK_a	Octanol-buffer partition coefficient
Adriamycin (ADR)	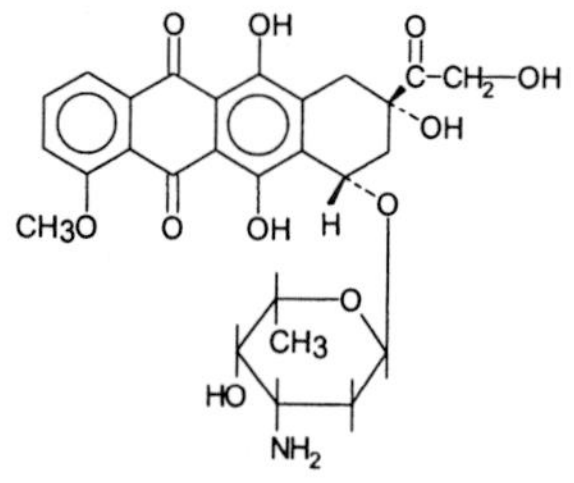 OH, O, C–CH_2–OH, OH, CH_3O, O, OH, H, O, CH_3, HO, NH_2	8.2	1.1
AD32	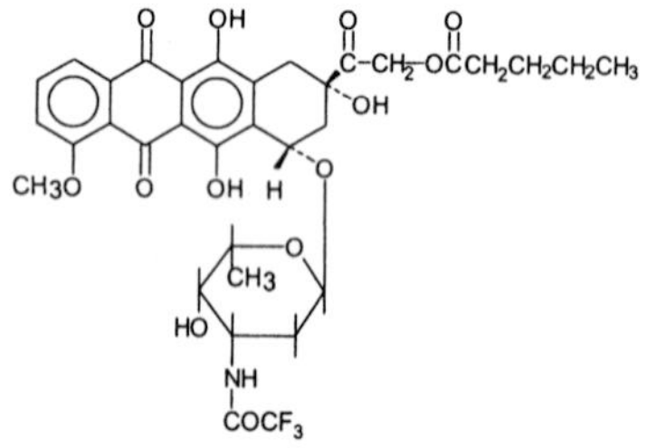 OH, O, C–CH_2–OCCH$_2$CH$_2$CH$_2$CH$_3$, OH, CH_3O, O, OH, H, O, CH_3, HO, NH, $COCF_3$	Uncharged	> 99.9
Chlorpromazine (CPZ)	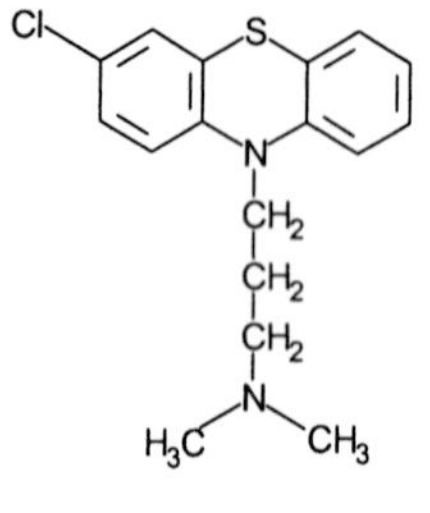 Cl, S, N, CH_2, CH_2, CH_2, N, H_3C, CH_3	9.4	5.2
Quinidine (QND)	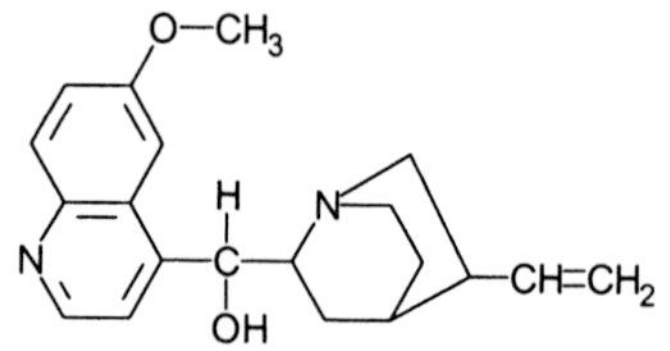 O–CH_3, N, H, C, OH, N, CH=CH_2	4.2, 8.6	1.8

tration) a decrease in the activation energy of ion movement to 29.3 kJ/mol. The decrease points to an enhanced release of the surface-bound water molecules by halothane at the pretransition temperature. Depression of the pretransition temperature and a decrease in the association energy among head groups from 9.7 kJ/mol for the control to 5.2 kJ/mol at 0.35 mM halothane also indicated a surface-disordering effect.

The interactions of catamphiphiles with three major phospholipid classes of mammalian plasma membranes have also been studied using the DSC technique [42]. Quite different interactions were observed. Whereas flunarizine significantly influenced the gel to liquid crystalline transition temperature of PS, it had little effect on that of PE (Figure 3.7). The liquid crystalline to inverted H_{II} phase transition of PE is, however, strongly induced in the presence of flunarizine. According to the authors, the results indicate different locations and ionization states for the drug in the phospholipid bilayer as well as the important role of the type of phospholipid in determining the ionization state of a drug. Both factors have implications for drug–membrane interactions. The results underline that type and degree of interaction are highly specific and depend on both the composition of the membrane and the structure of the drug.

Interesting results were obtained from a study of the interaction of adriamycin (ADR), an uncharged derivative, AD32, chlorpromazine, and quinidine in the presence of uncharged and charged phospholipids using DSC technique. Liposomes prepared from pure dipalmitoylphosphatidylglycerol (DPPG) or from binary mixtures of DPPG and dipalmitoylphosphatidylcholine (DPPC) showed that the modulation of bilayer properties by ADR was – as a result of electrostatic interactions – greatly enhanced in the presence of negatively charged phospholipid head groups. For the derivative AD32, a similar interaction with negative and neutral phospholipids was observed. At high drug concentrations, ADR as well as AD32 produced transitions with multiple peaks. In contrast, chlorpromazine and quinidine did not show this behavior. Comparison of the thermotropic effects on the membrane – such as lowering of T_t – by means of the octanol–water partition coefficient shows that additional factors are involved in the perturbation of the phase transition [43] (Tables 3.6 and 3.7).

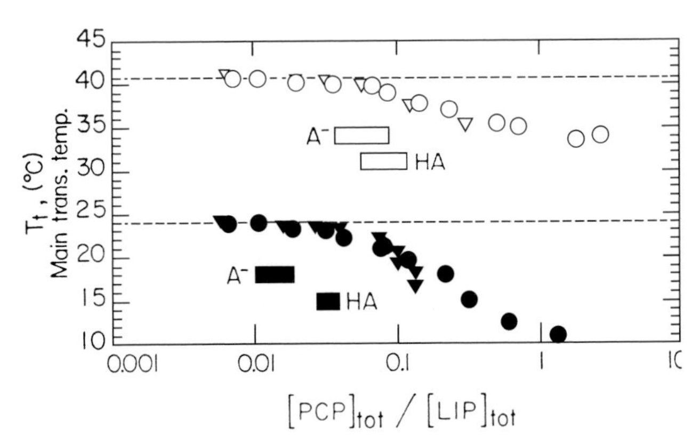

Fig. 3.8 Comparison of the effects of pentachlorophenol (PCP) on the gel-to-fluid transition of DPPC (open symbols) and DMPC (closed symbols) membranes. The circles represent the action of ionized PCP obtained at pH 10 and the triangles that of the unionized species at pH 3.4–3.5. Bars indicate the molar ratio (PCP-lipid) at which the pretransition disappears for the ionized (A^-) and unionized (HA) PCP. (Reprinted from Fig. 5 of ref. 44 with permission from Elsevier Science.)

Tab. 3.7 Thermodynamic data for DPPC and DPPG liposomes alone and in the presence of various drugs in PBS (pH 7.4) (adapted from Tab. 2 and 3 of ref. 43)

			Pretransition			Main transition		
Drug-lipid mixture	**Drug-lipid ratio**	**Total drug concentration (*M*)**	T_t^a (°C)	ΔH (*kcal mol*$^{-1}$)	**Cooperative unit (molecules)**	T_t (°C)	ΔH (*kcal/mol*$^{-1}$)	**Cooperative unit (molecules)**
DPPC[b]			34.7	1.0	380	41.5	7.9	300
+ADR	1:84	0.4×10^{-4}	34.4	1.0	370	41.5	7.0	390
+AD32	1:84	0.4×10^{-4}	33.3	0.6	590	41.4	7.7	300
+CPZ	1:84	0.4×10^{-4}	31.8	0.6	380	41.2	7.6	170
+QND	1:84	0.4×10^{-4}	34.5	1.0	340	41.5	7.7	340
+ADR	1:14	2.4×10^{-4}	34.6	1.0	380	41.6	7.4	370
+AD32	1:14	2.4×10^{-4}				40.3	7.0	90
+CPZ	1:14	2.4×10^{-4}				39.9	7.2	50
+QND	1:14	2.4×10^{-4}				40.6	8.0	70
DPPG-Na^+			32.8	0.3	1680	40.6	7.0	230
+ADR	1:84	0.4×10^{-4}	ME	0.2	1760	40.3	7.8	ME
+AD32	1:84	0.4×10^{-4}	ME	0.1	750	40.3	7.4	140
+CPZ	1:84	0.4×10^{-4}	30.1	0.2	1130	40.5	8.5	100
+QND	1:84	0.4×10^{-4}	32.0	0.2	2830	40.6	8.0	170
+ADR	1:14	2.4×10^{-4}				ME	7.4	ME
+AD32	1:14	2.4×10^{-4}				ME	7.9	ME
+CPZ	1:14	2.4×10^{-4}				38.9	8.5	30
+QND	1:14	2.4×10^{-4}				40.9	7.6	100

a) T_t values were reproducible to within ± 0.1 °C and ΔH values to within ± 0.3 kcal/mol.
b) The lipid concentration was 2.5 mg/ml or ~ 3.4×10^{-3} mM.
ME, multiple endotherms. PBS, phosphate buffer.

Furthermore, DSC can be useful for modeling the toxic effects of xenobiotics such as phenols and especially chlorinated phenols. A detailed study was performed by Smejtek *et al.* [44]. Among other observations, it was found that pentachlorophenol induces changes in the ξ-potential and the gel to fluid transition temperature in model lecithin membranes. The ionized form was more potent in abolishing pretransition. The unionized species induced an increase in melting transition width (Figure 3.8).

The effect of phenothiazine on membrane fluidity and competition between Ca^{2+} and phenothiazine for the anionic binding site at the phospholipid were the subject of another investigation based on DSC methodology [45]. Competition is also exhibited by other catamphiphiles and may play an important role in pharmacological activity.

Examination of the interaction of homologous series of phenylalkylamines and *N*-alkylbenzylamines with phospholipids showed a steady increase in T_t depression with increasing chain length. Surprisingly, no increase in T_t depression is observed with increasing chain length for *N*-alkyl-substituted benzylamines up to the hexyl derivative (Figures 3.9 and 3.10) [46]. An explanation will be given in Section 3.8.9.

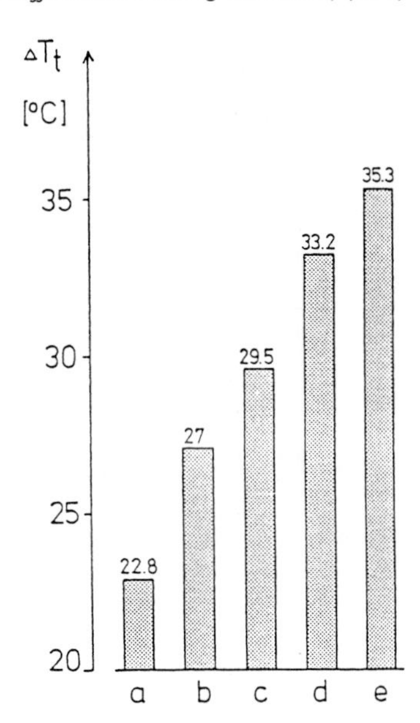

Fig. 3.9 Decrease in the transition temperature of DPPA by phenylalkylamines: (a) benzylamine, (b) phenylethylamine, (c) phenylpropylamine, (d) *p*-tolylethylamine, and (e) phenylbutylamine. Observed ΔT_t values are given on top of the columns. (Reprinted from Fig. 30 of ref. 46 with permission.)

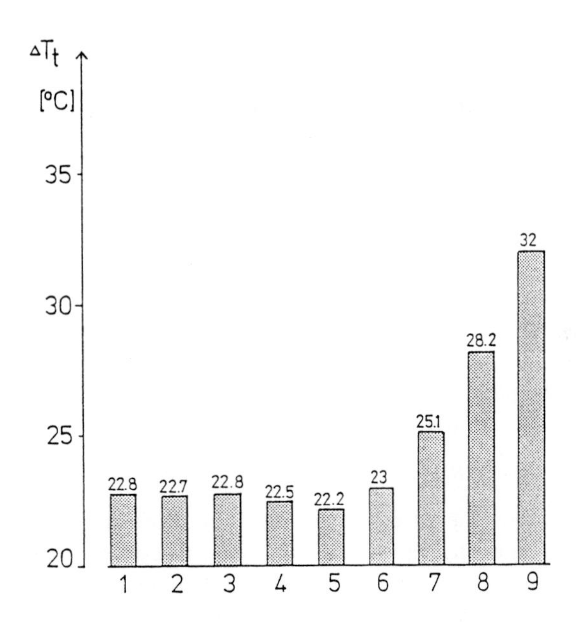

Fig. 3.10 Decrease in the transition temperature of DPPA by *N*-alkylbenzylamines unsubstituted in the aromatic ring. The numbers on the abscissa represent the number of methylene groups. The numbers on top of the columns are the observed ΔT_t. (Reprinted from Fig. 31 of ref. 46 with permission.)

In addition, the phase separation and domain formation that occurs in phospholipid mixtures (Figure 1.10) can be observed for phospholipid/drug mixtures using DSC (Figure 3.11) [12]. This is exemplified using the antiarrhythmic drug aprindine and DPPA bilayers. The control shows a single transition at about 65 °C. Upon addition of drug a new phase transition emerged at about 29 °C. With increasing drug concentration this signal became more and more intensive, and the original main transition signal at 65 °C disappeared. The difference between the control T_t and the

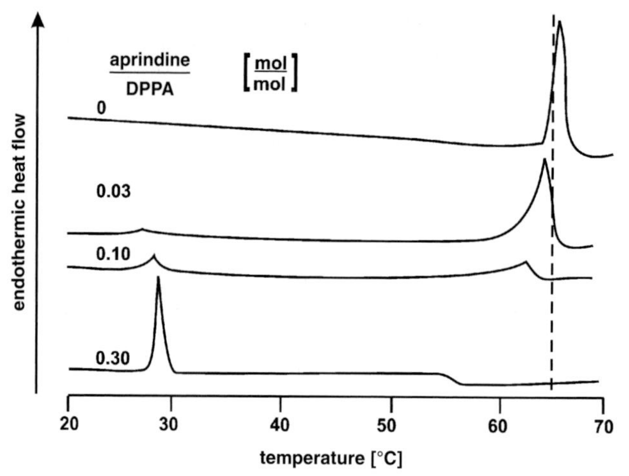

Fig. 3.11 Effect of aprindine on the transition temperature of dipalmitoylphosphatidic acid (DPPA). The amount of drug added is given relative to the amount of DPPA present. (Reprinted from Fig. 1 of ref. 12 with permission from Elsevier Science)

drug-induced T_t value, $\Delta T_{tind,}$ is characteristic for a certain drug but differs considerably between drugs (Table 3.8) [12]. The drug-induced signal is probably due to the formation of drug-containing DPPA domains that coexist with drug-free domains.

A highly significant correlation has been demonstrated between the reduction in phase transition temperature of DPPA in the presence of various amphiphilic drugs and the cardiodepressive effect of these drugs on guinea-pig hearts (IC_{50}) (see Section 5.1) [12]. Phenothiazines and other drugs that reverse multidrug resistance in tumor cells are other examples of compounds with phase separation and domain formation ability. Upon addition to liposomes composed of synthetic PS, the thermogram again shows the emergence of a new transition signal at a lower temperature, the position depending on drug structure [47] (Figure 3.12).

DSC has also been used to determine partition coefficients between phospholipids and water, avoiding radiolabeled solutes and high concentrations. One example is the partitioning of cardiac drugs into DMPC liposomes [48] (Table 3.9). The drug concentration in the liposomes was calculated using the following equation:

$$X = H/R(1/T - 1/T_0) \tag{3.4}$$

where H is the enthalpy of the pure liposome, R the gas constant, T_0 the melt onset of untreated liposomes, and T the melt onset of liposomes containing the drug. X can be converted from mole fraction to mass. Knowing the original amount of drug in solution, the partition coefficient, K, can be determined:

$$K = C_{org.}/C_{aq.}$$

where $C_{org.}$ stands for the drug concentration in the liposomes and $C_{aq.}$ for the one in the buffer phase.

A general model for the interaction of drugs and anesthetics with lipid membranes has been developed by Jørgensen *et al.* [49]. The situation is best described by a multistate lattice model for the main transition of lipid bilayers. The foreign mole-

Tab. 3.8 Catamphiphilic drugs: interaction with dipalmitoyl-phosphatidic acid and cardiodepression. (adapted from ref. 12)

Compounds	ΔT_t [a] (°)	IC_{50} [b] (μM)	AC_{50} [c] (μM)
Antiarrhythmic drugs			
Aprindine	36	40	0.5
Disopyramide	22	7000	85
Lidocaine	28	3250	20
Mexiletine	32	300	17
Procainamide	19	3000	340
Propafenone	27	17	1.2
Quinidine	30	45	22
β-Blockers			
Acebutolol	19	2600	110
Atenolol	15	6300	2000
Diacetolol	12	5100	64
Pindolol	25	580	45
Propranolol	37	22	4.4
Various			
2-Aminopyridine	13	1200	11000
Benzocaine	10	∅	240
Chloroquine	15	30	32
Chlorphentermine	34	80	43
Chlorpromazine	36	7	0.8
Dibucaine	48	35	0.2
Diltiazem	30	70	∅
Phentermine	28	700	160
Procaine	21	4000	170
Quinine	29	45	20
Tetracaine	42	100	0.5
Verapamil	34	150	∅

a) ΔT_t is the difference between T_t of the control and the drug-induced new T_t

b) IC_{50} is the concentration required to inhibit Ca^{2+} binding to phosphatidylserine monolayer by 50%

c) AC_{50} is the concentration required to elevate the threshold of alternating current to induce arrhythmia in isolated guinea-pig hearts by 50%.

cules are assumed to intercalate in the lattice as interstitials. The diversity in the thermodynamic properties of the model is explored by variation of the model parameters using computer simulation techniques that take into account thermal fluctuations.

The studied molecules include such diverse compounds as volatile anesthetics (halothane), local anesthetics of the cocaine type, calcium channel-blocking agents such as verapamil, antidepressants (chlorpromazine), and anticancer drugs such as adriamycin. It is argued that the factor of interest to the physiological effect may not

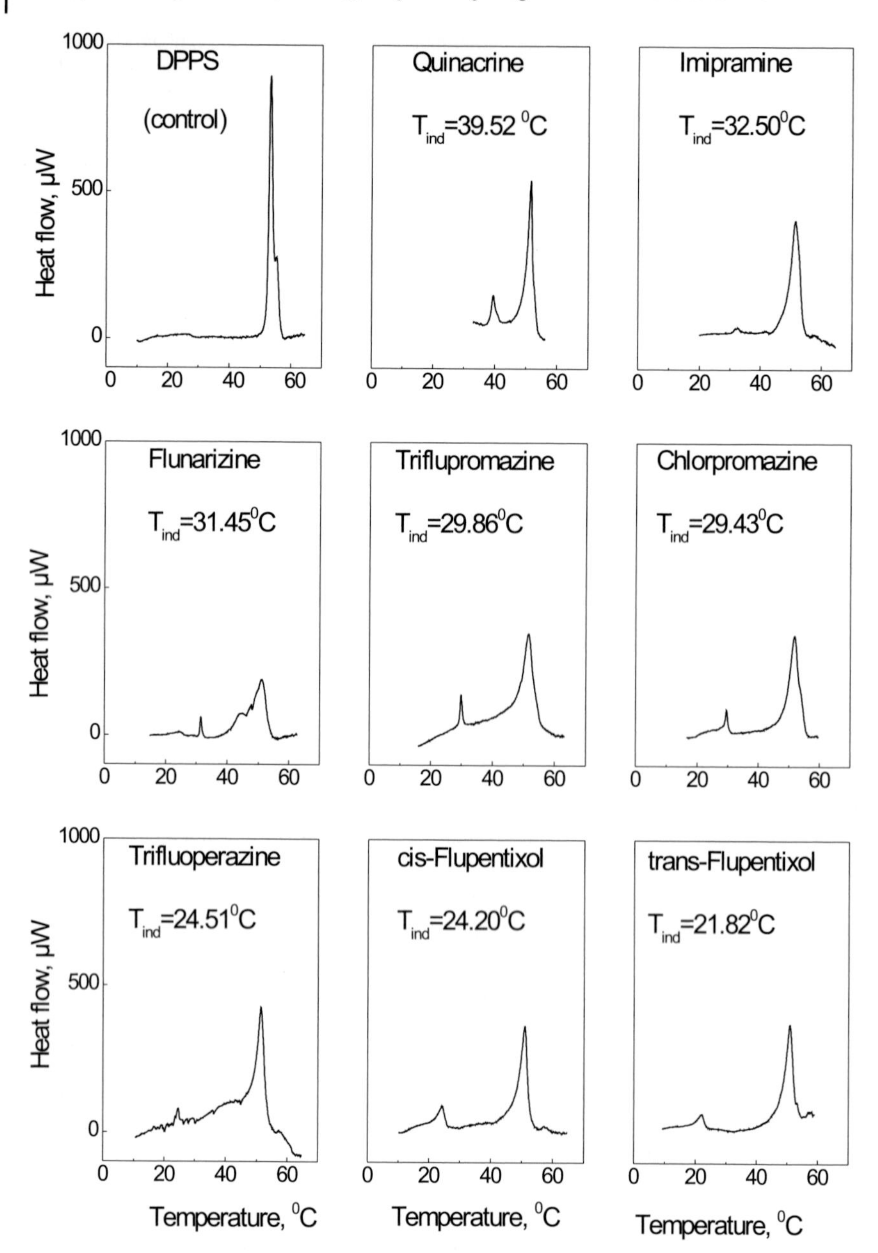

Fig. 3.12 Tracing of DSC thermograms of DPPS alone (control) and in the presence of MDR modifiers at a lipid-drug molar ratio of 1:0.05 (0.05 corresponds to the concentration of the drug in the ionized form). On the thermograms the small peaks on the left of the main phase transition peaks represent the drug-induced peaks. T_{ind} is the temperature that corresponds to the maximum excess specific heat of the drug-induced peak. The standard range in the scanning mode is 250 µW, i.e. 20.9 mJ g^{-1} °C^{-1}. (Reprinted from Fig. 3 of ref. 47 with permission from Bertelsmann-Springer.)

Tab. 3.9 Partition coefficients of cardiac drugs measured by DSC. (Reprinted from Tab. 1 of ref. 48, with permission from Elsevier Science)

Drug	***K***
Digoxin	0
Ouabain	0
Vardax	3.3
Piroximone	27
Sotalol · HCl	2.5
Propranolol	–
Lidocaine	26

be the global drug concentration as the dynamic lateral heterogeneity of the cooperative membrane assembly will lead to regions (domains) in which the local concentration is much higher than the global concentration.

Macroscopically, drug–membrane interactions manifest as changes in the physical and thermodynamic properties of the pure bilayer as varying amounts of the drug enter the membrane.

In terms of the characterization of the thermodynamic behavior, the phase diagram and the variation of response functions – for example specific heat and the membrane–water partition coefficients seen in crossing the phase diagram – are important indicators of the molecular interactions [49].

According to Jørgensen *et al.*, a foreign molecule can act as an interstitial (a) or substitutional (b) impurity. In addition, it is necessary to determine whether or not the concentration of the foreign molecule in the membrane is effectively constant [canonical or grand canonical ensemble (c)]. In case (a) the transition enthalpy, ΔH, does not vary. In case (b), ΔH decreases and in case (c) knowledge of the thermal variation of the partition coefficient is essential for the determination of the appropriate ensemble. Figure 3.13 shows the phase diagram for substitutional impurities (type I) in which the entropy of mixing normally leads to coexisting regions. In the case of interstitial impurities, both types of phase diagrams are possible. However, a wide region of coexistence can only occur if there are strong attractive interactions between interstitial molecules.

This microscopic interaction model can be used to explain more specific interactions between drug molecules and lipids. Such specific interactions could be a selective coupling between a drug molecule and a particular chain conformation of the lipid (kink excitation). This could have a dramatic effect on the fluctuation system. The drug molecule would then control the formation of interfaces between lipid domains and bulk phase in the neighborhood of the transition. First results on an extended model of this type [50] have confirmed this view and demonstrated that the partition coefficient can develop non-classical behavior by displaying a maximum near the transition. And such a maximum has in fact been observed experimentally

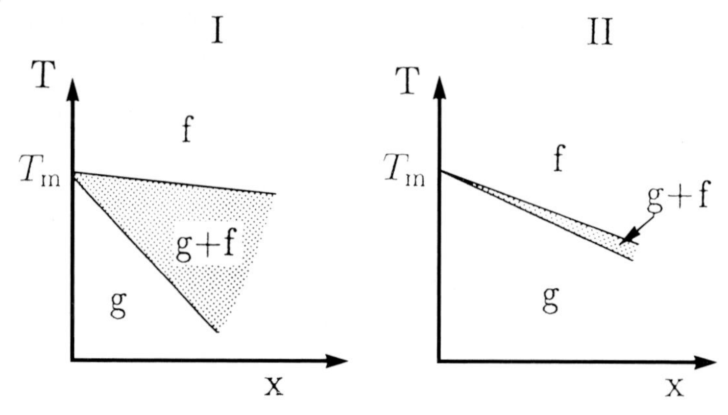

Fig. 3.13 Schematic drawing of two different types of phase diagram of a lipid membrane caused by foreign molecules. Type I: Substantial freezing-point depression and a wide coexistence region. Type II: Some freezing-point depression and a narrow coexistence region. (Reprinted from Fig. 2 of ref. 49 with permission from Elsevier Science.)

in membranes loaded with local anesthetics (cocaine derivatives) [51] or with certain insecticides (lindane) [52].

In an interesting study, the effect of five catamphiphiles on the transition temperature of three phospholipids with different head groups was evaluated [53]. It is important to remember that below the transition temperature phospholipid molecules in the gel phase are packed in a highly ordered organization. The hydrocarbon chains are in an *all-trans* conformation with greatly restricted motional freedom. Above the transition temperature, i.e. in the liquid crystalline state, they show increased motional freedom, and some of the C–C bonds assume the *gauche* conformation. At the same time, the thickness of the bilayer decreases and the surface expands, which leads to a loosening of the packing of the polar head group region. In phospholipids with the same fatty acids, the specific transition temperature depends mainly on the structure of the polar head group region. The results of the study showed that the transition from the gel to the liquid crystalline state for DPPA, DPPG, and DPPC occurred at different temperatures (Table 3.10), reflecting the in-

Tab. 3.10 Effect of different cationic amphiphilic drugs on the phase transition temperature T_t, of liposomes from different phospholipids. (Reprinted from Tab. 1 of ref. 53, with permission from Elsevier Science)

	Phase transition temperature (°C)		
	DPPA	**DPPG**	**DPPC**
Control	63	42	41
Phentermine	36	37	36
Lidocaine	35	38	36
Chlorphentermine	29	31	32
Propranolol	27	31	32
Chlorpromazine	27	30	29

fluence of the head group on T_t. Upon addition of the drugs to DPPA and also to DPPG a second transition signal at much lower temperature was induced. This is an indication of the formation of drug-containing domains in the phospholipid. This drug-induced transition was independent of the amount of drug added but varied with the drug applied. The most interesting observation was that, despite the large difference in their T_t values under drug-free control conditions, the T_t values induced by the various drugs were very similar for the three studied phospholipids (Table 3.10). This supports the authors' conclusion that the intercalation of drug molecules between the polar head groups of phospholipids almost completely eliminates the specific influence of the head groups of the studied phospholipids [53].

3.4 Fluorescence Techniques

Fluorescence probes are frequently used to study changes in membrane organization and membrane fluidity induced by anesthetics, various drugs, and insecticides. This technique measures fluidity as the rate and extent of phospholipid acyl chain excursion away from some initial chain orientation during the lifetime of the excited fluorescence state. Special techniques even allow the place of interaction to be localized, i.e. to the outer membrane region, the hydrophobic area, or the embedded proteins.

The effect of organophosphorous insecticides such as methylbromfenvinfos on membrane fluidity has been studied using the fluorescence anisotropy of 1,6-diphenyl-1,3,5-hexatriene (DPH), a probe known to be located in the hydrophobic core of the bilayer, and 1,3-bis(1-pyrene)propane (Py(3)Py), a probe distributed in the outer layer region [54]. DPH revealed a broadening of the transition profile and a solidifying effect in the fluid phase of DMPC and DPPC in the presence of 50 μM insecticide. An ordering effect of the insecticide in the fluid state was revealed by Py(3)Py. In addition, the pretransition in DPPC and DMPC vesicles was abolished by the insecticide. The addition of cholesterol decreased the influence of the insecticide. It was also observed that the influence on native membranes (erythrocytes, lymphocytes, brain microsomes, and sarcoplasmic reticulum) depended on the cholesterol content of the membranes.

In another study, the *in vitro* modulation of rat adipocyte ghost membrane fluidity by cholesterol oxysterols was investigated [55]. It was found that cholesterol oxysterols interact differently with rat adipocyte membranes. Cholestanone interacts predominantly with the phospholipids located at the inner leaflet (e.g. PE), whereas cholesterol interacts preferably with the phospholipids (PC) of the outer layer.

DPH probes have also been used to study the effect of the pyrethroid insecticide allethrin on DMPC [56].

The fluorescence probe 1-anilinonaphthalene-8-sulfonate was used to examine the binding of spin-labeled local anesthetics to membranes of human blood cells and rabbit sarcoplasmic reticulum. Two distinct fluorescence lifetimes are exhibited by this probe: the shorter life time represents the probe associated with pure lipid, the longer lifetime that associated with the protein region. It was found that spin-labeled local anesthetics quenched the fluorescence of both components. The results reflect the relative interaction of local anesthetics with membrane lipids and proteins. It was also found that positively charged local anesthetics produced a stronger quench on 1-anilinonaphthalene-8-sulfonate in protein regions, indicating an interaction of electrostatic nature between proteins and positively charged local anesthetics [57].

The modulation of synaptosomal plasma membranes (SPMs) by adriamycin and the resultant effects on the activity of membrane-bound enzymes have been reported [58]. Again DPH was used as fluorescence probe. Adriamycin increased the lipid fluidity of the membrane labeled with DPH, as indicated by the steady-state fluorescence anisotropy. The lipid-phase separation of the membrane at 23.3 °C was perturbed by adriamycin so that the transition temperature was reduced to 16.2 °C. At the same time it was found that the Na^+,K^+-stimulated ATPase activity exhibits a break point at 22.8 °C in control SPMs. This was reduced to 15.8 °C in adriamycin-treated SPMs. It was proposed that adriamycin achieves this effect through asymmetric perturbation of the lipid membrane structure and that this change in the membrane fluidity may be an early key event in adriamycin-induced neurotoxicity.

Fluorescence techniques have also been used to determine the localization of molecules in membranes. Using this technique, the localization of the linear dye molecule 3,3′-diethyloxadicarboxyamine iodide (DODCI) in lipid bilayer vesicles was determined as a function of lipid chain length and unsaturation. It was found that the fraction of the dye in the interior region of the membrane was decreased as a function of chain length in the order $C_{12} > C_{14} > C_{16} > C_{18}$. In unsaturated lipids it was $C_{14:1} > C_{14:0} > C_{16:1} > C_{16:0}$, which is in agreement with the general observation that the penetration of amphiphilic molecules into the interior of membranes increases with an increase in the fluidity of the membrane structure [59].

3.5
Fourier Transform Infrared Spectroscopy (FT-IR)

FT-IR-spectroscopy has been applied successfully to the study of the physicochemical behavior of phospholipid–water systems under various conditions. In particular, measurements on oriented phospholipid films obtained with polarized IR have provided detailed information at the molecular level. The spectra are characterized by the wavelength of the maximum of the absorption signal and the width and intensity of the signal as a function of the direction of the polarized light beam. Analysis of the absorption signals from the hydrophilic and hydrophobic regions of the phospholipids provides information about inter- and intramolecular interactions. Upon addition of amphiphilic drugs, local changes in the interaction between phospholipid molecules are observed, which reflect changes in phase transition temperature or the appearance of new phases. The application of FT-IR allows the determination of the orientation and degree of organization of substructures within the phospholipid. In addition, the localization of a drug within the bilayer seems possible. Detailed information on the organization of phospholipid membranes gained by FT-IR can substitute for information obtained from NMR studies. The latter provide information on interacting drug molecules, which is of great importance for the modeling of drug–membrane interactions.

FT-IR has been used particularly to study the effects of cations on phosphorus head groups and methylene groups of the acyl chain. These studies were stimulated by the importance of the biological role of the Ca^{2+}–phosphatidylserine interaction. Binding of Ca^{2+} to the phosphate ester groups leads to the dehydration of such groups. Binding of Ca^{2+} to the phosphate ester groups of 1,2-dioleoyl-*sn*-glycero-3-phospho-L-serine (DOPS) is not influenced by the presence of cholesterol. Instead, cholesterol perturbs the acyl chain packing, increases the degree of immobilization of the interfacial and phosphate ester, and increases the degree of dehydration of the head groups (Figure 3.14) [60]. The interaction of chemicals and drugs with phospholipids has also been studied by IR spectroscopy, e.g. the interaction of phenol, salicylic acid, and acetylsalicylic acid (ASA) [61]. IR spectra of hydrated samples of DPPC were measured in the absence and presence of these three compounds as a function of temperature and at different lipid–drug ratios. From the temperature dependence of the wavenumber of the CH_2-symmetric stretching vibrational mode (νCH_2), the authors calculated both the main transition temperature, T_t, and the pretransition temperature, T_p. As shown in Table 3.11, the three compounds had a similar effect on T_t but their effect on T_p was greater.

Another example is the determination of the interaction of D-propranolol with LUVs of DMPC by FT-IR and quasielastic light scattering (QLS) [62]. The results of the two complementary techniques showed that cationic D-propranolol interacts with DMPC. The FT-IR spectra indicate that the drug is localized at the aqueous interface of the bilayer. Figure 3.15 shows the temperature dependence of the wavenumber of the symmetric CH_2 stretching mode of DMPC LUV in the absence and presence of

Tab. 3-11 Transition temperatures of DPPC and mixtures of DPPC/PHE, DPPC/SA, and DPPC/ASA obtained from the temperature dependence of the ν_s (CH_2) wavenumber. (Reprinted from Tab. 1 of ref. 61, with permission from Elsevier Science)

Sample	Molar ratio	pH	T_t (°C)	T_p (°C)
DPPC		7.7	40.6	34
DPPC		12.4	36	–[a]
DPPC/PHE	30 : 1	7.7	37.6	24
DPPC/PHE	3 : 1	7.7	34	13
DPPC/PHE	1 : 1	7.2	32	~10
DPPC/PHE	3 : 1	12.4	36	–[a]
DPPC/SA	30 : 1	7.7	39.2	20
DPPC/SA	1 : 1	7.7	24	–[b]
DPPC/ASA	30 : 1	7.7	39	34
DPPC/ASA	1 : 1	7.7	33.2	16

a) No pretransition, T_p.
b) T_p not detected above 6°C.
PHE, Phenol; SA; salicylic acid; ASA; O-acetylsalicylic acid; DPPC; dipalmitoyl phosphatidylcholine.

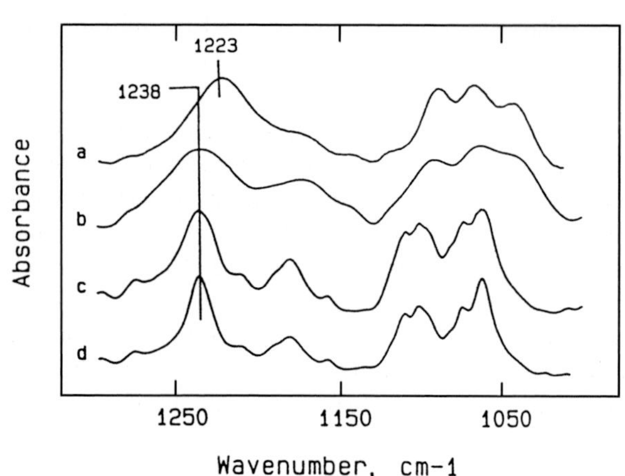

Fig. 3.14 Infrared spectra of the PO_2^- stretching (νPO_2^-) region for (a) fully hydrated pure DOPS in Tris buffer, (b) dried pure DOPS, (c) DOPS/Ca^{2+} (1:1, mol/mol), and (d) DOPS/CHOL/Ca^{2+} (1:1:1, mol/mol) in Tris buffer (pH 7.4) at 30 °C. (Reprinted from Fig. 4 of ref. 60 with permission from the American Chemical Society.)

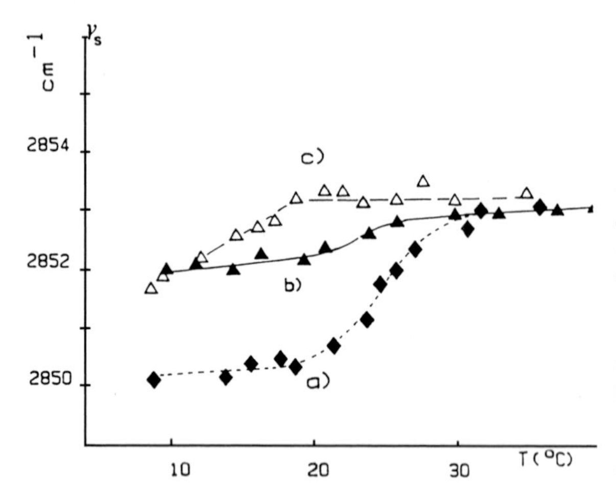

Fig. 3.15 Temperature dependence of the wavenumber of the symmetric CH_2 stretching mode of DMPC LUVs in the absence (a) and presence (b) of propranolol at pH 8.3 and (c) at pH 5.4. DMPC-drug = 2:1 (mol/mol). (Reprinted from Fig. 3 of ref. 62 with permission from Elsevier Science.)

the drug at different pH. QLS shows that, in addition to the shift in the transition temperature, the drug changes the gel phase of the membrane and induces a second transition phase.

An additional example for the usefulness of IR spectroscopy in studying drug interactions with phospholipid vesicles is the quantitative determination of acyl chain conformation in gramicidin–DPPC mixtures [63]. The technique provides a quantitative measure of the extent to which membrane-spanning peptides induce disorder of phospholipid gel phases and order in liquid crystalline phases.

3.6
Electron Spin Resonance (ESR)

ESR is another supplementary technique used to investigate membrane properties and drug–membrane interactions. It is especially useful for the determination of membrane disorder and localization of drug molecules in the membrane. An advantage of the technique is its sensitivity, which allows the study of such effects at molecular or submolecular levels. Various membrane parameters have successfully been studied by ESR spectroscopy of stable nitroxide radicals. Membrane order and dynamics have been evaluated as well as drug effects on membranes, including phase behavior and permeability. Spin-labeled proteins are used to throw light on the possible direct or indirect effects of drugs on membrane-bound proteins. The results published up to 1989 have been reviewed [64].

Spin labels are usually molecules that contain a nitroxide moiety with an unpaired electron localized on the nitrogen and oxygen atoms. These labels are specifically incorporated in the lipid part of the biological membrane. In this way, the properties of the different regions of the membrane can be studied. ESR measures the transition

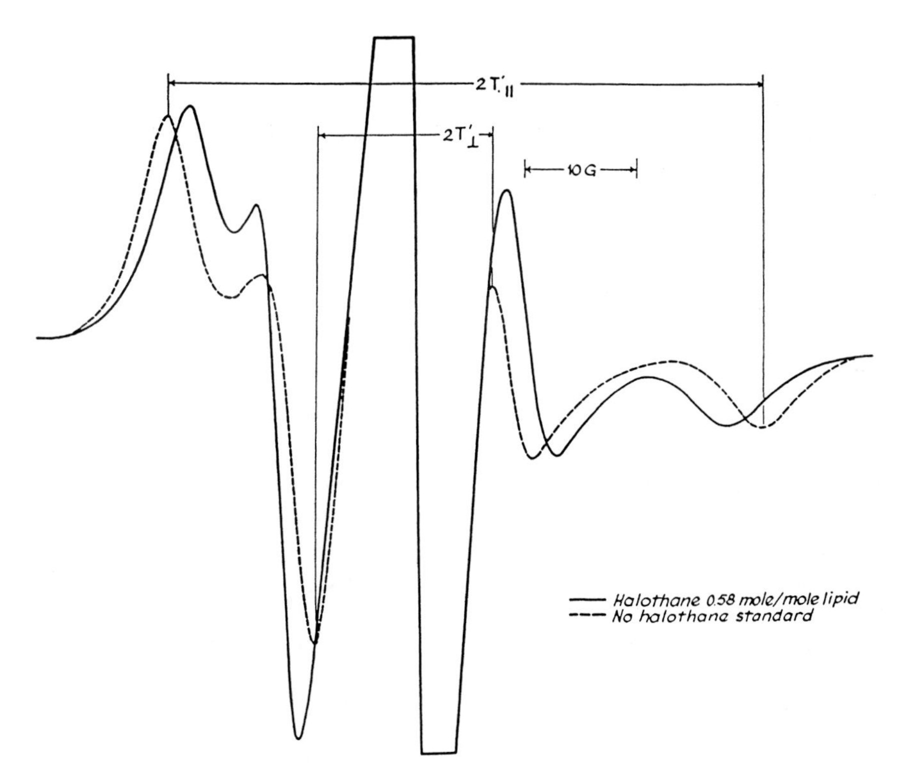

Fig. 3.16 ESR spectrum of the spin label Iβ, 1-stearoyl-2-(6-doxyl)myristoyl-glycero-3-phosphocholine (7,4) in phospholipid vesicles with and without halothane. $2T'_{\parallel}$ and $2T'_{\perp}$ are, respectively, the outer and inner hyperfine splitting in Gauss. (Reprinted from Fig. 1.1 of ref. 65 with permission from Wiley-Liss.)

of the electrons, which depends on the orientation, motion quantity, and magnetic properties of the environment of the nitroxide label.

ESR has been used in particular to study the effects of anesthetics on membrane structure and function. It has been found that many different parameters affecting membrane structure, order, and dynamics are changed simultaneously by anesthetics. Figure 3.16 shows the effect of halothane on the ESR spectrum of spin-labeled Iβ 1-stearoyl-2-(6-doxyl)myristoyl-glycero-3-phosphocholine in phospholipid vesicles with and without halothane, and Figure 3.17 shows how the degree of order of the membrane changes from as a function of increasing anesthetic concentration [65]. The parameters derived, the knowledge that amphipathic molecules are superior to apolar molecules, and the resulting differential effects of these classes of anesthetics on surface and core properties led to the conclusion that interfacial properties are crucial for the anesthetic effect [66]. However, the exact mechanism of action of anesthetics is still not known. A theoretical model has been described by Trudell [65] and is shown in Figure 3.18. It assumes an indirect influence of the anesthetic on the protein conformation as a result of perturbation of the phospholipid membrane.

This question of direct interaction with nerve proteins or indirect interaction via membrane perturbation has also been tackled by ESR spectroscopy. Two types of labeling have been used: fatty acids for lipid labeling and maleimide for frog nerve proteins. The anesthetics used were halothane as an example of a general anesthetic and procaine, lidocaine, and tetracaine as examples of local anesthetics. The latter interact primarily with head groups but can also merge into the hydrophobic hydrocarbon

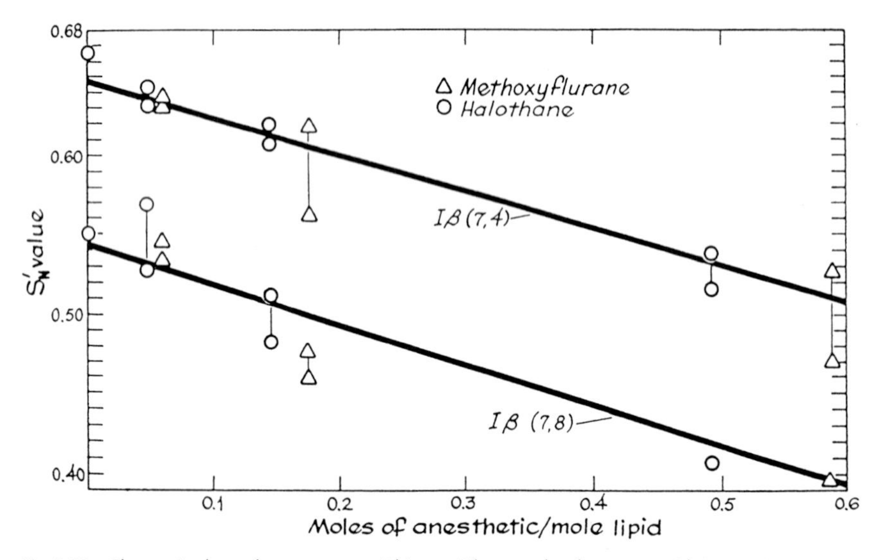

Fig. 3.17 Change in the order parameter (S'_n) of spin labels Iβ (7,4) and Iβ (7,8) 1-stearoyl-2-(10-doxyl) stearoyl-glycero-3-phosphocholine in phospholipid vesicles in the presence of varying amounts of halothane and methoxyflurane. Bilayer order decreases with increasing anesthetic concentration, as indicated by decreasing S'_n values. (Reprinted from Fig. 1.2 of ref. 65 with permission from Wiley-Liss.)

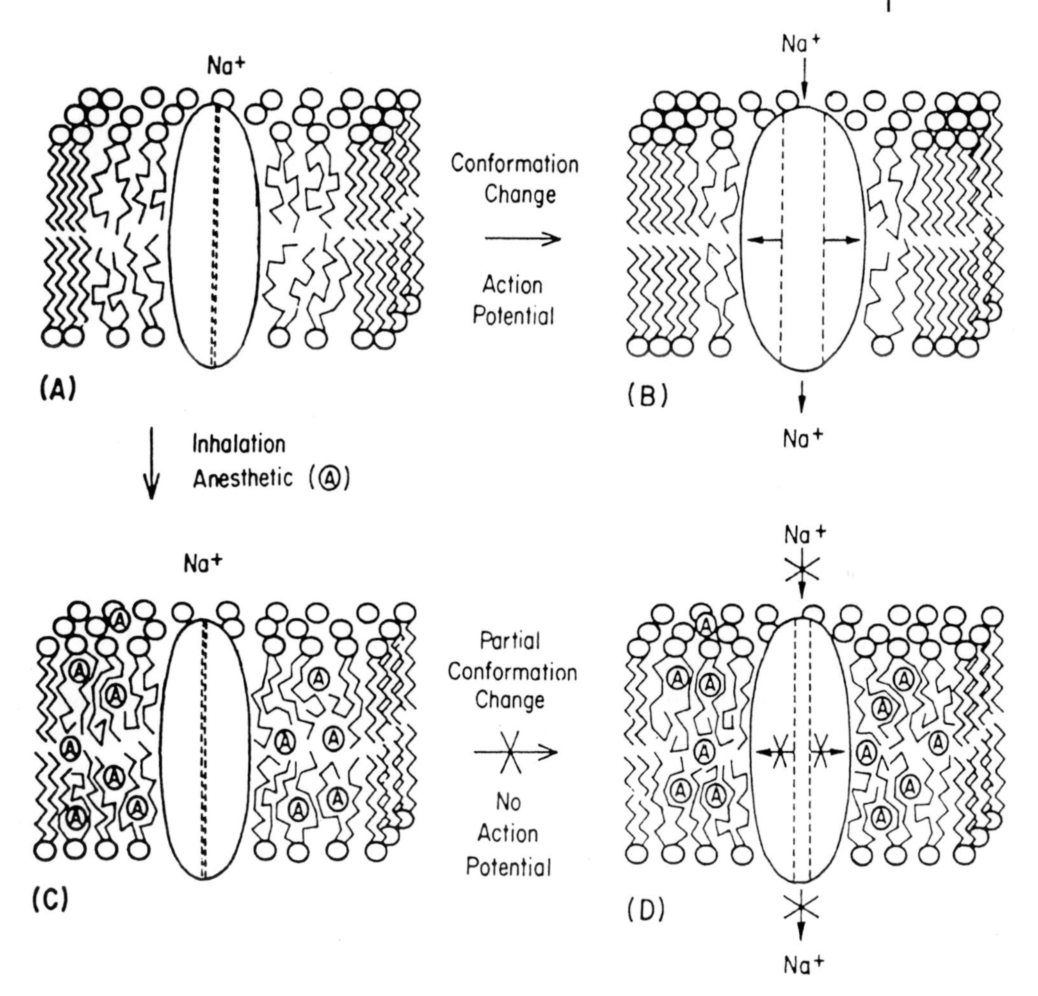

Fig. 3.18 (A) Schematic drawing of a phospholipid bilayer containing a membrane-solvated globular protein that has a sodium channel in the closed configuration. (B) The globular protein has expanded in conformation to allow a sodium ion influx. (C) Anesthetic molecules have fluidized the entire bilayer and destroyed the regions of solid phase. (D) Conversion of the high-volume fluid-phase lipids into low-volume solid phase is a high-energy process. Therefore, the protein is unable to expand or change conformation, and the excitation process does not occur. (Reprinted from Fig. 1.4 of ref. 65 with permission from Wiley-Liss.)

region. The authors conclude from the results "that the effect of halothane, lidocaine and tetracaine is efficiently transferred to the spin labeled membrane proteins via strong lipid protein interaction. The result supports the concept that the architecture and physiological activity of the membrane bound proteins are sensitive to changes in the physical state of membrane lipids" [67].

Another example is the perturbing effect of eight calcium channel blockers on membranes prepared from two different lipids [68]. The authors used total lipids from rat brain and synaptosomal membranes. The spin probe was 1-palmitoyl-2-stearoyl-phosphatidyl-choline labeled at the doxyl group at the carbon-16 position (16-PC). The apparent order parameter, S, is calculated from the apparent outer (A_{max}) and inner (A_{min}) splittings which were directly taken from the ESR spectra. It is used to describe the relative efficiency of the drugs in perturbing the lipid membrane.

$$S_{app.} = (A_{max} - B)\ 0.5407/C \tag{3.5}$$

$$B = A_{min} + 1.4\ [1 - (A_{max} - A_{min})/27.25 \tag{3.6}$$

$$C = (A_{max} + 2B)/3 \tag{3.7}$$

An example of an ESR spectrum is depicted in Figure 3.19 and shows the effect of nifedipine and verapamil. Nifedipine did not affect the ESR spectrum significantly, whereas the addition of verapamil led to a change in the ESR spectrum of liposomes, indicating a fluidization of the membrane [68]. Figure 3.20 shows that verapamil, gallopamil, mepamil, and diltiazem exert strong effects on the order parameter S, whereas the dihydropyridines nifedipine, nitrendipine, nimodipine, and niludipine show an insignificant or minor perturbation effect at a drug–lipid ratio of 1:2. This is true despite their high partitioning into lipid and biological membranes. According to studies by Herbette *et al.* [69], nimodipine is located at the water–hydrocarbon interface of the bilayer, while according to Bäuerle and Seelig [70] the uncharged molecule is homogeneously distributed across the hydrocarbon layer of the lipid mem-

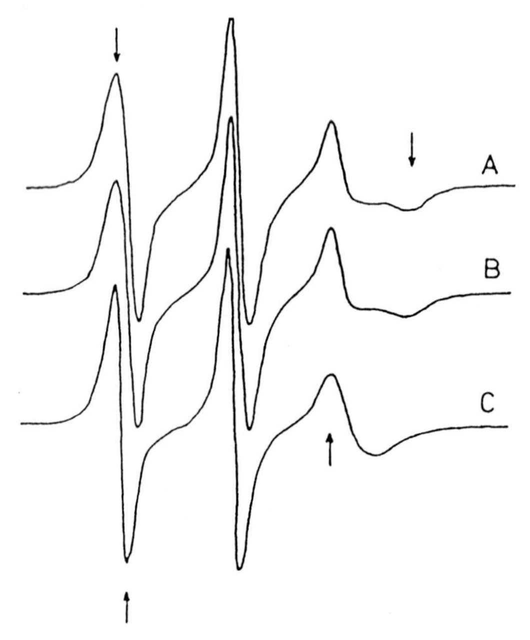

Fig. 3.19 ESR spectra of the 16-PC probe in total lipid (TL) liposomes. (A) control; (B) TL-nifedipine (2:1 molar ratio); (C) TL-verapamil (2:1 molar ratio). Spectrum width 6 mT. Temperature 37 °C. Arrows in spectrum (A) indicate A_{max} determination, arrows in spectrum (C) indicate A_{min}. (Reprinted from Fig. 2 of ref. 68 with permission from Elsevier Science.)

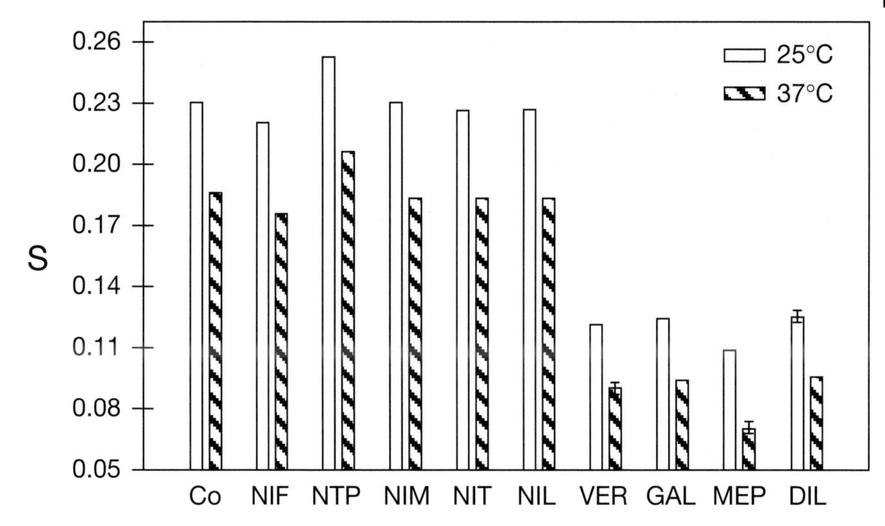

Fig. 3.20 Comparison of the perturbation effect of calcium channel blocker drugs on parameter S of 16-PC in liposomal membranes at 25 and 37 °C. Co, control; NIF, nifedipine; NTP, 2,6-dimethyl-4-(2-nitrophenyl)-3,5-pyridine dicarboxylic acid dimethyl ester; NIM, nimodipine; NIT, nitrodipine; NIL, niludipine; VER, verapamil; GAL, gallopamil; MEP, mepamil; DIL, diltiazem. (Reprinted from Fig. 3 of ref. 68 with permission from Elsevier Science.)

brane and does not interact with the head groups. In contrast, the effect of verapamil derivatives showed a significant disordering effect. This effect was similar to that produced by a temperature increase from 27 to 35 °C. Amphiphilic (charged) drugs are thought to interact with the polar head groups and at the same time to intercalate between the lipid acyl chains. This finding is supported by molecular modeling and NMR spectra [71], which have revealed that verapamil does indeed interact with the head groups and not with the hydrophobic hydrocarbon chain region (see Chapter 5).

It is interesting to note that the concentrations of verapamil that exert the perturbing effect are in the same range as concentrations that have been reported to have *in vitro* biological membrane effects. It is also remarkable that for this series of drugs no correlation has been found between the degree of perturbation and their log P [68].

3.7
Small-angle Neutron and X-ray Diffraction

X-ray diffraction [72], often in combination with DSC or NMR, is another useful tool for the study of drug–membrane interactions. It is especially useful to obtain information on the localization of drug molecules and on the conformational (phase) changes of membranes. The method is based on comparing the electron density profile of membranes to which drug has or has not been added. If partially hydrated probes are used, coherent Bragg-like scattering is obtained at reasonable resolution.

The observed differences provide information on the localization of the drug within the bilayer as well as on the perturbation of the membrane induced by the drug.

An example is the study of the topography of (–)-Δ^8-tetrahydrocannabinol (THC) and its 5′-iodo derivative in model membranes [73]. Electron density profiles of the lipid bilayer in the absence and presence of the cannabinoids were calculated using Fourier transform. Figure 3.21 shows the electron density profile for DMPC obtained at 22 °C; the space correlation is evident. Step-function equivalent profiles were then constructed to obtain an absolute electron density scale. The plot of the difference between electron density profiles resulted in improved representation of the site of increased electron density in the bilayer. The electron density profile differences inside the bilayer are given in Figure 3.22. They present the position of THC and the iodo atom of 5′-I-THC in the bilayer as shown in the figures. This is in agreement with the results of ^{2}H-NMR, which show that in THC the long axis of the

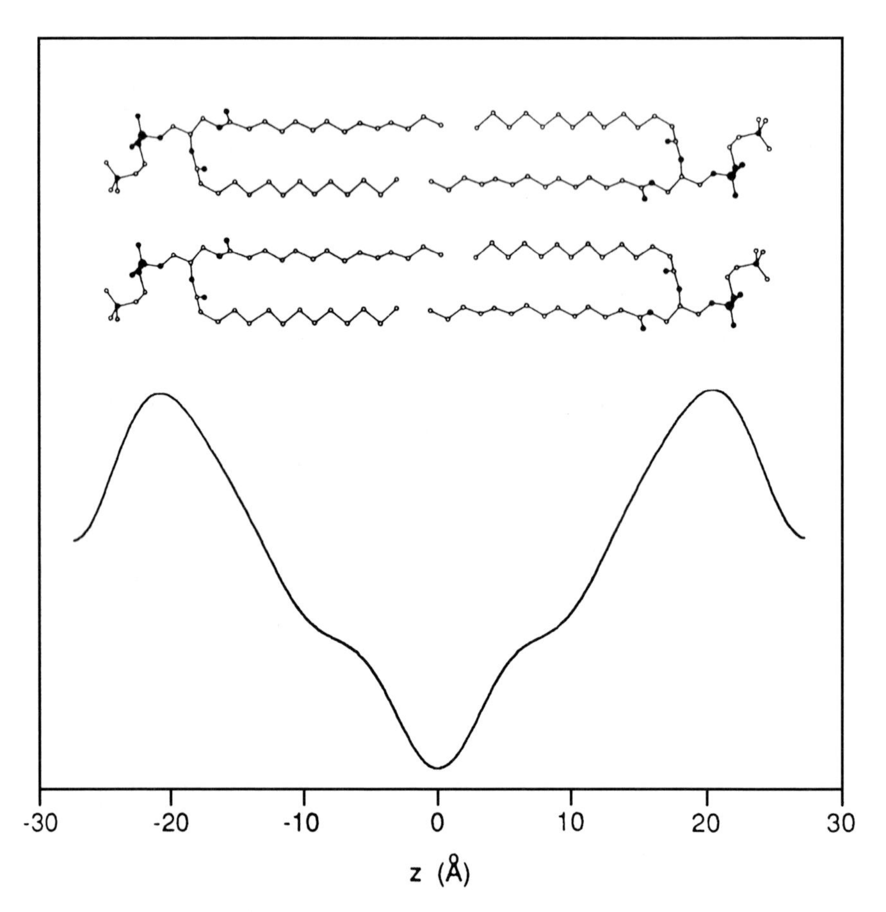

Fig. 3.21 Electron density profiles obtained by Fourier transformation of the small-angle diffraction intensities for DMPC at 22 °C. A molecular graphic representation of a DMPC bilayer above the electron density profile shows the space correlation in the dimension across the bilayer. (Reprinted from Fig. 5 of ref. 73 with permission from Elsevier Science.)

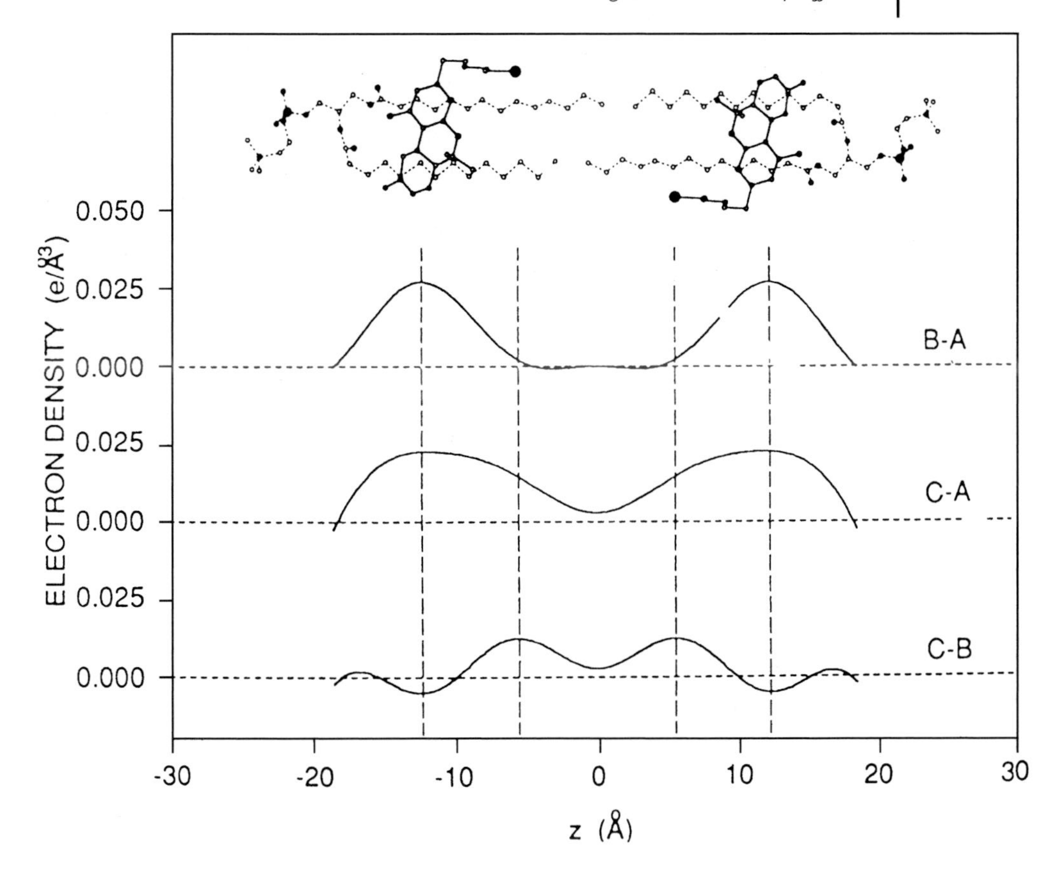

Fig. 3.22 Electron density profile differences inside the bilayer. Curve B–A is the difference between the profiles of DMPC + Δ^8-THC and DMPC, curve C–A is the difference between those of DMPC + 5′-I- Δ^8-THC and DMPC, and curve C–B is the difference between those of DMPC + 5′-I- Δ^8-THC and DMPC + Δ^8-THC. The outer pair of vertical dashed lines indicates the peaks in curve B–A and the inner pair indicates the peaks in C–B. They represent, respectively, the positions of the center of Δ^8-THC and the iodine atom of 5′-I-Δ^8-THC in the bilayer as shown by the graphic representations of the molecules above. (Reprinted from Fig. 8 of ref. 73 with permission from Elsevier Science.)

tricyclic system is oriented perpendicular to the bilayer chain [74]. Such an orientation would place the phenolic hydroxy group of THC near one of the carbonyl groups of DMPC, most likely one of the *sn*-1 chain, by formation of a hydrogen bond. The authors calculated that the position of the iodo group of 5′-I-THC requires that the five-membered chain is oriented toward the center of the bilayer and parallel to the lipid chain.

In this context it is interesting to note that there is evidence that the biological effect of cannabinoids is at least in part due to the interaction with membrane lipids. As shown by X-ray diffraction, it can be assumed that the free phenolic hydroxy group of biologically active cannabinoids is positioned near the bilayer interface. If the biological activity is due to interaction with membrane-integrated proteins, then it can be assumed from the results of the X-ray diffraction measurements that the binding site is also located near the membrane interface. In contrast, the inactive *O*-methyl analogs, which are positioned near the center of the bilayer, produce no significant perturbation of the membrane and membrane components. The example of the interaction of cannabinoids very clearly shows that the interaction of drugs with the highly structured matrix of the membrane leads to a specific orientation of the molecule within the membrane and not only to a partitioning. In this special case, the change in the hydrophobic hydrophilic center of gravity leads to a flip-flop in the orientation of 5'-I-THC and THC. The orientation may therefore directly or indirectly related to the biological effect.

Several other examples of drug–membrane interactions have been reported. Using X-ray diffraction techniques, interactions with tetracyclines [75], pindolol [76], and chlorpromazine [77, 78] have been described. In these studies, it was shown that in the presence of chlorpromazine the bilayer thickness or lipid head group separation in DPPC liposomes is only 30 Å, which is about 20 Å smaller than two fully extended DPPC molecules. Chlorpromazine produced an interdigitated phase, which is in agreement with the observed effect of chlorpromazine on the shape of erythrocytes.

An X-ray diffraction study on the effect of α-tocopherol on the phase behavior of dimyristoylphosphatidylethanolamine led to the following conclusions:

- α-Tocopherol is not randomly distributed within the phospholipid bilayer but exists in domains in the gel phase bilayer of the pure phospholipid below T_t.
- At higher temperatures, phospholipid molecules in the neighborhood of α-tocopherol are induced to form H_{II} phases with a decreased radius of curvature.
- In the H_{II} phase a defined stoichiometry of phospholipid and α-tocopherol molecules exists [79].

Another example worth mentioning is X-ray diffraction studies on amiodarone [80, 81], a drug that accumulates extensively in membranes (see Section 4.4). The effect of cholesterol on membrane structure has also been studied by X-ray diffraction. The results indicated the existence of microdomains in the DPPC–cholesterol mixed ripple phase [82].

3.8 Nuclear Magnetic Resonance (NMR)

Among the various techniques capable of evaluating the motional characteristics of biological membranes and the influence of ligands on membrane structure, NMR spectrometry plays a decisive role. High-resolution NMR spectra of model membranes were first obtained in 1966 [83]. Since then, almost every new NMR technique has been applied to study the dynamics and conformation of membranes, changes in membrane properties on addition of biological components and drugs, and the effect of membranes on drug orientation and conformation. Several reviews on the application of NMR to phospholipids have been published [84, 85]. NMR techniques provide detailed information about molecular conformation and ordering, and relaxation time measurements probe the amplitude and time scale of motions and allows interaction phenomena to be studied.

NMR spectra can be characterized by:

- the magnetic field at which resonance occurs, depending on the nucleus and the surrounding of the nucleus considered (^{1}H, ^{13}C, ^{31}P, etc.);
- The degree of spin–spin coupling produced by neighboring effects;
- The spin lattice relaxation rate, $1/T_1$, and the spin–spin relaxation rate, $1/T_2$, the latter expressed as the line width at half-maximal height of the resonance signal.

During the interaction between drug and phospholipid molecules, one or several of these parameters can change in a manner characteristic of both the drug and the "receptor" phospholipid membrane.

Additional information on molecular conformation can be obtained by NOE, transfer NOE or 2-D homonuclear correlated NMR spectroscopy (COSY), which is used to measure the distance between nuclei. For detailed information on the various techniques see refs. 86 and 87.

These methods enable changes in membrane organization to be detected and the localization of drug molecules in the membrane to be determined. They also provide information on substructures of drug molecules involved in the interaction and on possible changes in membrane and drug conformation. In addition, NMR techniques can be used to follow the rate of drug transfer into the membrane. The following nuclei are of special interest: ^{1}H, ^{2}H, ^{13}C, ^{19}F, and ^{31}P. ^{31}P-NMR techniques in solution as well as solid-state have been applied. The latter technique has been used particularly to study the interaction of peptides and proteins and has been reviewed recently [88].

Solid-state NMR allows a more direct approach to ligand–receptor interactions, normally with enhanced sensitivity, resolution and assignments, by specifically incorporating NMR isotopes (^{2}H, ^{13}C, ^{15}N, ^{19}F). Solid-state NMR can provide information on the orientational constraints of labeled groups in ligands and peptides caused by the spectral anisotropy of certain nuclei. Magic angle spinning (MAS) solid-state NMR methods have been applied to determine spin-coupled distances through dipolar coupling determinations, to high resolution (± 0.3Å) and chemical shifts to define the ligand-binding environment [89]. In addition, new methods enable the resolution of multiple spins, thus allowing structural determinations. Spectral editing permits the selective observation of specific ligands within a complex system [90].

3.8.1 Study of Membrane Polymorphism by ^{31}P-NMR

For a phase-separated region to exist, lipids have to move into and out of various phases. The lateral diffusion constant in liquid crystalline bilayers is about $10^{-8}\,cm^2/s$, which corresponds to an exchange frequency between lipid–lipid nearest neighbors of about $10^6/s$. A necessary precondition for the detection of phases by NMR technique is that the proportion of observable species in the phase is sufficiently large.

Not all NMR-active nuclei may therefore be suitable for the study of phase separation in phospholipids. Limited information can be gained on phospholipid phase transformation from ^{1}H- and ^{13}C-NMR because of problems in resolution. Only some signals, for example the choline methyl groups of phospholipids in the outer and inner leaflets of unilamellar bilayers, can be identified by ^{1}H- and ^{13}C-NMR when chemical shift reagents are used. However, ^{13}C-NMR can be applied to the study of phospholipid phase transition when the lipid is specifically enriched at the *sn*-2, carbonyl position [91].

Phosphorus-31 is present in biological membranes in addition to other nuclei and has special advantages. Phospholipid head groups contain an isolated $I = 1/2$ spin system that depends only on chemical shift anisotropy and dipolar proton–phosphorus interactions. It is therefore a useful probe for structure and motion. The chemical shift of ^{31}P depending on the local nuclear shielding varies with the orientation of the magnetic field with respect to the nucleus. The observed spectrum can therefore be measured over a wide range of about 100 ppm. As the chemical shift difference for ^{31}P is only ~4 ppm, the chemical shift anisotropy, because of orientational effects, controls the spectrum. A typical ^{31}P-NMR spectrum of polymorphic phases of phospholipid bilayers is depicted in Figure 3.23. It shows the usefulness of ^{31}P-NMR for the study of lipid polymorphism [92].

In micelles or reversed micelles, a single narrow-resonance signal is observed. If this lipid is incorporated into a bilayer, the motion becomes more restricted and a broad spectrum with a residual chemical shielding anisotropy of –30 to –50 ppm is observed. For details, the reader is referred to specific publications (see, for example, ref. 92). The hexagonal phase is characterized by a reduced residual shielding anisotropy and its sign is reversed [84, 85, 93]. The dynamics of phosphate head groups have been experimentally determined and compared with calculated values [94]. Transient NOE studies have been performed to identify the important proton species contributing to the ^{31}P–^{1}H dipolar interaction [95]. ^{2}H-NMR has been used to probe the membrane surface, to study the gel and liquid crystalline phases, and to determine the response of phospholipids in the gel and liquid crystalline states to membrane surface charges [96–98]. Many different effects and transition states have been studied and reported, for example the transitions of lamellar fluid to hexagonal and lamellar gel to fluid, and have been evaluated as a function of temperature and of membrane-affecting compounds such as acylglycerols, cholesterol, or polymyxin.

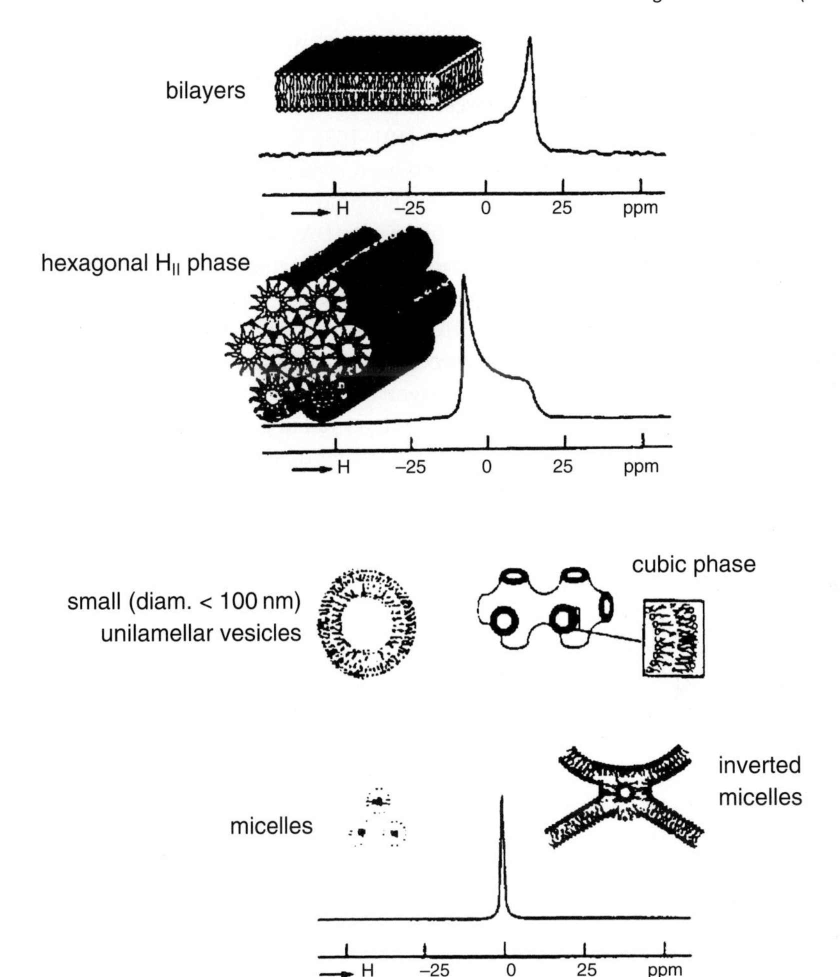

Fig. 3.23 NMR spectra for macroscopic polymorphic phases of phospholipids in bilayers, hexagonal H_{II}, and isotropic phases (small vesicles, micelles, inverted micelles, and cubic phases). Similar spectra are recorded for any one phase regardless of which phospholipid type forms the phase because the chemical shift differences for different phospholipids are smaller (2–4 ppm) than the anisotropy of chemical shift (~30–50 ppm). (Reprinted from Fig. 3 of ref. 92 with permission from Elsevier Science.)

3.8.2
Effect of Cholesterol and Diacylglycerols

The influence of cholesterol concentration at various temperatures on the thermotropic phase behavior and organization of saturated PE bilayers has been studied by combining DSC, FT-IR, and ^{31}P-NMR. It was found that incorporation of low levels of cholesterol into the bilayer caused a progressive reduction in the temperature, enthalpy, and overall cooperativity of the lipid hydrocarbon chain melting transition

[99]. The interaction of various cholesterol "ancestors" with lipid membranes on oriented bilayers has also been studied by ^{2}H- and ^{31}P-NMR [100].

The interaction of saturated diacylglycerols (DAGs) with phospholipid bilayers has been studied by various NMR techniques [101–103]. Complementary techniques, such as DSC and X-ray diffraction, have also been used [34]. It has been shown that DAGs induce a decrease in the area per phospholipid molecule and cause a parallel increase in lateral surface pressure of the bilayer. As DAGs with diC$_8$ are the most effective activators of PKC, it was concluded that the activation of the enzyme occurs via a transverse perturbation of the lipid bilayer structure [102]. ^{2}H-NMR spectra of DPPC in the absence and presence of DAGs of various chain length are depicted in Figure 3.24. DAGs with chain length longer than C$_{10}$ induced lateral phase separation. The modulation of the bilayer to hexagonal transition by DAGs has also been evaluated. It is found that their effect on the bilayer to H$_{II}$ phase transition temperature is generally at least an order of magnitude greater than their effects on the gel to liquid crystalline phase transition temperature. DAGs were better modulators than

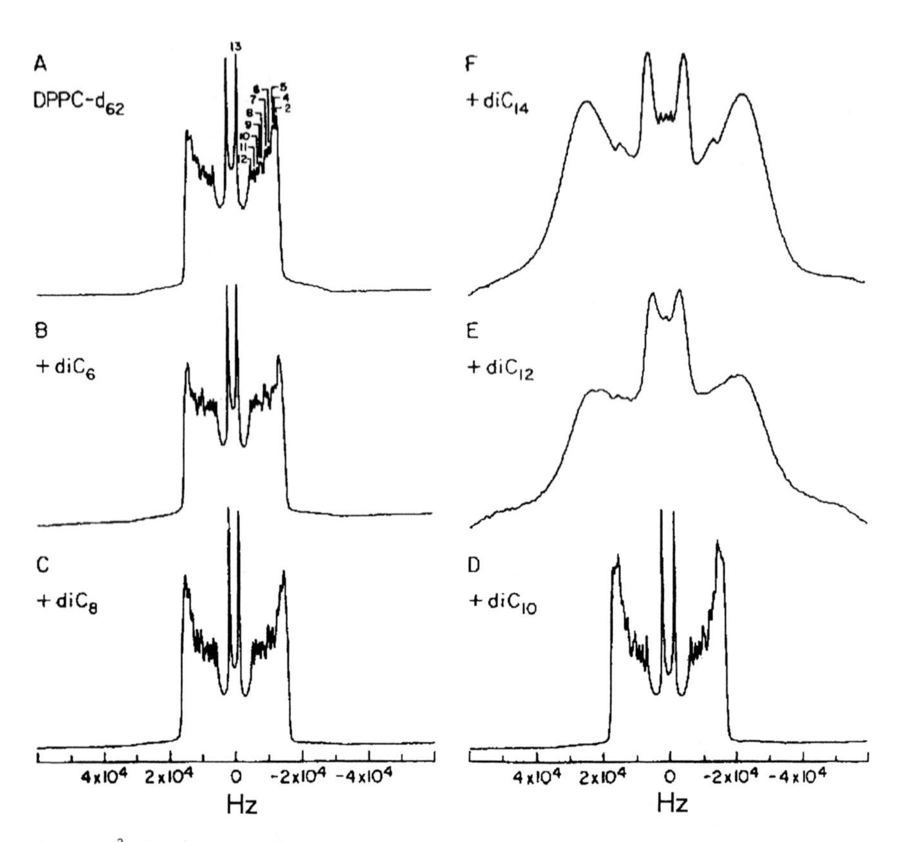

Fig. 3.24 ^{2}H-NMR spectra of DPPC-d$_{62}$ in the absence and presence of 25 mol% DAGs at 40 °C. Peaks 1 and 3 are resolved from peak 2 only at higher temperatures. (Reprinted from Fig. 1 of ref. 102 with permission from the American Chemical Society.)

monoacylglycerols, and triacylglycerols more effective than DAGs [104]. This is of interest because the effect of such additives on the rate of membrane fusion in model systems often correlates with the effect of these additives on the bilayer to hexagonal phase transition temperature [104]. The increase in the bilayer to H_{II} phase transition temperature generally inhibits cell fusion, and several compounds that stabilize bilayers show antiviral activity [37].

More recently, the effect of five DAGs, namely diolein, 1-stearoyl, 2-arachidonoyl-*sn*-glycerol (SAG), dioctanoylglycerol (diC$_8$), 1-oleoyl, 2-*sn*-acetylglycerol (OAG), and dipalmitin, on the structure of lipid bilayers composed of mixtures of PC and PS (4:1 mol/mol) was studied by ^{2}H-NMR [105]. To probe the surface region DPPC deuterated at the α- and β-position of the choline moiety was used. All DAGs, with the exception of dipalmitin, caused a continuous increase in the α-deuteron splittings and a parallel continuous decrease in the β-deuteron quadrupole splittings, indicating the introduction of conformational changes in the PC head group (Figure 3.25). In addition, DAGs at 25 mol% affected the spin–lattice relaxation time (T_2). Similar counterdirectional changes have been observed for charged catamphiphiles [106], charged phospholipids [107], and peptides [108]. The quadrupole splitting of the phospholipid head groups did not directly respond to the net membrane surface charge; rather, "the driving force for the changes in head group splittings is related to the intermolecular interactions within the local environment of neighboring lipids"

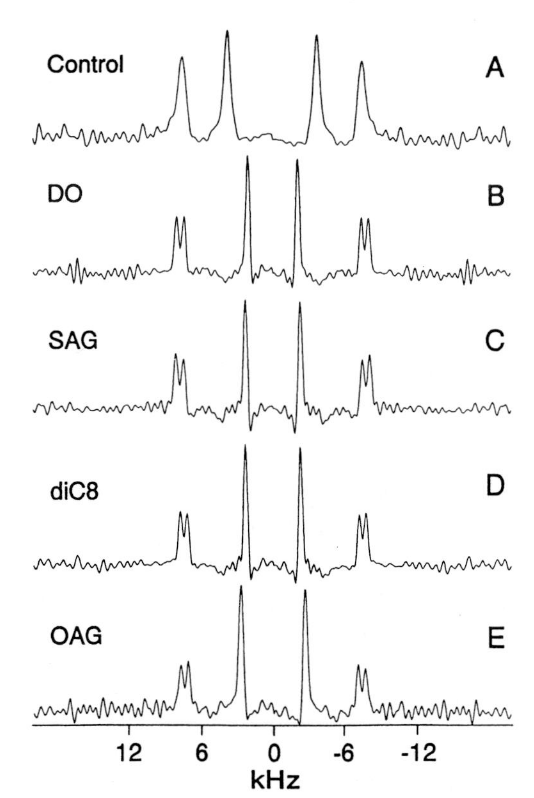

Fig. 3.25 De-paked ^{2}H-NMR spectra of DPPC-d_4 in PC/PS mixtures in the absence (A) or presence of (B–E) of 25 mol% DAG at 37 °C. (Reprinted from Fig. 2 of ref. 105 with permission from the Rockefeller University Press.)

[109]. These authors also discussed the possibility that the observed changes in Δν-values (quadrupole splitting) and the activation of PKC could be affected by another factor that could be the primary cause of both phenomena [105]. It has been shown that phoretin dramatically reduces the absorption of ^2H_2O by the membrane surface [110] and, in case of DAGs, 1,2-DPG reduced the degree of dehydration of the acyl carbonyl of DPPC in DPPC/1,2-DPG bilayers [111]. Thus, there exists the possibility that DAG-induced bilayer dehydration can affect both the change in phospholipid head group conformation and PKC activity (see Section 5.1).

3.8.3 Effect of Drugs

3.8.3.1 ^{31}P-NMR for the Study of Changes in Orientation of Phospholipid Head Group

The application of various NMR techniques to the study of the degree of drug–membrane interactions, the location of drug molecules within the membrane, the effects of drugs on surface charge, drug mobility, transmembrane transport, drug distribution in biological material, and the structural dependence of these processes is discussed below and illustrated by examples. The selection is admittedly subjective, but hopefully serves the purpose of covering those applications that could be of interest in drug development. This focus means that the discussion includes not only very recent papers, but also older ones that are appropriate to cover the various aspects and applications of NMR techniques.

Most of the examples published deal with the interaction of anesthetics with membranes, but calcium channel openers, β-blockers, antimalarial drugs, anticancer drugs, antibiotics, and insecticides have also been examined. The NMR resonance signals of atoms and groups within the lipid, and of drug molecules, have been followed and described as a function of the interaction. For the interaction of anesthetics and the involved mechanism of action, the reader's attention is directed to a review [112].

One important parameter that can be derived from ^{31}P-NMR spectra of phospholipids is the chemical shift anisotropy parameter, Δσ. This is defined as the width of the resonance signal of the low-frequency 'foot' at half-height (Figure 3.26). Chemical shift anisotropy is related to the average orientation of the phospholipid head groups relative to the normal plane of the bilayer and also to the molecular motion of the lipid molecules. As membrane-active compounds often change the average orientation of the lipid head groups as well as the phase transition temperature of the lipids from liquid crystalline to gel phase, Δσ is a sensitive parameter accounting for such changes.

The application of this technique is discussed using the example of the interaction of antimalarial drugs with DPPC [113].

Four cationic antimalarials, chloroquine (**1**), quinacrine (**2**), mefloquine (**3**), and quinine (**4**), have been studied using DPPC bilayer membranes and ^{31}P- and ^{2}H-NMR. Figure 3.26 shows the ^{31}P spectra of DPPC, fully hydrated above T_t, in the absence and presence of three of the drugs. Mefloquine exerts the strongest effect on the chemical shift anisotropy parameter (most effective broadening of the "foot"),

Chloroquine (1)

Quinacrine (2)

Quinine (3)

Mefloquine (4)

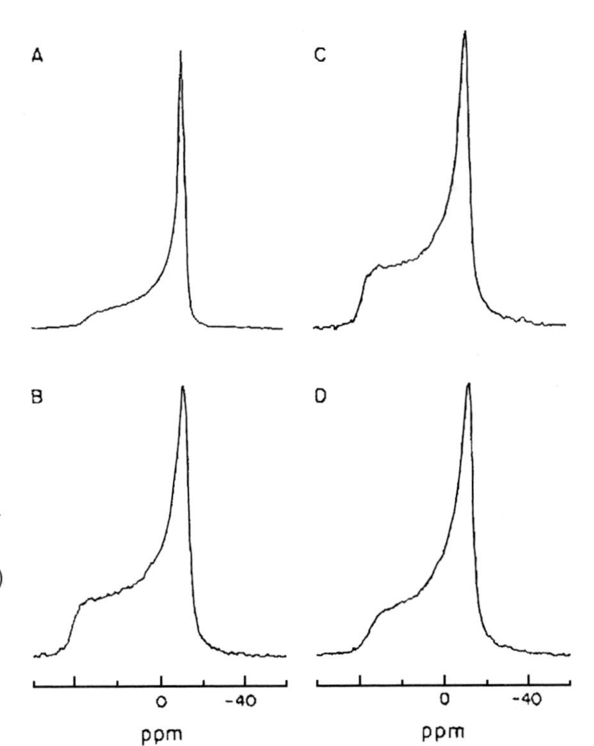

Fig. 3.26 ^{31}P-NMR spectra of aqueous dispersions of drug-DPPC mixtures (1:2, mol/mol) at 50 °C. (A) DPPC; (B) DPPC + mefloquine; (C) DPPC + quinine; (D) DPPC + quinacrine. (Reprinted from Fig. 2 of ref. 113 with permission from Elsevier Science.)

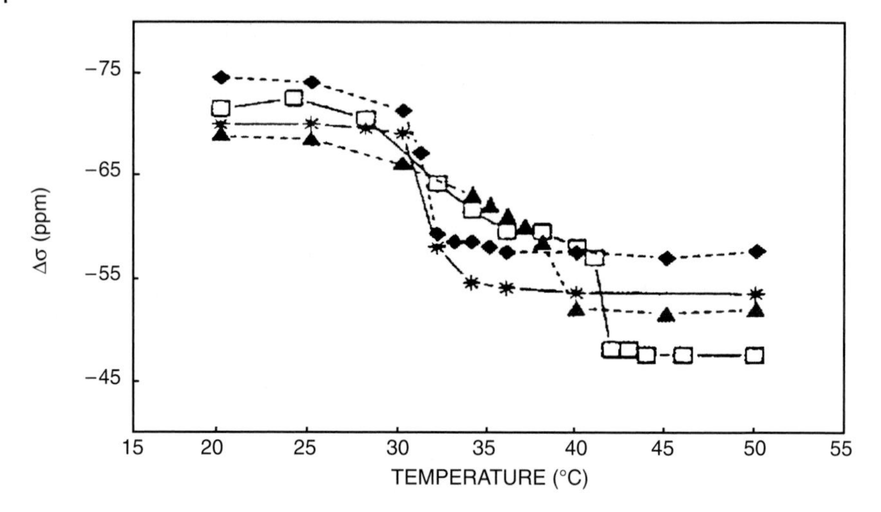

Fig. 3.27 Plot of $\Delta\sigma$ *vs.* temperature for aqueous dispersions of drug-DPPC mixtures (1:2, mol/mol). □, DPPC; ◆, DPPC + mefloquine; *, DPPC + quinine; ▲, DPPC + quinacrine. (Reprinted from Fig. 3 of ref. 113 with permission from Elsevier Science.)

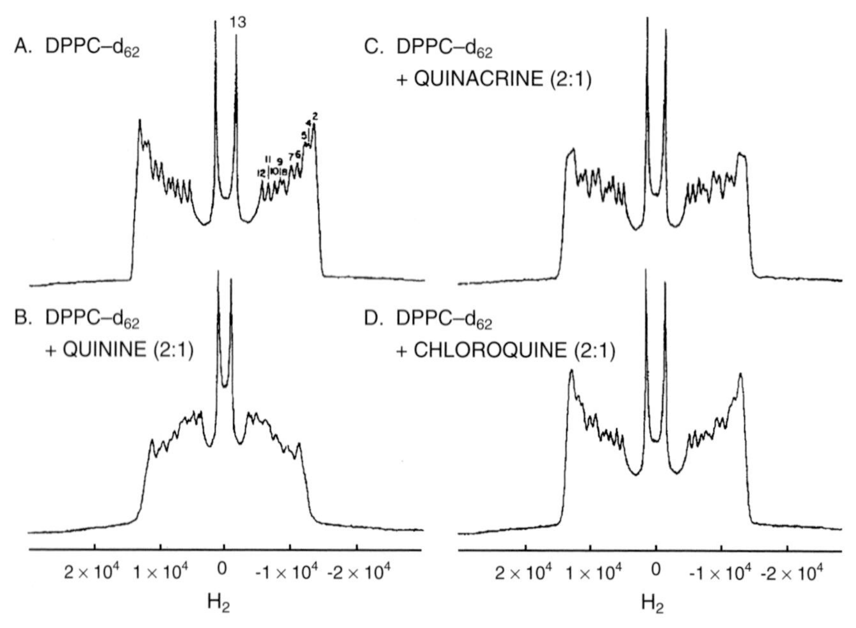

Fig. 3.28 ^{2}H-NMR of drug-DPPC mixtures (1:2, mol/mol) at 40 °C. (A) control (peaks 1 and 3 are resolved from peak 2 only at higher temperatures); (B) DPPC-d_{62} with quinine; (C) with quinacrine; (D) with chloroquine. (Reprinted from Fig. 4 of ref. 113 with permission from Elsevier Science.)

followed by quinine. The corresponding $\Delta\sigma$-values are 57 ppm and 54 ppm. It has previously been shown by computer simulation that changes of $\Delta\sigma$ above the phase transition temperature, T_t, are primarily due to changes in the orientation of the phosphate moiety of the phospholipid head groups relative to the bilayer normal. In combination with changes in the phase transition temperature $\Delta\sigma$ is an indicator of membrane-active compounds. The results indicate that mefloquine has a more pronounced effect on the orientation of head groups than quinine. In the case of chloroquine, no interaction was observed. The shapes of the spectra are typical of the bilayer conformation of lipid molecules, i.e. no change to the hexagonal phase has occurred. The temperature dependence of $\Delta\sigma$ of DPPC in the presence and absence of the antimalarials is shown in Figure 3.27. Mefloquine produced an increase in $\Delta\sigma$ to 57 ppm and a decrease in T_t of about 10 °C. In comparison, quinacrine caused a smaller but significant increase in $\Delta\sigma$ to 51.5 ppm and the decrease in T_t was 2.5 °C. The observed sharp rise in the absolute value of $\Delta\sigma$ (see Figure 3.27) upon lowering the temperature corresponds to T_t.

The drug-induced changes were confirmed by the results of ^{2}H-NMR experiments. For these studies, fully hydrated chain-perdeuterated phospholipid was used. The spectra in Figure 3.28 arise from the superposition of the axially averaged powder patterns produced by the various deuterons of the C^2H_2 and the terminal C^2H_3 groups. From each segment, the order parameter, S_{C2H}, can be derived from the peak-to-peak quadrupole splitting, $\Delta\nu$. It corresponds to the perpendicular orientation of the bond relative to the external magnetic field and is described by:

$$\Delta\nu^i = 3/4(e^2qQ/h)S^i_{C2H} \tag{3.8}$$

where $e^2qQ/h = 167$ kHz is the quadrupole coupling constant of a deuteron in the C^2H bond. The ^{2}H-NMR spectra of lipids deuterated at specific positions in the side chain show the presence of the order parameter profile along the side chain of phospholipids in a multilayer configuration. A plateau of relatively higher S_{C2H} is observed for C^2H_2 segments near the glycerol backbone [114]. Viewing the spectra in the absence and presence of the drugs confirms that DPPC-d_{62} remains in the basic bilayer structure. Quadrupole splitting was slightly decreased by quinacrine. A strong perturbation of the ^{2}H-NMR line shape was observed, however, in the presence of quinine and mefloquine at low molar ratios. In contrast, no interaction with the phospholipid even at high molar ratios was seen for chloroquine. The authors concluded that the results of ^{2}H- and ^{31}P-NMR reveal the ability of mefloquine, and to a slightly lesser extent quinine, to intercalate into the DPPC bilayer. This could account for the demonstrated ability of these drugs to penetrate the interior of the bilayer. In contrast, chloroquine and quinacrine show very little or no interaction with the head groups. According to the authors, the accumulation of mefloquine in the membrane could also explain the ability of mefloquine to overcome the resistance of chloroquine-resistant strains of *Plasmodia*.

A number of dyes belonging to the polyene class – oxonols, merocyanine, rhodamines, and cyanines – have been used in ^{31}P-NMR as potential-sensitive probes to investigate their effects on the membrane and their location in PC vesicles [115]. A substantial broadening of the ^{31}P resonances and a reduction in spin–lattice and spin–spin relaxation times was detected upon addition of the positively charged cyanine probes diS-C_3-(5) and diS-C_4-(5), but no detectable chemical shift. In contrast, the addition of anionic probes, including several oxonols and merocyanine 540, did not lead to line broadening, changes in relaxation rates, or chemical shift. This is probably due to repulsion between the anionic probes and the phosphate head groups. An increase in vesicle size detected by electron microscopy and possibly inhibition of local phosphate motion would explain ^{31}P resonance broadening and the reduction in relaxation rate induced by the two cyanines. It is suggested that the cyanine-induced increase in vesicle size is due to "an irreversible vesicle-fusion process possibly initiated by the screening of surface charge by the probes" [115].

3.8.4
Determination of Drug Transmembrane Transport

An interesting application of ^{31}P-NMR is the study of the transmembrane transport of drugs.

An example is presented in a paper by Huyinh-Dinh and coworkers [116]. These authors studied the transmembrane transport of lipophilic glycosyl phosphotriester derivatives of 3'-azido-3'-deoxythymidine (AZT), a drug used in the treatment of acquired immunodeficiency syndrome (AIDS). The usefulness of AZT is, however, limited by serious adverse reactions, especially bone marrow suppression. Previous studies by the same authors have shown that lipophilic phosphotriesters of thymidine are transported across unilamellar vesicles and that the length of the alkyl chain is of importance for drug–membrane interaction. The nucleoside and the hexadecyl moiety cannot be changed because they are essential for transport and antiviral effect. Therefore, it was suggested that changes at the sugar moiety could alter the physicochemical properties to improve selectivity. Experiments were performed with 6-substituted mannosyl phosphotriesters of AZT, derivative **5**, and a 1-substituted phosphotriester **6**. ^{31}P-NMR spectra were obtained in aqueous solution in the presence of LUVs and in the presence of LUVs and Mn^{2+} ions. These ions cannot permeate the LUV membranes, and their paramagnetism led to a large broadening of the resonance signals of those drug molecules that remained in the extravesicular volume. In consequence, only the resonance signals of molecules that had crossed

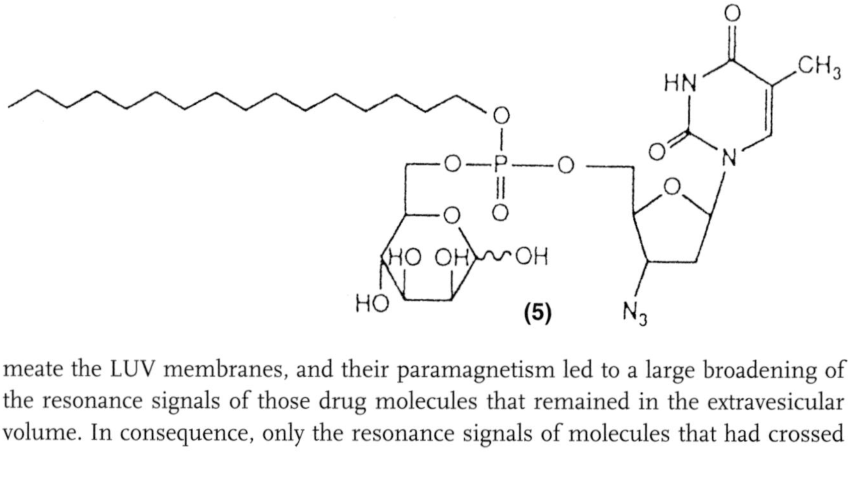

HN O CH3 O O P O O O N O HO OH OH HO N3

(5)

(6)

the membrane and were located within the vesicles were observed. It is obvious from Figure 3.29 that only molecules of derivative **6** were able to cross the membrane of the vesicles. The two phosphorus resonances of derivatives **5** and **6** in the absence of LUV correspond to the two diastereoisomers (Figure 3.29A). The chemical shifts of **5** and **6** are similar. The signal line width of **5** is, however, twice that of **6**. The line

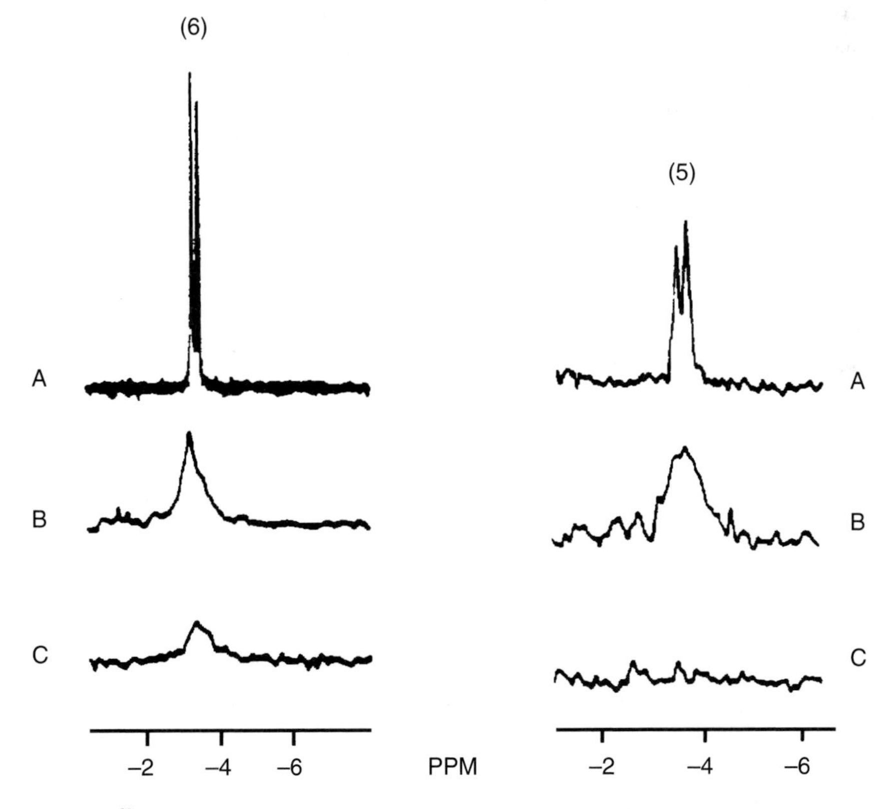

Fig. 3.29 ^{31}P-NMR (121 MHz) of phosphotriesters 5 and 6 in the absence (A) and presence (B) of LUVs and after addition of Mn^{2+} ions. (Reprinted from Fig. 2 of ref. 116 with permission from the American Chemical Society.)

width of AZT is similar to that observed for glycosyl phosphotriesters of thymidine or 5-fluoro-2'-deoxyuridine (data not shown). This indicates that these molecules form similar small micelles in aqueous solution. In contrast, the 6-mannosyl derivative (**5**) forms larger aggregates. The signals of both derivatives are broadened upon addition of LUV, that of **5** by a factor of 10. When Mn^{2+} ions were added, the resonance signals of **5** were completely broadened so that no signal could be observed. In contrast, the resonance signals of derivative **6** could still be observed. The intensity of the low-field resonance was decreased by a factor of 3, and that of the high-field resonance by a factor of 2, and the line width of the signals was not affected in the presence of Mn^{2+} ions. This shows that the phosphotriester, **6**, is partly transported into the intravesicular water–membrane interface. Its resonance signals are of reduced intensity but can still be detected. They are not influenced by Mn^{2+} ions. In contrast, derivative **5** interacts only with the outer layer of the membrane.

A combination of ^{1}H-NMR and the paramagnetic shift reagent praseodymium trichloride ($PrCl_3$) has been used to study the permeation of glycolic acid through LUVs of PC at various cholesterol concentrations and at different pH values [117]. The ^{1}H-NMR spectrum of glycolic acid in the presence of LUVs is shown in Figure 3.30. A single resonance at about 4 ppm can be observed. The addition of $PrCl_3$ led to a downfield shift of the glycolic acid outside the vesicles and a small upfield shift of the internal peak. Both peaks were broadened because of the slow exchange regime.

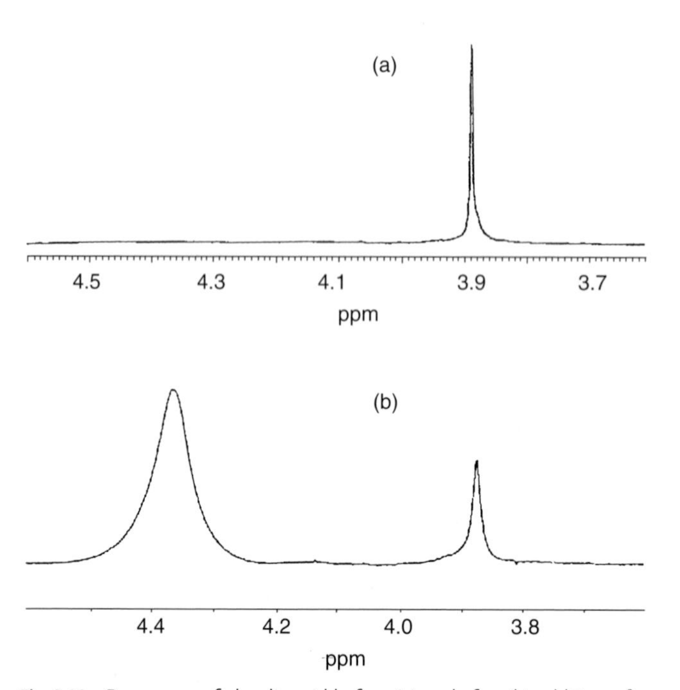

Fig. 3.30 Resonance of glycolic acid before (a) and after (b) addition of $PrCl_3$ as shift reagent to the vesicle system, pH 3.94, 30 °C, 100 nm vesicle diameter. (Reprinted from Fig. 2 of ref. 117 with permission from Elsevier Science.)

This increase in line width of the inner resonance as a result of exchange is described by the following equation:

$$\Delta\nu = k_{app.}/\pi \qquad (3.9)$$

where $\Delta\nu$ is the broadening in Hertz arising from the exchange process, and $k_{app.}$ (in s^{-1}) is the pH-dependent, vesicle concentration-independent rate constant of transport across the membrane. $\Delta\nu$ is obtained by subtracting the line width in the absence of exchange, ν_o, from the observed line width, ν. With decreasing temperature,

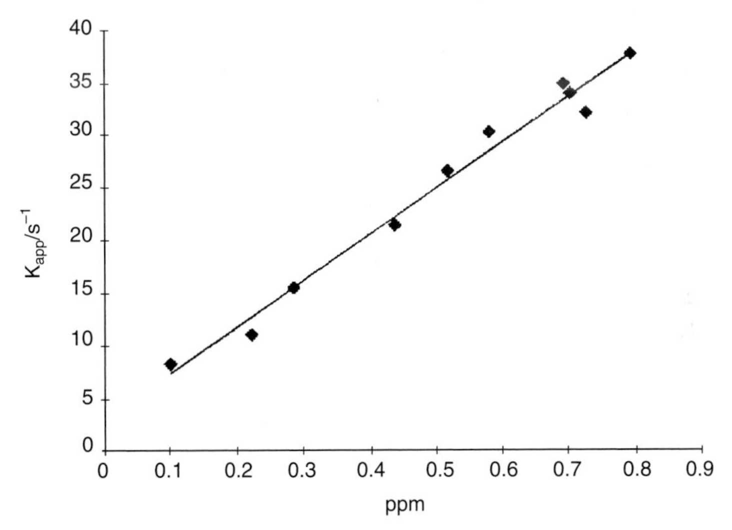

Fig. 3.31 Effect of pH (and therefore of α, the fraction of undissociated acid) on the apparent rate constant of permeation, $k_{app.}$. The slope is 43.7/s (SD = 2.2/s) and the intercept 3.1/s. (Reprinted from Fig. 4 of ref. 117 with permission from Elsevier Science.)

the line width decreased but became constant below 283 K ($\nu_o \gg 5$ Hz). The observed pH dependence of the permeation rate is given in Figure 3.31 and is described by Eq. 3.10

$$k_{app.} = k_a\alpha + k_b\,(1 - \alpha) \qquad (3.10)$$

where k_a is the rate constant of the uncharged acid, k_b is the rate constant of the negatively charged base, and α is the fraction of undissociated acid. This example of glycolic acid shows the ability of NMR techniques to evaluate the rate constants for permeation into LUVs under equilibrium conditions, and enables the study of the effects of pH, temperature, vesicle size and membrane composition (cholesterol) on the permeation rate of drug molecules.

A detailed review on NMR methods for measuring membrane transport has been published by Kuchel *et al.* [118] (see also ref. 119).

3.8.5
^{1}H-NMR in Combination with Pr^{3+} for the Study of Drug Location

Often ^{1}H-NMR is used in combination with paramagnetic probes of praseodymium cations, Pr^{3+}. Based on the bipolar interaction, Pr^{3+} allows the separation of the resonance signals of the intra- and extracellular choline head groups of phospholipids. Extravesicular Pr^{3+} is rapidly exchanged between the H_2O and phosphate sites of choline head groups in the outer layer of vesicles. The separation arises from the downfield shift of the extravesicular head group signal (O). The second signal originates from the intravesicular choline head group resonance (I). From the ratio of these signals, the size of the vesicles can be calculated [120].

Figure 3.32 shows the ^{1}H-NMR spectrum of DPPC vesicles in the presence of 5 mM Pr^{3+}. It shows the separation of resonance signals of the inner (*I*) and outer (*O*) choline head groups and the proton resonance of the lipid acyl chains (*H*) and of the terminal methyl group (*M*) [121]. Lysed vesicles were obtained by cycling the DPPC

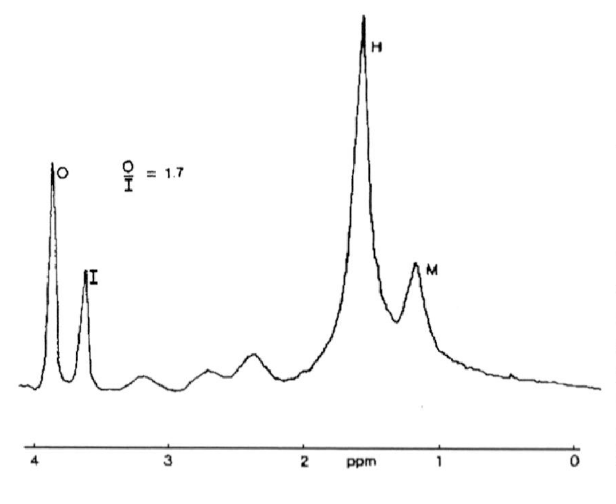

Fig. 3.32 ^{1}H-NMR of DPPC vesicles (10 mg/mL) in the presence of 5 mM Pr^{3+} showing separate signals from the inner (I) and outer (O) choline head groups, the lipid acyl chains (H), and terminal methyl groups (M). Chemical shifts are shown with respect to external TMS. (Reprinted from Fig. 1 of ref. 121 with permission from Elsevier Science.)

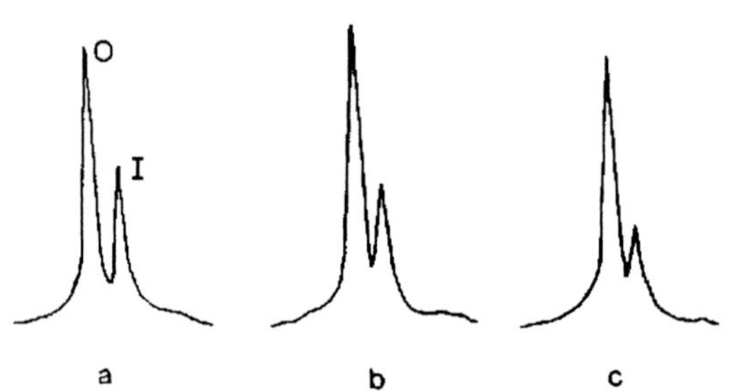

Fig. 3.33 ^{1}H-NMR spectra showing the increase in the O/I ratio of DPPC vesicles in the presence of 50 mM decan-1-ol. (a) After incubation, O/I ratio = 1.58. (b) After six cycles, O/I ratio = 1.92 (% lysis = 11.6). (c) After 10 cycles. O/I ratio = 2.25 (% lysis = 20.6). (Reprinted from Fig. 3 of ref. 121 with permission from Elsevier Science.)

vesicles from 60 °C to 30 °C to 60 °C, i.e. on cycling through T_t. After lysis, the interaction of Pr^{3+} with the inner head groups led to an increase in intensity of the outside signal and a decrease in intensity of the inside signal so that the O/I ratio increased. The ratio can be used to calculate the percentage of lysed vesicles. The authors have studied the effect of normal alcohols on vesicular permeability induced at the phase transition temperature. The incorporation of alcohols in the vesicles increased the degree of lysis on cycling through T_t. Compared with the control, an increase in the O/I ratio was observed after cycling through T_t, decan-1-ol being most effective (Figure 3.33).

In another ^{1}H-NMR study, the interaction of β-blockers with sonicated DMPC liposomes in the presence of Pr^{3+} was evaluated [122]. The presence of Pr^{3+} increased the splitting of the choline trimethylammonium group signals that arise from the phospholipid molecules located at the internal and external layer of the bilayer. The downfield shift of the external peak (E) is considerably stronger than the upfield shift of the internal peak (I) (Figure 3.34). The difference in chemical shift of the two sig-

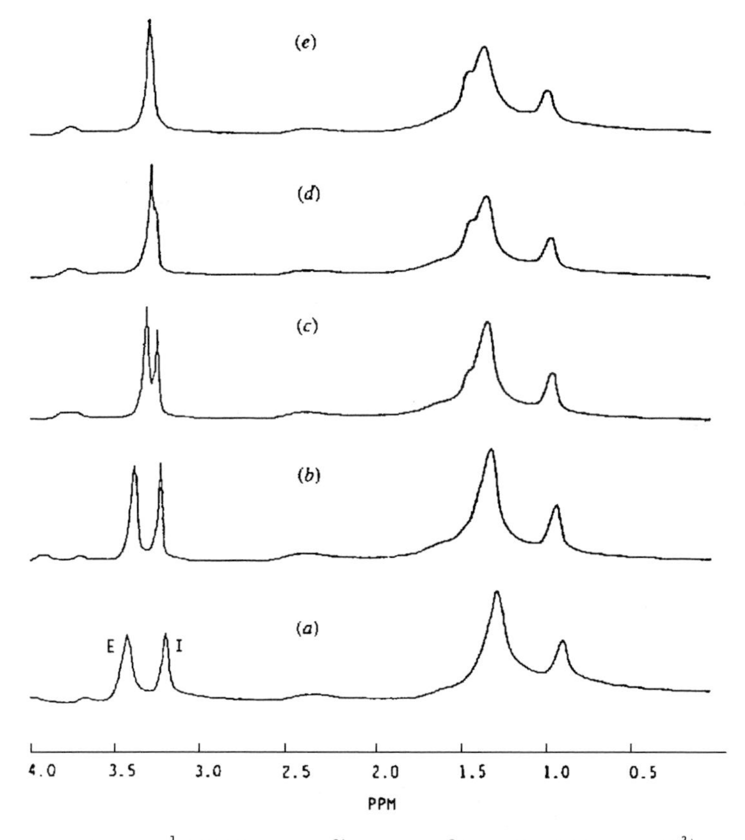

Fig. 3.34 ^{1}H-NMR spectra of liposomes of (a) DMPC (14.4 mM) + Pr^{3+} (2 mM) + propranolol at (a) 0 mM, (b) 0.5 mM, (c) 2 mM, (d) 4 mM, and (e) 5 mM. (Reprinted from Fig. 5 of ref. 122 with permission from Taylor & Francis.)

Tab. 3.12 Physicochemical properties of β-blockers of the propranolol type

Compound	p*Ka*	Log K'^{a}_{m}	Log P^{b}_{oct}
Propranolol (PPL)	9.45	2.62	3.56
Alprenolol (APL)	9.70	2.23	3.10
Oxprenolol (OPL)	9.50	1.54	2.18
Toliprolol (TPL)	9.60	1.54	1.93
Metoprolol (MPL)	9.70	1.23	1.88
Atenolol (ATL)	9.55	1.09	0.16

a) partition coefficient in L-α-dimyristoylphosphatidylcholine-phosphate buffer, pH 7.4.
b) Partition coefficient in octanol-buffer.

nals (ΔHz) increased linearly with the increasing Pr^{3+} concentration up to 10 mM (data not shown). Upon addition of the β-blockers the effect of Pr^{3+} was reversed, and propranolol had the strongest effect. The studied β-blockers are presented in Table 3.12, together with their partition coefficients in DMPC liposomes and in octanol. At 5 mM the splitting of the resonance signals was completely reversed by propranolol (Figure 3.34), whereas for oxprenolol and toliprolol the reversal was not completed even at 20 mM. The relative power of the β-blockers to reduce ΔHz follows a linear relationship with respect to the β-blocker concentration and is described by Eq. 3.11:

$$\log \Delta\mathrm{Hz} = D - KC \tag{3.11}$$

where the slope K is the displacement constant, C is the concentration (mM) of β-blockers, and D is the mean ΔHz (64.3 ± 2.4) at 2 mM Pr^{3+} in the absence of a β-blocker. Propranolol shows the strongest displacement effect. Additionally, a slight downfield shift of the methylene group signal of the acyl chain of DMPC is observed, indicating the penetration of propranolol into the liposomes. For the *meta*- and *ortho*-substituted β-blockers, a highly significant correlation between the displacement constant K and the phospholipid–buffer partition coefficient log K'_m has been found ($r = 0.95$, $n = 4$), compared with the regression coefficient with log $P_{oct.}$ ($r = 0.92$). β-blockers with *para*-substitution (metoprolol, atenolol) and low K'_m values deviate from this correlation. They exert stronger effects than would be expected on the basis of their low lipophilicity. The authors assume that β-blockers with *ortho*- and *meta*-substitutents are non-selective whereas those with *para*-substituents on the aromatic ring structure are involved in membrane interactions that lead to cardioselectivity. It is concluded that, in the case of non-selective β-blockers, partitioning into the membrane is decisive, whereas for the less lipophilic, selective derivatives polar group interaction at the membrane surface is dependent on the orientation of the β-blocker molecules. It can be concluded that the displacement of Pr^{3+} from membrane surfaces by β-blockers is related to their interaction with the polar head groups of the phospholipid and that *para*-substituted derivatives may adopt a different orientation. The results show that it is risky to rely on octanol–water partition coefficients to describe drug–membrane interaction and underline the great power of the NMR techniques to gain insight into such processes.

3.8.6 The Use of ^{2}H-NMR and ^{13}C-NMR to Determine the Degree of Order and the Molecular Dynamics of Membranes

Deuterium NMR (^{2}H-NMR) is a powerful technique to obtain information on both the degree of order and the molecular dynamics of liquid crystalline media. It has extensively been used on model as well as natural membranes. Deuterium NMR has been used as a probe to investigate chain packing in lipid bilayers, and the effects of hydrocarbons and alcohols and their location in the membrane have been determined [123].

The deuterium nucleus gives rise to a doublet in the resonance spectrum caused by the interaction between its quadrupole moment ($I = 1$) and the surrounding magnetic field gradient. In oriented samples, the separation between the two spectral lines observed follows the relationship in Eq. 3.12:

$$\nu_2 - \nu_1 = 3/4(e^2qQ/h)(3\cos^2\theta - 1) \tag{3.12}$$

where e^2qQ/h is the ^{2}H quadrupole coupling constant (~170 kHz) and θ the angle between the magnetic field and the axis of the molecular ordering [124]. All values of θ are possible in an aqueous dispersion of phospholipids. The spectrum of ^{2}H nuclei is a "power pattern." The two possible peaks correspond to $\theta = 90°$ (perpendicular edges) and the two shoulders to $\theta = 0°$ (parallel edges) [124] (Figure 3.35) [125].

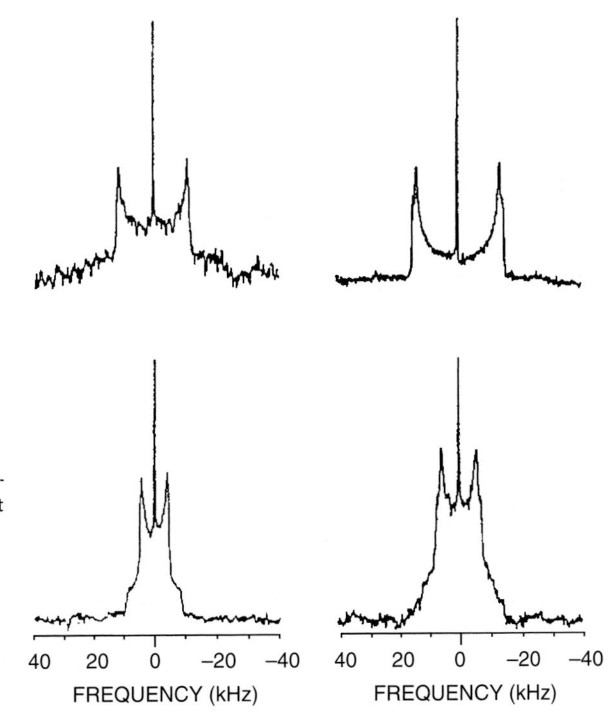

Fig. 3.35 ^{2}H-NMR spectra (30.7 MHz) of β-DTGL (7) labeled at C-1′ of the carbohydrate head group (left) and at C-3 of glycerol (right) in the absence (top) and presence (bottom) of tetracaine (8) at pH 9.5, 52 °C. (Reprinted from Fig. 2 of ref. 125 with permission from Elsevier Science.)

Solid-state ^{2}H experiments with drug–phospholipid preparations can be carried out by observing spectra from either the drug or the phospholipid. Depending on the degree of labeling, the orientation and position of the drug in the membrane can be determined. It can also be appropriate to combine ^{2}H labeling with ^{13}C labeling of, for instance, the drug molecules. It is known that the effects of anesthetics on glycolipids differ from those on phospholipids. Figure 3.35 gives as an example the ^{2}H-NMR spectrum of the glycolipid 1,2-di-*O*-tetradecyl-(β-D-glucopyranosyl)-*sn*-glycerol (β-DTGL) **(7)** labeled at C-1' of the carbohydrate head group (left) and C-3 of glycerol (right) in the absence (top) and presence (bottom) of tetracaine (TTC) **(8)** at pH 9.5 and 52 °C [125], where the tetracaine is almost totally uncharged. In the absence of tetracaine at 52 °C the spectrum describes an axially symmetric motion of the lamellar phase. In the presence of uncharged tetracaine, the quadrupole splitting of β-DTGL labeled at C-1' of the carbohydrate head group is reduced by more than a factor of 2 (Figure 3.35). A non-lamellar phase is induced. It can be assumed that the lipid is in the hexagonal phase, as observed for pure β-DTGL at elevated temperatures (58 °C), i.e. the lamellar structure of DTGL is very sensitive to tetracaine. In contrast, the interaction of tetracaine with PC with or without cholesterol results in a reduction of the lipid order parameters in the plateau and in the tail regions of the acyl chains. The effect is larger with the charged form of tetracaine [126].

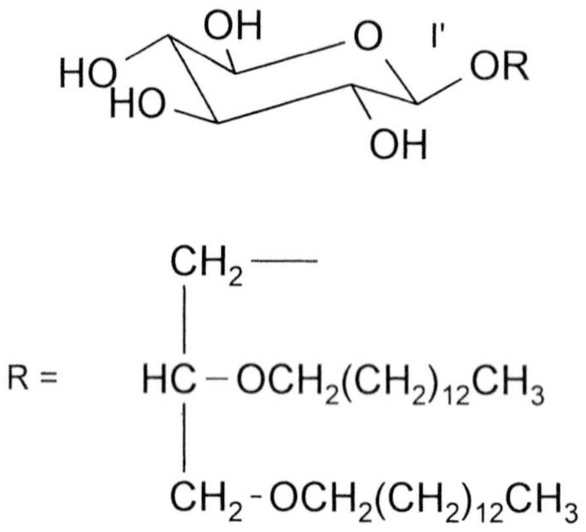

β-DTGL (7)

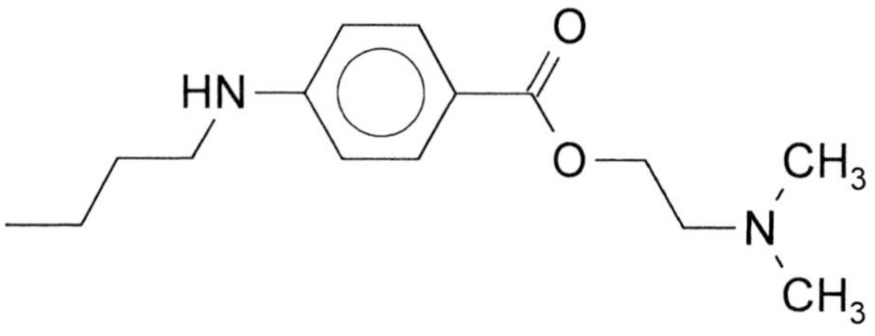

TTC (8)

^{2}H-NMR in combination with isothermal titration calorimetry (ITC) has been applied to investigate the effect of the neuronal marker dye FM1-43 [*N*-(3-triethylammoniumpropyl)-4-(*p*-dibutylaminostyryl)pyridinium-dibromide)] on the thermodynamics and lipid order of neutral and negatively charged phospholipid vesicles [127]. FM1-43 labels nerve terminals in an activity-dependent fashion. Whereas ITC allowed the measurement of the adsorption isotherm up to 100 μM dye, ^{2}H-NMR measurements gave useful information on the location of the dye in the membrane. The authors observed a non-linear dependence between the extent of adsorption and the free dye concentration. Two forces seemed to be involved in the adsorption process: the insertion of the non-polar part of the molecule into the hydrophobic membrane and – to some extent – electrostatic attraction-repulsion interactions of the cationic molecule. The authors were able to separate the two driving forces by employing the Gouy–Chapman theorem. The adsorption was corrected for the electrostatic effects and the dye–membrane interaction was then described by a simple partition equilibrium with a partition constant of 10^3–10^4. The partition enthalpy was determined to be $\Delta H = -2.0$ kcal/mol and the free energy of binding to be $\Delta G = -7.8$ kcal/mol. This result indicated that the insertion of FM1-43 is an entropy-driven process that follows the classical hydrophobic effect. The results of the ^{2}H-NMR experiments provided information on the structural changes of the lipid bilayer. FM1-43 not only disturbed the packing of the acyl chains but decreased the order of the fatty acyl chains. In addition, it changed the conformation of the PC head group. The $^-$P–N$^+$ dipole, being parallel to the membrane surface in the absence of dye, after dye insertion was rotated with its positive end toward the water phase. Compared with other catamphiphiles, the degree of rotation is, however, smaller, probably because of the two cationic charges in FM1-43, which can counteract each other in the rotation of the dipole.

3.8.7
Change in relaxation rate, $1/T_2$: a Method of Quantifying Drug–Membrane Interaction

The ^{1}H- and ^{13}C-NMR spectra of small molecules in the presence of proteins or phospholipids can reveal changes in the spin–lattice relaxation rate, $1/T_1$, and the spin–spin relaxation rate, $1/T_2$, the latter expressed as the line width of the resonance signal at half peak height. Changes in $1/T_2$ are related to a decrease in the rotational freedom of small molecules in the presence of a "receptor" with which they can interact [128, 129]. The transverse relaxation time, T_2, depends on both the internal and overall molecular mobility. Several mechanisms, e.g. dipole–dipole, spin rotation, quadrupole interaction, and chemical shift anisotropy, can be involved [130]. The changes in peak width, ν, at half peak height are proportional to the change in $1/T_2$. Expressed as a function of the drug to protein or drug to phospholipid ratio in the vesicles, $1/T_2$ can be used to quantify the degree of interaction, provided care is taken that no other factors have led to signal broadening. The relation follows Eq. 3.13:

$$T_2 = 1/(\pi \nu_{1/2}) \qquad (3.13)$$

For a large range of phospholipid concentrations, the broadening linearly depends on the lipid concentration. The slope of such plots is proportional to the "affinity" of the drug molecules or molecular substructures to the phospholipid. It can be taken as a measure of the degree of interaction.

Studies have been performed to elucidate the molecular basis of drug-induced phospholipidosis. It has been found that, in the case of drugs of the chlorphenter-

Tab. 3.13 Steroid structure of the pregnane analoges

Steroid	R1	R2	R3
1	3α-OH	5α-H	=O
2	3α-OH	5α-H	H2
3	3α-OH	5β-H	=O
4	3α-OH	5β-H	H2
5	3β-OH	5α-H	H2
6	3β-OH	5β-H	H2
7	3α-OH	5α-H	=O

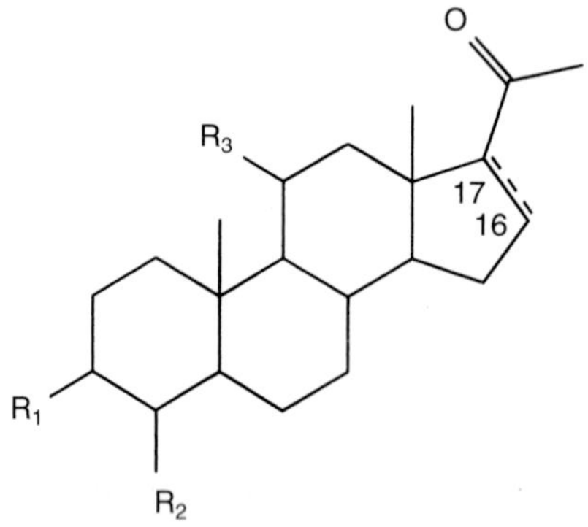

Tab. 3.14 ^{1}H- (left) and ^{2}H-NMR line width (right) ($\nu_{1/2}$), of resonances corresponding to the $C(O)CH_3$ protons of steroids incorporated in lecithin vesicles at different temperatures.

Steroid	1H $\nu_{1/2}$ (Hz)			2H $\nu_{1/2}$ (Hz)			Anesthetic activity (mg/kg)[a]
	37°C	50°C	70°C	37°C	50°C	70°C	
1	8.0	5.0	2.5	40	17	12	3.1
2	8.0	4.0	2.5	–	–	–	3.1
3	8.0	5.0	2.5	47	35	20	6.3
4	5.5	4.0	2.5	26	19	14	3.1
5	b	b	b	b	b	30	100
6	b	b	b	60	42	27	25
7	b	b	b	b	b	b	Inactive

a) Lowest dose producing loss of righting reflex.
b) resonance not detected

mine type and other catamphiphilic drugs, a relation between the degree of interaction and the disturbance of the phospholipid metabolism does exist (see also Section 4.4) [131].

The mobility of anesthetic steroids (Table 3.13) in model membranes has been examined by ^{1}H- and ^{2}H-NMR spectroscopy [132]. Line width broadening in the ^{1}H- and ^{2}H-NMR proton signals of the $C(O)CH_3$ substituent of the steroids in the presence of phosphatidylcholine was determined at various temperatures. A correlation with their anesthetic effect is assumed (Table 3.14). Resonance signals with relatively narrow line width were observed at 37 °C for the active anesthetics **1–4**. Upon raising the temperature, the resonance signals narrowed, which is consistent with an increase in mobility. For membrane preparations containing the inactive derivatives **5–7**, resonances (especially ^{1}H) could not be detected even at the highest temperature. This suggested that the intensive line width broadening of these steroids is due to extreme restriction of their mobility in the phospholipid bilayer. Only relatively narrow ^{1}H-NMR resonances of the $C(O)CH_3$ group of the steroids could be observed in the presence of the very intensive lecithin signals. To enable the observation of steroid signals of all derivatives, they were selectively deuterated (>95%) in the acetyl methyl position and the ^{2}H-NMR spectra recorded. Again, the resonance signals became smaller with increasing temperature, indicating increasing mobility, but remained relatively broad for the derivatives with low biological activity (Table 3.14). In this example, the interaction of drug molecules with phospholipids led to a decrease in biological activity.

Another example is depicted in Figure 3.36, which shows the broadening of the proton resonance signals of the NCH_3 group and the aromatic protons of trifluoperazine – a compound that also shows some ability to reverse multidrug resistance in tumor cell lines – as a function of bovine brain phosphatidylserine (BBPS). Figure 3.37 depicts plots of $1/T_2$ *vs.* BBPS concentration (the slope of which is a measure of drug–phospholipid interaction) for several catamphiphiles, and shows that interaction increases with increasing BBPS concentration. The interaction of trifluoperazine and other catamphiphiles is so strong that it cannot be reversed by addition of Ca^{2+} ions. Only in the case of lidocaine and verapamil, which show the weakest interaction, i.e. the shallowest gradient, is the interaction reversible by Ca^{2+}. In the case of amiodarone and some other catamphiphiles (see Figure 3.37), an increase in interaction was in fact observed upon addition of Ca^{2+}. This effect of increasing ion concentration is due to the strong hydrophobic interaction forces involved, especially in the case of amiodarone. This finding supports the results of NMR, X-ray diffraction, and modeling studies, which have shown that amiodarone is deeply buried in the hydrocarbon chains of the bilayer [14, 47] (see Section 4.2).

Detailed interaction studies have been performed with the neuroleptic drugs *cis*- and *trans*-flupentixol. These stereoisomers, like trifluoperazine, act as modifiers of multidrug resistance in tumor cell lines, in which it has been observed that the *trans* isomer is about three times as potent as the *cis* form. Surprisingly, it was found that, in the case of these two stereoisomers, the slope of the curve $1/T_2$ *vs.* lecithin concentration is different. The steeper slope for *trans*-flupentixol suggests that it interacts about twice as strongly with lecithin as does *cis*-flupentixol (Figure 3.38) [133].

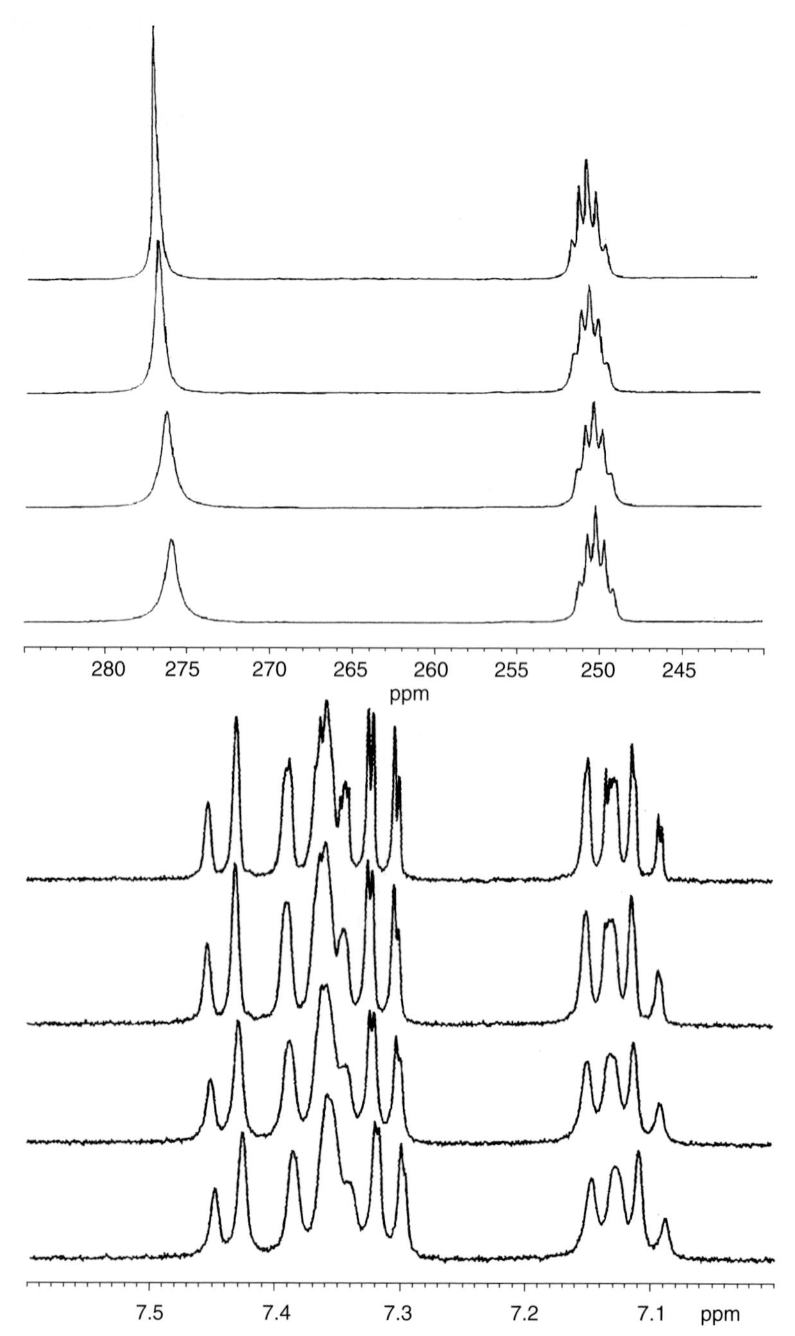

Fig. 3.36 Broadening of proton resonance signals of trifluoperazine (top, methyl groups ~2.9 ppm, standard DMSO 2.5 ppm; bottom, aromatic protons) as function of increasing bovine brain phosphatidylserine concentrations (BBPS) (control, 0.01, 0.03, 0.05 mg/ 500 μL). (Reprinted from Fig. 5 of ref. 47 with permission from Bertelsmann-Springer.)

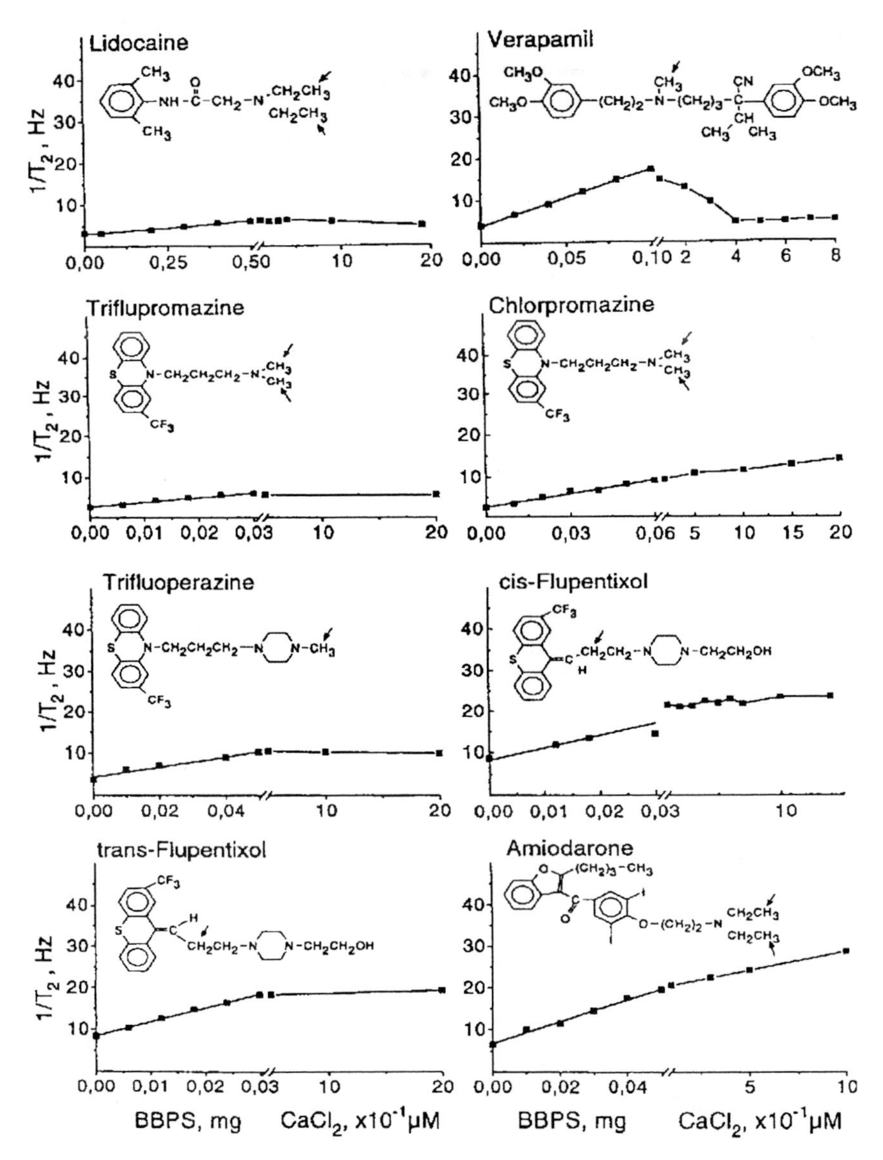

Fig. 3.37 Increase in drug proton relaxation rate ($1/T_2$) for a series of catamphiphiles as a function of increasing BBPS concentration alone and after addition of increasing concentrations of $CaCl_2$. Note the change in the abscissa. The spin systems measured are indicated by arrows on the corresponding structures. (Reprinted from Fig. 6 of ref. 47 with permission from Bertelsmann-Springer.)

This indicates a stereospecific interaction of these catamphiphiles with phospholipids. Both stereoisomers interacted more strongly with the negatively charged PS than with the neutral PC. In addition, the chemical shifts of the H-signals were different for *trans*- and *cis*-flupentixol [134]. The largest difference was observed for the H1 proton, indicating a reduction in electron density around this nucleus in the case of *trans*-flupentixol. To determine the main feature involved in the behavior of the hydrogens of these stereoisomers when interacting with phospholipids, the slopes of the interaction of *cis*-flupentixol with PC and BBPS were plotted against the slopes of the interaction of *trans*-flupentixol with the phospholipids. The structure of flupentixol and numbering of the protons is given in formula **9**. Figure 3.39 shows that H1 has the lowest slope and H10 the highest and that H9 lies in the middle between H1 and H10 on both plots. These hydrogens are in the region of the drug molecule where the stereoisomeric differences between *cis*- and *trans*- flupentixol occur. The coupling constants of H9 (7.3 Hz) and H10 (7.2 Hz) were used as constraints to calculate the conformations that best fit the experimental data. Experimental and computational results support the assumption that *cis*- and *trans*-flupentixol can interact in a stereodependent manner with both phospholipids [134].

Similar studies on *N*-alkyl-substituted benzylamines show a sudden increase in line width broadening at a chain length of about C_6 (Figure 3.40). The sudden increase in strength of interaction corresponding to a π-value of about 3 is paralleled by a strong increase in negative inotropic activity [135].

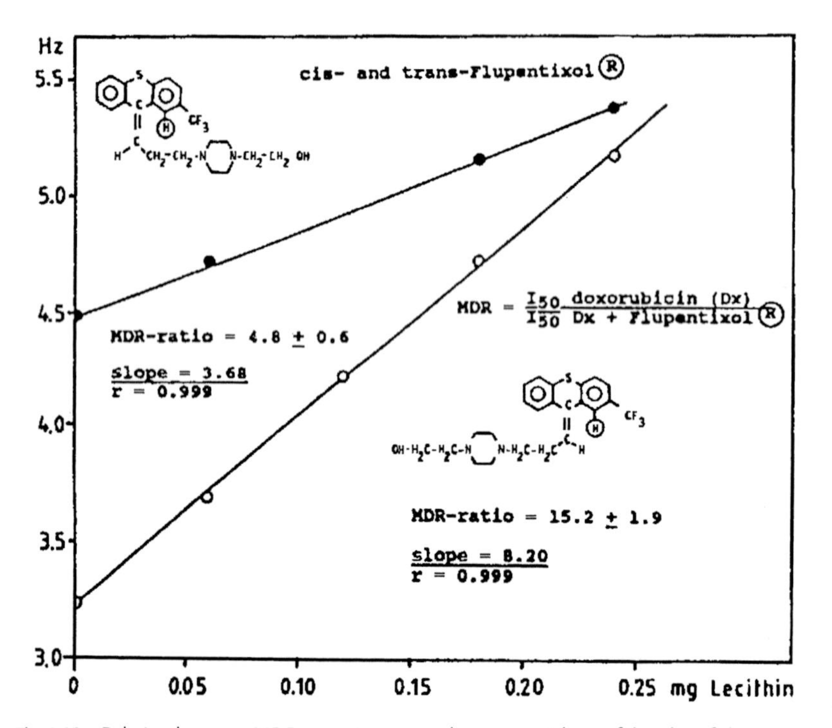

Fig. 3.38 Relation between MDR-reversing activity (ratio, reversal of resistance against doxorubicin) of *cis*- and *trans*-flupentixol and their degree of interaction with phospholipid liposomes (slope of the plot of changes in $1/T_2$ as a function of lecithin concentration). (Reprinted from Fig. 3 of ref. 133 with permission from Elsevier Science.)

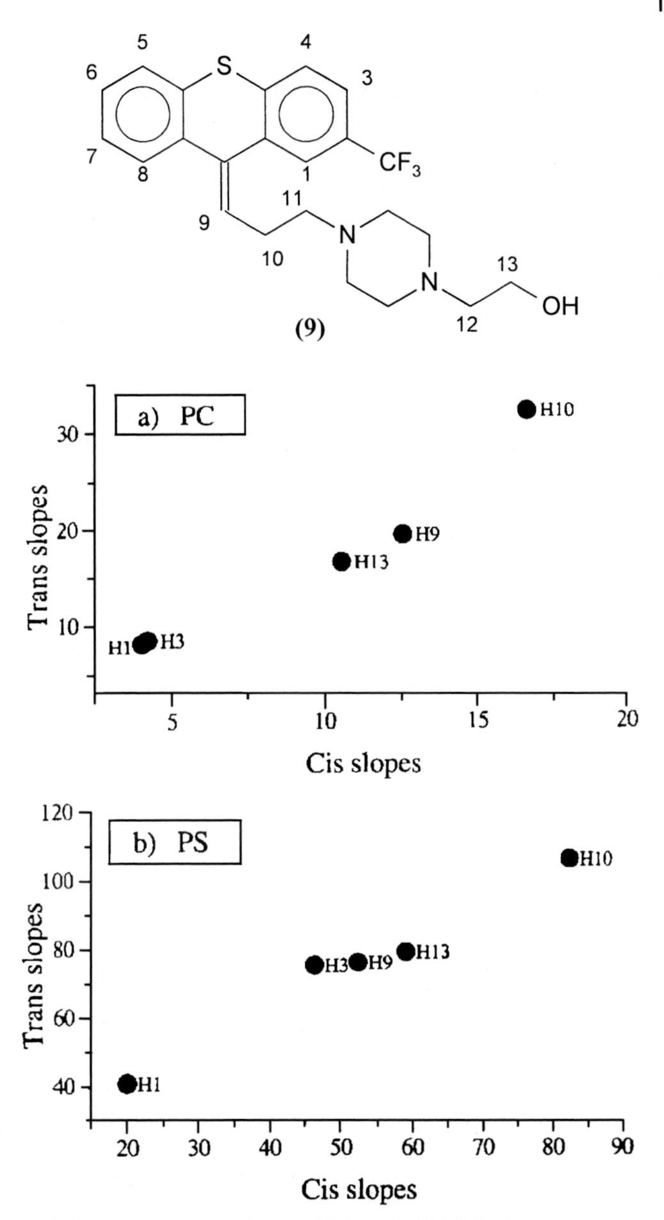

Fig. 3.39 Plots of proton $1/T_2$ slopes of *trans*-flupentixol *vs.* $1/T_2$ slopes of *cis*-flupentixol in (a) a phosphatidylcholine (DPPC) and (b) a phosphatidylserine (BBPS) lipid environment. (Reprinted from Fig. 3 of ref. 134 with permission from Wiley-VCH.)

The observation made in the case of the above two examples leads to another possible effect of drug–membrane interaction: a change in the preferred conformation or orientation of the drug molecule under the environmental conditions of the membrane.

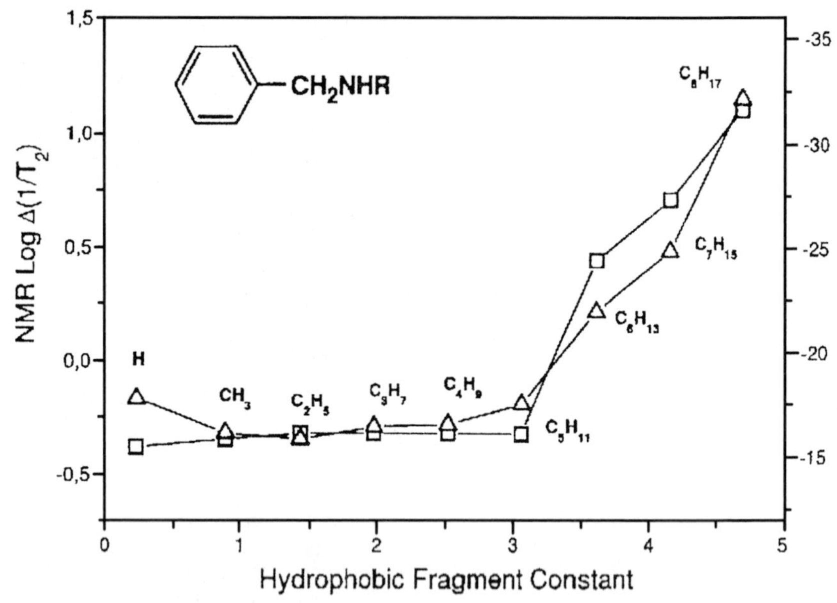

Fig. 3.40 Plot of NMR binding ($1/T_2$) (–Δ–) *vs.* *N*-alkyl hydrophobic fragment constants compared with an overlay plot of computed interaction energies *vs.* *N*-alkyl hydrophobic fragment constants (–□–) for benzylamines R = H to R = C_8H_{17}. (Reprinted from Fig. 3 of ref. 138 with permission from Wiley-VCH.)

3.8.8
NOE-NMR in the Study of Membrane-induced Changes in Drug Conformation

It is well known that the preferred minimal energy conformations of drug molecules can differ in the crystalline state, in solutions, and in the state bound to protein receptors. An early example of this is provided by the different conformations of acetylcholine bound to the nicotinic acetylcholine receptor derived from two-dimensional NOE spectra (Figure 3.41) [136]. It was observed that there were major conformational differences between the free and bound forms of acetylcholine. This applies especially to the arrangement of the N-C-C-O backbone, which changes from *gauche* in solution to nearly *trans* in the bound state. In the bound state the two electronegative oxygens are positioned on the same side of the acetylcholine molecule. The hydrophobic acetyl methyl and choline methyl form a continuous hydrophobic area over the rest of the molecule.

It should not therefore be surprising that a chiral matrix such as a phospholipid membrane can also act as a specific receptor and lead to a change in conformation of a ligand molecule.

An example of this is provided by the simple *N*-alkyl-substituted benzylamines mentioned above. A bilinear dependence of the change in phase transition temperature, displacement of Ca^{2+} ions, and change in $1/T_2$ relaxation rates on the chain length of the *N*-alkyl substitution was observed. There was no effect, or even a nega-

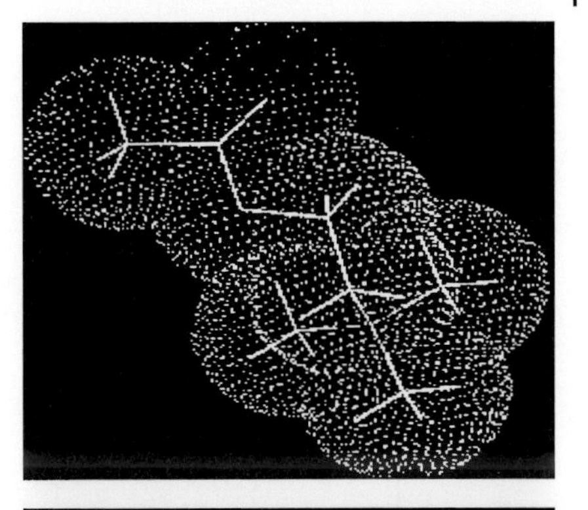

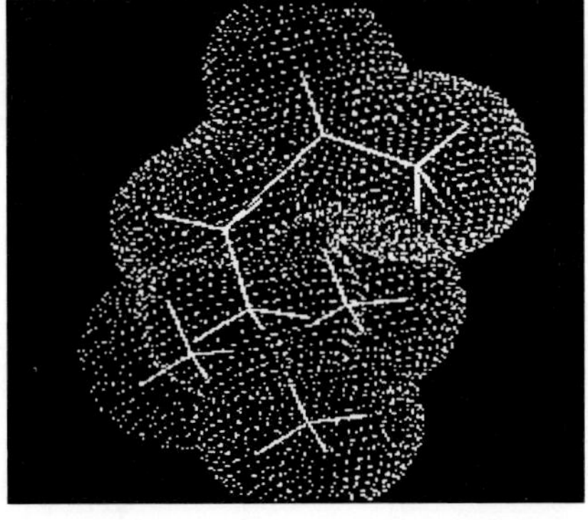

Fig. 3.41 Computer representation of the structure of acetylcholine in its free solution-state conformation, in which oxygen atoms appear on opposite sides of the molecule (top), and of acetylcholine in its receptor-bound state, in which oxygen atoms appear on the same side (bottom). (Reprinted from Fig. 4 of ref. 136 with permission from *Proc. Natl. Acad. Sci. USA.*)

tive effect, on the interaction with phospholipids up to a chain length of about C_5. Thereafter, a strong increase in interaction was observed. We speculated on a change in conformation from the extended to a folded conformation [137]. Transient Overhauser effect (TOE)–NMR experiments subsequently performed fully supported this speculation. With TOE, time-dependent changes in the magnetization of a particular nucleus after inversion of the magnetization of a neighboring nucleus can be measured. Two-dimensional TOE experiments with the derivatives were performed in the presence and absence of lecithin vesicles, and the results are shown in Figure 3.42. Up to a chain length of *n*-pentyl the extended conformation exists in the absence and presence of the phospholipid, whereas at longer chain lengths this form exists only in the absence of phospholipid (lecithin). In the absence of lecithin, a cross-peak occurs only between the *ortho* protons of the dichlorophenyl and the benzylic methylene. In the presence of phospholipid, however, additional cross-peaks, indicating proximity of the *N*-hexyl protons and the aromatic ring, are clearly present for the hexyl derivative. For the butyl derivative, however, no NOE effect between the *n*-butyl and the aromatic protons was detected.

Molecular modeling studies correlate very well with the experimental observations [138]. The plot of the computed interaction energies versus lipophilicity runs parallel to the plot of $1/T_2$ versus lipophilicity (Figure 3.40). The result confirms the proposed conformational change with lengthening of the *N*-alkyl groups of the benzylamines. It is an example of the general model proposed by Herbette and coworkers

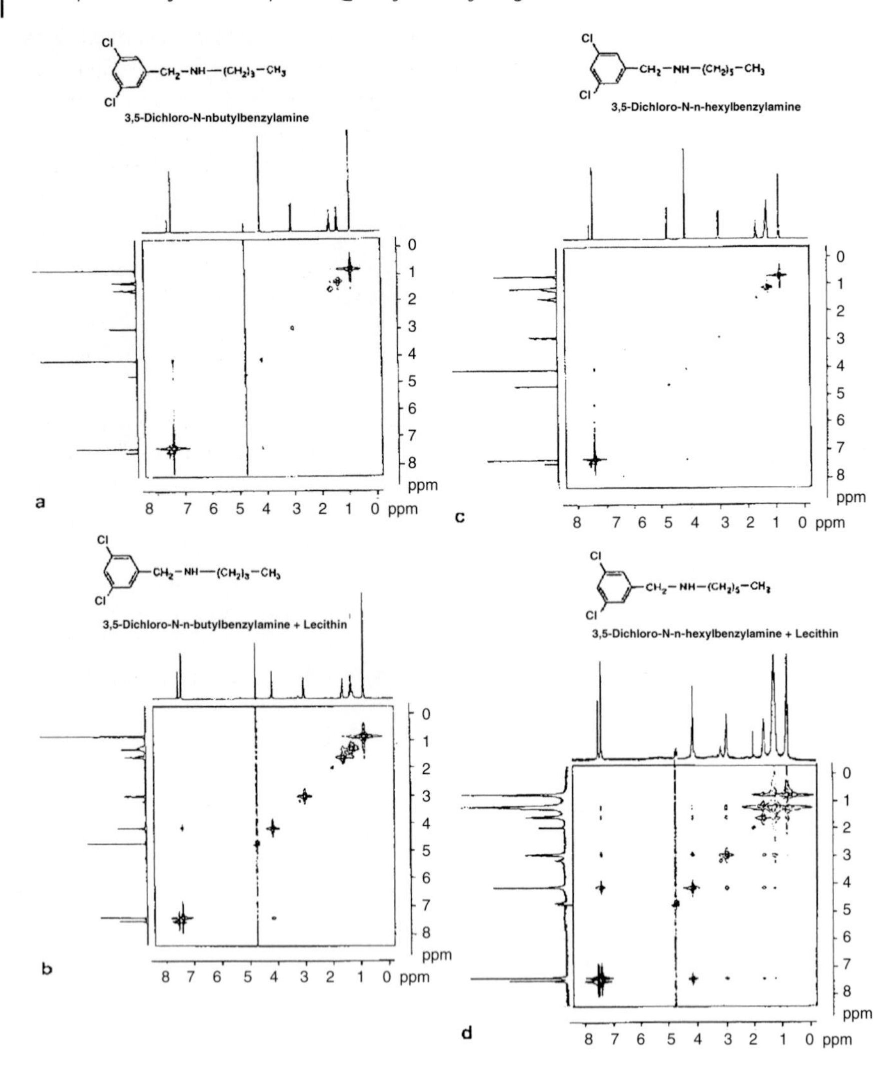

Fig. 3.42 NMR-NOE spectra of 3,5-dichlorobenzyl-*N*-butylamine and 3,5-dichlorobenzyl-N-hexylamine in deuterated phosphate buffer in the absence (a, c) and presence (b, d) of lecithin. (Reprinted from Fig. 3a-d of ref. 137 with permission from Wiley-VCH.)

[139]. For possible consequences of drug–membrane interactions on conformation and/or orientation, see also Chapter 5. The examples discussed provide support for the assumption that the active conformation can be induced by the lipid environment. In other cases, however, inactivity may be related to such changes. Using NOE-NMR, changes in conformation have also been shown for verapamil in the presence of lipid-mimetic solvents such as acetonitrile [140] and for the binding of verapamil and verapamil analogs to amphiphilic surfaces (phospholipid-coated polystyrene–divinylbenzene beads) [141].

Nuclear Overhauser effects in ^{1}H-NMR have also been used to determine the location of cationic forms of the local anesthetics procaine, tetracaine, and dibucaine within lipids and the induced dynamical perturbations of lipids. Small unilamellar phosphatidylcholine vesicles have been used as model membranes [142]. The strength of interaction between the drug and phospholipids followed the order dibucaine > tetracaine > procaine, in parallel with their decreasing mobility as determined from intermolecular NOE studies. However, only tetracaine induced a significant dynamic perturbation of the membrane vesicles, as revealed by the extent of transfer of the negative NOE from α-methylene protons to choline methyls, olefinic methines, acyl methylenes, and terminal methyl protons. The authors interpreted the results as follows: procaine interacts very weakly with lipids at the outer surface of the vesicles, whereas tetracaine binds to the lipids at both the outer and inner halves of the bilayer, "inserting its rod-like molecule in a forest of the acyl chains of the phospholipid," and dibucaine binds tightly to the polar head groups of the phospholipid, which reside only at the outer half of the vesicles. The authors concluded that the relative anesthetic potency of these drugs can be correlated with their ability to bind to lipids at the polar head group of the bilayer but not with their ability to affect membrane fluidity.

The localization of another local anesthetic, methoxyflurane ($CH_3OCF_2CHCl_2$), has also been investigated by 2-D NOE [143]. It has frequently been reported that anesthetics depress the temperature of order–disorder phase transition of lipid membranes. As the phase transition is related to the hydrocarbon chain conformation of phospholipid molecules, a direct interaction between the anesthetics and the lipid molecule has been assumed. However, the results of NOE-NMR at 25 °C clearly showed that a cross-peak is observed only between the choline methyl protons and the methoxy protons of the methoxyflurane. This is a clear indication of interfacial action of volatile anesthetics, which can lead to domain formation of the drug within the membrane (see also Section 5.1). The results were in agreement with the observation that, in the absence of anesthetics, phase transition occurs simultaneously at the same temperature, in the case of choline methyl and the acyl-chain methylene protons. However, the phase transition induced by anesthetics at reduced temperature occurred differentially in the lipid core and the choline group at the surface [144]. At higher temperatures (45 °C), well above T_t, no cross-peaks were observed between the anesthetic and DPPC protons. The authors speculated that the correlation times of each proton of DPPC became too short to be detected by 2-D NMR because the molecular motions of DPPC molecules are greatly increased by raising the thermal energy.

Calculations reported by Gaillard *et al.* [145] show that the lipophilicity of morphine glucuronides is conformation-dependent. These compounds exist as folded and extended conformers that differ in their partition behavior. They are hydrophilic in water and less polar in lipid environments such as membranes.

3.9
Circular Dichroism (CD)

Circular dichroism is another technique used to study drug–membrane interactions and possible changes in drug location and conformation. The interaction of the anticancer drugs doxorubicin, daunorubicin, idarubicin, and idarubicinol (for the formula of anthracyclines, see Section 5.2.2) with LUVs has been determined by CD in an attempt to understand the conformational aspects of the interaction [146]. The anthracycline chromophore is sensitive to ring interactions of molecules because of its abundance of π–π* and n–π* transitions. It is therefore very suitable for the application of absorption, fluorescence, and, especially, CD spectroscopy. Figure 3.43 shows the CD spectra of doxorubicin and daunorubicin in the absence and presence of LUVs containing different amounts of anionic phospholipid In the case of LUVs containing large amounts of PA (20%), a two-step binding of the molecules can be detected. The molecules interact via electrostatic forces involving both the negative charge of PA and the positive charge of the amino sugar of the anthracyclines. The CD spectrum of the couplet type characteristic of the free dimeric doxorubicin in aqueous phase disappears. It is replaced by a positive band of relatively high intensity (band I, Figure 3.43a). The isodichroic point at 450 nm indicates equilibrium between free and bound drug. The dihydroxyanthraquinone part of the molecule is situated outside the bilayer in the aqueous bulk phase. By increasing the molar ratio of anionic phospholipid to doxorubicin, the intensity of this positive band decreases and is shifted to longer wavelength (band II). The change in the spectrum indicates a deeper intrusion of the anthraquinone into the bilayer. This is paralleled by an in-

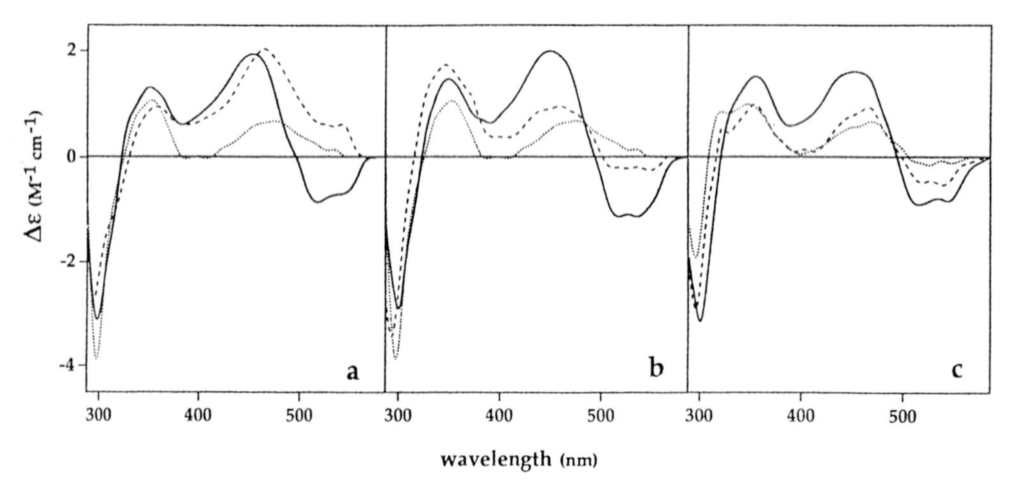

Fig. 3.43 Circular dichroism spectra of doxorubicin and daunorubicin in the absence of LUVs (——), in the presence of LUVs at a PC: anthracycline ratio of 1:2 (------), and in the presence of LUVs at a PC: anthracycline ratio of 1:5 (....). Anthracycline concentration = 5×10^{-4} M. (a) Doxorubicin plus LUVs containing 20% PA. (b) Doxorubicin plus LUVs containing 5% PA. (c) Daunorubicin plus LUVs containing either 5% or 20% PA. (Reprinted from Fig. 3 of ref. 146 with permission from Elsevier Science.)

crease in hydrophobic interactions in place of electrostatic interactions. If low amounts of PA are used (5%) no band I, characteristic of electrostatic interactions, is seen. The authors concluded that in a medium with a dielectric constant similar to that of methanol, doxorubicin and daunorubicin are embedded within the membrane at the level of the polar head groups. The electrostatic interaction is eliminated if the derivatives are more lipophilic, e.g. idarubicin. Neither idarubicin nor idarubicinol could be localized in the bilayer. However, from the fine structure in the absorption spectra the authors concluded that these compounds are positioned deeper within the bilayer than doxorubicin or daunorubicin.

The influence of cholesterol on drug–membrane interaction has also been studied, and it was found that the interaction of doxorubicin is not strongly influenced by cholesterol, whereas the interaction of the more lipophilic idarubicin depends strongly on the presence of cholesterol. It forms a complex comprising two or three molecules of idarubicin, one cholesterol molecule, and one or more molecules of PA. The recognition of cholesterol-rich membranes by idarubicin and idarubicinol, in contrast to doxorubicin and daunorubicin, could be of interest for drug design. Another interesting observation is the change in the orientation of the sugar moiety with respect to the ring system. The geometry of the glycosidic linkage can be defined by the torsion angle C(6a)–C(7)–O(7)–C(1'). In aqueous solution the mean plane of the sugar is perpendicular to the dihydroxyquinone plane. The geometry of this glycosidic linkage would have a great influence on the CD signal amplitude at 480 nm. At high molar phospholipid–drug ratios, the CD signal for the "membrane monomer" is positive. Its intensity is lower than that found for the monomer in aqueous solution and similar to that of the corresponding aglycone. The authors conclude that this "strongly suggests that in the membrane monomer the value of the C(6a)–C(7)–O(7)–C(1') torsion angle has changed when compared to the free drug molecule" [146].

3.10 UV Spectroscopy

A method that requires less expensive experimental equipment than NMR spectroscopy or X-ray diffraction is derivative transformation UV spectroscopy. It has been shown to be a reliable procedure for estimating drug partitioning into membranes [147]. The partitioning of the anticancer drug tamoxifen and its derivative 4-hydroxytamoxifen was determined. Figure 3.44 shows the second-derivative UV spectra of increasing 4-hydroxytamoxifen concentrations in DMPC liposomes. Interestingly, the partitioning of 4-hydroxytamoxifen increases with increasing drug concentration, whereas the partitioning of tamoxifen decreases with increasing concentration of the drug, as observed also for chlorpromazine and amphotericin B. The linear decrease in tamoxifen partitioning suggests that the lipid phase undergoes saturation or, alternatively, that the reduced accumulation is due to a less favorable membrane environment because of changes in the presence of the drug. The opposite behavior of 4-hydroxytamoxifen suggests that the hydroxyl group is responsible for the differential partitioning, leading to an increase in accumulation with increas-

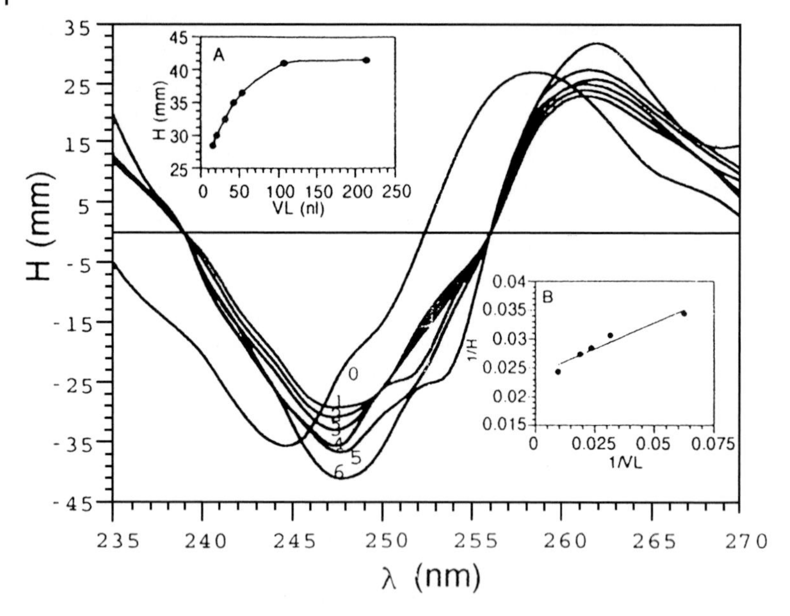

Fig. 3.44 Second-derivative absorption spectra of 10 μM 4-hydroxytamoxifen in DMPC liposomes at different lipid concentrations. The nominal concentration of DMPC in suspensions was 0 (curve 0), 21 (1), 28 (2), 42 (3), 56 (4), 70 (5) and 140 μM (6). Inset A shows the relationship between the peak height (H) derived from the data on main chart and the membrane volume. The double-reciprocal plot of these data is represented in inset B. (Reprinted from Fig. 1 of ref. 147 with permission from Academic Press.)

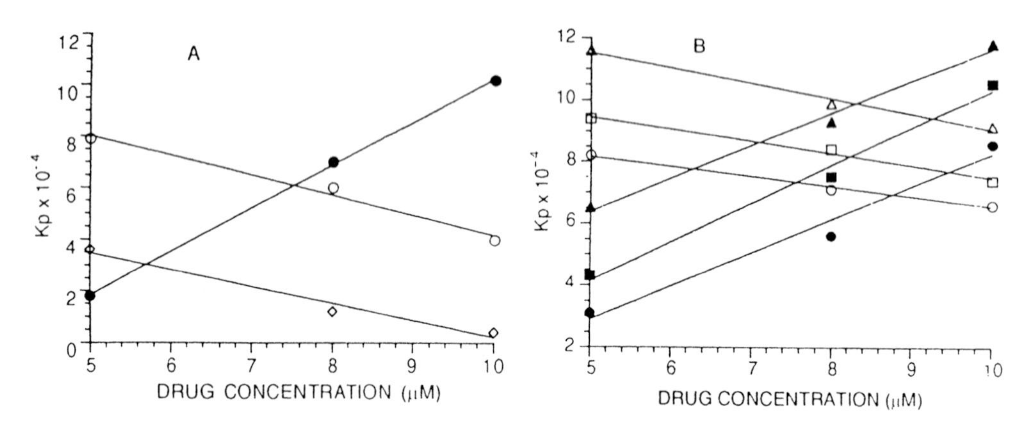

Fig. 3.45 Effect of drug concentration on the partition coefficients of tamoxifen (open symbols) and 4-hydroxytamoxifen (closed symbols) in DMPC bilayers (A) and in liposomes of sarcoplasmic reticulum lipids and native membranes (B). Note that the linear correlation between K_p and drug concentration is negative for tamoxifen and positive for 4-hydroxytamoxifen. (A) ○ and ● at 24 °C, ◇; at 37 °C: (B) ○ and ●, liposomes of sarcoplasmic reticulum lipids; □ and ■, native sarcoplasmic reticulum; △ and ▲, mitochondria. (Reprinted from Fig. 2 of ref. 147 with permission from Academic Press.)

ing drug concentration. This behavior can be explained by the polarity of the 4-hydroxytamoxifen molecule, which is able to form hydrogen bonds with phospholipid head groups or acyl groups at the surface of the bilayer. Consequently, the stability of the membrane will be disturbed, decreasing the packing and opening up space for the incorporation of more drug as its concentration increases. Furthermore, the authors could show that the degree of partitioning of the two drugs depends also on the composition of the liposomes. Compared with lipid dispersions, the drugs partition more strongly into native sarcoplasmic reticulum membranes and to an even greater extent into rat liver mitochondria (Figure 3.45). Maximum partitioning is observed in the range of the main phase transition. Cholesterol strongly affects the incorporation of drugs (Figure 3.46). This example demonstrates again that the partitioning of two analogs can have completely different effects on the membrane and in consequence on biological activity.

Similarly, using UV absorption spectroscopy Gracia and Prello [148] studied the influence of membrane chemical composition and drug structure on the localization of benzodiazepines at the lipid–water interface. Their results revealed that the benzodiazepines can be incorporated as an integral part of the bilayer and are not located only at the core, as reported from fluorescence polarization experiments [149]. The influence of lipid phase (gel or liquid crystalline), cholesterol content, lipid composition (egg phosphatidylcholine or DPPC), and structure of benzodiazepines determine their localization in the membrane. The strength of benzodiazepine–membrane interaction increases with a decrease in molecular order, molecular packing, and hydration,. The authors point to the pharmacological relevance of theses results because "the extent of partitioning of these drugs into biomembranes would be coupled to local oscillation of membrane dynamics which may be induced by physiological events."

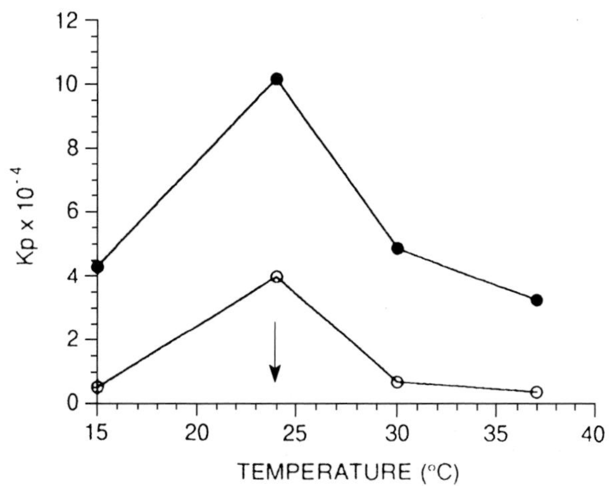

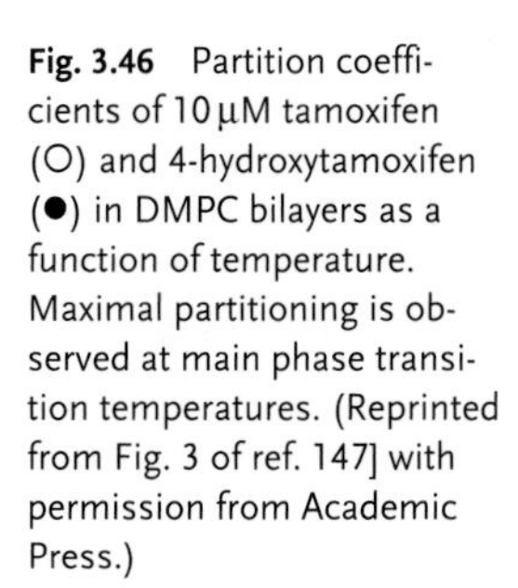

Fig. 3.46 Partition coefficients of 10 μM tamoxifen (○) and 4-hydroxytamoxifen (●) in DMPC bilayers as a function of temperature. Maximal partitioning is observed at main phase transition temperatures. (Reprinted from Fig. 3 of ref. 147] with permission from Academic Press.)

3.11 Combined Techniques for Studying Drug–Membrane Interaction

A single biophysical method is not always sufficient for a detailed analysis of the complexity of membrane dynamics and thermodynamics in the absence and presence of drug molecules. The methods discussed so far differ in their ability to describe the various aspects of membrane–drug interaction. HPLC techniques can be used to quantify the overall affinity of drugs to membranes and to determining log $P_{oct.}$ or log $K'_{oct.}$, or proportional log $K'_{membr.}$ values on IAM columns. Studies involving monolayers and calcium displacement can also be used to measure the relative affinity of drugs to phospholipid head groups and effects on membrane surface tension. Calorimetry, in particular, can supply data on changes in membrane fluidity, phase transition separation, and domain formation. The spectroscopic techniques supplement information on drug and phospholipid conformation, order, and dynamics as well as on drug orientation, location, and transfer rates. To obtain detailed insight into drug–membrane interaction, a combination of these techniques is advocated. Several applications have been published and shall be discussed using a few examples.

3.11.1 Combination of DSC and NMR

In addition to freeze–fracture electron microscopy, a combination of DSC, NMR, and the monolayer technique has been applied to study the various aspects of the interaction of the class IV calcium antagonist flunarizine [150] with a range of phospholipids. DSC shows only limited interaction of flunarizine with PC. The drug destabilizes the Lβ-phase without stabilizing the Lα-phase. In contrast, flunarizine influences not only the onset of phase transition but also the phase transition temperature and the completion of the transition of phosphatidylserine (PS), indicating

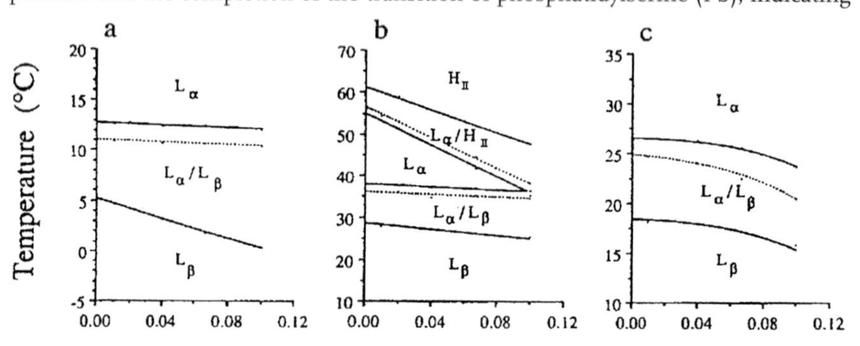

Fig. 3.47 Phase diagrams of flunarizine-dielaidoylphospholipid mixtures. (a) Phosphatidylcholine, (b) phosphatidylethanolamine, and (c) phosphatidylserine. (Reprinted from Fig. 3 of ref. 150 with permission from Academic Press.)

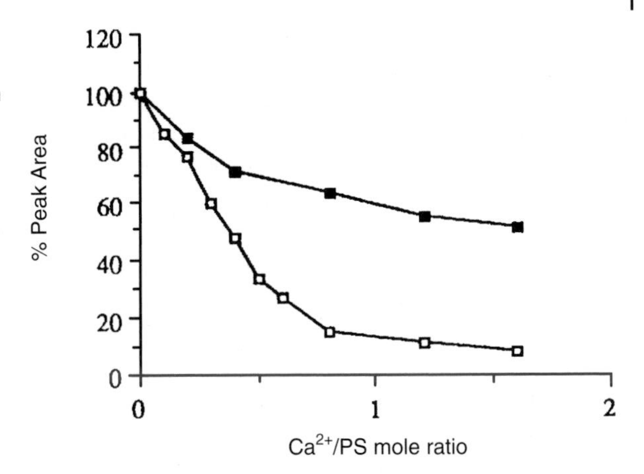

Fig. 3.48 ^{31}P-NMR study of the effect of flunarizine on the Ca^{2+}-induced reduction in the head group signal of dielaidoylphosphatidylserine: ❑, PS; ■, PS + flunarizine (1:1). (Reprinted from Fig. 2 of ref. 150 with permission from Academic Press.)

a significant interaction. With PE, flunarizine has no effect on the Lβ- to Lα-phase change. Instead, a strong concentration-dependent induction of the H_{II} phase is observed, which points to an expansion of the acyl chain region of the PE bilayer by the lipophilic molecules. The limited effect on the Lβ- to Lα-transition of DPPC, the complete absence of this effect in the case of PE, and the induction of the H_{II} phase in PE indicate the location of flunarizine in the hydrophobic core of these phospholipids. The strong effect on phase transition of PS suggests a more specific interaction of flunarizine with PS, probably with the head groups (Figure 3.47). This specific interaction follows also from studies on monolayers, for which the "cut-off" surface pressure for the interaction of flunarizine with the phospholipids has been evaluated. The cut-off point is the surface pressure at which flunarizine in an aqueous phase (0.5 μM) no longer penetrates into the monolayers of the three phospholipids. It shows that the strongest interaction occurs with PS (44.0 mN/m), the weakest with PE (33.4 mN/m), which is in good agreement with the DSC results. ^{31}P-NMR was then used to follow the displacement of flunarizine from the head groups by Ca^{2+}. ^{31}P-NMR enables the mobility of the phosphorus atoms in the head groups to be determined. The interaction of Ca^{2+} with the head groups of PS leads to immobilization of the head groups and, as a result, to broadening of the resonance signal so that the signal almost disappears. Figure 3.48 demonstrates this event in the absence and presence of flunarizine. It shows the ability of equimolar concentrations of flunarizine to prevent the binding of Ca^{2+} almost completely (see also ref. 42).

The effects of antiviral chemotherapeutic agents such as cyclosporin A, benzyloxycarbonyl-D-Phe-L-Phe-Gly, and amantadine on membrane properties have been studied using the combination of ^{31}P-NMR and DSC. It was found that benzyloxycarbonyl-D-Phe-L-Phe-Gly was most effective in raising the bilayer to H_{II} phase transition temperature. Cyclosporin A caused the greatest broadening of the ^{31}P-NMR signal. It was suggested that both effects are related to the inhibitory activity on membrane fusion and possibly also to their antiviral activity [151].

Another interesting example worthy of mention is the interaction of positively charged antitumor drugs with cardiolipin-containing DPPC vesicles [152], for which example ^{31}P-NMR and DSC measurements can be compared. The results of ^{31}P-NMR studies on DOPE–cardiolipin (2:1) liposomes suggested that adriamycin and 4'-*epi*-adriamycin "are capable of inducing phase separation under liquid crystalline conditions." In contrast, celiptium, 2-*N*-methyllelipticinium, and ethidium bromide could not induce phase separation. This finding is of particular interest because of the considerable evidence that cardiolipin–adriamycin interaction plays an important role in the inhibition of mitochondrial function, both *in vitro* and *in vivo*, and this could be related to anthracycline-induced cardiotoxicity.

A later publication from the same group [153] reporting on ^{31}P- and ^{2}H-NMR and DSC studies on zwitterionic and anionic phospholipids in the absence and presence of doxorubicin (adriamycin) described different results [152]. Doxorubicin had a stronger disordering effect on the membrane of lipid mixtures enriched with anionic lipids. However, "extensive segregation of DOPE and DOPA or DPS was not observed even under conditions of H_{II}-phase formation". According to the authors, the reason for this discrepancy was that in the earlier paper the phase separation was obtained with membranes "subject to gel–liquid crystalline phase transition", which was, however, discounted in the first paper.

The example shows the great importance of the experimental conditions (lipid composition, temperature, lipid–drug concentration ratio, etc.), which have a decisive influence on the quality, type, and strength of drug–membrane interactions. Any further interpretation with respect to biological activity *in vitro* and *in vivo* can become risky.

3.11.2
Combination of DSC and X-ray Diffraction

The thermodynamic and structural effects of diltiazem on lecithin (PC) liposomes have been studied by DSC and X-ray diffraction [154]. The authors found that at higher drug concentration the pretransition disappears, the main transition temperature decreases, and the lamellar thickness of the bilayer increases. The chains in the β-conformation were packed in a hexagonal undistorted lattice. There was a complete agreement in the observed data: all changes in the calorimetric and structural curves (X-ray) occurred at the same concentration, that is between 10^{-2} and 10^{-1} M.

The same two techniques have been used to study the effect of calcium channel-blocking drugs on the thermotropic behavior of DMPC [155]. The effect on the main phase transition of DMPC as a function of concentration of the drugs diltiazem, verapamil, nisoldipine and nimodipine was investigated. Nisoldipine and nimodipine had much greater effects on T_t than the other two compounds. They also markedly reduced the enthalpy of transition in an approximately linear manner, whereas verapamil and diltiazem showed no significant influence on enthalpy in the concentration range studied and may therefore form approximately ideal solutions in the lipid. The authors also determined the partition coefficients using the DMPC–buffer system. Their findings are summarized in Table 3.15. From the X-ray and neutron scattering studies performed, the authors concluded that the locations of nimodipine

Tab. 3.15 The effects of calcium channel-blocking drugs on the main phase transition of dimyristoylphosphatidylcholine (DMPC) adapted from ref. 155.

Drug (value of K_p)	Drug concentration (mM)	T_t (°C)	ΔH_{cal} (kcal/mol)	K^a	$\frac{\Delta H_{vH}}{\Delta H_{cal}}$ (CU^b)
Nimodipine	0.032	23.33	2.23	0.59	460
(128000)	0.043	22.77	2.61	0.36	700
	0.051	22.80	2.98	0.38	610
	0.064	21.50	1.50	0.46	580
	0.075	21.86	1.70	0.56	420
	0.085	21.46	1.40	0.60	700
	0.096	21.40	1.10	0.71	570
				0.52 ± 0.06	580 ± 50
Nisoldipine	0.015	23.85	4.73	0.52	260
(52800)	0.023	23.54	3.71	0.24	490
	0.013	23.75	3.89	0.60	310
	0.046	23.43	4.47	0.44	250
	0.067	22.79	3.40	0.37	360
	0.077	22.67	3.86	0.39	340
	0.087	22.58	2.66	0.47	410
	0.093	22.47	2.81	0.44	460
	0.105	21.92	2.17	0.47	450
	0.126	21.44	1.79	0.54	720
				0.45 ± 0.04	410 ± 50
DMPC	0	24.0	4.37	–	520
Diltiazem	0.090	23.82	4.52	0.52	400
(6030)	0.224	23.72	4.46	0.69	380
	0.448	23.82	4.77	0.59	320
	0.538	23.33	4.77	0.66	290
	0.627	23.05	4.41	0.62	390
	0.762	22.75	4.12	0.62	330
	0.896	22.73	4.38	0.65	360
	1.030	22.60	4.02	0.69	380
			4.43 ± 0.11	0.63 ± 0.02	360 ± 0.20
Verapamil	0.016	23.83	4.12	0.32	420
(21900)	0.040	23.69	3.77	0.42	460
	0.099	23.35	3.89	0.48	370
	0.199	22.85	3.57	0.57	360
	0.298	22.58	4.09	0.59	370
	0.397	22.19	4.48	0.58	300
	0.497	21.92	4.46	0.61	330
	0.596	21.69	3.81	0.68	360
	0.699	21.25	4.21	0.66	520
	0.900	20.88	3.94	0.71	730
	1.100	20.35	4.20	0.71	930
			4.05 ± 0.10	0.57 ± 0.04	470 ± 60

a) Distribution constant. **b)** Cooperativity unit.

and verapamil in the lipid bilayer was significantly different. Nimodipine is located at the hydrocarbon core to water interface of the lipid bilayer, reaching down only to the first methylene segments of the hydrocarbon core. This was also found in native membranes. On reducing the temperature the molecules were squeezed out of the region near to the hydrocarbon core. In contrast, it was found that verapamil, and to an even greater extent amiodarone, was located much deeper within the lipid structure.

The combination of X-ray and DSC has also been used to explore the effects of pindolol on model membranes. Changes in thermodynamic and structural parameters were observed. Depending on the temperature pindolol induced either a interdigitated 'gel' phase or, at high temperature, a lamellar phase. The data would seem to indicate that the non-polar moiety of pindolol penetrates deep into the bilayer interior while the hydrocarbon chains have to adopt a different conformation [156].

3.11.3
Combination of DSC and ESR

A combination of DSC and ESR spectroscopy has been applied to study the interactions of the angiotensin II non-peptide AT_1 receptor antagonist losartan with DPPC [157]. It is known that losartan (**10**) not only acts on AT_1 receptors but also interacts with biological membranes. This is of particular interest as it is thought that losartan binds to residues located within the transmembrane regions of the AT_1 receptor. Thus, the interaction of losartan with phospholipid bilayers was studied and its localization in the bilayer determined by ESR. The DSC results show a marked effect of losartan on cooperativity of the thermodynamic properties of DMPC and DPPC bilayers, especially DMPC. The transition signal is extensively broadened and shows a complex thermodynamic event characterized by three major transition peaks. Taken together with the observed temperature dependence of the spin-label ESR spectra, the small change in total calorimetric enthalpy indicated that this thermotropic behavior is related to the melting point of the lipid chain. Complex endotherms observed at high losartan concentrations have also been reported for the anionic lipid DMPG at low ionic strength [158]. It has recently been demonstrated that this corre-

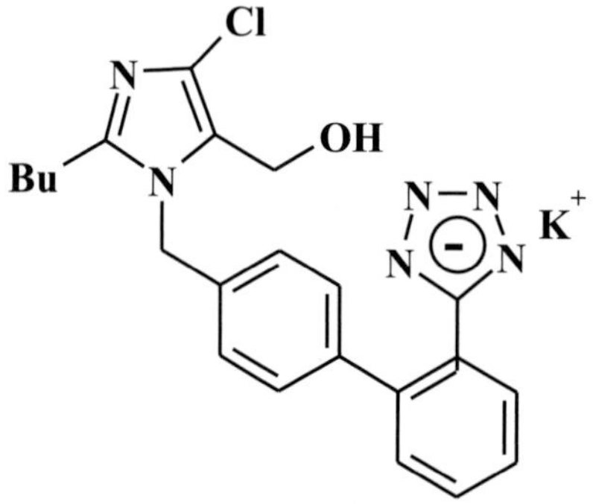

(10)

sponds to a reversible transition from a vesicular suspension to an extended bilayer network in the chain-melting region. As an explanation, it was proposed that solvent-associated or interfacial interactions may lead to a change in membrane curvature [159]. The induced structural transmission is indicated by the complex thermotropic behavior leading to 3 calorimetric peaks in the case of DMPC (Figure 3.49). In addition, the effect of losartan on chain motion and packing in bilayers of DMPC and DPPC was investigated by ESR spectroscopy using different positional isomers of DPPC (*n*-PCSL) spin probes. In agreement with the DSC results, the observed ESR spectra showed that losartan broadened the region of phase coexistence and shifted the chain-melting phase transition to a lower temperature. The decrease in chain mobility is induced in both the gel and fluid phase. However, in the fluid phase, this effect was observable with higher sensitivity with the spin label in the 5-position than with spin labels that were positioned further down the chain. This led to the as-

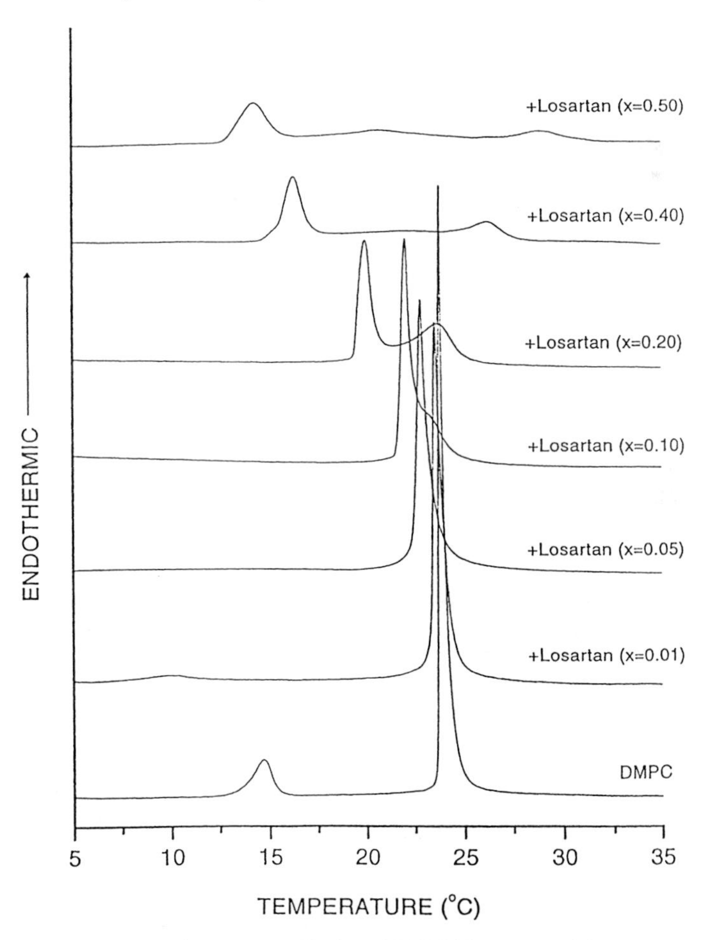

Fig. 3.49 DSC thermogram of hydrated DMPC bilayers containing losartan at different mole fractions, x, indicated. (Reprinted from Fig. 2 of ref. 157 with permission from Elsevier Science.)

sumption that losartan was located closer to the interfacial region than to the hydrophobic core of the DPPC bilayer. At higher losartan concentrations, however, the picture changed and indicated a different mode of intercalation: losartan seemed to be positioned deeper within the membrane. In summary, DSC showed that losartan can intercalate in the phospholipid bilayers. From the ESR spectra it can further be derived that a possible mode of intercalation is through the location of its polar hydroxyl group near the carbonyl groups of the phospholipids and the formation of hydrogen bonds. The authors suggest that the changes which can be induced by losartan in the vesicular structure should be considered in the pharmacological application of this antagonist. In addition to its antagonistic action at the receptor site, some of its effects may be mediated by intercalation with the membrane.

3.11.4
Combination of DSC and Fluorescence

The interaction of the local anesthetics dibucaine and tetracaine with biological membranes, e.g. the sarcoplasmic reticulum, and Ca^{2+},Mg^{2+}-ATPase has been studied by DSC and fluorescence techniques [160]. It was found that the temperature range of denaturation determined by DSC was 45–55 °C for the ATPase in sarcoplasmic reticulum membrane and 40–50 °C for the purified enzyme. Whereas millimolar concentrations of dibucaine and tetracaine reduced the denaturation temperature of ATPase by a few degrees, other local anesthetics, such as lidocaine and procaine (reported to have no effect on ATPase activity), did not significantly alter the DSC pattern of these membranes up to a concentration of 10 mM. These two drugs also show a lower partition coefficient in sarcoplasmic reticulum membranes. The intrinsic fluorescence of these membranes is, however, largely quenched by the small concentrations of dibucaine and tetracaine (Figure 3.50). This quenching of the intrinsic fluorescence can be fitted to the theoretical energy transfer prediction using the membrane–water partition coefficients, which were also determined in the study. It is suggested that the shift in denaturation temperature of the ATPase in the

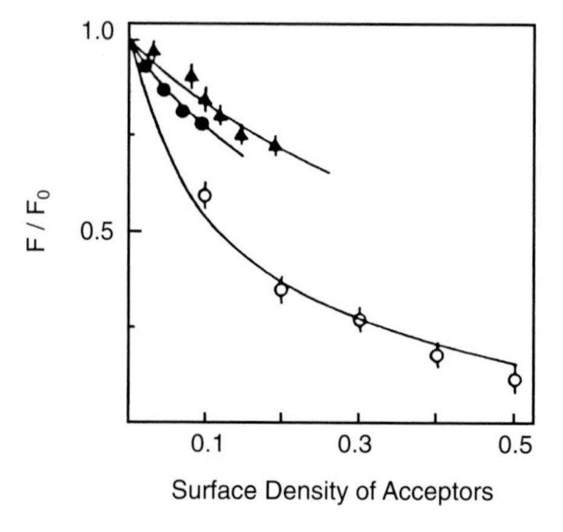

Fig. 3.50 Extent of quenching of the intrinsic fluorescence of sarcoplasmic reticulum membranes *vs.* the surface density of local anesthetics. The symbols correspond to dibucaine (▲) quinacrine (●), and tetracaine (○). The extent of quenching is expressed as the fluorescence in the presence of local anesthetics (F) over the fluorescence in their absence (F_o). (Reprinted from Fig. 6 of ref. 160 with permission from the American Chemical Society.)

presence of anesthetics, and most likely the inhibition of its activity, are related to the progressive disruption of the lipid annulus.

Sabra *et al.* [161] studied the suppression of unilamellar DMPC bilayer permeability in the main phase transition region by the insecticide lindane. The authors used a combined calorimetric and fluorimetric approach. An increase in lindane concentration was paralleled by a broadening of the transition signal and a decrease in transition temperature. No change in transition enthalpy was observed, indicating that lindane intercalated between the acyl chains without hindering their movement. The result can be interpreted as an enhancement of the thermal fluctuation of the bilayer by lindane. The results of the calorimetric measurements support the assumption that the bilayer is not disrupted by lindane, even at high concentrations. Fluorescence experiments were performed to measure the permeability changes of the unilamellar DMPC bilayer towards Co^{2+}. A head group-labeled lipid analog was used as fluorescent probe. The results showed that lindane sealed the bilayer to Co^{2+} penetration. The effect was concentration dependent and increased with increasing insecticide concentration. The authors compared their findings with the reported effects of anesthetics such as procaine, for which a decrease in fluctuation is paralleled

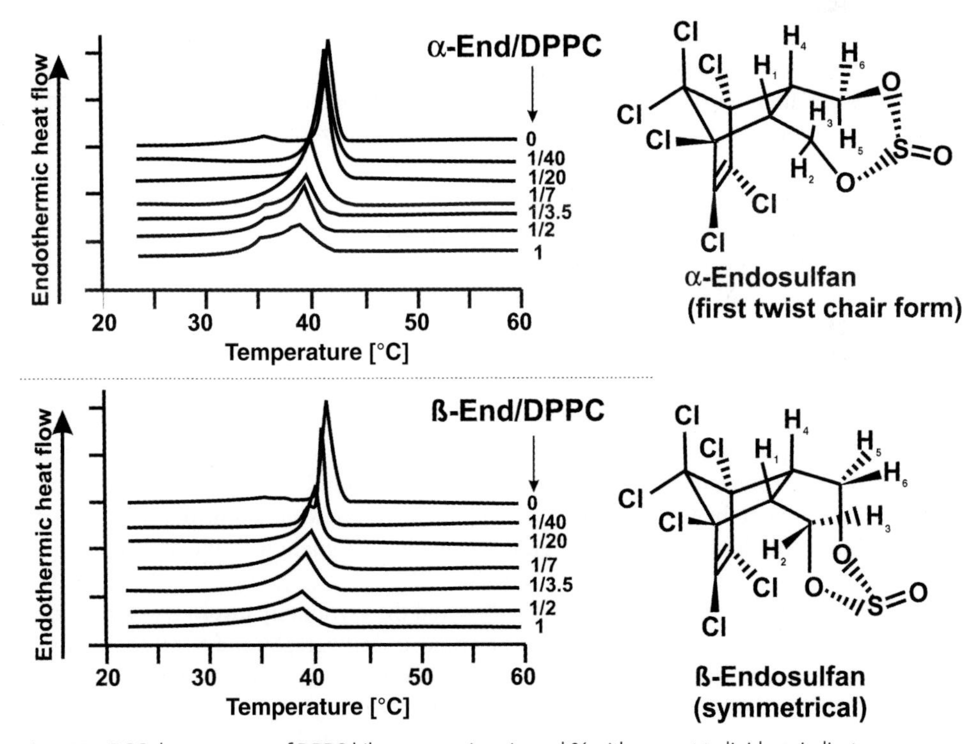

Fig. 3.51 DSC thermograms of DPPC bilayers in the absence and in the presence of varying concentrations of α-endosulfan (a) or β-endosulfan (b). The insecticide concentrations given in mol % with respect to lipid are indicated on the curves (left). The structures of endosulfan conformers are shown (right). (Adapted from Fig. 1 of ref. 162 and Fig. 1 of ref. 163.)

by an increase in permeability to ions, and with the effect of the hydrophobic *p*-dibutylbenzene, for which a decrease in permeability for ions was observed. The authors came to the conclusion "that the concept of 'small molecules' interacting with bilayers is too simple when it comes to explaining permeability concepts. One has to consider the individual molecular structure in detail when trying to predict physicochemical effects" [161].

Very recently, the perturbation induced by α- and β-endosulfan in lipid membranes has been studied by DSC and fluorescence technique [162]. Endosulfan is a broad-spectrum organochlorine insecticide that is used in the culture of a variety of cereals, fruits, vegetables, and cotton. The product is normally applied as a 7:3 isomeric mixture of α- and β-endosulfan. The data on membrane perturbation by the two isomers are another striking example that even small changes in molecular structure can lead to very different behavior in membrane interaction. Both isomers abolish the bilayer pretransition at a particular molar ratio of insecticide–lipid, shift the phase transition mid-point to lower temperature, and broaden the transition profile of DPPC (Figure 3.51a and b). In contrast to the β-form, however, the α-isomer, in addition to the main phase transition, induces a new phase transition signal, centered at 35.4 °C. (Figure 3.51a) This indicates a heterogeneous distribution of the α-isomer in a separate domain within the plane of the membrane (see Section 1.1.3.2). The β-form, on the other hand, undergoes homogeneous distribution. The DSC data show that the perturbations induced by the isomers are similar in some respects but different in others. To explore the indicated different localization of the isomers, fluorescence polarization techniques using special fluorescence probes have been applied [2-, 6-, and 12- (9-anthroyloxy) stearic acid and 16-(9-anthroyloxy) palmitic acid; 2-AS, 6-AS, 12-AS and 16-AP respectively]. In the case of the α-form, the results indicated an increase in the lipid structural order in the region probed by 2-AS and a decrease in the regions probed by 6-AS, 12-AS and 16-AP. In contrast, for the β-form, the results revealed a disordering effect in the upper region of the bilayer as detected by the 2-AS probe and ordering effects in the deeper regions, detected by the 6-AS, 12-AS, and 16-AP probes. In agreement with these results, the effect of β-endosulfan was decreased by the incorporation of cholesterol in DPPC bilayers, and completely abolished at 20 mol% cholesterol; however, even higher cholesterol concentrations did not prevent the membrane interaction of the α-form. The results suggest that the α-isomer interacts with the head group region and induces a phase separation that leads to membrane domains containing high local concentrations of the insecticide. This could lead to higher biological activity, in agreement with the observed higher degree of toxicity of the α-form.

The structures of β- and α-endosulfan and the asymmetry involved in the isomeric conversion have been determined by X-ray crystallography and NMR spectroscopy [163]. The α-isomer adopts an extended conformation, the twisted chair form, with the polar dioxathiepin-3-oxide at one pole and the lipophilic hexachloro group at the other pole, i.e. it possesses what can be called a "hydrophilic–hydrophobic dipole." The β-isomer, in contrast, assumes a symmetrical configuration (Figure 3.51) and is "apolar." This example again emphasizes that even minor structural or conformational changes in drug molecules can lead to very significant differences in type and degree of drug–membrane interaction, which may induce different biological activities. It supports the arguments against the use of isomeric mixtures in drug application.

3.11.5
Combination of FT-IR and NMR

The effect of temperature and pressure on the interaction with DPPC of the amphiphiles cholesterol and tetracaine on the conformational and dynamic properties of DPPC bilayers has been investigated using a combination of FT-IR, 2-D NMR, and time-resolved fluorescence anisotropy [164]. The authors studied the influence of pressure on DPPC–cholesterol and DPPC–tetracaine dispersions. In contrast to the effects of temperature, a change in pressure affected only the space available to the molecules and not their kinetic energy. Thus, the degree of intramolecular interaction can be affected by changes in pressure. The authors found a significant increase in the hydrophobicity of DPPC bilayers on incorporation of cholesterol. An increase in pressure up to 1 kbar suppressed water permeability much less than cholesterol embedded in fluid DPPC bilayers at 30 mol%. 2-D NOESY ^{1}H-NMR spectra of tetracaine in DMPC liposomes showed cross-peaks between H_2-phenyl and the $^{+}NH(CH_3)_2$ protons of tetracaine. This suggests that the tetracaine molecules assume a bent conformation in the lipid environment (DMPC/26 mol% tetracaine, pH 5.5). In addition, a weak cross-peak between the choline protons of DMPC and the $^{+}NH(CH_3)_2$ and phenyl protons of tetracaine was observed, suggesting that tetracaine is located in the interfacial region of the membrane. At pH 9 the uncharged form of tetracaine is found deeper within the interior of the membrane. This is indicated by the difference in pressure dependence of the δCH_2 band wavenumber of the DMPC bilayer in the absence and presence of tetracaine at pH 5.5 and 9.4, as observed by FT-IR. The pressure at which correlation field splitting becomes visible in DMPC was much higher in the presence of the uncharged form of tetracaine than for charged tetracaine, indicating that in the gel phase the acyl chains of DMPC are disordered more by uncharged than by charged tetracaine [164].

3.11.6
Combination of UV and ^{2}H-NMR

The partitioning of local anesthetics into membranes was studied. The surface charge effects on the phospholipid head groups were followed by a combination of UV and ^{2}H-NMR [165]. The liposomes used were composed of 1-palmitoyl-2-oleoyl-*sn*-glycero-3-phosphocholine (POPC) selectively deuterated at the α- and β-segments of the choline head group. The anesthetics studied were dibucaine and etidocaine in their protonated form. The intercalation of the drugs with the lipid molecules led to an increase in the surface area and to a positive electrical charge on the membrane. UV-binding isotherms were analyzed in such a way as to allow the determination of bilayer extension as well as the charge-induced concentration variation near the membrane surface. Intercalation was best described by a partition equilibrium. The surface partition coefficients measured at pH 5 were $660 \pm 80\ M^{-1}$ and $11 \pm 2\ M^{-1}$ for dibucaine and etidocaine respectively. The additional ^{2}H-NMR spectra showed that the head group conformation of the lipid molecules changed as a result of drug binding. The authors conclude that the observed membrane extension caused by even moderate amounts of drug may alter permeability and that drug absorption increases the membrane surface potential by some 20 mV, which will lead to a reduction in Ca^{2+} concentration at the membrane surface. Finally, the observed drug intercalation changed the phospholipid head group conformation and thereby the electric dipole field in the vicinity of the phosphocholine dipoles in the range of 50–100 mV. This would be sufficient to induce conformational transitions in the receptor protein [165].

3.11.7
Combination of DSC, FT-IR, and NMR

In an interesting study, three complementary techniques, namely DSC, FT-IR, ^{1}H-NMR, and deuterium NMR of side-chain perdeuterated DMPC (DMPC-d_{54}), were used to evaluate the interaction of reversible H^+,K^+-ATPase inhibitors [166] with DMPC. The aim of the study was to characterize the interaction between DMPC model membranes and two non-covalent inhibitors of gastric H^+,K^+-ATPase. Compound **11** partitions readily into the bilayer of DMPC and causes a slight disordering of the lipid hydrocarbon side-chain motion, a reduction in cooperativity, and a decrease in phase transition temperature. However, FT-IR and deuterium NMR showed that the bilayer structure remains intact up to high molar compound–lipid ratios. ^{1}H-NMR NOE measurements performed for **11** and analog **12** provided some insight into the orientation of the two compounds within the bilayer. Both drugs are largely cationic under physiological conditions (pK_a 9.54 for **11** and 8.63 for **12**). The spin probe data and intermolecular NOEs were unspecific for compound **11** but specific for compound **12** (SK&F 96464). For compound **11**, the NOEs in the spectral region – especially between 4 and 4.8 ppm – could not be reliably attributed to the effects of compound–lipid interactions. Only the resonances of the β-carbon of the glycerol moiety of DMPC at about 5.2 ppm were observed to be involved in NOEs with **11**. NOEs from the glycerol β-protons to several signals of **11** were observed, indicating considerable mobility of the compound in relation to the head group region.

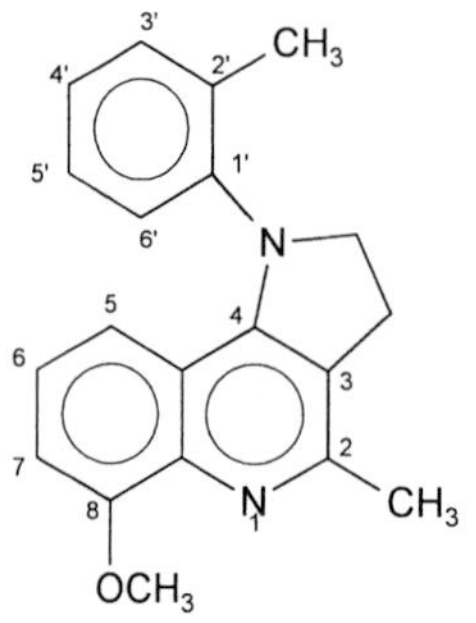

SK&F 96079 (11) 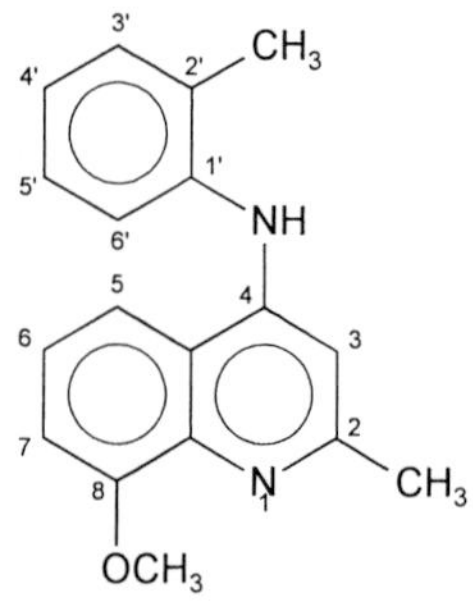 **SK&F 96464 (12)** 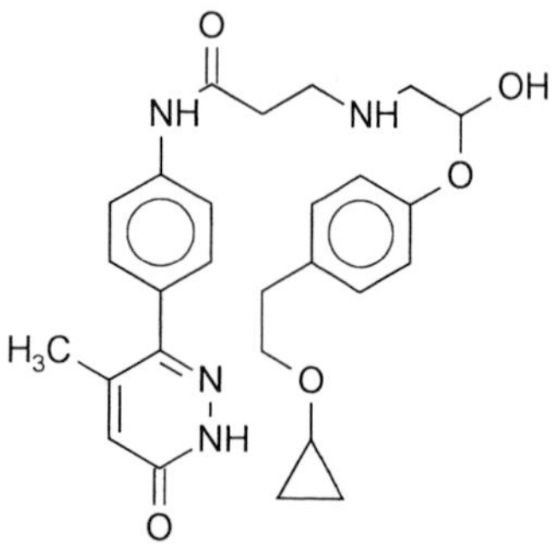 **SK&F 95018 (13)**

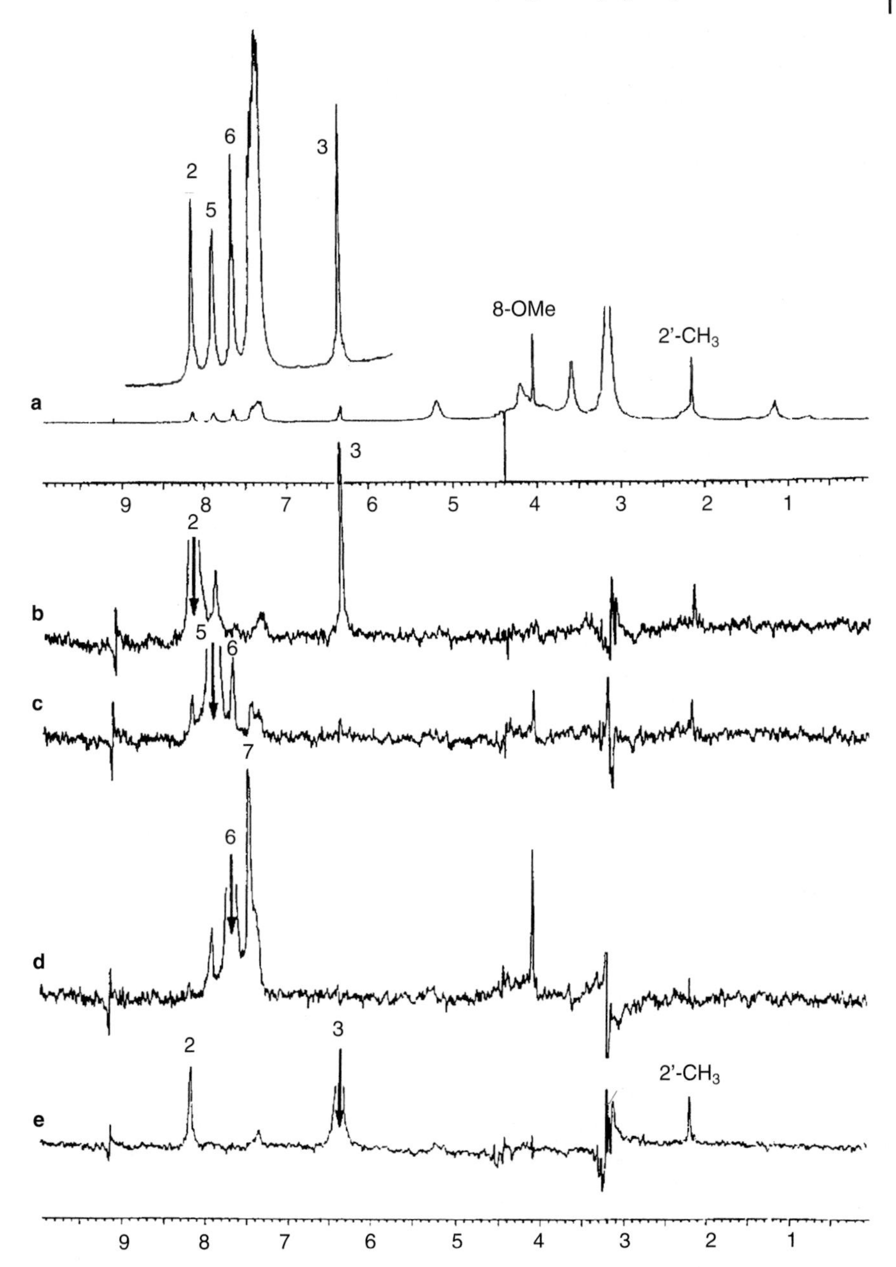

Fig. 3.52 (a) ^{1}H-NMR spectra (360 MHz, 25 °C) of a sonicated aqueous dispersion of 5 mg of DMPC-d_{54} and SK&F 96464 (12) in a 4:1 molar ratio. (b) to (e) correspond to NOE difference spectra resulting from 250 ms irradiation at the (b) H2, (c) H5, (d) H6, and (e) H3 signals of the compound. (Reprinted from Fig. 3 of ref. 166 with permission from Elsevier Science.)

In contrast, derivative **12** showed increased solubility in water and rather sharp resonance signals in the presence of DMPC, owing to rapid exchange between the free and membranous environment. The conformation of **12** was varied interactively using a Evans–Sutherland PS 300 to take into account the observed NOEs and the considerable shielding of the H2 and H3 protons and deshielding of H5 and H6 of the quinoline ring. Exposing H2, H3, and H5 to irradiation produced slight NOEs to the envelope of signals at about 7.3 ppm, which correspond to H7 and the protons attached to the *ortho*-toluidino group. Exposure of the H3 resonance to irradiation also produced a NOE to the 2'-CH_3 signal (see formula **12**). A conformation with a torsion angle of ~80° for the angle C_3–C_4–N–C_1' and of about –105° for C_4–N–C_1'–C_2' positioned the protons H2 and H3 well inside and the protons H5 and H6 outside the ring current shielding cone centered in the *ortho*-toluidine ring. This is in accordance with the NMR results. The related conformation obtained by rotation of the *ortho*-toluidine ring by 180° about the C_1'–N bond yields a conformation that is also significantly populated. For the two conformations assumed, the authors concluded, on the basis of the only slight NOEs observed between the compound and membrane signals, that a significant fraction of compound **12** is located in the aqueous phase. Typical NOE difference spectra resulting from protons of compound **12** which have been subjected to irradiation are given in Figure 3.52.

Under the conditions used in the experiment, **12** can form two favorable electrostatic interactions with DMPC head groups. Hydrogen bonds can occur between the amino protons of **12** and the *sn*-1 side chain carbonyl of a DMPC molecule and between the protonated quinoline ring of the molecule and a primary phosphodiester oxygen of the same DMPC molecule. Figure 3.53 shows the stereo plot of the two molecules in the bound conformation consistent with the intramolecular NOEs (nitrogen atoms in the drug formula are not shown in the stereo plots). These conformations are also consistent with the NOEs between the compound and the glycerol β-proton of DMPC. The two geometrically favorable electrostatic interactions possible for compound **12** seem to be the precondition for its specific orientation in the membrane. The authors further concluded that compounds of type **11**, which show only a slight effect on cooperativity of the gel to crystalline phase transition, should not possess the deleterious toxic properties, especially hemolytic, that have been observed for the amphiphilic derivative **13** (SK&F 95018). In the latter, disruption of model bilayers to non-ordered systems has been observed [167].

A combination of ^{2}H-NMR, CD, fluorescence, and electron absorption techniques has been used to investigate the interaction of the pentene antibiotic filipin with dimyristoylphosphatidylcholine [168].

3.12 Summary

The various methods available to study drug–membrane interactions are best applied to artificial membranes. However, it is important to be careful in drawing any conclusions regarding the pharmacological or biological relevance of the measured

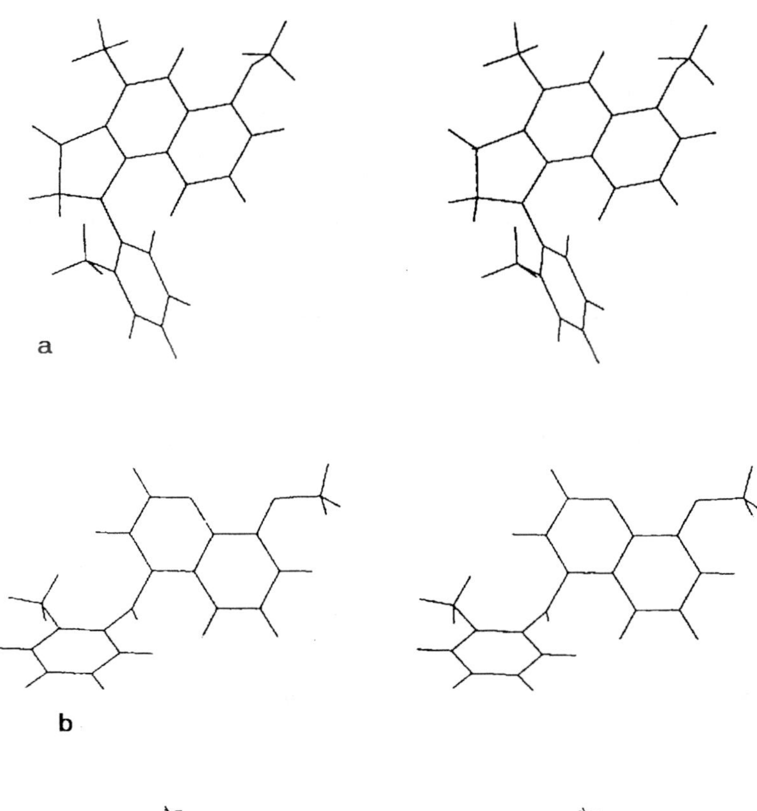

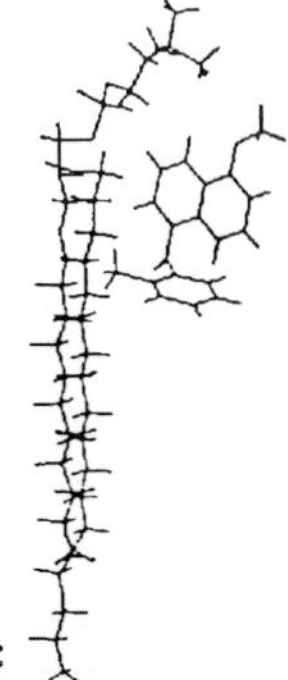

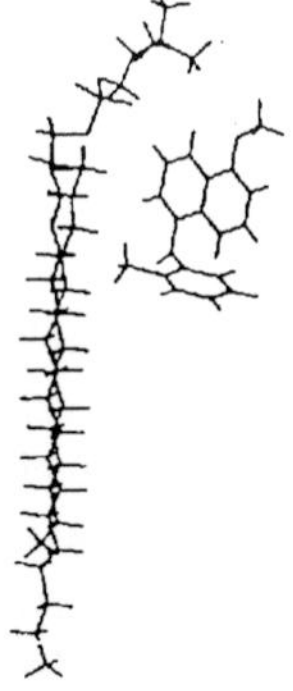

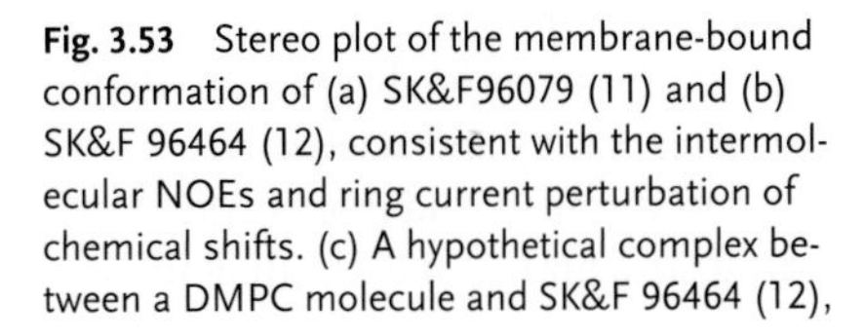

Fig. 3.53 Stereo plot of the membrane-bound conformation of (a) SK&F96079 (11) and (b) SK&F 96464 (12), consistent with the intermolecular NOEs and ring current perturbation of chemical shifts. (c) A hypothetical complex between a DMPC molecule and SK&F 96464 (12), consistent with fatty acid spin probe data. The structure is, according to the authors, intended to be only representative and not unique. (Reprinted from Fig. 5 of ref. 166 with permission from Elsevier Science.)

effects of drugs on artificial membranes. For example, changes in the phase transition of artificial bilayers observed upon addition of drugs affect the strength and type of interaction of drugs and phospholipid and are useful for the derivation of structure–interaction relationships.

Most biomembranes are in the liquid crystalline phase under physiological conditions and their complex composition makes it more difficult to measure phase transitions. It should be mentioned that the effect of drug molecules on fluidity and order of a membrane can even be reversed below and above T_t [169] and NMR measurements of drug–membrane interactions are normally performed above T_t.

Other effects caused by the intrusion of drug molecules into membranes, such as changes in surface charge, transmembrane potential, allosteric conformational changes in membrane-integrated proteins, etc., could well reflect biological membrane effects. Because of the microheterogeneity of biomembranes, it can also be assumed that catamphiphiles induce phase transitions of biological importance in particular restricted membrane regions (domains, lipid annuli). Drugs, such as flupentixol or amiodarone, that strongly interact with membranes cause complete separation of the control transition signal and the drug-induced phase transition signal. In between the two signals the thermogram returns to the baseline. This indicates that not all transition states between the maximal drug-induced phospholipid state and unaffected phospholipid molecules exist. The reduction in T_t is independent of the phospholipid–drug ratio. This means that the phospholipid–drug domains possess a constant composition. The domain becomes enlarged upon addition of drug (reflected by an increase in the area under the signal of the thermogram) but no change in structure or "quality" occurs.

The various NMR techniques discussed – alone or in combination with other analytic techniques – are most valuable tools for following the interaction of drug molecules with the hydrocarbon chains and polar head groups of bilayers. They allow us to obtain information on the preferred conformation and orientation of ligands and membrane constituents. The examples demonstrate that catamphiphiles can interact with polar head groups and that the preferred conformation of drugs and head groups may change. This could be crucial for drug activity or inactivity. The different methods, especially NMR, are also useful to determine transport kinetics and partitioning of drug molecules into bilayers. The phase separation and domain formation seen by DSC in negatively charged lipids is probably due to a change in head group structure, whereas changes in phase transition temperature reflect changes in membrane fluidity and the cooperativity of the lipid chains. The type and strength of interactions depend on the structure and physicochemical properties of both the drug molecule and the phospholipids which have a direct influence on drug accumulation within a particular membrane and therefore on drug activity and selectivity.

It is hoped that the following chapters will help to stimulate and intensify research of the effects of membrane composition and drug structure on their mutual interactions.

References

1 Unger, S.H., Chiang, G.H., *J. Med. Chem.* **1981**, *24*, 262–270.

2 Thurnhofer, H., Schnabel, J., Lika, G., Pidgeon, C., Hauser, H., *Biochim. Biophys. Acta* **1991**, *1064*, 275–286.

3 Pidgeon, C., Marcus, C., Alvarez, F., in *Applications of Enzyme Biotechnology*, T.O. Baldwin, J.W. Kelly (Eds.), Plenum Press, New York **1992**, pp. 201–220.

4 Kaliszan, R., Kaliszan, A., Wainer, I.W., *J. Pharm. Biomed. Anal.*, **1993**, *11*, 505–511.

5 Kaliszan, R., Nasal, A., Bucinski, A., *Eur. J. Med.Chem.* **1994**, *29*, 163–170.

6 Hinderling, P.H., Schmidlin, O., Seydel, J.K., *J. Pharmacokinet. Biopharm.* **1994**, *12*, 263–287.

7 Seydel, J.K., Albores Velasco, M., Coats, E.A., Cordes, H.-P., Kunz, B., Wiese, M., *Quant. Struct.–Act. Relat.* **1992**, *11*, 205–210.

8 Yang, Q., Liu, X.-Y., Umetani, K., Kamo, N., Miyke, J., *Biochim. Biophys. Acta*, **1999**, *1417*, 122–130.

9 Sicre, P., Cordoba, J., *J. Colloid Interface Sci.*, **1989**, *132*, 95–99.

10 Lüllmann, H., Plösch, H., Ziegler, A., *Biochem. Pharmacol.* **1980**, *29*, 2969–2974.

11 Hwang, S.-B., Chen, T.Y., *J. Med. Chem.* **1981**, *24*, 1202–1211.

12 Girke, St., Mohr, Kl., Schrape, S., *Biochem. Pharmacol.* **1989**, *38*, 2487–2496.

13 Klein, Ch.D.P., Klingmüller, M., Schellinski, Ch., Landmann, S., Hauschild, St., Heber, D., Mohr, Kl., Hopfinger, A.J., *J. Med. Chem.* **1999**, *42*, 3874–3888.

14 Seydel, J.K., Coats, E.A., Pajeva, I.K., Wiese M., in *Bioactive Compound Design, Possibilities for Industrial Use*, M.G. Ford, R. Greenwood, G.T. Brooks, R. Franke (Eds.), Bios Scientific Publishers, Oxford **1996**, pp. 137–147.

15 Sackmann, E., *Ber. Bunsenges. Phys. Chem.* **1978**, *82*, 891–909.

16 Ushida, K., Yao, H., Ema, K., *Phys. Rev. E: Stat. Phys., Plasma, Fluids, Rel. Interdiscip. Top.* **1997**, *56*, 661–666.

17 Cruzerro-Hansson, L., Ipsen, J.H., Mouritzen, O.G., *Biochim. Biophys. Acta* **1989**, *979*, 166–176.

18 Menger, F.M., Wood, M.G., Zhou, Q.Z., Hopkins, H.P., Fumero, J., *J. Am. Chem. Soc.* **1988**, *110*, 6804–6810.

19 Brumm, T., Jørgensen, K., Mouritzen, O.G., Bayer, Th.M., *Biophys J.* **1996**, *70*, 1373–1379.

20 Caffray, M., Moynihan, D., Hogan, J., *Chem. Phys. Lipids* **1991**, *57*, 275–291.

21 Caffray, M., Hogan, J. *Chem. Phys. Lipids* **1992**, *61*, 1–109.

22 Tilcock, C.P.S., Bally, M.B., Farren, S.B., Gruner, S.M. *Biochemistry* **1984**, *23*, 2696–2703.

23 Cheetham, J.J., Chen, R.J.B., Epand, R.M. *Biochim. Biophys. Acta* **1990**, *1024*, 367–372.

24 Wachtel, E., Borochov, N., Back, D., *Biochim. Biophys. Acta* **2000**, *1463*, 162–166.

25 MacMullen, T.P., McElhaney, R.N., *Curr. Opin. Colloid Interface Sci.* **1997**, *1*, 83–90.

26 Gingell, D., Gingsberg, L. in *Cell Surface Reviews*, Vol. 5, G. Poste, G.L. Nicolson (Eds.), North-Holland, Amsterdam **1978**, pp. 791–833.

27 Choi, S., Ware, W., Lauterbach, St. R., Phillips W.M. *Biochemistry* **1991**, *30*, 8563–8568.

28 Wachtel, E.J., Borochov N., Bach, D., *Biochim. Biophys. Acta* **1991**, *1066*, 63–69.

29 Lohner, K., *Chem.Phys.Lipids* **1991**, *57*, 341–362.

30 Tenchov, B., *Chem. Phys. Lipids* **1991**, *57*, 165–177.

31 Seddon, J.M., Hogan, J.L., Warrender, N.A., Pebay-Peyroula., E., *Progr. Coll. Polym. Sci.* **1990**, *81*, 189–197.

32 Epand, R.M., Epand, R.F., Lancaster, C.R.D., *Biochim. Biophys. Acta* **1990**, *945*, 161–166.

33 Basáñez, G., Nieva, J.L., Rivas, E., Alonso, A., Goñi, F.M., *Biophys. J.* **1996**, *70*, 2299–2306.

34 Siegel, D.P., Bansbach, J., Yeagle, P.L., *Biochemistry* **1989**, *28*, 5010–5019.

35 Turnois, H., Fabrie, C.H.J.P., Burger, K.N.J., Mandersloot, J., Hilgers, P., Van Dalen, H., De Gier, J., De Kruijff, B., *Biochemistry* **1990**, *29*, 8297–8307.

36 Epand, R.M., Epand, R.F., McKenzie, R.C., *J. Biol. Chem.* **1987**, *262*, 1526–1529.

37 Epand, R.M., Lobl, Th., Renis, H.E., *Biosci. Rep.* **1987**, *7*, 745–749.

38 Borchard, K., Heber, D., Klingmüller, M., Mohr, K., *Biochem. Pharmacol.* **1991**, *42* (Suppl.), 61–65.

39 Bauer, M., Megret, C., Lamure, A., Lacabanne, C., Fauran-Clavel, M.-J., *J. Pharm. Sci.* **1990**, *79*, 897–901.

40 Moret, C., *Neuropharmacology* **1985**, *24*, 1211–1219.

41 Yoshida, T., Taga, K., Okabayashi, H., Kamaya, K., Ueda, I., *Biochim. Biophys. Acta* **1990**, *1028*, 95–102.

42 Thomas, P.G., Verkleij, A.J., *Biochim. Biophys. Acta* **1990**, *1030*, 211–222.

43 Constantidines, P.P., Inouchi, N., Tritton, Th.R., Sartorelli, A.C., Sturtevant, J.M., *J. Biol. Chem.* **1986**, *261*, 10196–10202.

44 Smejtek, P., Barstad, A.W., Wang, S., *Chem. Biol. Interact.* **1989**, *71*, 37–61.

45 Kim, N.H., Roh, S.B., *Korean J. Pharmacol.* **1990**, *26*, 77–82.

46 Hanpft, R., *Dissertation*, Universität Kiel, **1987**.

47 Pajeva, I.K., Wiese, M., Cordes, H.-P., Seydel, J.K., *J. Cancer Res. Clin. Oncol.* **1996**, *122*, 27–40.

48 Redman-Frey, N.L., Antinore, M.J., *Anal. Chim. Acta 251*, 79–81 (**1991**)

49 Jørgensen, K., Ipsen, J.H., Mouritsen, O.G., Bennett, D., Zuckerman, M.J., *Biochim. Biophys. Acta* **1991**, *1062*, 227–238.

50 Jørgensen, K., Mouritsen, O.G., Bennett, D., Zuckerman, M.J., *Ann. N.Y. Acad. Sci.* **1991**, *625*, 747–750.

51 Singer, M.A., Jain, M.K., *Can. J. Biochem.* **1980**, *58*, 815–821.

52 Antunes-Madeira, M.C., Madeira, V.M.C., *Biochim. Biophys. Acta* **1985**, *820*, 165–172.

53 Hanpft, R., Mohr, K., *Biochim. Biophys. Acta* **1985**, *814*, 156–162.

54 Blasik, J., Pol. J., *Environ. Study* **1996**, *5*, 31–35.

55 Lau, W.F., Das, N.P., *Experientia* **1995**, *51*, 731–737.

56 Moya-Miles, M.R., Munoz-Delgado, E., Vidal, C.J., *Biochem. Mol. Biol. Int.* **1995**, *36*, 1299–1308.

57 Chan, S.D., Wang, H.H., *Biochim. Biophys. Acta* **1984**, *770*, 55–64.

58 Deliconstantinos, G., Kopeikina-Tsiboukidon, L., Villiotou, W., *Biochem. Pharmacol.* **1987**, *36*, 153–161.

59 Krishna, M.M.G., Periasama, N., *Biochim. Biophys. Acta* **1999**, *1461*, 58–68.

60 Choi, S., Ware, W., Lauterbach, St. R., Phillips, W.M., *Biochemistry* **1991**, *30*, 8563–8568.

61 Casal, H.L., Martin, A., Mantsch, H.H., *Chem. Phys. Lipids* **1987**, *43*, 47–53.

62 Cao, A., Hantz-Brachet, E., Azize, B., Taillandier, E., Perret, G., *Chem. Phys. Lipids* **1991**, *58*, 225–232.

63 Davies, M.A., Brauner, J.W., Schuster, H.F., Mendelsohn, R., *Biochem. Biophys. Res. Commun.* **1990**, *168*, 85–90.

64 Ondrias, K., *J. Pharm. Biomed. Anal.* **1989**, *7*, 649–675.

65 Trudell J.R., in *Drug and Anesthetic Effects on Membrane Structure and Function*, R.C. Aloia, C.C. Curtain, L.M. Gordon (Eds.), Wiley-Liss, New York **1991**, pp. 1–13.

66 Ueda, I., in: *Drug and Anesthetic Effects on Membrane Structure and Function*, R.C. Aloia, C.C. Curtain, L.M. Gordon, (Eds), **1991**, pp. 15–33

67 Grof, P., Belagyi, J., *Biochim. Biophys. Acta* **1983**, *734*, 319–328.

68 Ondrias, K., Ondriasova, E., Stasko, A., *Chem. Phys. Lipids* **1992**, *62*, 11–17.

69 Herbette L.G., Chester, D.W., Rhodes, D.G., *Biophys. J.* **1986**, *49*, 91–94.

70 Bäuerle, H.-D., Seelig, J., *Biochemistry* **1991**, *30*, 7203–7211.

71 Seydel, J.K., Coats, E.A., Cordes, H.-P., Wiese, M., *Arch. Pharm.* **1994**, *327*, 601–610.

72 Cunningham, B.A., Bras, W., Quinn, P.J., Lis, L.J., *J. Biochem. Biophys. Methods* **1994**, *29*, 87–111.

73 Mavromoustakos, Th., Yang, D-P., Charalambous, A., Herbette, L.G., Makriyannis, A., *Biochim. Biophys. Acta* **1990**, *1024*, 336–344.

74 Makriyannis, A., Banijamali, A., Jarell, H.C., Yang, D.-P., *Biochim. Biophys. Acta* **1989**, *986*, 141–145.

75 Suwalsky, M., Neira, F., Sanchez I., *Z. Naturforsch.* **1991**, *46c*, 133–138.

76 Colotto, A., Mariani, P., Grazia Ponzi Bossi, M., Rustichelli, F., Albertini, G., Amaral, L.Q., *Biochim Biophys. Acta* **1992**, *1107*, 165–174.

77 McIntosh, T.J., McDaniel, R.V., Simon, S.A., *Biochim. Biophys. Acta* **1983**, *731*, 109–114.

78 Suwalski, M., Gimenez, L., Saenger, V., Neira F., *Z. Naturforsch.* **1988**, *43c*, 742–748.

79 Wang, X., Takahashi, H., Hatta, I., Quinn, P., *Biochim. Biophys. Acta* **1999**, *1418*, 335–343.

80 Chatelain, P., in *Drug and Anesthetic Effects on Membrane Structure and Function*, R.C. Aloia, C.C. Curtain, L.M. Gordon (Eds.), Wiley-Liss, New York **1991**, pp. 183–202.

81 Trumbore, M., Chester, D.W., Moring, M., Rhodes, D., Herbette, L.G., *Biophys. J.* **1988**, *54*, 535–543.

82 Adachi, T., Takahashi, H., Hatta, I., *Physica B.* **1995**, *213/214*, 760–762.

83 Chapman, D., Penkett, S.A., *Nature* **1966**, *211*, 1304–1305.

84 Cullis, P.R., de Kruijff, B., *Biochim. Biophys. Acta* **1979**, *559*, 399–420.

85 De Kruijff, B., Cullis, P.R., Verkleij, A.J., Hope, M.J., van Echteld, C.J.A., Taarschi, T.F., van Hoogevest, P., Killian, J.A., Rietveld, A., van der Steen, J.J.H.H.M., in *Progress in Protein–Lipid Interactions*, Vol. 1, A.Watts, J.J.H.H.M. de Pont (Eds.), Elsevier, Amsterdam **1985**, pp. 89–142.

86 Fribolin, H., *Ein- und zweidimensionale NMR-Spektroskopie*, VCH Verlagsgesellschaft, Weinheim **1988**.

87 Martin, G.E., Zektzer, A.S., *Two-dimensional NMR Methods for Establishing Molecular Connectivity*, VCH, Weinheim **1988**.

88 Watts, A., *Biochim. Biophys. Acta* **1998**, *1376*, 297–318.

89 Watts, A., *Curr. Opin. Biotechnol.* **1999**, *10*, 48–54.

90 Lee, Y.K., Kurur, N.D., Helmle, M., Johannessen, O.G., Nielsen, N.C., Levitt, M.H., *Chem. Phys. Letts.* **1994**, *242*, 304–309.

91 Wittebrot, R.J., Blume, A., Huang, T.-H., Das Gupta, S.K., Griffin, R.G., *Biochemistry* **1982**, *21*, 3487–3502.

92 Watts, A., Spooner P.J.R., *Chem Phys. Lipids* **1991**, *57*, 195–211.

93 Seelig, J., in *Modern Trends of Colloid Science in Chemistry and Biology*. H.-F. Eicke (Ed.), Birkhäuser Verlag, Basle **1985**, pp. 261–273.

94 Dufourc, E.J., Mayer, Chr., Stohrer, J., Althoff, G., Kothe, G., *Biophys. J.* **1992**, *61*, 42–57.

95 Milburn, M.P., Jeffrey, K.R., *Biophys. J.* **1990**, *58*, 187–194.

96 Watts, A., *Studia Biophys.*, **1988**, *127*, 29–36.

97 Davis J.H., *Biophys. J.* **1979**, *27*, 339–358.

98 Macdonald, P.M., Leise, J., Marassi, F., *Biochemistry* **1991**, *30*, 3558–3566.

99 McMullen, T.P.W., Levis, R.N.A.H., McElhaney, R.N., *Biochim. Biophys. Acta* **1999**, *1416*, 19–134.

100 Krajewski-Bertrand, M.-A., Milon, A., Nakatani, Y., Ourisson G., *Biochim. Biophys. Acta* **1992**, *1105*, 213–220.

101 De Boeck, H., Zidovetzki, R., *Biochemistry* **1989**, *28*, 7439–7446.

102 De Boeck, H., Zidowetzki, R., *Biochemistry* **1992**, *31*, 623–630.

103 Hamilton, J.A., Vural, J.M., Carpentier, Y.A., Deckelbaum, R.J., *J. Lipid Res.* **1996**, *37*, 773–782.

104 Epand, R.M., Epand R.F., Lancaster, C.R., *Biochim. Biophys. Acta* **1988**, *945*, 161–166.

105 Goldberg, E.M., Lester, D.S., Borchardt, D.B., Zidovetzki, R., *Biophys. J.* **1995**, *69*, 965–973.

106 Seelig, J., MacDonald, P.M., Scherer, P.G., *Biochemistry* **1987**, *26*, 7535–7541.

107 Scherer, P.G., Seelig, J., *EMBO J.* **1987**, *6*, 2915–2922.

108 Dempsey, C,. Warrs, A., *Biochemistry*, **1987**, *26*, 5803–5811.

109 Marassi, F.M., Macdonald, P.M., *Biochemistry* **1992**, *31*, 10031–10036.

110 Bechinger, B., Seelig, J., *Biochemistry* **1991**, *30*, 3923–3929.

111 López-Gracia, F., Villaín, J., Gómez-Fernández, J.C., Quinn, P.J., *Biophys. J.* **1994** *66*, 1991–2004.

112 Aloia, R.C., Curtain, C.C., Gordon, L.M. (Eds.), *Drug and Anesthetic Effects on Membrane Structure and Function*, Wiley-Liss, NewYork **1991**.

113 Zidowetzki, R., Sherman, I.W., Atiya, A., de Boeck, H., *Mol. Biochem. Parasitol.* **1989**, *35*, 199–208.

114 Seelig, J., *Q. Rev. Biophys.* **1997**, *10*, 353–418.

115 Bammel, B.P., Brand J.A., Simmons, R.B., Evans D., Smith, J.C., *Biochim. Biophy. Acta*, **1987**, *896*, 136–152.

116 Henin, Y., C. Gouytte, C., Schwartz, O., Debouzy, J.-C., Neumann, J.-M., Huynh-Dinh, T., *J. Med. Chem.* **1991**, *34*, 1830–1837.

117 Males, R.G., Herring, F.G., *Biochim. Biophys. Acta* **1999**, *1416*, 333–338.

118 Kuchel, Ph.W., Kirk, K., King, G.F., in *Subcellular Biochemistry*, Vol. 23, *Physicochemical Methods in the Study of Biomembranes*, H.J. Hilderson, G.B. Ralston (Eds.), Plenum Press, New York **1994**, pp. 247–313.

119 Bulliman, B.T., Kuchel, Ph.W., Chapman, B.E., *J. Magn. Res.* **1989**, *82*, 131–183.

120 Hutton, W.C., Yeagle, P.L., Martin, R.B., *Chem. Phys. Lipids* **1977**, *19*, 255–265.

121 Hunt, G.R.A., Kaszuba, M., *Chem. Phys. Lipids* **1989**, *51*, 55–65.

122 Betagery, G.V., Theriault, Y., Rogers, J.A., *Membrane Biochemistry* **1989**, *8*, 197–206.

123 Boden, N., Jones, S.A., Sixl, F., *Biochemistry* **1991**, *30*, 2146–2155.

124 Rice, M.D., Wittebort, R.J., Griffin, R.G., Meirovitch, E., Stimson, E.R., Meinwald, Y.C., Freed, H.J., Scheraga, H.A., *J.Am. Chem. Soc.* **1981**, *103*, 7707–7710.

125 Auger, M., Smith, I.C.P., Jarrell, H.C., *Biochim. Biophys. Acta* **1989**, *981*, 351–357.

126 Auger, M., Jarrell, H.C., Smith, I.C.P., *Biochemistry* **1988**, *27*, 4660–4667.

127 Schote, U., Seelig, J., *Biochim. Biophys. Acta* **1999**, *1415*, 135–146.

128 Jardetzky, O., Roberts, G.C.K., *NMR in Molecular Biology*, Academic Press, New York **1981**.

129 Seydel, J.K. *Trends Pharmacol. Sci.* **1991**, *12*, 368–371.

130 Craig, D.J., Higgins, K.A., in *Annual Reports on NMR-Spectroscopy* 22, A.G. Webb (Ed.), Academic Press, London **1990**, pp. 61–138.

131 Seydel, J.K., Wassermann, O., *Naunyn-Schmiedeberg's Arch. Pharmacol.* **1973**, *279*, 207–210.

132 Makriyannis, A., DiMeglio, Chr. M., Fesik, W., *J. Med Chem.* **1991**, *34*, 1700–1703.

133 Seydel, J.K., Cordes, H.-P., Wiese, M., Chi, H., Schaper, K.-J., Coats, E.A., Kunz, B., Engel, J., Kutscher, B., Emig, H., in *QSAR: Rational Approaches to the Design of Bioactive Compounds*, C. Silipo, A. Vittoria (Eds.), Elsevier, Amsterdam **1991**, pp. 367–376.

134 Pajeva, I.K., Wiese, M., *Quant. Struct.–Act. Relat.* **1997**, *16*, 1–10.

135 Seydel, J.K., Cordes, H.-P., Wiese, M., Chi, H., Croes, N., Hanpft, R., Lüllmann, H., Mohr, K., Patten, M., Padberg, Y., Lüllmann-Rauch, R., Vellguth, S., Meindl, W.R., Schönberger, H., *Quant. Struct.–Act. Relat.* **1989**, *8*, 266–278.

136 Behling, R.W., Yamane, T., Navon, G., Jelinski, L.W., *Proc. Natl. Acad. Sci. USA* **1988**, *85*, 6721–6725.

137 Seydel, J.K., Albores Velasco, M., Coats, E.A., Cordes, H.-P., Kunz, B., Wiese, M., *Quant. Struct.–Act. Relat.* **1992**, *11*, 205–210.

138 Coats, E.A., Wiese, M., Chi, H-L., Cordes, H.-P., Seydel, J.K., *Quant. Struct.–Act. Relat.* **1992**, *11*, 364–369.

139 Mason, R.P., Rhodes, D.G., Herbette, L.G., *J. Med. Chem.* **1991**, *34*, 869–877.

140 Tetreault, St., Ananthanarayanan, V.S., *J. Med. Chem.* **1993**, *36*, 1017–1023.

141 Retzinger, G.S., Cohen, L., Lau, S.H., Kezdy, F.J., *J. Pharm. Sci.* **1986**, *75*, 976–982.

142 Kuroda, Y., Fujiwara, Y., *Biochim. Biophys. Acta* **1987**, *903*, 395–410.

143 Yokono, S., Ogli, K., Tsukamoto, I., Yokono, A., Miura, S., *Ann. N.Y. Acad. Sci.* **1991**, *625*, 751–755.

144 Yokono, S., Sheeh, D.D., Ueda, I., *Biochim. Biophys. Acta* **1981**, *645*, 237–242.

145 Gaillard, P., Carrupt, P.-A., Testa, B., *Bioorg. Med. Chem. Lett.* **1994**, *4*, 737–742.
146 Gallois, L., Fiallo, M., Garnier-Suillerot, A., *Biochim. Biophys. Acta* **1998**, *1370*, 31–40.
147 Custodio, J.B.A., Almeida, L.M., Madeira, V.M., *Biochem. Biophys. Res. Commun.* **1991**, *176*, 1079–1085.
148 Gracia, D.A., Prello, *M.A., Biochim. Biophys. Acta* **1999**, *1418*, 221–231.
149 Kurikungel H., Barain, P., Restall, C., *Biochim. Soc. Trans.* **1992**, *20*, 1575–1581.
150 Thomas, P.G., *Cell Biol. Int. Rep.* **1990**, *14*, 389–397.
151 Epand R.M., Epand, R.F., McKenzie, R.C., *J. Biol. Chem.* **1987**, *262*, 1526–1529.
152 Nicolay, K., Saureau, A.-M., Tocanne, J.F., Brasseur, R., Huart, P., Ruyschaert, J.-M., De Kruijff, B., *Biochim. Biophys. Acta* **1988**, *940*, 197–208.
153 de Wolf, F.A., Nicolay, K., de Kruijff, B., *Biochemistry* **1992**, *31*, 9252–9262.
154 Ambrosini, A., Dubini, B., Leone, L., Ponzi Bossi, M.G., Russo, P., *Mol. Cryst. Liq. Cryst.* **1990**, *179*, 317–334.
155 Bae, S.-J., Kitamura S., Herbette, L.G., Sturtevant, J., *Chem. Phys. Lipids* **1989**, *51*, 1–7.
156 Colotto, A., Mariani, P., Ponzi Bossi, M.G., Rustichelli, F., Albertini, G., Amaral, L.Q., *Biochim. Biophys. Acta* **1992**, *1107*, 165–174.
157 Theodoropoulou, E., Marsh, D., *Biochim. Biophys. Acta* **1999**, *1461*, 135–146.
158 Heimburg, T., Biltonen, R.L., *Biochemistry* **1994**, *33*, 9477–9488.
159 Schneider, M.F., Marsh, D., Jahn. W., Kloesgen, B., Heimburg, T., *Proc. Natl. Acad. Sci. USA* **1999**, *96*, 14312–14317.
160 Gutierrez-Merino, C., Molina, A., Escudero, B., Diez, A., Laynez. J., *Biochemistry* **1989**, *28*, 3398–3406.
161 Sabra, M.C., Jørgensen, K., Mouritsen, O.G., *Biochim. Biophys. Acta* **1996**, *1282*, 85–92.
162 Videira, R.A., Antunes-Madeira, M.C., Madeira, V.M.C., *Biochim. Biophys. Acta* **1999**, *1419*, 151–163.
163 Schmidt, W.F., Hapeman, C.J., Fettinger, J.C.F., Rice, C.P., Bilboulian, S., *J. Agric. Food Chem.* **1997**, *45*, 1023–1026.
164 Reis, O., Zenerino, A., Winter, R., in *Biological Macromolecular Dynamics*, S. Cusak, H. Büttner, M. Ferrand, P. Langan, P. Timmnis (Eds.), Academic Press, London **1997**.
165 Seelig, A., Allegrini, P.R., Seelig, J., *Biochim. Biophys. Acta* **1988**, *939*, 267–276.
166 Reid, D.G., MacLachlan, L.K., Mitchell, R.C., Graham, M.J., Raw, M.J., Smith, P. A., *Biochim. Biophys. Acta* **1990**, *1029*, 24–32.
167 Jones, H.B., Reid, D.G., Luke, J.S., *Toxicol. In Vitro* **1989**, *3*, 299–309.
168 Milhaud, J., Mazerski, J., Bolard, J., Dufoure, E.J., *Eur. Biophys. J.* **1989**, *17*, 151–158.
169 Surewicz, W.K., Leyko, W., *Biochim. Biophys. Acta* **1981**, *643*, 387–397.

4
Drug–Membrane Interaction and Pharmacokinetics of Drugs

Joachim K. Seydel

4.1
Drug Transport

Drug action may be divided into two steps:
1) Transport, usually passive, of drugs to the site of action. The concentration of drugs at the specific receptor depends on both the dose and the pharmacokinetic properties of the drug molecules.
2) A specific interaction between the drug molecule and a receptor.

This pharmacokinetic phase/pharmacodynamic phase model of dose–response can be described physiologically by considering each component of the phase as being separated by lipid membrane barriers that divide essentially aqueous environments.

Normally, drugs reach their target organ via the blood. Therefore, the drug molecules first have to enter the circulation, which requires the passage through barrier membranes in the gastrointestinal tract. This process is called resorption or absorption.

Generally, the following mechanisms can be involved in the transport of drugs across biological membranes:
- simple diffusion;
- supported diffusion;
- active transport;
- transport by pinocytosis;
- transport through pores.

The speed of absorption depends on both the properties of the membrane and the structure or physicochemical properties of the drug. As most drugs are administered by the oral route, the discussion in this context is restricted to the membranes of the intestinal system. The gastrointestinal absorption of drugs is the most significant factor in their bioavailability. The barrier the drug has to overcome in the stomach and intestinal tract is a monolayer epithelium. Therefore, the most important factor influencing oral drug absorption is the permeability of the monolayer of intestinal epithelial cells lining the gastrointestinal tract. As active transepithelial resorption is the exception rather than the rule, and as no watery pores exist in the intact mucosa, drugs, if they are to cross the epithelium of the intestine by passive diffusion, including paracellular and transcellular permeation, must have certain lipophilic properties. On the other hand, for the drug to partition into the watery fluid of the stomach or intestine requires a certain degree of solubility in water. It is therefore not surprising and very well documented that the speed of resorption or the percentage absorbed in general follows a non-linear dependency with lipophilicity, normally being expressed as the octanol–water partition coefficient.

To elucidate the role of hydrophobic bonding, a detailed study on the kinetics of intestinal absorption has been performed on sulfonamides. It was concluded that transport across the microvillus membrane occurs via “kinks” in the membrane “which are pictured as mobile structural defects representing mobile free volumes in the hydrocarbon phase of the membrane and whose diffusion coefficient is fairly fast ($\sim 10^{-5}$ cm^2/s)” [1]. The thermal motion of the hydrocarbon chain leads to the formation of “kinks.” It was also postulated that a transient association of the drug molecules with proteins on the surface of the microvillus membrane is involved in the formation of the activated complex in the absorption process [1].

In vivo drug absorption is influenced not only by the hydrophilic/lipophilic properties, charge, and volume, but also by the metabolic stability of the drug. Resorption from the mucosa of the intestine is favored compared with the resorption from the stomach. The reason for this is the unique construction of the luminal cell membrane, which results in an enlarged mucosa surface of the intestine.

Examples of structure–resorption correlations using octanol–water partition coefficients as descriptors have been published and are discussed here using several examples. These correlations with log $P_{oct.}$ are generally useful. This is especially true as long as neutral drugs are involved, but values often are imprecise in case of charged drugs.

It is commonly assumed that transfer processes can be modeled by bulk phase thermodynamics and that surface or interfacial effects are negligible. These assumptions may, in the case of partitioning into amphiphilic structures formed by micelles or bilayer membranes, not always be appropriate. These interfacial ‘solvents’ have a large surface to volume ratio, similar to interfacial solvents used in reversed-phase liquid chromatography. The partitioning into such phases is the basis of the chromatographic separation.

The characteristics of membrane permeation are partition, including affinity, location, specific interaction with certain phospholipids, and diffusion kinetics. Because of the complex events involved during drug absorption *in vivo*, true membrane permeability modeling cannot always be expected. Therefore, many attempts have been made to develop suitable *in vitro* systems to study the permeation process and its dependence on membrane composition and drug physicochemical properties.

A planar bilayer membrane of egg phosphatidylcholine–decane, for example, has been used to study the permeation of monocarboxylic acids. A highly significant correlation between their permeability and log $P_{hexadecane}$ was found ($r = 0.996$) [2]. Later, the same authors studied the diffusion of 22 small non-electrolytes through planar lipid bilayer membranes and compared their findings with their partition coefficient in four organic solutes (hexadecane, olive oil, octanol, ether). Permeabilities and partition coefficients covered a range of about six powers of 10. The best correlation again was with hexadecane ($r = 0.995$), followed by olive oil ($r = 0.990$) and ether ($r = 0.918$). For octanol the correlation coefficient was significantly less ($r = 0.841$).

The role of bilayer composition and temperature in drug permeability has been studied using as an example the anthracyclines doxorubicin (**1**), daunorubicin (**2**), and pirarubicin (**3**) (see Chapter 5, p. 273). The kinetics of uptake into large unilamellar vesicles (LUVs) in the presence of DNA encapsulated inside the LUVs as a driving force has been investigated using fluorescence techniques [3]. DNA was incorporated as a model receptor to mimic drug binding inside the liposomes. The permeability coefficient of the neutral form of the drugs, PM^{o}, decreased as a function of increasing amounts of negatively charged phospholipid in the bilayer. The explanation given is that "the kinetics of passive diffusion of the drugs depends on the amount of neutral drug embedded in the polar head group region, which decreases as the quantity of negatively charged phospholipids increases" [3]. As expected, the permeability coefficient decreased when the amount of cholesterol in the bilayer increased because the membrane became more rigid. Two steps are involved in the interaction of the anthracyclines with the bilayer because of the pH effect on the ionizable anthracyclines:

1) an electrostatic interaction between the positively charged amino groups of the drug and the negatively charged phospholipid;
2) a hydrophobic interaction with the hydrocarbon chain of the bilayer.

Depending on the pK_a of the drug and the lipophilicity of the neutral form, the separation of these two steps is more or less pronounced. Finally, the neutral molecules embedded in the polar head groups "flip-flop" through the bilayer. The diffusion of drug, therefore, depends on the amount of neutral drug in the polar head group region and the kinetics of translocation. As was to be expected, the authors found – using the neutral drug concentration in the bulk medium for the calculation of $P_{M^{o}}$ – a dependence of $P_{M^{o}}$ on the presence of anionic phospholipids in the bilayer. Despite the degree of homology in the structure of the three studied anthracyclines, the authors found a decrease in the activation energy with increasing amounts of phosphatidic acid only for doxorubicin (E_a = 57 kJ/mol), being in the range of 100 ± 15 kJ/mol otherwise. Another important observation was that for pirarubicin the activation energy increased with increasing cholesterol. Previously, it has been shown that to cross the hydrophilic/hydrophobic interface of a phospholipid membrane a polar solute has to lose its hydration water molecules. This process should determine the activation energy of the transport process. For a given species, activation energies are generally almost identical for permeation across all membranes independent of the membrane composition. Thus, it seems that the loss of hydration water (dehydration) is not the rate-limiting step in the case of doxorubicin and pirarubicin.

The influence of chain packing and permeant size on the permeability of lipid bilayers in gel and liquid crystalline DPPC has systematically been explored [4] by a combination of NMR line-broadening and dynamic light-scattering methods. Seven monocarboxylic acids of varying chain length and degree of chain branching were used. The observed permeability coefficients, P_M, were compared with those predicted by the bulk solubility–diffusion model, P_o, in which the bilayer is presented as a slice of bulk hexadecane. The ratio of the two permeability coefficients was defined as the permeability decrement f ($f = P_M/P_o$) and accounts for the decrease in permeability coefficients due to the chain ordering effect. The chain ordering correction factor was shown "to depend exponentially on the ratio of permeant cross-sectional area to the mean free surface area of the membrane" [4]. Thus, the authors were able to derive a model that combines the effect of bilayer chain packing and permeant size on the permeability of bilayer membranes (Eqs. 4.1 and 4.2).

$$f = f_o \exp\left(-\lambda a_s / a_f\right) \tag{4.1}$$

where a_s is a characteristic cross-sectional area of solute, f_o and λ are constants independent of permeant size and bilayer packing structure and

$$a_f = A - A_o = A_o(1/\sigma - 1) \tag{4.2}$$

Similar to the definition of free volume, the mean free surface area per lipid molecule is related to the area occupied by one phospholipid molecule, A. A_o(= 40.8 Å) is the area of a phospholipid molecule in the crystal and σ (= A_o/A) is the reduced surface density. For the surface density data see ref. 5.

A kinetic model has been derived taking into account both the membrane volume and partition kinetics [6]. Two rate constants are involved, one corresponding to solute transport from the solution to the membrane (k_1) and one from the membrane to the solution (k_2). It is shown that only in the case that k_1 is much smaller than k_2 is equilibrium for the partitioning between membrane and solution attainable. In other cases, it is unattainable until the solute concentrations in both solution phases have reached an identical value. For very high values of the ratio k_1/k_2, the time span reduces with decreasing volume fraction of the donor phase. In contrast, the time required is shortest if the ratio of k_1/k_2 is very small. The values relating to the solution phase are at a maximum when the volume of a donor solution phase is equal to that of the acceptor solution phase. The time for reaching equilibrium for each phase increases, because k_2 decreases concomitantly, with k_1 remaining constant. In a later paper, the same authors extended their kinetic model of membrane transport by including possible asymmetrical membranes [7]. They consider continuous elution of the sink side of the membrane [8] and many other factors such as solid gel *vs.* liquid crystalline state or saturated *vs.* unsaturated phospholipid composition of the membrane.

Problems of the relationships between intestinal drug absorption kinetics and partitioning into the oily phase (tributyrin, tricaprylin) have also been discussed [9]. Different theories have been developed to explain the observed data and especially the non-linear correlations often found. As suggested, the rate constant, k_1, of a drug partitioning into the intestinal wall increases with an increase in the partition coefficient $K_{app.}$ until it becomes limited by extraction into the bloodstream on the opposite side of the membrane [10] or by the growing influence of the unstirred water layer [11, 12]. In the case of highly lipophilic drugs, a distribution phase inside the intestinal wall has been observed and described by a two-compartment model [13, 14]. In this case the "true" absorption rate constant, k_2, is much smaller than the penetration rate constant k_1. The "true" absorption constant appears to have an optimal partition coefficient, the value of which depends on the nature of the lipophilic phase chosen to simulate the lipidic components of the membranes [9]. In order to achieve better correlations and predictions on drug absorption, factors such as blood perfusion and interaction with mucus have to be considered [9]. A review on methodologies for the measurement of passive and transporter-mediated transmembrane solute fluxes has recently been published [15].

Very recently, drug–liposome partitioning has been described as a tool for the prediction of absorption in the human intestinal tract [16]. The pH-metric log P titration method was used to measure the distribution coefficients of 21 structurally unrelated ionizable drugs (Table 4.1). The distribution coefficients into sonicated small unilamellar vesicles (S-SUV) of purified soybean phosphatidylcholine (Epikuron 200) and the determined solubility at pH 6.8 of the drugs were used to describe the absorption of these drugs. The absorbed fraction covered a range from <5% to almost complete absorption. Parallel log $P_{oct.}$ and log $D_{oct.}$ were determined. Again, large differences were found in the interactions of the drug molecules with octanol and phospholipids, especially for the ionized forms. For the neutral form of acids the following relation was observed:

$$\log P_{SUV} = 0.37 \log P_{oct.} + 2.2 \qquad r^2 = 0.89 \tag{4.3}$$

The liposomal system seemed to be less discriminating than octanol. Compounds with low log $P_{oct.}$ showed high partitioning in the liposomal system for the neutral form of the acids. The range covered by the neutral form was from –1.8 to +4.2 for log $P_{oct.}$ and from 1.7 to 4.3 for log $P_{SUV.}$ The correlation between log P_{suv} and log $P_{oct.}$ for the ionized form of the acids was poor:

$$\log P_{SUV} = 0.33 \log P_{oct.} + 2.6 \qquad r^2 = 0.72 \tag{4.4}$$

The partition coefficients for the anionic species were at least two orders of magnitude higher in the liposome system than in octanol.

For the neutral form of 13 bases, again a very poor correlation between partitioning into octanol and liposomes was observed. The range found for the partition coefficients was 2.0 to 6.0 in octanol and +1.0 to ≤ 5.0 in the liposomal system (–2 is the observable minimum according to the authors). Poor correlations were also found for the cationic forms of the bases ($r^2 = 0.49$). Generally, the differences (log $P_{oct.}$ – log P_{SUV}) for the neutral forms are smaller than the differences observed for the charged species.

The authors defined a new "absorption potential", AP_{SUV}, to describe human intestinal absorption data. It is defined in Eq. 4.5 and listed in Table 4.1.

$$AP_{SUV} = \log [\text{distribution at pH}_{6.8} \times (\text{solubility at pH}_{6.8} \times (V/\text{dose})] \tag{4.5}$$

where V is the volume of intestinal fluid.

No correlation between log D_{SUV} and intestinal absorption was found. However, a sigmoidal correlation was seen when AP_{SUV} was used instead. Only acetylsalicylic acid and allopurinol were outliers. According to the authors, this may indicate additional paracellular diffusion through the membrane promoted by the low molecular weight of these compounds. In contrast, no correlation was found between log $D_{oct.}$ or $AP_{oct.}$ and the percentage of intestinal absorption.

Tab. 4.1 Partitioning of ionizable drugs (see pK_a) into small unilamellar vesicles (SUVs) of DOPC and into octanol for uncharged and charged species, determined by the potentiometric titration technique, compared with their intestinal absorption (%) at the doses indicated. (Adapted from Tab. 2 of ref. 16)

No.	Compound	Log $P_{suv.}$ neutral	Log $P_{suv.}$ ion	Log $P_{oct.}$ neutral	Log $P_{oct.}$ ion	pK_a 37°C	Log $D_{suv.}$ pH 6.8	Log $D_{oct.}$ pH 6.8	Solubility pH 6.8 (mg/ml)	Dose (mg)	$AP_{SUV.}$	$AP_{oct.}$	Intestinal absorption (%)
1	Acetylsalicylic acid	2.4	1.6	1.1	−2.0	3.41	1.6	−2.0	0.8	500	1.2	−2.4	90
2	Acyclovir	1.7	2.0	−1.8	−2.0	2.16	1.7	−1.8	0.8	200	1.7	−1.3	20
						9.04							
3	Allopurinol	2.5	2.7	0.2	−0.9	9.00	2.5	0.1	0.4	300	2.0	−0.6	90
4	Amiloride[a]	1.8	1.6	0.1	−0.7	8.35	1.6	−0.6	0.1	5	2.3	0.4	50
5	Atenolol	2.2	1.0	0.5	−1.5	9.25	1.0	−1.3	10.0	50	2.7	0.4	56
6	Diclofenac	4.3	2.9	4.2	0.3	4.01	2.9	1.4	0.8	50	3.5	2.3	99
7	Famotidine	2.3	1.7	−0.8	−0.2	6.56	2.1	−1.0	0.8	40	2.8	−0.5	45
						11.02							
8	Fluoxetine	3.0	2.2	4.5	1.5	9.62	2.1	1.9	2.5	30	3.5	3.3	80
9	Furosemide	3.0	1.9	2.6	−2.0	3.61	1.9	−1.0	0.8	40	2.6	−0.1	65
						10.24							
10	Ibuprofen[a]	3.8	2.1	4.0	−0.1	4.45	2.1	1.6	2.5	200	2.6	1.6	80
11	Miconazole	3.7	2.9	4.9	1.2	6.12	3.6	4.5	0.02	250	1.9	3.1	25
12	Moxonidine[b]	1.8	1.3	0.9	−0.2	7.36	1.5	0.4	0.8	0.3	4.3	3.1	99
13	Nizatidine	3.0	2.8	−0.2	−0.9	2.44	2.9	−0.3	10.0	300	3.8	0.5	99
						6.75							
14	Olanzapine	3.7	2.7	3.0	0.0	5.44	3.0	2.0	0.02	10	2.7	1.7	75
			2.3			7.80							
15	Paromomycin	1.7	1.2	−2.0	−2.0	5.99	−0.6	−2.0	10.0	250	0.4	−1.0	3
						7.05							
						7.57							
						8.23							
						8.90							
16	Propranolol	3.2	2.5	3.4	0.5	9.14	2.5	1.2	7.5	80	3.9	2.9	99
17	Rifabutine	3.4	3.5	4.5	2.7	6.90	3.2	4.3	0.1	150	2.4	2.8	53
			2.7		2.0	9.37							
18	Terbinafine	5.0	3.0	6.0	2.3	7.05	4.6	5.5	0.02	250	2.9	3.5	80
19	Xipamide	3.3	1.7	2.8	−1.7	4.58	1.7	0.5	0.8	20	2.7	1.5	70
						10.47							
20	Zidovudine	1.9	2.4	−0.5	−2.0	9.45	1.9	−0.7	7.5	100	3.2	1.6	90
21	Zopiclone	1.8	1.4	1.5	−0.8	6.76	1.6	1.3	0.1	8	2.1	1.5	80

a) All data at 25 °C.

b) SUV data at 25 °C for DOPC-liposomes.

Very recently, these partition data of Balon *et al.* [16] were analyzed for their structural dependence by Schaper *et al.* [17]. Using multiregression analysis, it was found that the distribution into soybean small unilamellar phospholipid liposomes increased with increasing size/bulk, described by the polarizability, α, and decreased with polarity, characterized by the sum of structural partial charge Σq^{+}. Additionally, an influence of H-bond acceptor strength was found for the neutral and positively charged species. All descriptors were calculated by the program package HYBOT/HYBOT-PLUS 98 [18].

Surprisingly, the regression coefficients obtained with the significant parameters were almost identical for neutral and positively and negatively charged forms of the drug molecules.

Therefore, it was possible to perform a correlation analysis of the combined data set of acids and bases by using the following equation:

$$\log D_{pH} = \log (f_{ui}10^{\log P_{ui}} + f_{(-)}10^{\log P(-)} + f_{(+)}10^{\log P(+)}) \quad (4.6)$$

By fitting a non-linear model equation with identical regression coefficients in the three subequations with identical regression coefficients for each of the bulk descriptors α_{ui}, $\alpha_{(-)}$, and $\alpha_{(+)}$, three new subequations were obtained. Finally, with identical regression coefficients in the three subequations for the descriptors α and Σq^{+}, with the exception of ΣCa_{ui} and $\Sigma Ca_{(+)}$, Eqs. 4.7a–c were derived.

$$\log P_{ui} = 3.356\ (\pm 0.559) + 0.075\ (\pm 0.015)\ \alpha_{ui} - 0.197\ (\pm 0.074)\ \Sigma Ca_{ui} - 1.063\ (\pm 0.320)\ \hbar q^{+}_{ui} \quad (4.7a)$$

$$\log P_{(-)} = 1.232\ (\pm 0.528) + 0.075\ (\pm 0.015)\ \alpha_{(-)} - 1.063\ (\pm 0.320)\ \Sigma q^{+}_{(-)} + 0.942\ (\pm 0.492)\ I\_on_{(-)} \quad (4.7b)$$

$$\log P_{(+)} = 2.755\ (\pm 0.930) + 0.075\ (\pm 0.015)\ \alpha_{(+)} - 0.078\ (\pm 0.076)\ \Sigma Ca_{(+)} - 1.063\ (\pm 0.320)\ \Sigma q^{+}_{(+)} \quad (4.7c)$$

$n = 54 \qquad r = 0.921 \qquad s = 0.360 \qquad Q = 0.893$

The value in brackets are the 95% confidence intervals, and Q is the correlation coefficient of the plot $y_{obs.}$ *vs.* $y_{pred.}$ by the leave-one-out procedure.

At first glance the results seem to contradict the "piston theory" of Avdeef (see Chapter 2). However, closer inspection of the equations reveals large differences in the intercepts, indicating large differences in the "basic lipophilicity" or, as the authors called it, the "liposome-philicity" of the different species. This, then, indicates differences in the depth of partitioning into the bilayer.

The equation derived by Schaper *et al.* [17] is the first general equation with high predictive power describing the partitioning of bases as well as acids and their ionized forms. Also, for the first time it was found that specific N-H acidic drugs show an opposite effect to the partitioning of ionized drugs. The latter generally show decreased partitioning. The authors assume "that anions with delocalized negative charge like anions of pyrimidones/pyrimidinols have the possibility to undergo some yet unknown special interaction with phospholipid molecules" [17] (see the indicator variable, $I\_on_{(-)}$ in Eq. 4.7b).

The permeability of SUVs for amino acids has been followed as a function of surface charge and cholesterol content. Increasing the negative charge by incorporating dicetylphosphate in the DMPC liposomes led to an enhanced permeability of positively charged amino acids (e.g. lysine). Increasing the cholesterol content decreased the permeation of net charged amino acids (e.g. alanine) [19].

The importance of hydrogen bonds for drug–membrane interaction and transport has been stressed recently [20]. The authors analyzed the data of Vaes *et al.* [21] regarding the partitioning of polar and non-polar chemicals into SUVs of L-α-dimyristoylphosphatidylcholine and into octanol. Descriptors used were a bulk descriptor, α, and the H-bond acceptor and donor strength, ΣC_a, ΣC_d. Regression analysis led to Eq. 4.8:

$$\log P_{DMPC} = -0.106 + 0.273\alpha - 1.030\Sigma C_a + 0.169\Sigma C_d \quad (4.8)$$

$$n = 19 \qquad r = 0.975 \qquad s = 0.275 \qquad Q = 0.966$$

The partitioning increased with α and decreased with stronger HB donor and HB acceptor effect ΣC_d (negative values) and $\Sigma C_{a.}$ respectively]. In the corresponding equation describing log $P_{oct.}$, the term describing H-bond donor strength was not significant.

The importance of hydrogen bond contribution to permeability and absorption processes was also studied on a heterogeneous set of chemicals [22] as well as for the transport of a series of passively absorbed drugs in humans [23].

The examples underline the importance of molecular bulk and mainly the hydrogen bond accepting strength for the interaction of chemicals and drugs with phospholipids.

4.1.1 Absorption Models

4.1.1.1 Caco-2 Cells as an Absorption Model

Caco-2 cells, especially the human intestinal epithelial cell line, have been proposed and used for the simulation and prediction of intestinal drug absorption after oral administration. These membranes of cells have useful properties for correlation with in vivo data such as enzymatic and transporter systems [24].

Several independent correlations between drug uptake in cell monolayers and data on intestinal absorption *in vivo* have been established. Such studies have shown that cell line monolayers can be used to predict absorption *in vivo* and to identify drugs that are not well absorbed. Cell line studies have the advantage of needing only small amounts of drugs. However, differences in the absorption properties of different regions in the gastrointestinal tract cannot be considered. In addition, the relatively low capacity of the metabolic and transport system is a limitation. This can lead to an underestimation of carrier-mediated uptake of drugs in the intestine. Another disadvantage is that these studies are labor intensive.

As permeability is a function of lipophilicity, molecular size, H-bonding capacity, and charge, and as charge is included in lipophilicity (log *P*) when the distribution coefficient D instead is used and molecular size and H-bonding are components of lipophilicity, permeability becomes solely a function of molecular size and H-bonding [25–27]. These two properties, besides solubility and dissolution, are assumed to play an important role in oral bioavailability. Using linear combinations of suitable molecular size and hydrogen bonding descriptors, i.e. without knowledge of the distribution coefficient, log D, it was shown that reasonable estimates of permeability can also be achieved for non-congeneric series of compounds. The results demonstrate the importance of hydrogen-bonding descriptors for the description of membrane permeation. The introduction of a new descriptor, Cad (the sum of absolute values of free energy H-bond factors, characterizing the total H-bond ability of a compound) (see also refs. 28 and 29), facilitates the prediction of permeability properties. Cad is calculated from the molecular structure of the solute and is strongly correlated with the H-bond descriptors, Ca+Cd. Figure 4.1 shows the obtained sigmoidal correlation of permeability and total hydrogen-bonding capacity (Cad) for a series of 17 compounds (Table 4.2) [25].

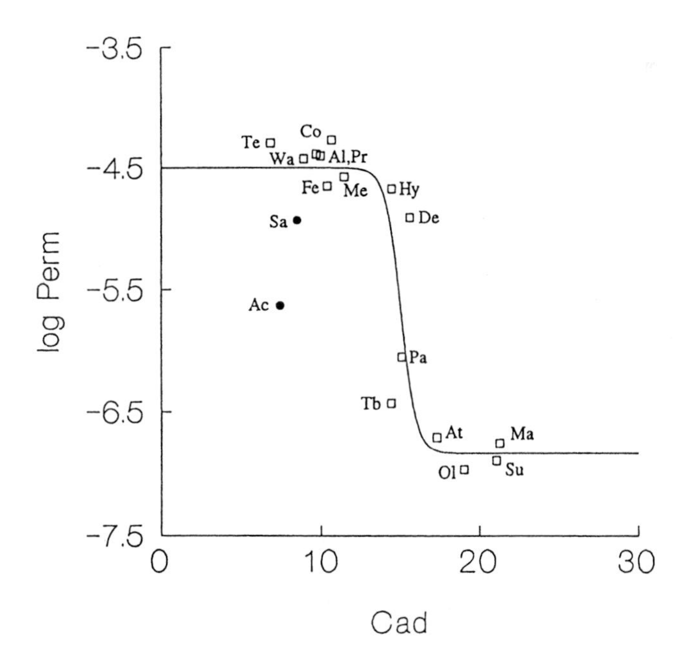

Fig. 4.1 Correlation of permeability and total hydrogen-bonding capacity for a series of structural heterogeneous compounds: corticosterone (Co), testosterone (Te), propranolol (Pr), alprenolol (Al), warfarin (Wa), metoprolol (Me), felodipine (Fe), hydrocortisone (Hy), dexamethasone(De), salicylate (Sa), acetylsalicylate (Ac), practolol (Pa), terbutaline (Tb), atenolol (At), mannitolol (Ma), sulphasalazine (Su), and olsalazine (Ol). (Reprinted from Fig. 4 of ref. 25 with permission from Wiley-VCH.)

Tab. 4.2 Hydrogen bonding descriptors for the selected compounds. (Reprinted from Tab. 4 of ref. 25 with permission from Wiley-VCH)

Symbol	HA	HA_i	HD	HD_i	HT	HT_i	HB	HBOND[a]	Ca[b]	Cd[b]	Cad[b]	SP[c]	Λ[d]
Sa	3	4	2	1	5	5	3	5	3.7	4.8	8.5	37	– 1.9
Ac	4	5	1	0	5	5	4	5	4.3	3.1	7.4	52	– 4.2
Ma	6	6	6	6	12	12	6	12	9.9	11.4	21.3	70	– 8.6
Tb	3	3	4	5	7	8	4	8	6.3	8.1	14.5	43	–
Al	2	2	2	3	4	5	3	5	6.1	3.8	9.9	27	– 6.0
Pr	2	2	2	3	4	5	3	5	5.9	3.8	9.7	27	– 6.4
Pa	4	4	3	4	7	8	5	8	9.3	5.7	15.0	49	– 8.2
At	4	4	4	5	8	9	5	9	9.3	8.0	17.3	55	– 8.8
Me	3	3	2	3	5	6	4	6	7.6	3.8	11.4	36	– 7.3
Te	3	3	1	1	4	4	2	3	4.6	2.2	6.8	29	– 6.5
Ol	8	10	4	2	12	12	8	11	9.4	9.7	19.1	94	–
Wa	5	7	1	0	6	7	4	5	6.7	2.2	8.9	50	– 7.1
Co	5	6	2	1	7	7	4	6	7.6	3.1	10.7	55	– 9.4
Hy	5	6	3	3	8	9	5	8	9.1	5.3	14.4	67	–10.0
Fe	5	5	1	1	6	6	5	6	8.7	1.7	10.4	49	– 7.6
De	5	5	3	3	8	8	6	8	10.3	5.3	15.6	64	–10.3
Su	8	9	3	2	11	11	9	11	14.1	7.0	21.1	106	–

a) Calculated with HBOND
b) Calculated with HYBOT. The sign of Cd has been reversed.
c) Calculated with MOLOC.
d) Calculated from the relationship of 1–octanol-water log P values vs. van der Waals volume V, of different n–alkanes as the distance on the ordinate.
Abbreviations and Symbols: HA, number of hydrogen bond acceptors (lone-pair count) in neutral form (HA_i for state at pH 7.4); HB, total number of atoms capable of forming hydrogen bonds; HD, number of hydrogen bond donors in neutral form (HD_i for state at pH 7.4); HT, total number of potential hydrogen bonds for neutral form (HT_i for state at pH 7.4); Λ, polarity term, mainly hydrogen bonding; SP, polar part of the surface area.

In another study, Caco-2 cell cultures were used for the characterization of the intestinal absorption of antibiotics [30]. In this study, two markers with low lipophilicity and paracellular permeability – fluorescein and [^{3}H]-mannitol – were used to assess the reproducibility of the assay conditions. In parallel, the transepithelial electrical resistance of the cell monolayer was monitored. According to the authors, 21-day-old cultures show a significantly better coefficient of variation than younger cell cultures (15 days) and are therefore more suitable for determining paracellular permeability. The results agreed with those published by other authors comparing *in vivo* absorption and absorption in Caco-2 cells for various drugs [24]. The tested antimicrobials could be grouped according to their apparent permeability coefficient $P_{app.}$. Group 1 drugs had a $P_{app.}$ of $< 0.2 \times 10^{-6}$ cm/s, a very poor absorption, and a bioavailability of < 1%. Group 2 drugs had $P_{app.}$ values between 0.2×10^{-6} and 2×10^{-6} cm/s and bioavailability between 1% and 90%. Group 3 drugs had $P_{app.}$ values $> 2 \times 10^{-6}$ cm/s and a bioavailability over 90% (Table 4.3).

Tab. 4.3 Human oral bioavailability together with mean (± SE) P_{app} values for various antibiotics. (Reprinted from Tab. 2 of ref. 30 with permission from Elsevier Science)

Antibiotic	Route of administration	Human oral bioavailability (%)[a]	P_{app} (cm/s) × 10^{-6b}
Vancomycin	i.v.	< 1	< 0.1
Teicoplanin	i.v./i.m.	< 1	< 0.1
Rifamycin-SV	i.v	< 1	0.17 ± 0.016
Cephaloridin	i.m	< 3	0.18 ± 0.017
Penicillin-G	i.v/i.m	20–30	0.24 ± 0.033
Chlorotetracycline	p.o.	30	0.48 ± 0.045
Erythromycin	p.o.	30–60	0.47 ± 0.181
Ampicillin	p.o.	30–70	1.84 ± 0.743
D-Cycloserine	p.o.	100	3.62 ± 0.105
Novobiocin	p.o.	90	3.98 ± 0.142
Rifampicin	p.o.	98–100	5.79 ± 0.053
Rifapentine	p.o.	100	11.80 ± 1.397
Trimethoprim	p.o.	100	21.12 ± 0.384

a) The range of values is reported when available.
b) Point estimations are not available for P_{app} values $< 0.1 \cdot 10^{-6}$ (cm/s) due to the quantitation limit of the corresponding antibiotic assay.

A theoretical model using Molsurf parameterization and PLS statistics was developed to predict Caco-2 cell permeability [31]. In this study, the compounds investigated were the same as those used by van de Waterbeemd *et al.* [25]. Various PLS models were derived. A good correlation between calculated and observed permeability data in Caco-2 cells [32] was observed. The compounds could be classified into those with poor absorption and those with acceptable permeability. Again, hydrogen bonding has the greatest favorable impact followed by lipophilicity and the presence of "surface electrons", i.e. valence electrons not highly bonded to the molecule. Classifying hydrogen-bond acceptors into oxygen and nitrogen types improved the predictive power for Caco-2 cell permeability. The correlation was sigmoidal [31].

In drug research, particular attention is paid to the study and prediction of the membrane permeability of peptides. The reason for this is the low success rate of peptide drugs in clinical performance because of the poor bioavailability of most pharmacologically active peptides. In addition to their metabolic instability, the transport of peptides across the intestinal epithelium is very limited [33, 34]. It has been suggested that the membrane transport of several compounds depends on the total number of hydrogen bonding sites in the drug molecule. This would indicate a significant role for desolvation energy in drug transport. To investigate this aspect further, the intestinal absorption for a series of radiolabeled peptides was determined in rats by comparing their biliary and urinary excretion. It was found that the absorption of the model peptides was inversely correlated with the number of calculated hydrogen bonding sites. The results generally were similar to those previously obtained *in vitro* using Caco-2-cells. However, it seems that a significant molecular size component is also involved, and it has to be admitted that not all assumed hydrogen bonds are energetically equivalent [35].

Shortly after this study the influence of peptide structures and hydrogen bonding on their transport across Caco-2 cell monolayers was again studied and the strong influence of peptide bond modifications became obvious [36]. The investigators chose peptides with less hydrogen bonding functionality in the side chain (Tables 4.4 and 4.5). A satisfying correlation was found between the permeability coefficients, log $P_{eff.}$, and the number of potential hydrogen bonding sites for peptides I to V as well as other peptides ($r = 0.97$).

Tab. 4.4 Structures of the peptide used to study the influence of bond modification on transport across Caco-2 cells [36]

$CH_3CONR_1CH(CH_2C_6H_5)CONR_2CH(CH_2C_6H_5)CONR_3CH(CH_2C_6H_5)CONHR_4$

Compound	R_1	R_2	R_3	R_4
I	H	H	H	H
II	H	H	CH_3	H
III	H	CH_3	CH_3	H
IV	CH_3	CH_3	CH_3	H
V	CH_3	CH_3	CH_3	CH_3

Tab. 4.5 Caco-2 cell permeability and octanol-buffer partition coefficients (P) for the studied peptides (see Tab. 4.4). (Adapted from Tab. 2 of ref. 36)

Peptide	Permeability coefficient[a]		
	$P_{monolayer}$[b]	P_{eff}	Log P[c]
I	0.67 (0.03)	0.66 (0.03)	2.30 (0.02)
II	2.88 (0.63)	2.78 (0.63)	2.63 (0.01)
III	6.11 (0.52)	5.68 (0.52)	2.53 (0.01)
IV	16.7 (1.7)	13.8 (1.7)	2.92 (0.01)
V	33.9 (2.9)	23.8 (2.8)	3.24 (0.02)

a) $cm/s \cdot 10^{-6}$ mean (± SD).
b) Calculated from P_{eff} using $P_{aq} = 8 \cdot 10^{-5}$ cm/s and $P_{filter} = 1 \cdot 10^{-3}$ cm/s.
c) Mean (± SD).

Several years later, this problem was again tackled using the calculated dynamic molecular surface properties of 19 oligopeptides divided into three homologous subgroups. Both the polar and non-polar parts of surface area were considered, and a strong relationship with the transepithelial permeability of the studied peptides was found, hydrogen bonding and hydrophobicity being the most important properties [37].

The effect of size and charge on the passive diffusion of peptides across Caco-2 cell monolayers via paracellular pathways as well as on retention time has been studied using the IAM technique [38]. It was found that a positive net charge of the hydrophilic peptides favors permeation across the intestinal mucosa via the paracellular pathway. With an increase in size, the molecular sieving of the epithelial membrane dominates the transport of peptides and the influence of the net charge becomes less important. The permeability coefficient, $P_{app.}$, of the capped amino acids and the model peptides can be calculated by:

$$P_{app.} = (\Delta Q/\Delta t)/A \times c(0) \tag{4.9}$$

where $\Delta Q/\Delta t$ is the linear appearance rate of mass in the receiver solution and A is the cross-sectional area (i.e. = 4.72 cm^2). The molecular radii of the peptides were calculated from their diffusion coefficients at 2 °C using the NMR technique. $c(0)$ is the intestinal solution concentration in the donor compartment at $t = 0$. The results are summarized in Tables 4.6 and 4.7.

Tab. 4.6 Capped amino acids, tripeptides, and hexapeptides and their physicochemical properties. (Adapted from Tab. 2 of ref. 38)

Compound	Radius[a] (Å)	MW	Lipophilicity[b] (Log K'_{IAM})	Net charge
Ac-Asp-NH_2	2.3	174	−1.48	Negative
Ac-Gly-Asp-Ala-NH_2	2.5	302	−1.62	Negative
Ac-Trp-Ala-Gly-Gly-Asp-Ala-NH_2	3.3	616	−0.82	Negative
Ac-Lys-NH_2	1.8	187	−0.92	Positive
Ac-Gly-Lys-Ala-NH_2	2.8	315	−0.97	Positive
Ac-Trp-Ala-Gly-Gly-Lys-Ala-NH_2	3.0	629	−0.11	Positive
Ac-Asn-NH_2	1.7	173	−1.20	Neutral
Ac-Gly-Asn-Ala-NH_2	2.6	301	−1.24	Neutral
Ac-Trp-Ala-Gly-Gly-Asn-Ala-NH_2	3.1	615	−0.37	Neutral

a) Stokes-Einstein radius calculated from the diffusion coefficient in D_2O.
b) Capacity factor determined from partitioning of the solute between 0.02 M phosphate buffer, pH 7.4, and an immobilized artificial membrane of PC analogs.

Tab. 4.7 Capped amino acids, tripeptides, and hexapeptides and their transport characteristics across Caco-2 cell monolayers. (Adapted from Tab. 3 of ref. 38)

	Apparent Permeability Coefficient · 10^{-8} (cm/s)		
Compound	**X = Asp**	**X = Asn**	**X = Lys**
Ac-X-NH_2	10.04 ± 0.43	ND	17.16 ± 1.58
Ac-Gly-X-Ala-NH_2	7.95 ± 1.03	25.79 ± 4.86	9.86 ± 0.18
Ac-Trp-Ala-Gly-Gly-X-Ala-NH_2	3.19 ± 0.27	5.12 ± 0.31	4.94 ± 0.33

ND = not determined.

Tab. 4.8 PAMPA flux and its relation to % absorption. (Reprinted from Tab. 1 of ref. 39, with permission from the American Chemical Society)

	Compound	*A* %[a]	c1	Log *D*	flux pH 6.5	c2	flux pH 7.4	c3	c4
1	Acetylsalicylic acid	100	h	–2.57	22	m	15	m	m
2	Alprenolol	93	h	1.00	31	h	27	h	h
3	Atenolol	54	m	–2.14	8	l	8	m	m
4	Ceftriaxone	1	l	–1.23	<5	l	<5	l	l
5	Cephalexin	91	h	–1.00	<5	l			l
6	Chloramphenicol	90	h	1.00	57	h	52	h	h
7	Corticosterone	100	h	1.89	59	h	51	h	h
8	Coumarin	100	h	1.39	84	h	66	h	h
9	Dexamethasone	100	h	1.74	39	h	37	h	h
10	Diltiazem	92	h	2.22	40	h	33	h	h
11	Guanabenz	75	h	1.67	21	m	30	h	h
12	Hydrocortisone	89	h	1.53	59	h	52	h	h
13	Imipramine	99	h	2.52	17	m	24	m	m
14	Metoprolol	95	h	0.07	20	m	25	h	h
15	Olsalazine	2	l	–4.5	<5	l	<5	l	l
16	Propranolol	90	h	1.54	17	m	28	h	h
17	Salicylic acid	100	h	–2.14			8	m	m
18	Sulfasalazine	13	l	–0.13	<5	l	<5	l	l
19	Sulpiride	35	m	–1.15	<5	l	15	m	m
20	Terbutaline	73	h	–1.4	10	m	15	m	m
21	Testosterone	98	h	3.31	56	h	45	h	h
22	Theophylline	98	h	–0.02	8	m	12	m	m
23	Tiacrilast	99	h	–1.05	7	m	5	l	m
24	Verapamil	95	h	1.91	65	h	28	h	h
25	Warfarin	93	h	0.12	57	h	16	m	h

a) *A* %, human absorption values; c1, classification human absorption; log *D*, distribution coefficients (pH 7.4 octanol-water); c2, classification permeation at pH 6.5; c3, classification pH 7.4; c4, combined classification pH 7.4 and 6.5 (c2 and c3). Classification scale: l, low; m, medium; h, high. Flux measurements were performed in triplicate. Standard deviations in all cases were less than 5% relative to the flux value.

4.1.1.2 Parallel Artificial Membrane Permeation Assay (PAMPA)

A new physicochemical high-throughput screening system (PC-HTS) using a parallel artificial membrane permeation assay (PAMPA) for the description of passive absorption processes has been developed by Kansy *et al.* [39]. PAMPA is based on a 96-well microtiter plate assay that allows for the measurement of hundreds of compounds per day. The authors used hydrophobic filter material as a support. Bilayers composed of lecithin in an inert organic solvent were fixed on this support, and the permeation of the compounds through this closed organic layer was determined. The obtained flux values through the formed bilayer were compared with documented human absorption data. It was possible to separate and classify the permeants into three groups: those that were well absorbed (absorption 70–100%, corresponding PAMPA flux 30–100%), intermediately well-absorbed compounds (absorption 30–70%; PAMPA flux 5–30%), and compounds with low absorption (absorption 0–30%; PAMPA flux <5%). Using this classification scheme, the *in vivo* absorption could correctly be predicted for 80% of the compounds, taking into account the flux at different pH values (Table 4.8). The results show the great potential of PAMPA in screening large series of compounds and to provide information on lipophilicity, ionization state, and solubility simultaneously without single compound measurements.

4.1.1.3 Surface Plasmon Resonance Biosensor Technique

Very recently, a so-called "surface plasmon resonance" (SPR) biosensor technique has been developed that allows the differentiation and prediction of the degree of

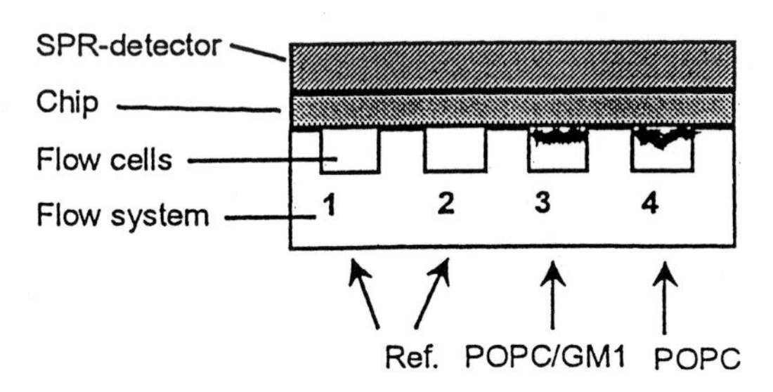

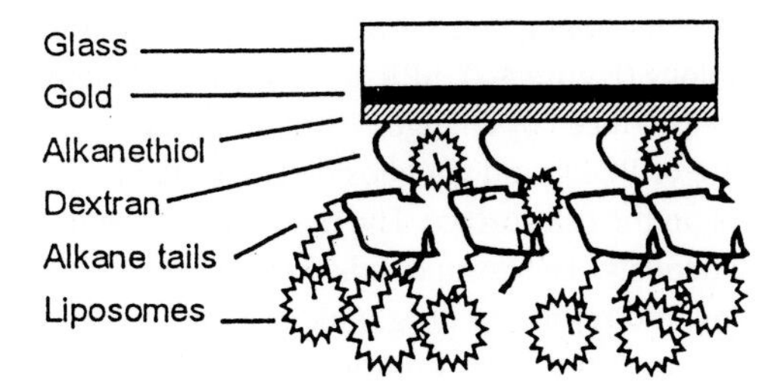

Fig. 4.2 (a) Changes in SPR as a result of changes in mass/refractivity index at the sensor surface indicate the binding of the drug candidate. In a BIACORE system, individually and serially addressable flow cells allow simultaneous analysis from four positions on the surface. The drug is injected over the four flow cells, and the SPR system generates four sensograms. The signal from the reference surface is subtracted from that of the liposome surface to give a differential sensogram. (b) Different liposomes are captured on alkane tails covalently attached to a dextran matrix. (Reprinted from Fig. 1 of ref. 40 with permission from the American Chemical Society.)

Tab. 4.9 Drug classification data and biosensor results from POPC and POPC/GM1 liposomes. (Reprinted from Tab. 1 of ref. 40 with permission from the American Chemical Society)

Drug	Fa (%)[a]	Fa_Cl[b]	Transp.[c]	POPC/GM1[d]	POPC[d]	MW
Desipramine	100	h	t	1199	1009	266
Verapamil	100	h	t	1056	780	455
Propranolol	100	h	t	839	683	259
Alprenolol	96	h	t	480	297	249
Sulfasalazine	12	l	t	348	241	398
Pindolol	92	h	t	314	142	248
Oxprenolol	97	h	t	277	120	265
Naproxen	100	h	t	183	109	230
Metoprolol	95	h	t	179	82	267
Carbamazepine	100	h	t	144	70	236
Terbutaline	73	h	t	117	47	225
Ketoprofen	100	h	t	101	62	254
Hydrochlorothiazide	55	m	t	95	60	298
Furosemide	50	m	t	92	56	331
Sulpiride	36	m	t	78	47	341
Atenolol	54	m	t	38	24	266
Antipyrine	100	h	t	12	19	188
Tranexamic acid	55	m	pt	14	20	157
Amoxicillin	90	h	ct	9	17	365
Foscarnet	17	l	p	62	32	126
Raffinose	0.3	l	p	34	24	504
Lactulose	0.6	l	p	29	22	342
Mannitol	26	l	p	29	22	182
Urea	100	h	p	26	23	60
Creatinine	100	h	p	13	22	113
L-Leucine	100	h	c	11	18	131
D-Glucose	100	h	c	10	18	180

a) F_a (%), fraction absorbed in humans.
b) *Fa*_Cl, Fa classification: h = high, m = medium, l = low.
c) Transp., transport mechanism: t, passive transcellular; p, passive paracellular; c, carrier mediated.
d) POPC/GM1 and POPC: binding level for POPC/GM1 and POPC liposomes.

transcellular (t), paracellular (p) and actively absorbed (c) drug fractions from the human intestine (Table 4.9) [40]. The method uses liposomes attached to a surface sensor. The interaction between drug and liposomes is directly measured by SPR technology (Figure 4.2). SPR is sensitive to changes in refractivity index at the sensor surface produced by changes in mass. Liposomes were prepared either of POPC or a mixture of POPC and ganglioside GM1 (95:5). Ganglioside GM1 was used to mimic sugars on the cell surface. The method can use up to three liposome preparations simultaneously. The compounds were injected and the SPR signals produced by the binding and release of drug molecules were monitored. Steady state was rapidly obtained (ca. 450 s). The liposome preparations are stable for up to 2 weeks. New preparations can be used after washing off the previous liposomes from the surface. Up to 100 substances can be measured within 24 h when testing a single concentration. Among the 27 compounds tested, the majority was absorbed to more than 70%.

4.1.1.4 The Use of IAM Columns

Fast test systems for studying drug–membrane interaction in relation to drug transport become even more important in combination with combinatorial chemistry and high-through put screening methods. IAM columns could be very helpful in this respect. HPLC screening of drug–membrane interaction has been evaluated and reported using as an example β-adrenoceptor antagonists on several IAM columns [41]. The columns were packed with single- (IAM.PC) or double-chain phospholipids (IAM.PC.DD), in which the chain was linked by ester or ether groups to the glycerol-backbone of PC. The silica surface was end-capped with methylglycolate; this is indicated by IAM.PC.MG. The best correlations between retention time and pharmacokinetic parameters were obtained with ester-linked IAM.PC.MG columns (Table

Tab. 4.10 Comparison of regression equations (n, number of data points; r, correlation coefficient; s, standard error) relating pharmacokinetic parameters data (X) for β-adrenoceptor blocking compounds to retention factors (k') obtained using an ^{ester}IAM.PC.MG and IAM.PC.DD column. (Reprinted from Tab. 3 of ref. 41 with permission from Elsevier Science)

Log $K'_{IAM.PC.MG}$				**Log $K'_{IAM.PC.DD}$**		
Log X	n	r^a	s	n	r	s
Log r_τ	7 (9)	0.842 (0.834)	0.305 (0.444)	7	0.810	0.332
Log f_b	8 (13)	0.775 (0.829)	0.377 (0.367)	8	0.744	0.399
Log r_a	6	0.952	0.149	6	0.932	0.176
Log K_{BC}	8 (10)	0.865 (0.896)	0.202 (0.180)	8	0.894	0.181
Log V_{uss}	7 (9)	0.876 (0.858)	0.247 (0.233)	7	0.847	0.274
Log K_p	8 (10)	0.830 (0.870)	0.514 (0.447)	8	0.751	0.609
Log r	8 (9)	0.812 (0.822)	0.588 (0.575)	8	0.672	0.746
Log $B_{\%}$	6 (8)	0.801 (0.800)	0.320 (0.274)	6	0.776	0.337

a) The numbers in the parentheses are from Chapter 3 [5]. The log K' values obtained on the ^{ester}IAM.PC.MG column utilized a mobile phase with 10% (v/v) acetonitrile in 0.1 M sodium phosphate buffer at pH 7.0. The log K' values obtained on the IAM.PC.DD column utilized a mobile phase with 10% (v/v) acetonitrile in 0.01 M PBS at pH 7.0. Log X represents the logarithm of the pharmacokinetic parameters: log r_τ, the ratio of fraction of drug bound and unbound to tissue; log f_b, the fraction of drug bound in plasma; log r_a, the ratio of the fraction of drug bound and unbound to albumin; log K_{BC}, the true red cell partition coefficients; log V_{uss}, the steady-state volume of distribution referenced to the unbound drug in plasma; log K_p, the partition coefficient of drug between plasma protein and plasma water; log r, the ratio of the fraction of drug nonrenally and renally eliminated; log $B_{\%}$, the percentage binding of drug to serum proteins. All pharmacokinetic data from Chapter 3 [6].

Tab. 4.11 Lipophilicity parameters and measured and calculated IAM capacity factors for various compounds. (Reprinted from Tab. 1 of ref. 42 with permission from Elsevier Science)

	Log P	Log k_w	Log k_{cal}
Benzoic acid	1.88	–0.222	0.458
Caffeine	0.07	0.021	–1.028
Furosemide	2.29	1.615	0.795
Griseofulvin	1.95	1.975	0.516
Hydrocortisone	1.86	1.503	0.442
Indomethacin	4.23	2.364	2.388
Isosorbide dinitrate	1.12	–0.146	–0.165
Ketoprofen	2.79	1.341	1.206
Minoxidil	1.23	0.432	–0.075
Naproxen	2.82	0.981	1.230
Salicylic acid	2.27	0.394	0.779
Theophylline	–0.25	–0.101	–1.189

4.10), probably because of the deeper partitioning of the drugs into this stationary phase.

IAM column chromatography has also been used to predict the transdermal transport of drugs [42]. The retention time, log k_w (capacity factor extrapolated to 100% aqueous phase at pH 5.5, IAM column) and log $P_{oct.}$ were compared for the studied drugs (Tables 4.11 and 4.12). The coefficients of permeability through human skin, K_p, were not correlated with either log k_w or log $P_{oct.}$. The authors had, however, pre-

Tab. 4.12 Water solubilities, total transdermal fluxes, permeability coefficients, and differences between experimental and calculated IAM values for various compounds. (Reprinted from Tab. 2 of ref. 42 with permission from Elsevier Science)

	C_w [a]	J [b]	Log K_p [c]	Δ Log k_w
Benzoic acid	0.5	720	–0.40	–0.680
Caffeine	22	1.5	–4.72	1.049
Furosemide	0.54	0.21	–3.97	0.820
Griseofulvin	0.014	0.24	–2.32	1.459
Hydrocortisone	0.28	0.42	–3.38	1.061
Indomethacin	0.016	0.25	–2.36	–0.024
Isosorbide dinitrate	1.09	4.8	–2.91	0.019
Ketoprofen	2.97	12	–2.95	0.135
Minoxidil	1.9	0.81	–3.93	0.507
Naproxen	0.016	4.8	–1.08	–0.249
Salicylic acid	0.002	1.9	–0.58	–0.385
Theophylline	7.5	1.08	–4.40	1.189

a) C_w (water solubility) in units of mg/ml.
b) J (total transdermal flux) in units of mg cm^{-2} h^{-1}.
c) K_p in units of cm/s.

viously derived a correlation between these two parameters for the partitioning of neutral compounds [43].

$$\log k_w = 0.816\ (\pm 0.035)\ \log P_{oct.} - 1.055\ (\pm 0.140) \qquad (4.10)$$

$n = 10 \qquad r = 0.993 \qquad s = 0.11$

This relationship was used to calculate log k_w and the difference from the observed log k_w values was used to describe the skin permeability, K_p (Δlog k_w and log $P_{oct.}$ were not significantly intercorrelated, $r = 0.61$).

$$\log K_p = -2.419\ (\pm 0.276)\ \Delta\log k_w - 2.206\ (\pm 0.174) \qquad (4.11)$$

$n = 10 \qquad r = 0.952 \qquad s = 0.517$

The only exceptions were griseofulvin and hydrocortisone, both neutral compounds but still deviating from the correlation described in Eq. 4.10. These compounds show extra interactions with phospholipids [42].

4.1.1.5 Partitioning into Immobilized Liposomes

Drug partitioning has also been determined on liposomes and biomembranes immobilized in gel beads by freeze–thawing [44]. Drug partitioning was expressed as capacitor factor, K_s, normalized with respect to the amount of immobilized phospholipid. Lipids used were PC, PS, egg phospholipid, lipids extracted from human red cell membrane vesicles, vesicles from cytoskeleton-depleted human red cell membrane vesicles, and ghosts from human red cell membranes. Interestingly, all the different liposomal and biomembrane compositions gave similar results for the partition behavior of the studied β-blockers, phenothiazines, and benzodiazepines. The results obtained from liposomes of pure PC were the only exception. Mixed liposomes seem to better mimic biomembrane properties. For several drugs after oral administration in humans a non-linear relation was obtained between percent absorption and log K_s obtained in vesicles. Almost no absorption was found below K_s values of 0.6 or above 3.0. Maximum absorption occurred between K_s 1.5 and 2.5.

Recently, the partitioning of oligopeptides has been determined in two chromatographic systems [45]. IAM chromatography and immobilized liposome chromatography (ILC) were used. The relationship was analyzed by the partial least-squares technique between these two sets of experimental data and three different sets of calculated descriptors. The results of all three models were statistically significant and showed that negative charge and large hydrogen-bonding capacity prevent partitioning into membranes. In contrast, the hydrophobic properties of oligopeptides support partitioning.

Compounds that show low but intrinsic absorption can be optimized by various galenic techniques and procedures. However, those which possess no absorption ability at all cannot be optimized by such procedures. New strategies have been developed for novel drug delivery systems to control drug release, transport, and absorption across mucosal membranes. A special class of modifiers are amphiphilic compounds. The cationic *N*-trimethyl-chitosan (TMC) has been shown to be a novel type of macromolecular penetration enhancer. It increases the paracellular permeability of the intestinal epithelia and thereby enhances the paracellular uptake of hydrophilic and macromolecular drugs. It has been demonstrated that the degree of quaternization is an important factor in the ability of TMC to open tight junctions in the mucosal membrane. Only TMCs with a high degree of substitution are effective, and they show no toxicity when applied to Caco-2 cells [46].

The correlation between the structures of catamphiphilic drugs and their binding to brush border membranes isolated from rat small intestine has also been reported [47]. It was found that quaternary ammonium compounds such as tetramethylammonium and choline compete for the binding site. The competition effect increased with increasing chain length of the hydrocarbon chain of the quaternary amines. In imipramine-related compounds the ranking in binding affinity was *N*-didesmethylimipramine (primary amine) > desimipramine (secondary) > imipramine (tertiary) > methylimipramine (quaternary).

Tab. 4.13 *n*-Decane-, 1-hexadecene-, and 1,9-decadiene-water partition coefficients (molar concentrations) for various solutes at 25 °C. (Reprinted from Tab. 1 of ref. 49 with permission from Bertelsmann-Springer)

		Partition coefficient		
	Permeant	**Decane-water**	**Hexadecene-water**	**Decadiene-water**
1	Water	$(2.0 \pm 0.2) \times 10^{-5}$	$(5.9 \pm 0.2) \times 10^{-5}$	$(1.2 \pm 0.1) \times 10^{-4}$
2	Formic acid	$(1.8 \pm 0.1) \times 10^{-4}$	$(3.0 \pm 0.6) \times 10^{-4}$	$(6.2 \pm 0.4) \times 10^{-4}$
3	Acetic acid	$(3.1 \pm 0.5) \times 10^{-4}$	$(7.8 \pm 0.2) \times 10^{-4}$	$(1.3 \pm 0.2) \times 10^{-3}$
4	Acetamide	9.3×10^{-6}	$(3.8 \pm 0.5) \times 10^{-5}$	$(1.3 \pm 0.2) \times 10^{-4}$
5	Butyric acid	$(9.5 \pm 0.4) \times 10^{-3}$	$(1.8 \pm 0.1) \times 10^{-2}$	$(3.9 \pm 0.1) \times 10^{-2}$
6	Adenine	$(3.7 \pm 0.1) \times 10^{-7}$	$(3.1 \pm 0.5) \times 10^{-6}$	$(5.8 \pm 1.7) \times 10^{-6}$
7	Benzoic acid	$(4.9 \pm 0.3) \times 10^{-2}$	$(1.5 \pm 0.1) \times 10^{-1}$	$(3.1 \pm 0.1) \times 10^{-1}$
8	*p*-Toluic acid	$(2.0 \pm 0.1) \times 10^{-1}$	$(5.2 \pm 0.1) \times 10^{-1}$	$(9.0 \pm 0.1) \times 10^{-1}$
9	α-Hydroxy-*p*-toluic acid	$(5.0 \pm 0.2) \times 10^{-5}$	$(2.5 \pm 0.1) \times 10^{-4}$	$(7.3 \pm 0.7) \times 10^{-4}$
10	α-Chloro-*p*-toluic acid	$(2.8 \pm 0.3) \times 10^{-1}$	$(4.2 \pm 0.1) \times 10^{-1}$	$(5.3 \pm 2.4) \times 10^{-1}$
11	α-Cyano-*p*-toluic acid	$(3.7 \pm 0.1) \times 10^{-3}$	$(9.6 \pm 1.9) \times 10^{-3}$	$(1.7 \pm 0.1) \times 10^{-2}$
12	α-Methoxy-*p*-toluic acid	$(4.3 \pm 1.2) \times 10^{-2}$	$(6.4 \pm 0.5) \times 10^{-2}$	$(1.1 \pm 0.1) \times 10^{-1}$
13	α-Naphthoic acid	$(6.9 \pm 0.4) \times 10^{-1}$	2.0 ± 0.2	2.2 ± 0.2
14	β-Naphthoic acid	2.0 ± 0.1	4.1 ± 0.1	5.7 ± 0.2
15	α-Carboxy-*p*-toluic acid	$(1.4 \pm 0.2) \times 10^{-5}$	$(4.0 \pm 0.2) \times 10^{-5}$	$(1.2 \pm 0.2) \times 10^{-4}$
16	α-Carbamido-*p*-toluic acid	$(1.8 \pm 0.3) \times 10^{-6}$	$(7.0 \pm 0.6) \times 10^{-6}$	$(4.0 \pm 0.7) \times 10^{-5}$
17	9-Anthroic acid	1.2 ± 0.1	3.0 ± 0.1	6.4 ± 0.5
18	2′,3′-Dideoxyadenosine	$(3.6 \pm 0.2) \times 10^{-6}$	$(5.0 \pm 0.1) \times 10^{-5}$	$(6.4 \pm 1.1) \times 10^{-5}$
19	2′-Deoxyadenosine	$(8.0 \pm 0.4) \times 10^{-8}$	6.0×10^{-7}	$(2.4 \pm 0.9) \times 10^{-6}$
20	Prednisolone	$(1.5 \pm 0.2) \times 10^{-5}$	$(1.6 \pm 0.1) \times 10^{-4}$	$(7.9 \pm 1.0) \times 10^{-4}$
21	Hydrocortisone	$(6.6 \pm 0.1) \times 10^{-5}$	$(5.1 \pm 0.8) \times 10^{-4}$	$(1.75 \pm 0.05) \times 10^{-3}$
22	Hydrocortisone-21-pimelamide	–	$(2.44 \pm 0.01) \times 10^{-4}$	$(1.3 \pm 0.1) \times 10^{-3}$

Tab. 4.14 Permeability coefficients (P_m) for various non-electrolyte solutes across egg lecithin bilayer membranes at 25 °C. (Reprinted from Tab. 2 of ref. 49 with permission from Bertelsmann-Springer)

	Permeant	V (Å^3)^a	P_m (cm/sec)
1	Water	20.6	$(1.9 \pm 0.9) \times 10^{-3}$
2	Formic acid	38.5	$(2.9 \pm 0.1) \times 10^{-3}$
3	Acetic acid	55.5	$(5.0 \pm 0.2) \times 10^{-3}$
4	Acetamide	59.7	$(2.9 \pm 0.3) \times 10^{-4}$
5	Butyric acid	89.5	$(1.0 \pm 0.2) \times 10^{-1}$
6	Adenine	107	$(1.38 \pm 0.02) \times 10^{-5}$
7	Benzoic acid	108	$(5.7 \pm 0.5) \times 10^{-1}$
8	*p*-Toluic acid	125	1.1 ± 0.2
9	α-Hydroxy-*p*-toluic acid	133	$(1.6 \pm 0.4) \times 10^{-3}$
10	α-Chloro-*p*-toluic acid	139	$(6.4 \pm 0.1) \times 10^{-1}$
11	α-Cyano-*p*-toluic acid	144	$(2.7 \pm 0.5) \times 10^{-2}$
12	α-Methoxy-*p*-toluic acid	148	$(3.5 \pm 0.1) \times 10^{-1}$
13	β-Naphthoic acid	149	$(1.7 \pm 0.2) \times 10^{1}$
14	α-Naphthoic acid	149	2.3 ± 0.6
15	α-Carboxy-*p*-toluic acid	152	$(1.8 \pm 0.3) \times 10^{-4}$
16	α-Carbamido-*p*-toluic acid	157	$(4.1 \pm 0.4) \times 10^{-5}$
17	9-Anthroic acid	190	3.2 ± 0.8
18	2′,3′-Dideoxyadenosine	195	$(6.3 \pm 0.1) \times 10^{-5}$
	2′,3′-Dideoxyinosine		$< 10^{-6}$
	2′,3′-Dideoxyguanine		$< 10^{-6}$
19	2′-Deoxyadenosine	203	$(9.4 \pm 0.7) \times 10^{-7}$
20	Prednisolone	309	$(1.5 \pm 0.6) \times 10^{-4}$
21	Hydrocortisone	316	$(5.6 \pm 0.3) \times 10^{-4}$
22	Hydrocortisone-21 -pimelamide	452	$(1.8 \pm 0.5) \times 10^{-4}$

a) Molecular volume calculated by the atomic increment method.

4.1.2 Computational Methods, QSAR

Because of the importance of drug absorption and permeation of membranes, various methodologies have been applied to explain observed differences in permeability in terms of the physicochemical properties of the drug molecules, to compare *in vitro* and *in vivo* data, and, finally, to derive meaningful models to predict *in vivo* data, thus avoiding experimental animal studies and saving money. As discussed above, several kinetic models have been derived to describe the permeation process of small molecules across a bilayer. In these studies, the permeation sites of different permeants with moderately polar or polar properties were determined [48] and described by the partition coefficients obtained in various partitioning systems. Hexadecane–water and olive oil–water partitioning led to a significant correlation with permeability ($r = 0.95$). The correlation with log $P_{oct.}$ and log P_{ether} was poor ($r = 0.75$, $r = 0.74$ respectively). In addition, the molecular volume contributed significantly to the permeability properties. According to the authors, the results are consistent with the solubility–diffusion model, which explains the dependence on both the hydrophilicity and molecular volume of non-electrolytes. It has been concluded that the solubility of the compounds in the membrane is best described when using slightly polar alkanes as models. In a later, similar study on 22 heterogeneous compounds, 1,9-decadiene was selected as the most suitable solvent for simulating the permeability of these non-electrolytes through egg phospholipid bilayers. The partition coefficients of the 22 solutes in three alkanes are given in Table 4.13, and the determined permeability coefficients in Table 4.14. The best description was obtained by applying Eq. 4.12:

$$\log P_m\delta/K_{w\rightarrow hc} = \log D_0 - n\log V \tag{4.12}$$

where P_m is the permeability coefficient, δ is the effective thickness of the barrier domain, $K_{w\rightarrow hc}$ is the bulk solvent–water partition coefficient, D_o is a constant depending on the microviscosity in the barrier domain, and V is the molecular volume of the different permeants (Figure 4.3) [49]. In many examples, log $P_{oct.}$ alone failed to describe gastrointestinal drug absorption. Inclusion of molecular volume or weight and hydrogen bonding led to an improved description of the biological data. A huge database has been set up by C. Hansch *et al.* (Pomona College) including numerous

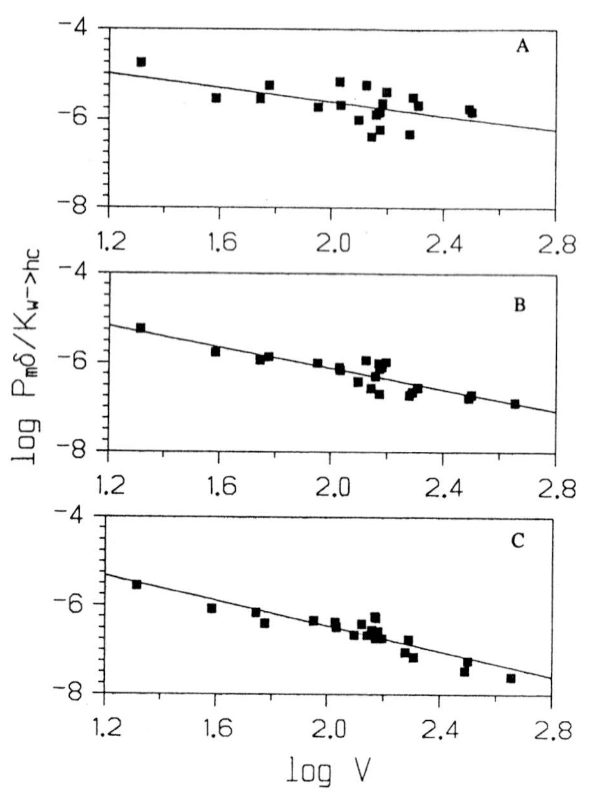

Fig. 4.3 Plots of log $P_M/K_{w\,6\rightarrow hc}$ *vs.* log *P* of permeant volume (A^3/molecule), using (A) *n*-decane, (B) 1-hexadecane, and (C) 1,9-decadiene as reference solvents. (The lines are least-squares fits.) (Reprinted from Fig. 4 of ref. 49 with permission from Bertelsmann-Springer.)

examples dealing with different types of membranes and various classes of drugs and chemicals. The examples document the important contribution of lipophilicity for the absorption process, in addition to volume, shape, and hydrogen bonding.

Examples that underline this statement come from the study on the permeability of 22 small non-electrolytes through lipid bilayer membranes discussed above [48], and a QSAR study has been performed (kindly supplied by C. Hansch, Data bank, Pomona College, USA):

$$\log \text{Per} = 1.068\ (\pm 0.080)\ \log P' - 1.133\ (\pm 0.541)\text{MGVOL} + 1.940\ (\pm 0.467) \quad (4.13)$$

$n = 22$ $r = 0.989$ $s = 0.287$ $Q^2 = 0.969$ (codeine was omitted)

where P' is the partition coefficient in hexadecane–buffer. Compounds, permeability coefficients, and physicochemical properties are listed in Table 4.15.

The permeation of toad urinary bladder by 22 non-electrolytes is another example [50]:

Tab. 4.15 Physicochemical properties and observed and predicted (Eq. 4.13) permeability coefficients for a series of compounds (source: Pomona databank).

		3 LOG PER M	1 YPRED	2 DEV	4 LOG P'	7 MGVOL
1	H_2O	–2.469	–2.922	0.454	–4.377	0.167
2	HF	–3.509	–3.943	0.435	–5.377	0.126
3	NH_3	–0.886	–1.133	0.247	–2.658	0.208
4	HCl	0.462	0.374	0.088	–1.222	0.231
5	HCOOH	–2.137	–2.653	0.516	–3.959	0.324
6	CH_3NH_2	–1.097	–0.868	–0.229	–2.260	0.349
7	$HCONH_2$	–4.000	–3.921	–0.079	–5.102	0.365
8	HNO_3	–3.036	–2.889	–0.147	–4.161	0.341
9	NH_2CONH_2	–5.398	–5.582	0.184	–6.553	0.465
10	HSCN	0.415	0.929	–0.514	–0.495	0.427
11	CH_3COOH	–2.161	–2.084	–0.078	–3.276	0.465
12	$C_2H_5NH_2$	–0.921	–0.629	–0.292	–1.886	0.490
13	$HOCH_2CH_2OH$	–4.056	–3.727	–0.328	–4.770	0.508
14	CH_3CONH_2	–3.770	–3.627	–0.143	–4.678	0.506
15	C_2H_5COOH	–1.456	–1.563	0.107	–2.638	0.606
16	$CH_3CH(OH)CH_2OH$	–3.553	–3.272	–0.281	–4.194	0.649
17	$HOCH_2CH(OH)CH_2OH$	–5.268	–4.946	–0.322	–5.699	0.707
18	C_3H_7COOH	–1.022	–1.105	0.083	–2.060	0.747
19	$HOCH_2CH_2CH_2CH_2OH$	–3.569	–3.616	0.047	–4.367	0.790
20	C_6H_5COOH	–0.260	–0.477	0.217	–1.276	0.932
22	Salicylic acid	–0.114	–0.486	0.373	–1.222	0.990
23	Codeine	–0.854	–2.028	1.174	–1.377	2.206

PER, permeability coefficient; P', partition coefficient in hexadecane-buffer; MGVOL, molecular volume.

$$\log \text{Per} = 1.173\ (\pm 0.310)\ \log P_{\text{olive oil}} - 2.891\ (\pm 1.241)\ \log \text{MW} + 0.192\ (\pm 0.094)\ \text{H}_b + 9.003\ (\pm 0.363) \quad (4.14)$$

$n = 22$ $r = 0.920$ $F_{3,\,18} = 33.3$ $s = 0.519$

where H_b is the number of possible hydrogen bonds.

In a recent interesting study, the effective intestinal permeability ($P_{\text{eff.}}$), which describes both the rate and extent of intestinal drug absorption (human jejunum, *in vivo*), was correlated with experimentally and theoretically derived property parameters [51]. pK_a, $\log P_{\text{oct.}}$, and $\log P_{\text{ion}}$ were determined and $\log D$ values at pH 5.5, 6.5, and 7.4 calculated for the 22 structurally diverse compounds. Finally, theoretical parameters used were number of hydrogen bond donors (HBD) and polar surface area (PSA). PSA was defined as the part of the surface area associated with oxygen, nitrogen, sulfur, and the hydrogens bonded to any of these atoms. The best model obtained after PLS analysis of the 13 passively and transcellularly absorbed compounds included the parameters HBD, PSA, and $\log D_{5.5}$ or $\log D_{6.5}$. Similar, good models for describing human *in vivo* $P_{\text{eff.}}$ values were obtained by using HBD and PSA alone or HBD, PSA, and CLOGP.

$$\log P_{\text{eff.}} = -0.010\text{PSA} + 0.192\log D_{5.5} - 0.239\text{HBD} - 2.883 \quad (4.15)$$

$n = 13$ $r^2 = 0.93$ $Q^2 = 0.90$

$$\log P_{\text{eff.}} = -0.011\text{PSA} - 0.278\text{HBD} - 2.546 \quad (4.16)$$

$n = 13$ $r^2 = 0.85$ $Q^2 = 0.82$

$$\log P_{\text{eff.}} = 0.162\text{CLOGP} - 0.010\text{PSA} - 0.235\text{HBD} - 3.067 \quad (4.17)$$

$n = 13$ $r^2 = 0.88$ $Q^2 = 0.85$

No significant differences were observed when $\log D_{6.5}$ was used instead of $\log D_{5.5}$. Therefore, no conclusions could be drawn about whether pH 5.5 (which corresponds to the pH in the microenvironment in the unstirred layer in neighborhood of the intestinal wall) or pH 6.5 (which corresponds to the pH of the lumen) is better suited for modeling $P_{\text{eff.}}$

In contrast to the sigmoidal relationships derived from permeability studies in Caco-2 cells [25], these models and relations assume that a linear relationship exists. No dependence of permeability on the molecular weight (MW) is observed. According to the authors, the reason for this could be the small range in MW covered in this data set (188–455 Da). The linear relation could result from the fact that for drugs with low permeability paracellular absorption is of little importance *in vivo*, whereas for drugs with high permeability the unstirred water layer is no barrier *in vivo* [52].

To validate the derived models, the authors used those models to predict $\log P_{\text{eff.}}$ for a set of 34 compounds for which the F_a values (fraction of drug absorbed) after oral administration in humans have been published [32, 39, 53]. A good correlation with the absorption data was found when, for simplicity, only HBD and PSA were included. Ten of these compounds were also included in the original data set of 13. First, the molecular descriptors were calculated and it was tested whether they matched the property space defined by the original data set was tested. Six compounds were outside the confidence limits of the PCA model. Log $P_{eff.}$ for 34 compounds was calculated using Eq. 4.15, and the compounds could be classified into drugs with low ($F_a < 20\%$, log $P_{eff.} < -5$), medium ($F_a = 20–80\%$, log $P_{eff.}$ –5 to –4), and high ($F_a = 80\%$, log $P_{eff.} > -4$) absorption. A plot shows that most of the compounds were correctly classified and that there is a sigmoidal correlation between F_a and log $P_{eff.}$. The main outlier was cephalexin, which is considered to undergo active transport (see ref. 39).

A new theoretical descriptor, the dynamic polar molecular surface area, PSA_d, has been calculated for a series of β-receptor blocking agents and used to describe their transepithelial permeability, P_c, in Caco-2 cells *in vitro* [54] (Tables 4.16 and 4.17). PSA_d was calculated from all low-energy conformations identified in molecular mechanics calculations in a vacuum and in simulated chloroform and water environments. The effect of the environment on PSA_d was small, with the exception of a very flexible compound with larger MW and several functional groups capable of forming hydrogen bonds. The large variation in epithelial permeability (0.104×10^{-6} to 242×10^{-6} cm/s) could be convincingly explained using $PSA_{d,\,vacuum}$ as a single descriptor. Again a sigmoidal correlation is found (Figure 4.4). The insert in Figure 4.4 shows the correlation between log P_c and $PSA_{d,\,vacuum}$ ($n = 9$, $r^2 = 0.907$, $Q^2 = 0.81$) if the outlier H216/44 is excluded ($r^2 = 0.968$, $Q^2 = 0.942$). However, correlations with lower statistical significance were obtained when the calculated octanol–water distribution coefficient, log D_{calc}, or log k', the experimentally determined immobilized liposome chromatography retention, were used instead. H216/44 is a large flexible compound containing several functional groups capable of forming hydrogen bonds (note the large difference between $PSA_{d,\,CHCl3}$ and $PSA_{d,\,water}$, Table 4.17).

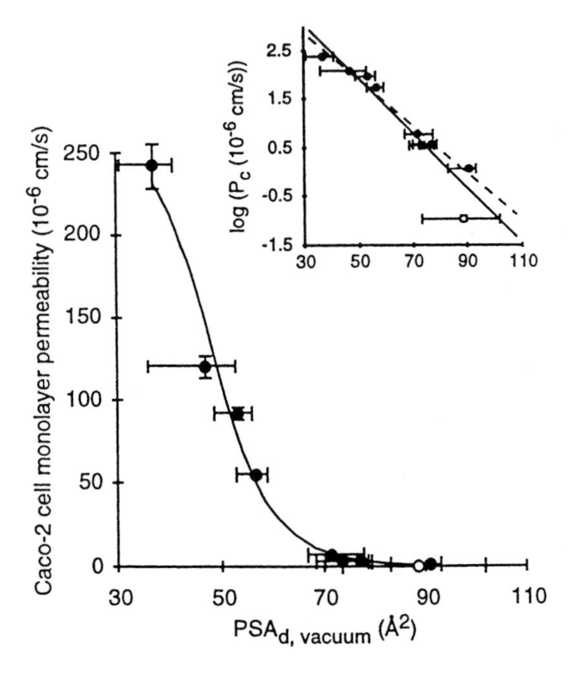

Fig. 4.4 Relationship between $PSA_{d,\,vacuum}$ of nine compounds and their Caco-2 cell monolayer permeability coefficients. The insert shows the correlation between log P_c and $PSA_{d,\,vacuum}$ The dashed line describes the resulting linear correlation (H216/44 is excluded, open circle). Error bars represent standard deviation of P_c and range of $PSA_{d,\,vacuum}$. (Reprinted from Fig. 2 of ref. 54 with permission from the American Chemical Society.)

Tab. 4.16 Structures of β-adrenoreceptor-blocking agents. (Reprinted from Tab. 1 of ref. 54 with permission from the American Chemical Society)

OH H N R_1 O R_4 R_2 R_3

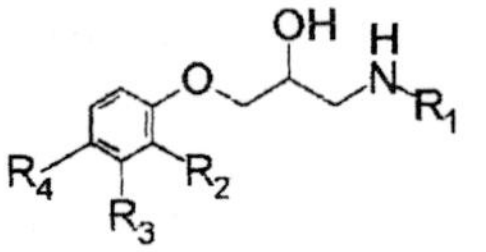

Compound	R_1	R_2	R_3	R_4	pK_a	MW
A H 216/44	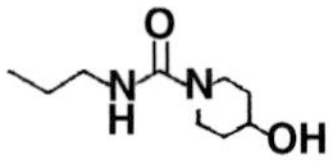 O N H N OH	H	H	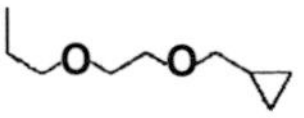 O O	8.5	479
B Atenolol	$CH(CH_3)_2$	H	H	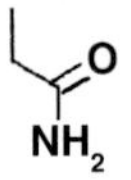 O NH_2	9.6	266
C H 95/71	$CH(CH_3)_2$	H	H	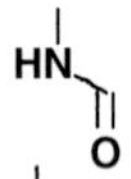 HN O	9.5	252
D Practolol	$CH(CH_3)_2$	H	H	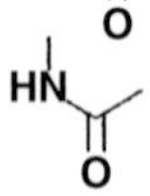 HN O	9.5	266
E H 244/45	$CH(CH_3)_2$	H	H	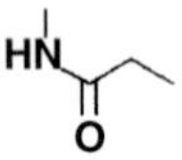 HN O	9.5	280
F Pindolol	$CH(CH_3)_2$	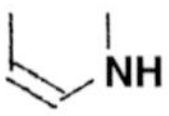 NH		H	9.7	248
G Metoprolol	$CH(CH_3)_2$	H	H	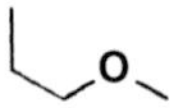 O	9.7	267
H Oxprenolol	$CH(CH_3)_2$	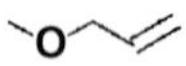 O	H	H	9.5	265
I Alprenolol	$CH(CH_3)_2$	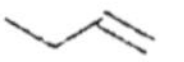	H	H	9.6	249

Tab. 4.17 Structural descriptors, physicochemical properties, and Caco-2 cell monolayer permeability coefficients of the β-adrenoreceptor antagonists (see Tab. 4.16). (Reprinted from Tab. 2 of ref. 54, with permission from the American Chemical Society)

Compound	$PSA_{d,vacuum}$ (Å^2)	$PSA_{d,\ chloroform}$ (Å^2)	$PSA_{d,\ water}$ (Å^2)	$SA_{d,\ vacuum}$ (Å^2)	Log D_{calc}	Log D_m	Log K_s [a] (M^{-1})	P_c [b] (10^{-6} cm/s)
A H216/44	88.7	70.8	116.6	574.8	−0.61	−0.10	1.09	0.104 ± 0.016
B Atenolol	90.9	91.6	92.6	370.4	−2.31	−2.04	0.37	1.02 ± 0.10
C H95/71	77.0	77.7	77.7	347.8	−1.34	−1.56	0.62	3.75 ± 0.34
D Practolol	73.4	74.3	74.1	372.3	−1.35	−1.31	0.72	3.46 ± 0.53
E H244/45	71.6	72.7	72.9	396.9	−0.82	−1.08	0.68	6.03 ± 0.26
F Pindolol	56.5	57.4	57.5	340.7	−0.63	−0.55	1.61	54.7 ± 0.6
G Metoprolol	53.2	53.9	53.9	395.4	−1.11	−0.42	1.15	91.9 ± 4.0 [c]
H Oxprenolol	46.7	49.1	49.6	373.4	−0.41	0.08	1.63	119.6 ± 6.7
I Alprenolol	37.1	38.6	39.1	365.4	0.40	0.85	2.44	242 ± 14 [c]

a) $n = 3$; relative SD ≤ 10%. b) $n = 6-8$. c) From ref. 55.
PSA_d, dynamic polar molecular surface area; K_s, specific capacity factor, for details see ref. 55.

In addition, the results of absorption studies for a series of β-adrenoceptor blockers *in vitro* using Caco-2 monolayers and rat intestinal segments have been used to test new descriptors [55]. For this purpose, the authors calculated dynamic molecular surface properties considering all low-energy conformations. Molecular mechanics were used to consider the flexibility of the molecules, and the van der Waals volume (vdW) and water-accessible surface areas were also calculated (Table 4.18 and 4.19). When the dynamic polar vdW surface areas were used in regression analysis to describe cell permeability data obtained in Caco-2 cells and in rat ileum, excellent correlations were obtained ($r^2 = 0.99$ and 0.92 respectively).

However, these molecular modeling techniques, like molecular dynamics calculations and dynamic polar surface calculations, require expensive computer time. Hydrogen bond formation and molecular volume turned out to be the most important

Tab. 4.18 Dynamic molecular surface area properties[a] of the β-adrenoreceptor blocking agents. (Reprinted from Tab. 4 of ref. 55 with permission from the American Chemical Society)

	Dynamic surface area (Å^2)			
	Polar		Total	
Compound	van der Waals'	Water accessible	van der Waals'	Water accessible
Atenolol	96.0	156.8	345	571
Practolol	77.7	107.0	344	577
Pindolol	54.4	72.5	314	520
Metoprolol	50.6	56.2	364	600
Oxprenolol	43.0	40.4	342	544
Alprenolol	34.9	30.7	337	535

a) The dynamic surface areas were calculated from the surface areas of all low-energy conformations according to a Boltzmann distribution.

Tab. 4.19 Permeability coefficients of the β-adrenoreceptor antagonists in Caco-2 monolayers and excised segments of rat ileum and colon. (Reprinted from Tab. 5 of ref. 55 with permission from the American Chemical Society)

	Caco-2 cell monolayers:[a]	Rat intestinal Segment:[b] P_{app} (cm/s × 10^{-6})	
Compound	P_c (cm/s × 10^{-6})	Ileum	Colon
Atenolol	1.02 ± 0.10	5.0 ± 1.0	1.92 ± 0.20
Practolol	3.27 ± 0.04	5.5 ± 1.3	2.44 ± 0.55
Pindolol	50.9 ± 3.2	24.9 ± 4.5	34.7 ± 9.7
Metoprolol	91.9 ± 4.0	40.5 ± 9.8	96 ± 22
Oxprenolol	129 ± 6.2	50 ± 25	62 ± 17
Alprenolol	242 ± 14	68 ± 11	116 ± 18

a) Cellular permeability (P_c), calculated from apparent permeabilities (P_{app}; $n = 4$) at two different stirring rates.
b) $n = 3$–5.

descriptors of oral absorption data and can easily be determined or calculated. Therefore, these parameters seem to be most suitable for the prediction of absorption and bioavailability.

4.2 Drug Distribution

4.2.1 Distribution into the Brain Compartment

Drug distribution is a decisive factor in drug action, selectivity, and toxicity. Drugs absorbed from the intestinal tract into the blood circulation undergo redistribution into the various tissues and body compartments. The distribution processes into the body fluids and compartments depend again on the physicochemical properties of both the drugs and the membranes and on the equilibrium between free drug and the fraction bound to serum proteins. The protein concentration in the serum, for example, exceeds that in the cerebral extracellular fluid about 2000-fold. Therefore, in general, a good correlation is found between the drug concentration in the cerebrospinal fluid (CSF) and its free concentration in the plasma. This, however, will depend not only on the steady-state plasma concentration but also on the off-rate from plasma protein binding for a particular drug [56]. Similarly, the metabolic rate can depend on the off-rate of the protein–drug complex [57, 58]. Protein binding can also be the reason for a non-linear correlation between the brain penetration and lipophilicity of a drug, the reason being the linear relation between lipophilicity and the degree of protein binding.

The distribution of drugs depends on both the physicochemical properties of the drug molecules and the composition of tissue membranes. These factors can either result in a uniform or uneven distribution of drugs into the various body compartments and fluids. In the extreme, distribution may tend toward an accumulation of drugs in particular tissues or to an almost complete exclusion of the drug from a particular compartment in a defined length of time. One unique compartment that has to be considered in this respect is the brain, which is separated from the capillary system of the blood by the blood–brain barrier, whose membrane has a special structure. It consists of a cerebral capillary network formed by a capillary endothelium that consists of a cell layer with continuous compact intercellular junctions. It has no pores, but special cells, astrocytes, which support the stability of the tissues, are situated at the bases of the endothelial membrane separating the brain and CSF from the blood. The astrocytes form an envelope around the capillaries.

This complex barrier has to be crossed by drug molecules via passive diffusion or active transport in order to reach the brain compartment. This is required for all drugs acting on the central nervous system (CNS). The degree of uptake into the CNS or CSF can be quite different despite the similar mechanism involved in diffusion. This can be explained by the much greater surface of the blood–brain membrane compared to the surface of CSF and/or the existence of specific carrier proteins in the CNS.

The permeability coefficient, P (cm/s), is defined by the following operational equation for non-directed flux from the blood to the brain across the cerebral capillaries. It can be determined by measuring the concentration of a drug in the plasma and brain as a function of time [56, 59]:

$$dC_{brain}/dt = PAfC_{plasma} \tag{4.18}$$

where C_{plasma} is the net plasma concentration at any time t after drug administration, C_{brain} is the brain parenchymal concentration (excluding intravascular material), A is the surface area of the capillary bed (about 180 cm^2/g brain for gray matter), and f is the steady-state fraction of drug unbound and non-ionized in plasma.

It should be kept in mind that the effective delivery of a drug to the brain is a function not only of the blood–brain barrier but also of the relative affinities of the drug for the brain and blood compartment.

A variety of approaches have been developed to describe and predict the delivery of drugs into the brain. The early examples were restricted to small series of small and mostly homogeneous non-electrolytes and showed acceptable correlations with their octanol–water partition coefficients [59].

However, some years later, it was found that the octanol–water partition coefficient alone was generally not sufficient to describe the delivery of drugs into the brain. More and more exceptions were detected. One example is the diffusion of some highly lipophilic vinca alkaloids whose permeability coefficients were observed to be smaller than expected from their high octanol–water partition coefficients [56, 60, 61]. The authors concluded "that the additive rules that govern bulk phase octanol–water partition coefficients [62] break down for large molecules in which distances between hydrophobic and ionic regions within the molecule become significant with respect to the distance of the membrane-water interface and of the thickness of the biological membrane (about 100 Å)". Young *et al.* [63] developed a new physicochemical model that correlated the brain penetration capacity of a series of centrally acting H_2-receptor histamine antagonists with the difference in their partition coefficients in the octanol–water and cyclohexane–water systems ($\Delta\log P$). Whereas $\log P_{oct.}$ ($r = 0.43$) alone could not describe the penetration behavior, $\log BB = \log (C_{brain}/C_{blood})$, significant correlations were obtained using $\log P_{cyclohex}$ ($r = 0.732$) alone and, to an even greater extent, using the difference between $\log P_{oct.}$ and $\log P_{cyclohex}$ (see also Chapter 2).

$$\log BB = -0.485\ (\pm 0.160)\ \Delta\log P + 0.899\ (\pm 0.500) \tag{4.19}$$

$$n = 20 \qquad r = 0.831 \qquad s = 0.439 \qquad F = 40.23$$

Analyzing the same data set, van de Waterbeemd and Kansy [64] derived a significant correlation when using $\log P_{cyclohex.}$ and a calculated volume descriptor V_M (cm^3) ($r = 0.934$, $s = 0.29$):

$$\log BB = 0.338 \log P_{cyclohex.} - 0.00618\ V_M + 1.359 \tag{4.20}$$

Thus, only one experimental $\log P$ value was necessary. The same authors demonstrated that a significant correlation between the penetration coefficient, log BB, and the total H-bonding capacity, $\Lambda_{cyclohex.}$, defined by Eq. 4.21 and the molecular volume exists. The solute parameter, $\Lambda_{cyclohex.}$, can be replaced by the calculated hydrophilic fraction of the van der Waals surface area, SP, so that log BB can be estimated without any experimentally determined descriptor (Eq. 4.23):

$$\Lambda_{cychex.} = -\log P_{cyclohex.} + 0.039\ V_M + 1.098 \tag{4.21}$$

and

$$\log BB = -0.338\ (\pm 0.032)\ \Lambda_{cyclohex.} + 0.007\ (\pm 0.001)\ V_M + 1.730 \tag{4.22}$$

$n = 20 \quad r = 0.934 \quad s = 0.29 \quad F = 58$

$$\log BB = -0.021\ (\pm 0.003)\ SP - 0.003\ (\pm 0.001)\ V_M + 1.643\ (\pm 0.465) \tag{4.23}$$

$n = 20 \quad r = 0.835 \quad s = 0.48 \quad F = 19.5$

The rational behind the usefulness of $\Delta\log P$ in the description of log BB is that it represents a measure of hydrogen-bonding ability. The hydrogen-bonding ability increases with increasing $\Delta\log P$, i.e. the more hydrophilic the compound, the more it will partition into the more watery blood system and not in the more lipophilic brain tissue. It has also been shown that $\Delta\log P$ is mainly a measure of the solute hydrogen bond acidity, whereas the hydrogen bond basicity seemed to play a less important role [65].

For a set of histaminergic H_1-receptor agonists it was found later that using the difference between $\log P_{oct.}$ and $\log P_{cyclohex}$ led to an overestimation of the brain penetration by a factor of 2–5 [66]. According to van de Waterbeemd and Kansy [64], the application of the molar volume, V_M, and polar surface area, SP, led to an even larger overestimation of the brain penetration of H_1-receptor agonists. This is in contrast to the satisfactory performance of these variables in the case of H_2-receptor antagonists.

In the case of a structurally inhomogeneous series of antihistamines, the sedative side-effects were found to be much better described by their octanol–water distribution coefficient at pH 7.4, log *D*, than by $\Delta\log P_{oct.\text{-}alk.}$ or their hydration capacity, Λ_{alkane} [67].

The problem is that predicting log BB outside of the parameter space of small test series is very risky. As pointed out by Abraham *et al.* [26], $\log P_{cyclohex.}$ and also log BB depend on the volume of an alkane. However, according to Eq. 4.20, log BB cannot be less than 1.359 units for an alkane. For alkanes $\Lambda_{cyclohex.}$ is zero [65] and Eq. 4.22 would predict log BB = 1.94 for methane.

This is because the definition of Eq. 4.21 seems not to apply in this case. Thus, for methane. with $\log P_{cyclohex.} = 1.33$ and $V_M = 30.3\ cm^3$. the log BB value calculated by Eq. 4.20 is 1.62 log units but the real value is only 0.04 log units [26]. Therefore, the authors attempted to evaluate systematically those factors influencing the distribution of solutes between blood and brain, starting with their general solvation equation:

$$\log \mathrm{SRP} = c + \mathrm{r}R_2 + s\pi^{\mathrm{H}}{}_2 + a\alpha^{\mathrm{H}}{}_2 + b\beta^{\mathrm{H}}{}_2 + \nu V_{\mathrm{x}} \tag{4.24}$$

where SRP is a solubility-related property for a series of solutes in a given system, and the solute descriptors are R_2, an excess molar refractivity, $\pi^{\mathrm{H}}{}_2$ the solute dipolarity/polarizability, $a\alpha^{\mathrm{H}}{}_2$ and $b\beta^{\mathrm{H}}{}_2$, the hydrogen bond acidity and basicity, respectively, and V_{x}, the characteristic volume of McGowan [68].

For an extended data set including the 20 derivatives of Young *et al.* [63], the authors derived Eq. 4.25 (8 out of 65 compounds were excluded as outliers for various reasons):

$$\log \mathrm{BB} = -0.038 + 0.198R_2 - 0.68s\pi^{\mathrm{H}}{}_2 - 0.75\ a\alpha^{\mathrm{H}}{}_2 - 0.698b\beta^{\mathrm{H}}{}_2 + 0.995V_{\mathrm{x}} \tag{4.25}$$

$n = 57 \qquad r = 0.952 \qquad s = 0.197 \qquad F = 99.2$

Parameters such as dipolarity/polarizability, hydrogen bond acidity, hydrogen and bond basicity favor distribution into blood, and the volume of McGovern, a measure of solute size, and R_2 favor distribution into brain.

In a first step, the two data sets were treated separately, and both yielded highly significant correlations, the regression coefficients remaining almost constant. This was true despite the fact that in one data set V_{x} (cm^3/mol/100) covered a range of 0.085–1.095 and in the second data set a range from 1.138 to 3.178 units. This supports the assumption that Eq. 4.25 can generally be applied to predict the penetration ability of drugs into the brain before synthesizing the drug. This is especially true as these authors described methods that help to estimate the solute descriptors through fragment schemes.

A different approach to estimate the ability of drugs to penetrate the blood–brain barrier was published by Seelig *et al.* [69]. They determined the lipophilicity, $\log k'_{\mathrm{w}}$, of the compounds by polycratic reversed-phase HPLC (C_{18} silica column, varying methanol–buffer ratios) and the surface activity of 24 drugs with or without CNS activity. The surface activity was quantified by their Gibbs adsorption isotherms characterized by three parameters: the onset of surface activity, the critical micelle concentration, and the surface area requirement of the drug at the air–water interface. When plotting the critical micelle concentration (CMC) as a function of the concentration of surface activity onset, C_{o}, at pH 8 for cationic and at pH 6.8 for anionic compounds, three hydrophobic regions were indicated. Accordingly, the drugs can be classified into three groups:

1) Those that are highly lipophilic and bind to other amphiphilic sites such as proteins. These drugs show a low concentration in the endothelial membrane and may also accumulate in the membrane interior. They, therefore, fail to enter the CNS.
2) Those that are less lipophilic and possess smaller cross-sectional areas at the air–water interface. These drugs interact readily with lipid membranes and their high $\mathrm{p}K_{\mathrm{a}}$ values allow a reprotonation at the CNS side of the membrane,
3) Those that are highly hydrophilic compounds and cross the blood–brain barrier only at concentrations that are sufficiently high to shift the equilibrium from the free to the membrane-bound state.

Later, van de Waterbeemd *et al.* [70] demonstrated for a set of 125 CNS-active and CNS-inactive compounds covering a large range of lipophilicity (log $D = -4.80$ to $+6.54$) and molecular size (MW 138–631) that it is possible to discriminate between the two classes of compounds. Log D values for optimal brain uptake should be between 1 and 4. It was, however, not necessary to determine the experimental log D value when MW and H-bonding were combined. The ratio of the principal axes (length/width) of the compounds should be below 5. It was also shown that the calculated polar surface area, SP, of drugs is a suitable descriptor of a drug's H-bonding potential. The authors stated that the constraints on a drug's physicochemical properties are somewhat more restrictive for penetration into the brain than for oral absorption. In this case, the MW should be below 450 and the polar surface area below 90 $Å^2$. Thus, a combination of MW and the total number of potential H-bonds in the drug's ionized state, the major components of lipophilicity and permeability, are useful for estimating CNS activity of new drug candidates.

A computational method for the determination of blood–brain partitioning of organic solutes using free energy calculations has been developed by Lombardo *et al.* [71]. It avoids the risk of breaking down complex molecules into smaller fragments, which may lead to non-additivity and may not always be correct in differing stereochemical and conformational environments. Log P values are composed of polarity and volume terms. From this knowledge the authors concluded that a solvation model that incorporates these factors should be suitable to calculate log P values expressed as solvation free energies. They calculated the free energies of solvation in water ($\Delta G^{o}{}_{w}$) and in hexadecane ($\Delta G^{o}{}_{HD}$) for 57 compounds included in the data sets discussed above [26, 63]. The free energy was computed with the AMSOL 5.0 program using the solvation model AM1-SM2.1 for water and AM1-SM4 for n-hexadecane [72, 73]. Because of the high flexibility of some of the derivatives – prior to computing – the initial conformations of solutes were generated by a Monte Carlo conformational search [74]. All compounds were optimized in the unionized state in the gas phase. The obtained relationship is given in Eq. 4.26:

$$\log \mathrm{BB} = 0.054\ (\pm 0.005)\ \Delta G^{o}{}_{w} + 0.43\ (\pm 0.07) \tag{4.26}$$

$$n = 55 \qquad r = 0.82 \qquad s = 0.41 \qquad F = 108.3$$

Two compounds had to be excluded from the data set. They were outliers (1.3 log units deviation) as also found in the analysis by Abraham *et al.* [26]. It must also be remembered when considering the derived equations that log BB is a complex parameter that encompasses brain partitioning and permeability and may also depend on other processes such as metabolism, active transport, and so forth. Thus, the standard deviations of these determinations may fall within the range of the mean values. The authors could demonstrate that the derived equation could also estimate log BB outside of the training data set. Therefore, the value of the above correlation with the solvation free energy lies in its power to rank compounds for their ability to cross the blood–brain barrier before synthesis. Interestingly, it was also shown by Lombardo *et al.* [71] that the calculated $\Delta G^{o}{}_{w}$ correlated well with the determined permeability coefficient, PC, using endothelial cell monolayers from bovine brain microvessels for 10 drugs taken from the literature [75]. ΔG^{o}_{w} was the only significant parameter in the correlation.

$$PC = 3.33\ (\pm 0.46)\ \Delta G^{o}_{W} + 100.4\ (\pm 8.0) \tag{4.27}$$

$n = 10 \quad r = 0.93 \quad s = 14.4 \quad F = 53.1$

A new method for predicting hydrogen bonding capacity using GRID calculations has recently been published [76]. In this calculation, an amide-NH probe was used to explore the hydrogen bond acceptor regions, a carbonyl probe to detect hydrogen bond donor regions, and a water probe for the detection of both. The surface and volume maps were calculated for three contour levels because of their dependence on the selected level. Using the data set of Young *et al.* [63] on the brain partitioning behavior of 20 compounds the newly developed descriptors were tested for their performance. In all regression equations derived, the best correlations were obtained with the surface maps at –2.0 kcal/mol using the water probe:

$$\log C_{brain}/C_{blood} = -0.00823\ (S\_2\_W) + 1.2689 \tag{4.28}$$

$n = 20 \quad r = 0.85 \quad s = 0.0012 \quad F = 46.9$

Similar to the theoretical model of Palm *et al.* [54], based on dynamic surface properties for the prediction of drug absorption into human intestinal Caco-2 cell lines the authors used their molecular dynamics/GRID approach to correlate the absorption of the same set of six β-adrenoceptor antagonists with the coefficients obtained by the water probe at contour level –2 and –3 kcal/mol.

$$\log P_{c}\ \text{Caco-2 (cm/s)} = -0.0276\ (S\_3\_W) + 0.3096 \tag{4.29}$$

$n = 6 \quad r = 0.93 \quad s = 0.405 \quad F = 24.0$

Very recently, new descriptors have successfully been derived from 3-D molecular fields. These descriptors were correlated with the experimental permeation data by discriminant partial least-squares methods. The training set consisted of 44 compounds. The authors were able to deduce a simple mathematical model that allows external prediction. More than 90% of blood–brain permeation data were correctly predicted [77].

4.2.2
Distribution, Localization, and Orientation of Drugs in Various Tissues and Membranes

As already stated, the distribution pattern of drugs in various tissues depends on both the structure and properties of the drug molecules and the composition of the membranes.

A model-based dependence of human tissue–blood partition coefficients of chemicals on lipophilicity and tissue composition was recently described [78]. For 36 neutral chemicals, the partitioning between seven different tissues and blood in humans was modeled, considering accumulation in the membrane, protein binding, and distribution in the aqueous phase as relevant processes. The tissue–blood partition coefficients were described by a non-linear function of lipophilicity of the chemicals and tissue composition. The equation derived is suitable for the estimation of partition coefficients for physiologically based pharmacokinetic models. It is, however, limited to non-ionizable compounds and is tested in a log P range $-2 < \log P < 5$.

Considering the different composition of phospholipids in membranes, it would be expected that the distribution of drugs into tissues and their localization therein would differ, and that the partition coefficients in membranes, log P_M, would deviate in size and ranking from those determined in bulk octanol–water. Log P_M can be strongly affected by the presence of charged head groups in the phospholipids, especially in the case of amphiphilic drugs.

Several *in vitro* studies have been undertaken in which the partitioning of compounds into membranes with varying lipid composition was determined. For example, the partitioning of the anticancer drug teniposide showed a strong dependence on the type of the phospholipid head group. The partition coefficient into multilamellar vesicles of DOPC was 4290 at 37 °C. It was reduced by the inclusion of additional phospholipids (33 mol%) in the order cardiolipin (Cl) > phosphatidylglycerol (PG) > PS > PE (Table 4.20). There also was a strong dependence on degree of saturation of the lipid chain and on cholesterol content [79].

Tab. 4.20 Role of lipid head group in partitioning of teniposide into phospholipid mixtures. (Reprinted from Tab. 1 of ref. 79, with permission from Elsevier Science)

Lipid	Mole ratio	K_p
DOPC	–	4288 ± 206
DOPC/DOPE	2:1	2275 ± 108
DOPC/DOPS	2:1	2132 ± 274
DOPC/DOPG	2:1	1478 ± 195
DOPC/Cl	2:1	1368 ± 29
POPC	–	2817 ± 243
POPC/POPE	1:1	1165 ± 60

Partitioning was evaluated at 37 °C. The partition coefficient (K_p) is presented as the mean ± SE.

The importance of PS in membranes for the distribution process can be derived from a study by Smejtek *et al.* [80] on aminopyridines. Aminopyridines are known to facilitate synaptic transmission at low calcium concentrations, an effect associated with the block of K^+ channels. For a series of aminopyridines, the zeta-potential (ζ) of PS vesicles in the presence of such aminopyridines was determined. The zeta-potential was determined from the electrophoretic mobility, μ:

$$\mu = \varepsilon\varepsilon_0\zeta/\eta \tag{4.30}$$

where ε is the relative dielectric constant, ε_0 the permittivity, η the viscosity, and ζ the zeta-potential value. This was done to relate the strength of association with the biological effect of the aminopyridines. The association constants of the studied

Tab. 4.21 Association constants of aminopyridines and calcium for phosphatidylserine membranes. (Reprinted from Tab. 1 of ref. 80 with permission from Elsevier Science)

	Membrane-active ion	Assoc. const. (M^{-1})
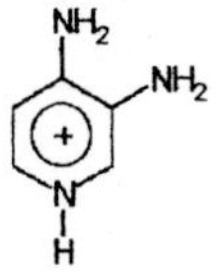	3,4-Diaminopyridine	6.5
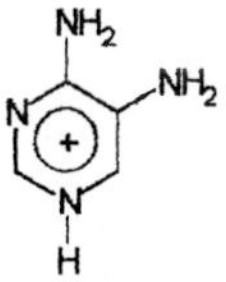	4,5-Diaminopyrimidine[a]	3.8
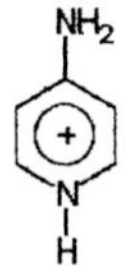	4-Aminopyridine	2.6
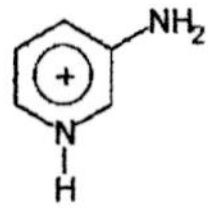	3-Aminopyridine[a]	1.8
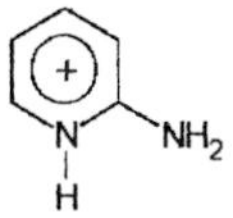	2-Aminopyridine[a]	1.6
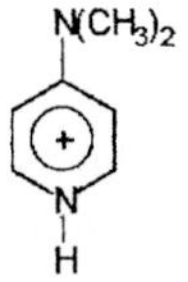	4-Dimethylaminopyridine	0.5
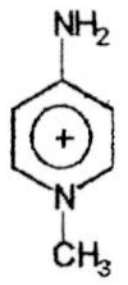	4-Aminopyridine methiodide	0.2
+2	Calcium	12.1

a) Average from measurements at two pH values.

Tab. 4.22 Erythrocyte (E/M) and tissue (T/M) medium ratios of the drugs in human erythrocytes, rat aortas and left atria at pH 7.2, at a medium concentration of 10^{-6} mol/L. (Reprinted from Tab. 3 of ref. 81 with permission from the American Society for Pharmacology and Experimental Therapeutics)

Drugs	E/M	T/M (aortas)	T/M (left atria)
Flunarizine	196 ± 3	682 ± 52	784 ± 155
R 56865	37 ± 2	230 ± 47	402 ± 37
Nitrendipine	18 ± 1	77 ± 2	96 ± 4
Verapamil	6 ± 1	23 ± 2	57 ± 6
Diltiazem	3 ± 0.3	11 ± 0.8	28 ± 2
Lidocaine	1.5 ± 0.5	4.5 ± 1.1	5.0 ± 0.5

The values represent mean ± SD ($n = 6$). For further details, see text.

aminopyridines as well as of Ca^{2+} for PS is given in Table 4.21. It is suggested from CNDO/2 calculations that the binding of aminopyridines to PS membranes increases with the density of excess charge in the protonated aminopyridine ring. The association constant correlates with the biological effect (see Section 5.1).

The superior description of drug–membrane interaction and distribution into tissues by the partition coefficients determined with liposomes compared with those in octanol–buffer is supported by several papers. The partitioning into left atria, AT, aorta and erythrocytes has been determined (Table 4.22) [81, 82] for six different drugs. Regression equations were derived using log $P_{oct.}$ and the IC_{50} for the displacement of Ca^{2+} from PS monolayers by the drug molecules. The authors also calculated membrane concentration and membrane partitioning ratios for the six drugs. High membrane concentrations with respect to the medium concentration were observed [81]. Later, using five compounds of the same series, the change in proton relaxation, $1/T_2$, was used as an independent variable based on binding experiments with phospholipids, measured by NMR spectroscopy [83]. Whereas the degree of Ca^{2+} displacement (IC_{50}) and the change in $1/T_2$ significantly correlated with the tissue distribution, the correlation with log $P_{oct.}$ showed only low predictive power (Q^2).

$$\log AT = -1.189\ (\pm 0.512) \log IC_{50\ Ca^{2+}} + 2.882\ (\pm 0.545) \quad (4.31)$$

$$n = 5 \quad r = 0.97 \quad s = 0.23 \quad F = 54.6 \quad Q^2 = 0.83$$

$$\log AT = 0.861\ (\pm 0.33) \log 1/T_2 + 1.428\ (\pm 0.34) \quad (4.32)$$

$$n = 5 \quad r = 0.98 \quad s = 0.21 \quad F = 68.8 \quad Q^2 = 0.89$$

$$\log AT = 0.594\ (\pm 0.52) \log P_{oct.} - 0.259\ (\pm 1.95) \quad (4.33)$$

$$n = 5 \quad r = 0.85 \quad s = 0.47 \quad F = 10 \quad Q^2 = 0.27$$

Support for these results comes from a paper in which the partitioning of 17 drugs into adipose tissue was examined [84]. For these drugs, log $P_{oct.}$, log $D_{oct.\ at\ pH\ 7.4}$, and the retention time on an immobilized artificial membrane, log k'_{IAM}, were determined. The adipose storage index (ASI) was defined as:

$$ASI = C_{ad.\ max} / D_{ad.\ max} \quad (4.34)$$

where $C_{ad.\ max}$ is the maximum concentration in adipose tissue after a single dose and $D_{ad.\ max}$ is the hypothetical (averaged) concentration of evenly distributed drug at $t = t_{ad.\ max}$. This can be calculated from the mass balance of a kinetic distribution experiment or, if elimination is slow, the value can be approximated to $t = 0$. ASI correlated best with the difference between log $D_{oct.}$ and log k'_{IAM}:

$$ASI = 1.81 \log D_{oct.} - \log k'_{IAM} + 0.40 \quad (4.35)$$

$n = 17 \qquad r^2 = 0.83 \qquad F = 72.3$

Tab. 4.23 Adipose storage index, capacity factors derived from the IAM column, and physicochemical parameters for 17 diverse drugs. (Reprinted from Tab. 1 of ref. 84 with permission from the American Chemical Society)

No.	Name	ASI[a]	Log $D_{7.4}$[b]	Log P[c]	Log K_{IAM}[d]	Log $D_{7.4}$ – log K_{IAM}[e]	pK_a[f]	Nature of drug[g]
1	Haloperidol	0.1	2.27	4.30	1.71	0.56	8.4	Basic
2	Clozapine	0.3	1.04	3.90	1.66	–0.62	7.5	Basic
3	Imipramine	0.3	2.70	4.60	1.44	1.26	9.5	Basic
4	Desipramine	0.3	1.48	4.00	1.44	0.04	10.2	Basic
5	Phenoxybenzamine	0.34	2.07	–	1.72	0.35	10.3	Basic
6	Indomethacin	0.4	1.30	3.10	1.05	0.25	4.5	Acidic
7	Chlorpromazine	0.5	3.39	5.00	2.06	1.33	9.3	Basic
8	Pentobarbital	1.1	1.40	1.40	1.38	0.02	8.0	Acidic
9	Nitrendipine	1.2	0.97	0.97	1.36	–0.39		Basic
10	Ketanserine	1.9	1.92	3.01	1.47	0.45	9.3	Basic
11	Clobazam	2.3	1.90	1.90	0.73	1.17		Neutral
12	Diazepam	4.6	2.80	2.80	0.98	1.82		Neutral
13	Phenytoin	5	2.50	2.50	0.51	1.99	8.3	Acidic
14	Thiopental	5	2.80	2.80	0.21	2.59	7.5	Acidic
15	(1,1-Bis(4-chlorophenyl)-2,2-dichloroethylene)	7.5	5.90	5.90	1.57	4.33		Neutral
16	*N*-Acetyldesiprimine	7.8	3.91	3.90	0.79	3.11		Neutral
17	Amiodarone	8.1	5.66	6.70	1.85	3.89	8.4[h]	Basic

a) Adipose tissue storage index.
b) Log (*n*-octanol-water distribution coefficient at pH 7.4).
c) Log (*n*-octanol-water partition coefficient of neutral form).
d) Log (IAM column capacity factor).
e) Log (*n*-octanol-water distribution coefficient at pH 7.4) – log K_{IAM}.
f) pK_a of compound.
g) Nature of compound.
h) pK_a estimated on the basis of the Hammet equation for tertiary amines ($9.61–3.30\Sigma\sigma^*$, using σ^* for the phenoxyethyl substituent of 0.384).

Both descriptors alone led to correlations with lower statistical significance. The data set included basic, acidic, and neutral drugs (Table 4.23). According to the authors, the difference log $D_{oct.}$ – log k'_{IAM} describes "the equilibrium of the drug between *n*-octanol and phospholipid membrane phase."

Similarly, the partitioning of nitroimidazoles into octanol–water, log $P_{oct.}$, and into four liposome preparations with different lipid composition, log K_M, was determined and regression equations were derived to explain the observed variation in the pharmacokinetic parameters of these drugs [85]. The log K_M ranged from 1.5 to 0.5 and was at least two- to threefold greater than log $P_{oct.}$ (Table 4.24). In addition, the hydrophobic substituent constants of functional groups in various partitioning systems

Tab. 4.24 Partition coefficients into various liposome systems (log K_M) and in *n*-octanol-saline (log $P_{oct.}$) at 30 °C[a] of Nitroimidazoles. (Adapted from Tab. 2 of ref. 85)

	Log K_M for liposome composition[b]				
Nitroimidazole	**I**	**II**	**III**	**IV**	**log $P_{oct.}$**
SR-2555	0.50	0.08	–0.32	–0.23	–1.64
Azomycin riboside	0.86	0.65	0.29	0.25	–1.34
SR-2508	0.66	0.37	0.16	0.14	–1.34
Desmethylmisonidazole	0.69	0.49	0.17	0.21	–0.85
RO-07-0741	0.80	0.60	0.42	0.10	–0.52
Misonidazole	0.92	0.74	0.12	0.14	–0.37
Azomycin	1.13	0.83	0.54	0.40	0.16
Iodoazomycin riboside	1.29	0.92	0.44	0.34	0.32
RO-07-2044	1.47	1.06	0.62	0.50	0.46

a) The maximum relative standard deviation (RSD) was 10 %, although in most cases the RSD was < 4.4 %.
b) I, DMPC; II, DPPC; III, DMPC:CHOL (1:1 mole ratio); IV, DMPC/CHOL/DCP (7:2:1 mole ratio). DCP, dicetylphosphate.

were determined and showed mostly negative and different values (Table 4.25). Correlation of the various log K_M values with pharmacokinetic parameters was optimal for K_M values derived from cholesterol-free liposomes. The best predictive power for the plasma clearance was obtained with $K_{M\ DMPC}$, for the acute LD_{50} with $K_{M\ DPPC}$. Oral bioavailability was better described with all four K_m values determined in liposome preparations than with log $P_{oct.}$

The same authors also reported on the correlation found between the radiosensitizing effects of five nitroimidazoles in either murine EMT-6 or Chinese hamster V79 tumor cell cultures and log $P_{oct.}$ or log K_M. For all five partition coefficients, no significant correlation was observed for EMT-6 cells. However, highly significant correlations were obtained for V79 cells with all four log K_M values, including those in liposome preparations with cholesterol added. No significant correlation was obtained with log $P_{oct.}$ (log $K_{M\ I}$ to log $K_{M\ IV}$: $r = 0.92, 0.95, 0.947, 0.937$) (log $P_{oct.}$: $r = 0.588$).

The example of the anticancer drugs tamoxifen and 4-hydroxytamoxifen shows that merely the absence or presence of certain substituents can change the distribu-

Tab. 4.25 Hydrophobic substituent constants of functional groups of nitroimidazoles in liposome (I–IV) systems and in the *n*-octanol-saline ($P_{oct.}$) (see Tab. 4.24). (Adapted from Tab. 4 of ref. 85)

Substituents (R) at ring nitrogen atom	I	II	III	IV	$P_{oct.}$
H	0	0	0	0	0
$CH_2CHOHCH_2OCH_2CF_3$	0.34	0.23	0.08	0.10	0.30
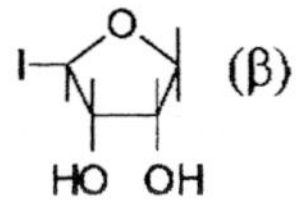 (β) HO OH	0.16	0.09	–0.10	–0.06	0.16
$CH_2CHOHCH_2OCH_3$	–0.21	–0.09	–0.42	–0.26	–0.53
$CH_2CHOHCH_2F$	–0.33	–0.23	–0.12	–0.30	–0.68
$CH_2CHOHCH_2OH$	–0.44	–0.34	–0.37	–0.19	–1.01
$CH_2CONHCH_2CH_2OH$	–0.47	–0.46	–0.38	–0.26	–1.50
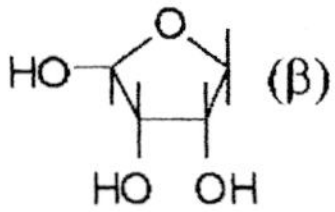 HO– (β) HO OH	–0.17	–0.18	–0.25	–0.15	–1.50
$CH_2CON(CH_2CH_2OH)_2$	–0.63	–0.75	–0.86	–0.63	–1.80

tion behavior [86]. For tamoxifen the partition coefficient for DMPC bilayers and liposomes of lipids extracted from native sarcoplasmic membranes linearly decreased with increasing drug concentration, as observed also for chlorpromazine, amphotericin B, and teniposide [74, 87, 88]. This may be due to saturation of the lipid phase under the experimental conditions (see Section 3.10) For 4-hydroxytamoxifen, however, a linear increase in K_p (partition coefficient for DMPC liposomes) with increasing drug concentration is observed. As the two drugs differ only in the 4-OH group, the authors suggest that the hydroxy group is responsible for the difference in the partition coefficient and its dependence on drug concentration. The polar molecule can form hydrogen bonds with the aqueous phase and therefore K_p will be smaller than for tamoxifen. However, the hydroxy group can positionally orient the amphipathic molecule in the phospholipid bilayer with the polar hydroxy group directed to the aqueous phase and the hydrophobic ring system interacting with the hydrophobic area of the phospholipid. The charge interaction of the hydroxy group with the polar head groups may then perturb the hydrogen bonding at the bilayer surface and destabilize it. This, in turn, weakens the lipid packing and facilitates the incorporation of the drug with increasing drug concentration.

This example underlines the finding that the interaction of drug molecules during transport across membranes involves not only electronic and steric effects, but also the effect of molecular shape and volume on the architecture and organization of membranes, and the energetics of the transfer process. Even if partitioning in the octanol–water system is highly correlated with that into liposomes, the contribution of the enthalpy and entropy to the free energy of partitioning can be very different [89].

It is therefore not surprising that in many cases a suitable preparation of liposomes is a better model for describing drug distribution into specific tissues than log $P_{oct.}$. To select the best liposome preparation remains, however, a problem until we know more about the individual composition of biological membranes in different tissues and their possible interaction modes with certain drug structures.

4.2.3
Distribution *in vivo*

In explaining the variance in pharmacokinetic parameters, log $P_{oct.}$ is not always a sufficient descriptor. The failure to correlate the kinetics of distribution and the storage of barbiturates in adipose tissue with log $P_{oct.}$ can serve as an example. Again, other descriptors have been investigated [90].

Detailed *in vivo* studies on the tissue distribution in rabbits of a set of 10 drugs have been reported together with their partition coefficients, K_p, in various tissues and partition coefficients log P_{app} in four solvent systems: octanol, benzene, chloroform, and triolein [91]. The results are summarized in Table 4.26. The K_p values in the studied tissues, including bones, showed a large variation in ranking and a significant variation from tissue to tissue. It was found that the tissue-to-plasma partition coefficient of the non-ionized form of the drugs, K_{pfu}, correlates best with log $P_{oct.}$ of the non-ionized form.

Tab. 4.26 Tissue-to-plasma partition coefficients (K_p) of basic drugs for various rabbit tissues[a], octanol-buffer partition coefficient, log P_{app}, and unbound fraction of drug in serum, f_p (Adapted from Tab. 2 of ref. 91)

Key	Substrate	K_p[b]									
		Lung	Brain	Heart	Gut	Muscle	Adipose	Skin	Bone	Log $P_{app.}$	f_p
1	Pentazocin	32.1	5.1	6.4	4.3	6.4	2.5	5.2	4.5	2.18	0.40
2	Nitrazepam	1.8	2.1	1.4	2.2	1.7	2.3	1.6	1.2	2.21	0.17
3	Haloperidol	53.5	8.2	14.3	10.8	7.2	27.6	6.2	5.4	2.69	0.23
4	Biperiden	131.0	25.7	34.4	22.5	8.5	120.0	9.9	5.2	2.83	0.39
5	Diazepam	8.4	3.2	6.0	6.7	3.5	12.2	1.6	1.0	2.99	0.091
6	Promethazine	151.4	20.0	35.0	32.9	15.4	132.5	13.5	9.5	3.10	0.22
7	Trihexyphenidyl	74.3	21.2	22.8	21.5	13.2	76.4	8.1	7.9	3.17	0.37
8	Chlorpromazine	64.0	9.3	14.0	11.1	5.2	40.9	5.4	4.3	3.28	0.095
9	Clotiazepam	11.0	3.2	2.6	3.6	1.6	5.9	1.4	1.0	3.49	0.03
10	Clomipramine	144.3	10.6	40.8	29.2	6.2	86.2	5.6	5.7	3.58	0.067

a) Results are given as the mean at 16 h after the beginning of the infusion studies when the rabbits were sacrificed for tissue sampling. At least three rabbits were used to determine the values. All SE values were within 10% of the mean.

b) The value of K_p is the ratio of tissue concentration to the arterial blood concentration. Log P_{app} is the octanol-water partition coefficient at pH 7.4 and f_p, the unbound fraction of the drugs in serum.

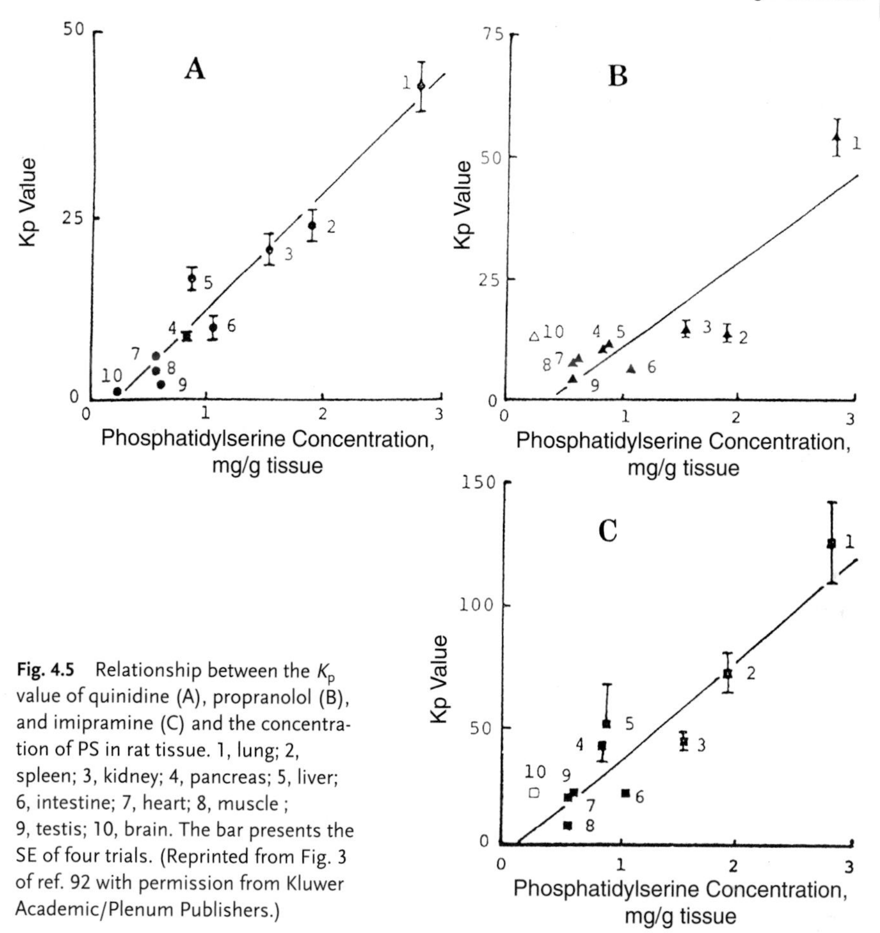

Fig. 4.5 Relationship between the K_p value of quinidine (A), propranolol (B), and imipramine (C) and the concentration of PS in rat tissue. 1, lung; 2, spleen; 3, kidney; 4, pancreas; 5, liver; 6, intestine; 7, heart; 8, muscle ; 9, testis; 10, brain. The bar presents the SE of four trials. (Reprinted from Fig. 3 of ref. 92 with permission from Kluwer Academic/Plenum Publishers.)

In another *in vivo* and *in vitro* study, the importance of the PS content of tissues to the distribution pattern of three weekly basic drugs was investigated in 10 different tissues [92]. The distribution of quinidine, propranolol, and imipramine was determined under steady-state conditions and was expressed as tissue-to-plasma partition coefficients, K_p. The PS concentration, in the various tissues was determined by 2-D thin-layer-chromatography, and a linear correlation was obtained for each drug when plotting K_p against the tissue PS concentration, with the exception of the brain compartment. An example is given in Figure 4.5. It shows the influence of the PS concentration for the distribution pattern of these drugs. The authors also determined the binding constant, *nK*, *in vitro* to five phospholipids including PS. The association constant to PS was the highest and again correlated for single drugs with the K_p values for the distribution of these drugs *in vivo* ($r = 0.93$). (The authors did not use the logarithmic form of K_p!):

$$K_p = 14.3 \times (\log nK) \times (\text{PPS conc.}) - 8.09 \tag{4.36}$$

where n is the number of binding sites and K the association constant.

Tab. 4.27 Binding parameters of propranolol, imipramine and quinidine to individual standard phospholipids[a] (Adapted from Tab. 3 of ref. 92)

	Propranolol			Imipramine			Quinidine		
Affinity	*K*	*n*	*nK*	*K*	*n*	*nK*	K^b	n^c	nK^d
PS			35.1			813			20.2
High	0.238	98.4		5.58	148		0.415	45.0	
Low	0.0111	731		0.0161	936		0.0048	309	
PG			14.9			287			9.28
High	0.571	10.2		1.71	158		0.302	20.2	
Low	0.0073	1240		0.0150	1160		0.0121	263	
PI			14.7			70.8			8.18
High	0.116	118		0.850	82.5		0.368	20.7	
Low	0.0034	300		0.0017	389		0.0015	372	
PC			1.10			1.31			0.13
High	0.657	1.31		0.381	3.14		0.446	0.13	
Low	0.0372	6.68		0.0025	45.7		0.0231	3.00	
PE			0.20						0.02
High	0.0150	12.5		0.211	0.99	0.25	0.0280	0.865	
Low				0.0081	5.15				

a) Binding parameters were determined using the *n*-hexane-pH 4.0 buffer partition system at 25 °C. Each binding parameter was calculated by a non-linear least-squares analysis method.
b) Association constant (μM^{-1}).
c) Number of binding sites (nmol/mg lipid).
d) Binding ability (mL/mg lipid). $nK = n_1K_1 + n_2K_2$.

Tab. 4.28 Comparison of membrane-predicted and literature values of steady-state volumes of distribution (V_{USS}) of β-blockers. (Reprinted from Tab. 6 of ref. 93, with permission from Elsevier Science).

β-Blocker	Predicted (V_{USS}, L)					Literature values [94]
	n-Octanol-buffer system	I	II	III	IV	
Propranolol	864	1606	941	567	1271	1950
Alprenolol	708	226	514	483	636	316
Metoprolol	156	226	316	540	202	240
Pindolol	133	295	326	109	186	200
Nadolol	–	123	120	–	156	186
Acebutolol	226	100	124	150	141	126
Atenolol	103	185	74	122	82	79
Average, % error	59	33	39	78	23	

I, DMPC liposomes; II, DPPC liposomes; III, DMPC/CHOL (1:1 mole ratio) liposomes; IV, DMPC/CHOL/DCP (7:2:1 mole ratio) liposomes. DCP, dicetylphosphate.

This relation was used to estimate the K_p values of desipramine for different tissues based on the *in vitro* association constant of desipramine, which was determined as 155 mL/mg lipid. The results obtained were satisfying.

Liposomes as a possible distribution model in QSAR studies were proposed by Betageri and Rogers [93] and used to describe pharmacokinetic data of seven β-blockers [94]. The partitioning of the β-blockers into octanol and four liposome preparations with different phospholipid and cholesterol composition was determined and compared in regression analyses with various pharmacokinetic parameters. An example is given in Table 4.28 for the steady-state volume of distribution. The best prediction was obtained when liposomes consisting of DMPC/CHOL/DPC (7:2:1 mole ratio) were used (23% average error), the worst with log $P_{oct.}$ (average error 59%).

A further example of the different distribution behavior of drugs in octanol and tissue is the partitioning into sarcoplasmic reticulum of 1,4-dihydropyridines and an additional series of drugs, including antiarrhythmics [95] (Tables 4.29 and 4.30).

Tab. 4.29 1,4-Dihydropyridine partition coefficients into biological membranes and octanol-buffer. (Reprinted from Tab. 1 of ref. 95 with permission from the American Chemical Society)

Drug	Biological membranes[a] (sarcoplasmic reticulum)	Octanol-buffer
Bay P 8857	125 000	40
Iodipine	26 000	
Amlodipine	19 000	30
Nisoldipine	13 000	40
Bay K 8644	11 000	290
Nimodipine	6 300	730
Nifedipine	3 000	

a) Similar values were obtained with cardiac sarcolemmal lipid extracts, indicating a primary interaction of the drug with the membrane bilayer component of these biological membranes.

Tab. 4.30 Drug partition coefficients into biological membranes and octanol-buffer. (Reprinted from Tab. 2 of ref. 95 with permission from the American Chemical Society)

Drug	Biological membranes (sarcoplasmic reticulum)	Octanol-buffer
Amiodarone	921 000	350
Beta X-61	12 500	120
Beta X-67	3 200	250
Propranolol	1 200	18
Beta X-57	350	3
Cimetidine	300	1
Timolol	16	0.7

Two facts immediately become obvious:

1) The partitioning into lipids made of biological membranes can by far exceed that into octanol–water.
2) The ranking can be different.

The larger partition coefficient into lipids is because only the neutral form can efficiently partition into octanol. Ordered phospholipid bilayers can, however, take up both forms: neutral and charged. The influence of anionic and cationic charge on the ability of amphiphilic drugs to partition and bind to DPPC liposomes was investigat-

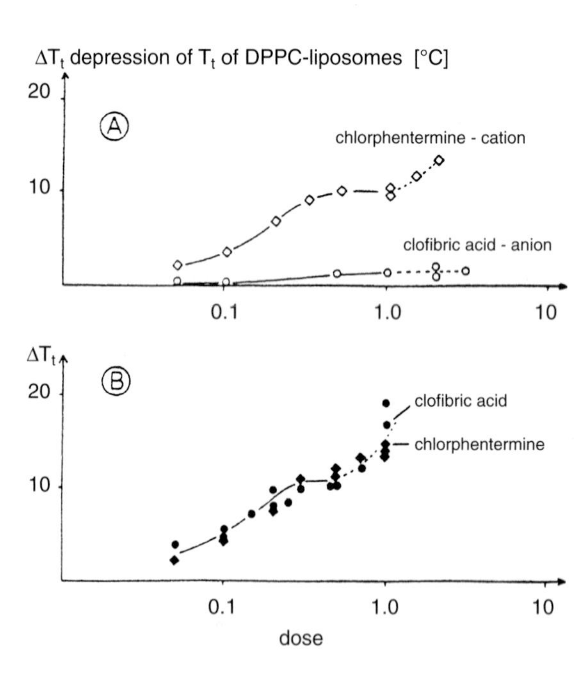

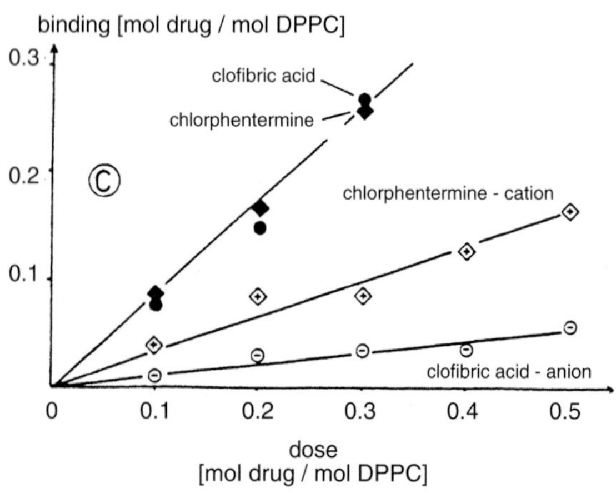

Fig. 4.6 (A and B) Dependence of the drug-induced reduction of the transition temperature (ΔT_t) on the amount of chlorphentermine and clofibric acid added. The dissociation equilibrium of the drugs was shifted to the indicated forms by adjusting the pH of the liposome suspension to pH 6 and over pH 9 respectively (A). Dose-effect curves of the uncharged forms (pH » 10 and pH 4.5, respectively. Dotted line indicates drug amounts at which morphologic alterations occurred in the liposome suspension (B). (Reprinted from Fig. 2 of ref. 96 with permission from Elsevier Science). (C) Liposomal binding of chlorphentermine and clofibric acid depending on the total amount of drug added to the liposome suspension (mean values of triplicate determinations). Maximum deviation from the mean was 10% of the mean. (Reprinted from Fig. 3 of ref. 96 with permission from Elsevier Science)

ed [96] using clofibric acid and chlorphentermine as examples. These compounds possess identical aromatic ring structures. When charged, however, their side chain bears a negative or positive charge respectively. Drug binding to DPPC was studied spectrophotometrically, and the drug effect on the phase transition temperature, T_t, was analyzed by DSC. Chlorphentermine had a considerably stronger effect on both binding and depression of T_t than did clofibric acid. It was also shown that clofibric acid had a lower intrinsic activity to reduce T_t. However, when the ionization equilibrium was shifted toward the uncharged form, both drugs had the same effect on T_t and showed the same degree of binding (Figure 4.6A–C). The results can be explained by the different capacity of the cationic and anionic forms to interact with the head groups of DPPC and to reach the interior of the bilayer.

Sometimes, the charged form partitions into phospholipids even more favorably than the neutral form. The charged form of amlodipine shows a relatively low partitioning into octanol ($\log P = 1.48$) compared with the uncharged dihydropyridine nimodipine ($\log P = 2.41$). The high partitioning of the partly ionized amlodipine at pH 7.4 into the lipid membrane ($\log P_M = 4.28$) – which exceeds the partition coefficient, $\log P_M$, into sarcoplasmic reticulum of the neutral nimodipine ($\log P_M = 3.70$) about fourfold – can be explained by hydrophobic interaction and additional favorable charge interaction with the negatively charged oxygen of the phospholipid head group [95]. This underlines that drug interactions with both model and biological membranes are complex events and cannot be modeled adequately by isotropic systems such as octanol–water.

X-ray diffraction studies have been performed to examine and compare amlodipine's crystal structure and its location with that of the uncharged nimodipine. The results showed that the dihydropyridine ring of amlodipine is more planar than that of nimodipine, that the torsion angle between the dihydropyridine and aryl rings is greater, and that the protonated amino group of amlodipine is directed away from the dihydropyridine ring structure. The membrane electron density profile structure showed that the time-averaged location of amlodipine, despite its positive charge, was near the hydrocarbon to water interface, as observed for uncharged dihydropyridines. In contrast to uncharged dihydropyridines, however, amlodipine's location was determined by an ionic interaction between its protonated amino group and the negatively charged phospholipid head group region and additional hydrophobic interactions with the acyl chain region near the glycerol backbone, similar to other dihydropyridines (Figure 4.7) [95, 97].

The "non-specific" interaction of amlodipine with the membrane may thus lead to a defined and specific location, orientation, and conformation required for binding to a transmembrane receptor protein. It may help to understand and explain its pharmacodynamic and pharmacokinetic profile, which shows a slow onset and a long duration of activity compared with uncharged drugs of this class [95].

It needs to be emphasized that the results from calcium channel antagonists cannot be generalized with respect to the relative order and degree of partitioning into octanol, liposomes, and native membranes. The partition coefficients, K_M, and $D_{oct.}$ of dopamine antagonists in brain membranes and liposomes formed by the extracted lipids were also found to vary. In the case of these drugs, $D_{oct.}$ was larger

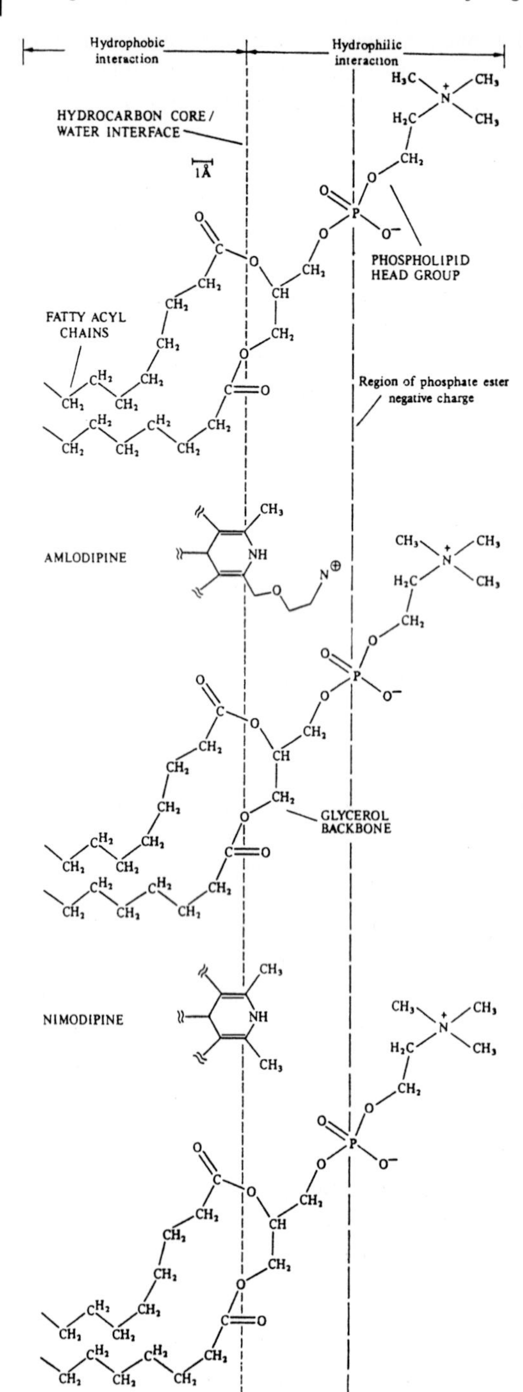

Fig. 4.7 Location of amlodipine within the membrane bilayer derived from its center-of-mass location and crystal structure. Its location near the hydrocarbon core-water interface can facilitate both a hydrophobic interaction with the phospholipid acyl chain and an ionic interaction between the protonated amino function of the drug and the charged anionic oxygen of the phosphate head group. Nimodipine structure and location are consistent with only hydrophobic interactions with the phospholipid acyl chains. No electrostatic interaction with the head groups of PI was noted. (Reprinted from Fig. 2 of ref. 95 with permission from the American Chemical Society.)

than K_M for the native membranes and, except for pimozide, a calcium channel blocker, K_M was larger in liposomes than in the native membrane [98]. In this study, thermodynamic parameters were also determined for the transfer of dopamine antagonists from buffer to membrane or liposomes.

4.3 Uptake into and Distribution within Bacterial Cells

In contrast to mammalian cells, the membrane of bacterial cells is much more complex and, as in the case of *Escherichia coli* or mycobacteria, it is asymmetric (see Section 1.2.2). The reason for this is that these small cellular life forms depend on diffusion of nutrients and metabolites. All substrates going in and out of the cell must diffuse through their cell walls. This might be one reasons why the surface area to volume ratio is important for bacterial cell shapes. This ratio is determined by the structure of their outer cell wall. To cross such a barrier, mainly by passive diffusion, chemotherapeutics must have other properties in addition to those necessary for suitable pharmacokinetics in the host, as in most cases, the target of the chemotherapeutics is within the cytoplasm.

Depending on the physicochemical and structural properties of the drugs, including molecular volume or weight, the following possibilities for the transport across the membrane exist:

- Diffusion through pores. This pathway is limited to small and hydrophilic drugs. The membrane pores have an exclusion limit of ~600 Da.
- Diffusion through the hydrophilic outer core and the hydrophobic inner bilayer in case of the Gram-negative bacteria or a highly lipophilic outer core and a cell wall skeleton of mainly peptidoglycans in the case of mycobacteria.
- Self-promoted uptake, as known to be the case for polycationic drugs like polymyxin, gramicidin and so forth.

An example for the pore pathway is discussed using the example of sulfonamides. For this class of compounds it is generally accepted that the degree of ionization determines the antibacterial activity, the ionized form being more potent then the neutral form. Total ionization, however, leads to decreased activity in whole cells because cell wall permeation becomes the rate-limiting step [99–101]. The possible effect of molecular size on cell wall permeation has been studied on a series of substituted 5-sulfanilamido-1-phenylpyrazoles that show only a small variation in pK_a values but a large difference in molecular weight [102].

The variation in the determined minimum inhibitory concentrations (MICs) against *E. coli* could be explained by the variation in size or molar refractivity of the substituents in *o*-position, MR_o, of the phenyl ring, and the MW.

$$\log 1/\mathrm{MIC} = -0.139\ (\pm 0.020)\ \mathrm{MR_o} - 0.0172\ (\pm 0.0061)\ \mathrm{MW} + 6.01\ (\pm 2.13) \tag{4.37}$$

$$n = 18 \quad r = 0.948 \quad s = 0.21 \quad F = 79 \quad Q^2 = 0.88 \quad (\text{correlation } \mathrm{MR_o/MW},\ r = -0.02)$$

A similar equation is obtained after replacing MW by the surface area.

$$\log 1/\mathrm{MIC} = -0.139\ (\pm 0.032)\ \mathrm{MR_o} - 0.0012\ (\pm 0.00057)\ \text{surface area} + 0.539\ (\pm 0.316) \tag{4.38}$$

$n = 18 \quad r = 0.936 \quad s = 0.251 \quad F = 52.9 \quad Q^2 = 0.81$
(correlation $\mathrm{MR_o}$/surface $r = -0.03$)

The importance of MW in explaining antibacterial activity had not been described before. To obtain support for the assumption that MW accounts for permeation problems for this series of sulfonamides with higher MW and for the observed inactivity of the COOH-substituted derivatives, a selected number of derivatives was tested in a cell-free system for inhibitory activity against *E. coli* pteroate synthase, the target of sulfonamides. This led to an interesting result: MW or surface is no longer a significant descriptor for activity and the COOH-substituted derivative is no longer an exception.

$$\log 1/I_{50} = -0.126\ (\pm 0.019)\ \mathrm{MR_o} - 0.437\ (\pm 0.084) \tag{4.39}$$

$n = 15 \quad r = 0.97 \quad s = 011 \quad F = 208 \quad Q^2 = 0.92$

This strongly supports the assumption of pore diffusion being a limiting factor for sulfonamides with a MW of ≥ 300. The activity, on the other hand, depends only on the steric influence of the *o*-substituents of the 5-sulfanilamido-1-phenylpyrazoles. The regression coefficient with $\mathrm{MR_o}$ in the two derived equations is constant in both systems and accounts for the negative steric effect of the *o*-substituents for the interaction with the active site at the synthase.

Similarly, in the isomeric 3-sulfanilamido-1-phenyl-pyrazole series, MW and calculated surface are no longer significant in the regression equation to describe the cell-free (I_{50}) activities.

$$\log 1/\mathrm{MIC} = 0.560\ (\pm 0.253)\ \sigma - 0.00132\ (\pm 0.00048)\ \text{surface area} \tag{4.40}$$

$n = 19 \quad r = 0.854 \quad s = 0.210 \quad F = 21.64 \quad Q^2 = 0.63$

$$\log 1/\mathrm{I}_{50} = 0.288\ (\pm 0.085)\ \sigma - 0.827\ (\pm 0.039) \tag{4.41}$$

$n = 15 \quad r = 0.896 \quad s = 0.069 \quad F = 52.9 \quad Q^2 = 0.73$

In contrast to the 5-sulfanilamido-1-phenylpyrazoles, σ-Hammett is of importance for the explanation of the variation in activity. It accounts for the variation in pK_a in this series. Steric effects were not involved and molecular modeling showed that the *o*-substituents in the 5-substituted series indeed occupied a different region. This can explain the observed difference in the influence of steric effects within the two series of sulfonamides. The important message is therefore that these compounds reach the cytoplasm by diffusion through the pores of the membrane and that MW or the occupied surface is the limiting factor. The inactivity of the COOH-substituted derivatives is due to their negative charge, which hinders diffusion through the negatively charged surface of the bacteria.

4.3.1 Diffusion Through the Outer Asymmetric Core of *E. coli*

It has been shown that lipophilic solutes permeate very slowly through membranes of Gram-negative bacteria because of the hydrophilic outer leaflet of the bacterial membranes. The dependence of rate of diffusion and the final equilibrium distribution of lipophilic drugs such as fluoroquinolones and tetracyclines which possess multiple protonation sites has been reviewed [103].

In a systematic study, the partition behavior of various lipophilic drugs between lipopolysaccharides of different phenotypes and glycerophospholipids was determined [104]. The inhibitory effect on *E. coli* cultures of a series of lipophilic rifampicin derivatives was studied using growth kinetic techniques [105]. It was observed that the length of the lag phase until onset of inhibition depended not only on the intrinsic activities (Figure 4.8) but also on the drug concentration (Figure 4.9, left). The concentration-dependent onset of inhibition indicates a diffusion-controlled mechanism due to drug–membrane interactions. If the series was studied at an identical drug concentration of 2 μM the onset of inhibition varied as a function of their lipophilicity. The largest delay in onset observed was 4 h, corresponding to about 16 bacterial generation times. The assumption of a rate-limiting step in membrane diffusion was further supported by the results of experiments in which a cell

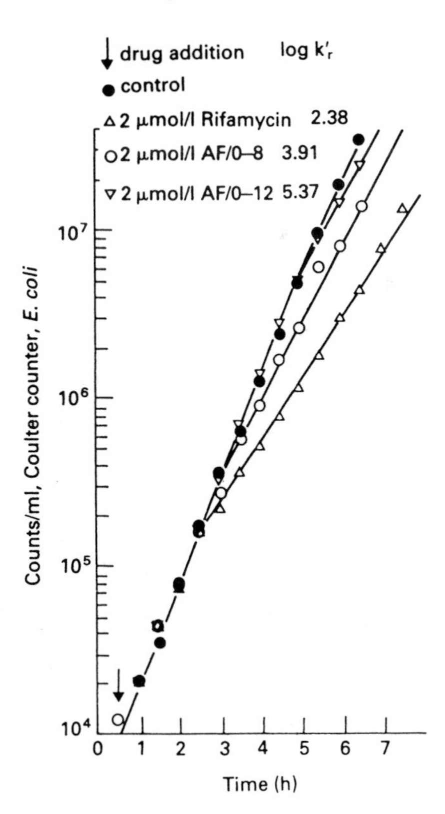

Fig. 4.8 Typical generation rate curves of *E. coli* at 37 °C in the absence (+) and presence of 2 μM of various rifampicin derivatives possessing different lipophilic properties (log k_r). The variation in lag phase (onset of inhibition) is clearly shown. (Reprinted from Fig. 1 of ref. 105.)

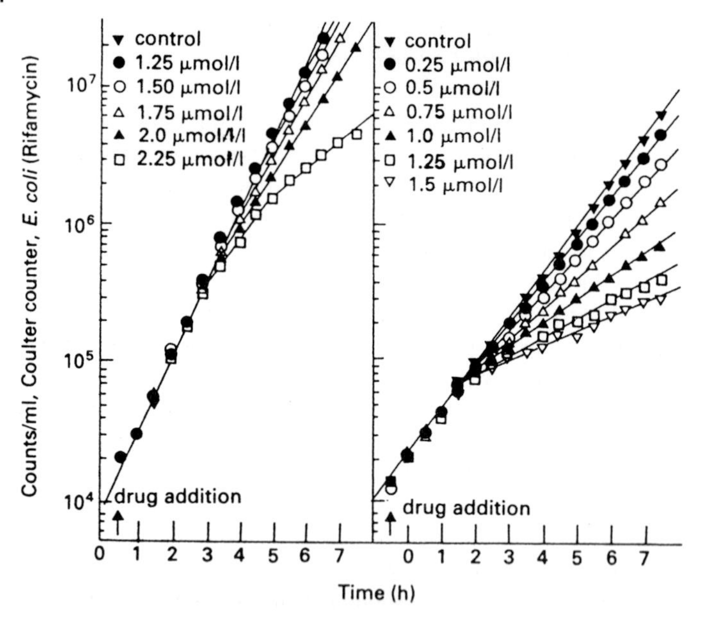

Fig. 4.9 Typical generation rate curves of the cell-wall deficient *E. coli* mutant (right) in comparison with a wild-type *E. coli* strain (left) in the absence and presence of rifamycin at the concentration indicated. Note the concentration-independent onset of inhibition on the right-hand fig. (Reprinted from Fig. 2 of ref. 105.)

wall-deficient mutant was used as a test organism (Figure 4.9, right). These mutants are characterized by structural defects in cell wall lipopolysaccharides (LPS) and phospholipids. The concentration-dependent lag phase disappeared (Figure 4.9). Because of the many variables involved it is not easy to deduce a mathematical model to describe the observed delay in onset of inhibition as a function of structure. Besides, the variation in lipophilicity of the drug molecules, i.e. in the diffusion properties, the intrinsic activity (MIC) varied within the series. As the onset of inhibition varied, the number of bacterial cells increased during the experiment and therefore the cell surface is not constant. To reduce the number of variables in the experimental set-up, the authors determined the concentration, $C_{3\text{h}}$ necessary to achieve an onset of inhibition 3 h after drug addition as the biological activity parameter. They did this for all 14 rifampicin derivatives studied. Using this procedure the number of bacterial cells (cell surface) remained constant. To account for the differences in intrinsic activity, the drug concentration was divided by the corresponding MIC value. The concentration corrected by this procedure, $C_{3\text{h}}$, was used as the dependent variable and correlated with the lipophilicity (log k') of the derivatives.

A highly significant bilinear dependence was obtained with an optimal lipophilicity, $\log k'_{\text{O}} = 3.18$:

$$\log 1/C_{3\text{h}} = 0.697\ (\pm 0.33)\ \log k' - 0.897\ (\pm 0.30)\ \log\ (0.0028\ k' + 1) + 4.37\ (\pm 0.49) \tag{4.42}$$

$n = 14 \qquad r = 0.857 \qquad s = 0.17$

The implication for the therapeutic application of these rifampicins is quite obvious. Highly lipophilic derivatives may fail in therapy despite low *in vitro* MIC values (high intrinsic activity) because the time needed for the diffusion into the bacterial cell might be too long compared with their biological half-life (concentration–time profile in the serum).

In the case of charged amphiphilic drugs, diffusion through Gram-negative bacterial cell walls may result in an even more complex type of drug–membrane interaction leading to partial resistance. This behavior is exhibited by, for example, 2,4-diamino-5-benzylpyrimidines, inhibitors of bacterial dihydrofolate reductase [106]. A new series of 2,4-diamino-5-benzylpyrimidines was synthesized [107, 108] and was found to show increased inhibitory activity against isolated dihydrofolate reductase (DHFR) compared with trimethoprim. The inhibitory activity against whole-cell *E. coli* was, however, disappointingly low. Having a molecular weight of about 600, these derivatives cannot use the pore pathway, in contrast to the smaller and hydrophilic trimethoprim and can reach the intracellular cytoplasm only by diffusion through the highly hydrophilic and negatively charged outer core of *E. coli*. This outer core consists of LPS, which could hinder the passage of the charged and lipophilic benzylpyrimidines.

This phenomenon has already been described [109, 110] for a large but heterogeneous series of chemotherapeutics and Gram-negative mutants and reviewed by Vaara [111].

For the newly synthesized more homogeneous series of 5-benzylpyrimidines, the antibacterial activity was determined systematically against four *E. coli* mutants (F614, F612, F588, F515) with various defined degrees of cell wall defects and against the corresponding isolated enzymes (Table 4.31) [106]. A steady increase in antibacterial activity ($1/I_{50}$) with increasing cell wall defects was observed (F614/F612 > 588 > 515). The tested compounds are catamphiphiles. Under physiological conditions (pH 7.4) one of the N atoms of the pyrimidine moiety is almost 50% ionized (pK_a trimethoprim 7.3) and the substituents in the 4'-position are lipophilic. It could, therefore, be speculated that the major factor in the binding of benzylpyrimidines to cell wall components is the interaction of the positively charged N atom of the pyrimidine ring with the negatively charged phosphate or carboxylate groups of LPS. Figure 4.10 shows the molecular structure of LPS Re of *E. coli* F515 [112]. The number of phosphate groups decreases from the smoother mutants (F614/612) to deep rough mutants (F588/515). The charge interaction will be reinforced by hydrophobic interaction of the substituents in the 4-position of the benzyl moiety and the hydrophobic area of lipid A of LPS. This means that the strength of interaction is determined not only by the overall lipophilicity of the substituents, but also by the area of the lipophilic substituents reaching down to lipid A. Charge interaction can be considered to be constant within the series (no change in pK_a as a function of the substituents in the 4-position of the benzyl moiety). The strength of interaction

Tab. 4.31 Code, structure, I_{50}-values against various *E. coli* mutants, and physicochemical properties of the tested 2,4-diamino-5-benzylpyrimidines. (Reprinted from Tab. 2 of ref. 106)

Compound	Formula	I_{50} F515 (μmol/L)	I_{50} F588 (μmol/L)	I_{50} F612 (μmol/L)	I_{50} F614 (μmol/L)	MW	Log k'	Log P_{ClogP}	Log P_{VG}	Log $k'_{calc.}$
TMP	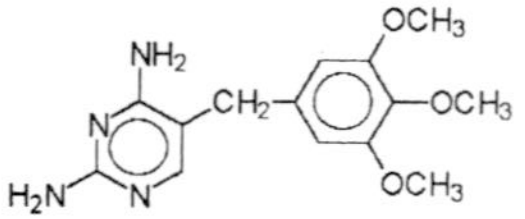 NH_2, N, H_2N, N, CH_2, OCH_3, OCH_3, OCH_3	0.80 ± 0.07	1.79 ± 0.09	1.34 ± 0.23	1.31 ± 0.10	290.32	0.55	0.91	1.05	0.46
K-96	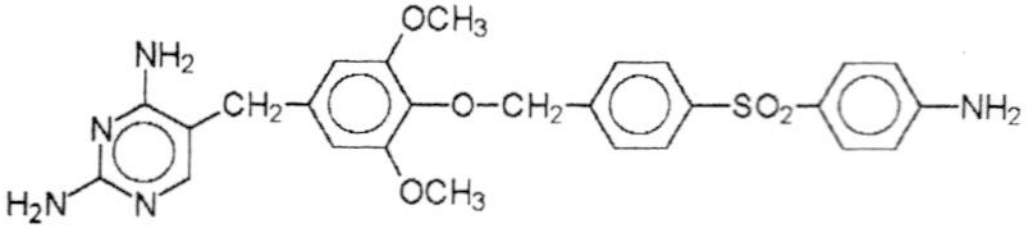 NH_2, N, H_2N, N, CH_2, OCH_3, OCH_3, O–CH_2, SO_2, NH_2	At 10 μmol/L about 20% inhibition	At 10 μmol/L about 20% inhibition	ND	ND	521.61	ND	NC	NC[b]	NC
K-107	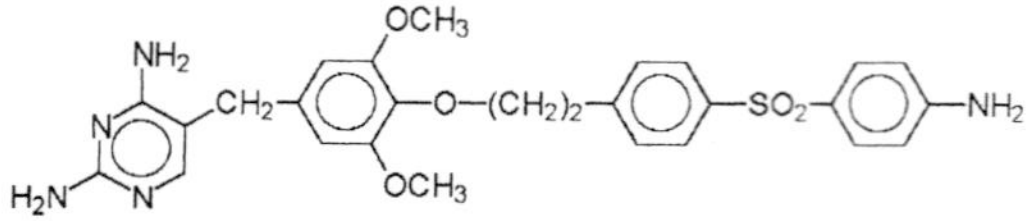 NH_2, N, H_2N, N, CH_2, OCH_3, OCH_3, O–$(CH_2)_2$, SO_2, NH_2	3.60 ±0.15	10.92 ± 1.38	ND	ND	550.65	1.64	1.89	2.53	1.98
K-130	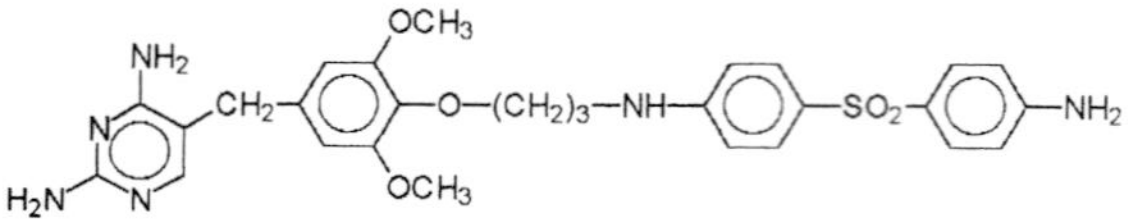 NH_2, N, H_2N, N, CH_2, OCH_3, OCH_3, O–$(CH_2)_3$–NH, SO_2, NH_2	1.22 ±0.08	8.10 ± 0.59	14.67 ±1 .62	13.73 ± 1.12	564.68	2.09	1.99	2.58	2.06
K-150	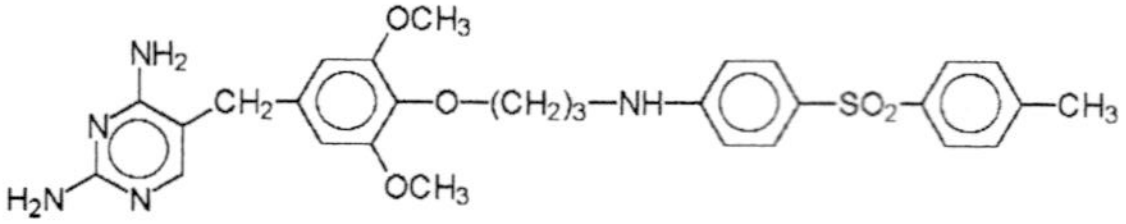 NH_2, N, H_2N, N, CH_2, OCH_3, OCH_3, O–$(CH_2)_3$–NH, SO_2, CH_3	0.84 ±0.06	14.15 ± 1.15	20.35 ± 0.69	16.22 ± 1.12	563.69	3.80	3.50	3.83	3.50
KC-142	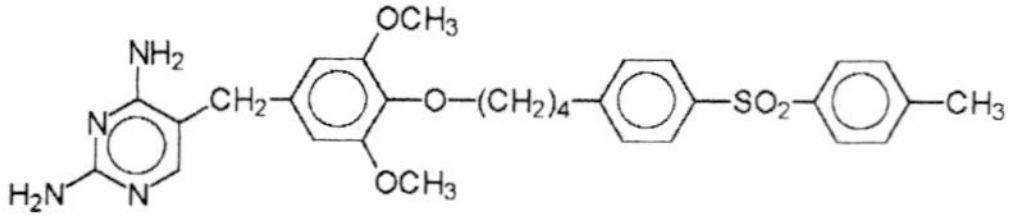 NH_2, N, H_2N, N, CH_2, OCH_3, OCH_3, O–$(CH_2)_4$, SO_2, CH_3	2.45 ±0.59	[a]	[a]	[a]	562.68	ND	4.28	5.16	4.63

Compound	Formula	I_{50} F515 (μmol/L)	I_{50} F588 (μmol/L)	I_{50} F612 (μmol/L)	I_{50} F614 (μmol/L)	MW	Log k'	Log P_{ClogP}	Log P_{VG}	Log $k'_{calc.}$
KC-1303	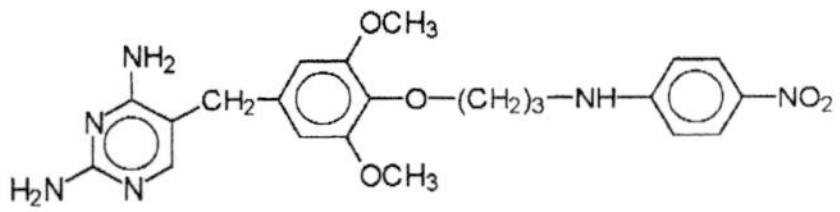 NH_2, H_2N, N, N, CH_2, OCH_3, OCH_3, $O-(CH_2)_3-NH$, NO_2	0.39 ±0.04	1.17 ± 0.18	2.05 ± 0.48	2.83 ± 0.24	454.59	2.39	2.35	2.49	2.18
KC-1307	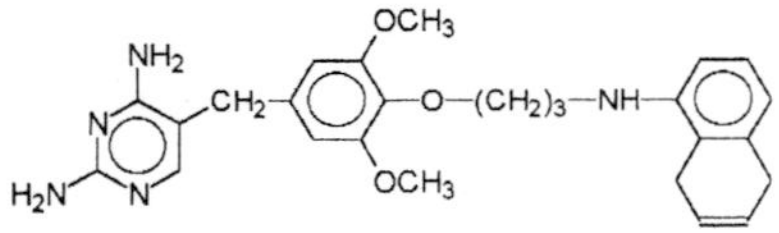 NH_2, H_2N, N, N, CH_2, OCH_3, OCH_3, $O-(CH_2)_3-NH$	1.94 ±0.26	9.59 ± 0.99	18.55 ± 2.40	13.44 ± 1.28	463.58	ND	3.05	3.76	3.24
KC-1308	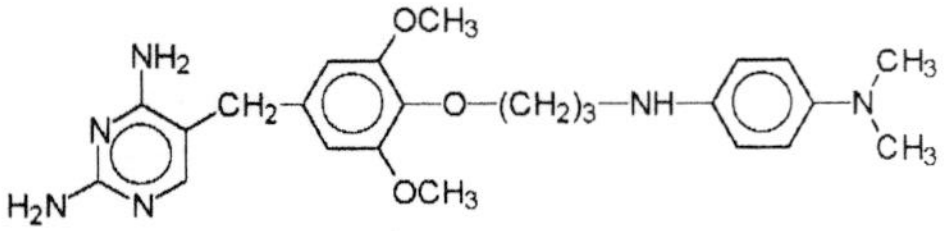 NH_2, H_2N, N, N, CH_2, OCH_3, OCH_3, $O-(CH_2)_3-NH$, N, CH_3, CH_3	2.57 ±0.03	3.16 ± 0.32	2.22 ± 0.47	3.29 ± 0.41	452.56	2.02	2.16	2.80	2.26
KC-1310	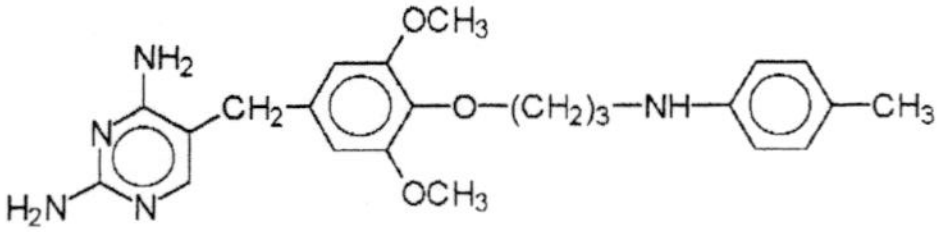 NH_2, H_2N, N, N, CH_2, OCH_3, OCH_3, $O-(CH_2)_3-NH$, CH_3	1.07 ±0.15	6.34 ± 0.47	ND	ND	423.52	2.69	2.46	3.01	2.52
KC-1311	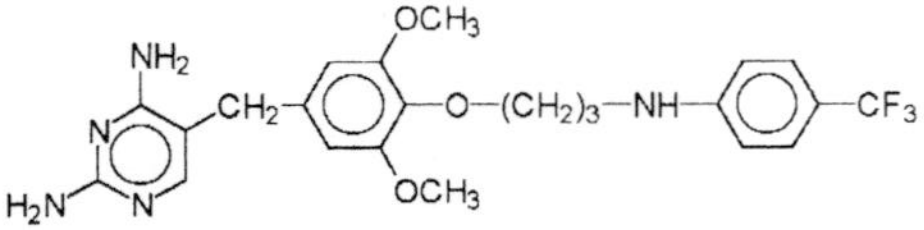 NH_2, H_2N, N, N, CH_2, OCH_3, OCH_3, $O-(CH_2)_3-NH$, CF_3	0.75 ±0.03	7.23 ± 0.60	14.36 ± 3.24	22.10 ± 1.92	477.48	ND	3.37	3.42	3.20
GH-101	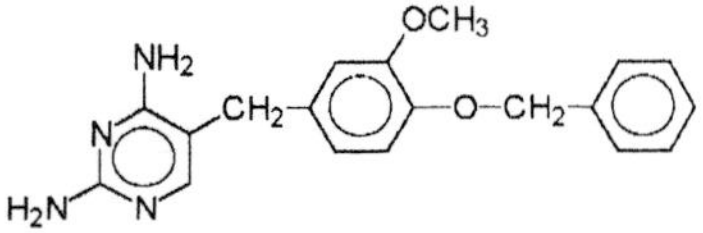 NH_2, H_2N, N, N, CH_2, OCH_3, $O-CH_2$	3.02 ±0.14	16.01 ± 1.02	30.38 ± 3.88	35.36 ± 2.62	336.40	ND	2.82	3.07	2.74

Tab. 4.31 Code, structure, I_{50}-values against various *E. coli* mutants, and physicochemical properties of the tested 2,4-diamino-5-benzylpyrimidines. (Reprinted from Tab. 2 of ref. 106)

Compound	Formula	I_{50} F515 (μmol/L)	I_{50} F588 (μmol/L)	I_{50} F612 (μmol/L)	I_{50} F614 (μmol/L)	MW	Log k'	Log P_{ClogP}	Log P_{VC}	Log $k'_{calc.}$
GH-306	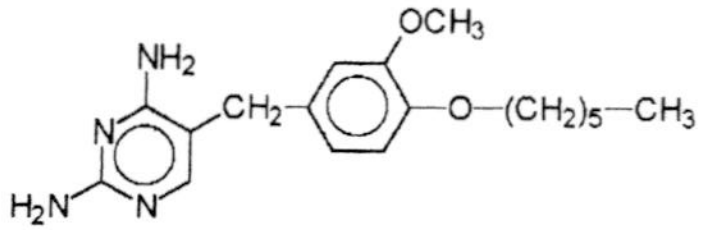	3.03 ± 0.28	9.59 ± 0.76	32.00 ± 5.42	42.72 ± 2.27	330.44	3.00	3.69	3.30	3.79
RO-18958	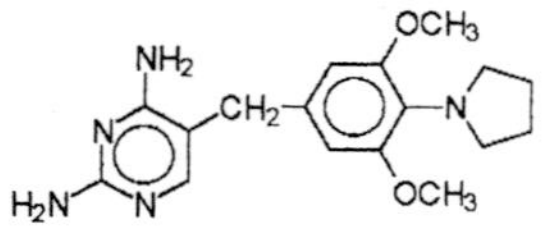	0.32 ± 0.02	1.30 ± 0.16	4.57 ± 0.92	5.28 ± 0.39	353.43	2.24	2.60	2.12	2.09
A-44733	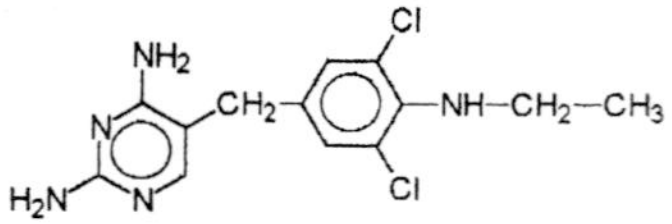	11.69 ± 1.86	22.03 ± 2.97	55.09 ±11.40	73.74 ± 6.49	312.22	2.22	2.32	2.76	2.31

a) inactive up to 15 μmol/L.
ND, not determined.
NC, not calculated.

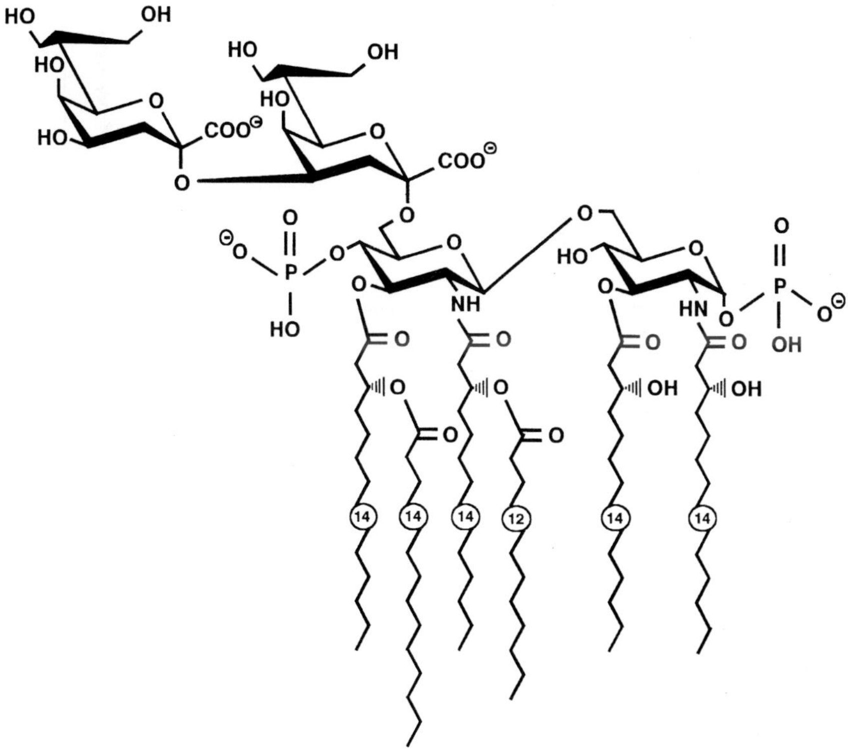

Fig. 4.10 Chemical structure of the LPS configuration of *E. coli* F515.

should then depend primarily on the lipophilicity of the substituents and/or their molecular weight. This is supported by the derived regression analysis. When the ratios of the I_{50} values of the smooth mutant and the deep rough mutant were correlated with the descriptor of lipophilicity, log k', a statistically significant correlation was found (Eq. 4.43).

$$\log I_{50F614}/I_{50F515} = 0.362\ (\pm 0.142)\ \log k' + 0.09\ (\pm 0.152) \qquad (4.43)$$

$n = 8 \qquad r^2 = 0.86 \qquad s = 0.15 \qquad F = 38.9 \qquad Q^2 = 0.77$

More detailed information was obtained from NMR binding measurements. Changes in $1/T_2$ relaxation times and changes in the chemical shift of the resonance signals of the benzylpyrimidines were observed in the presence of increasing LPS concentration. Whereas no interaction can be seen for trimethoprim and K-130 in the case of LPS derived from the deep rough mutant F515, increasing interactions occurred between K-130, GH-306, and LPS derived from the deep rough *E. coli* mutant F588 and the rough mutant F470. Trimethoprim again shows no interaction. K-130 and GH-306 varied greatly in their antibacterial effects against these mutants. The interaction with LPS-470 was so strong that the drug–LPS complex started to precipitate during the measurement. There was no change in the quality of the mag-

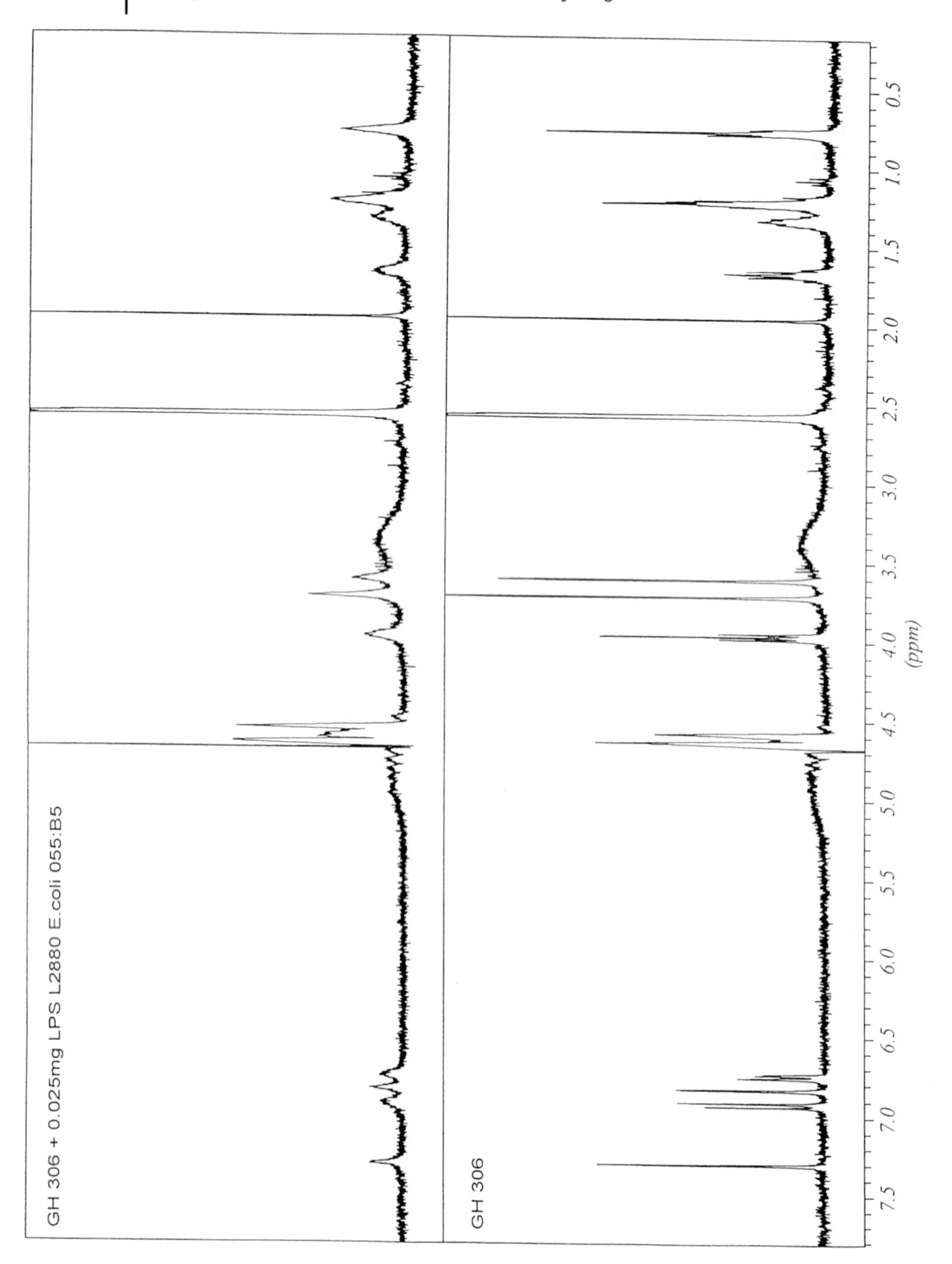

Fig. 4.11 NMR spectrum of GH-306 (0.5 μM) in the absence and presence of 25 μg of *E. coli* O55:B5 L2880-derived LPS. The relaxation rate of the internal standards dimethylsulfoxide and acetonitrile remained unchanged. (Reprinted from Fig. 8 of ref. 106 with permission from Elsevier Science.)

netic field, as indicated by the unchanged resonance signal of the internal standards (dimethylsulfoxide and acetonitrile). If LPS derived from a wild-type *E. coli* (serotype 055:B5) was used instead, the LPS remained in solution upon addition of the drugs K-130 and GH-306 owing to its long hydrophilic O-specific side chain. A broadening of all resonance signals of K-130 is already observed at a very low LPS concentration (0.025 mg/mL). Under identical conditions the broadening of the resonance signals is even more pronounced for GH-306 (Figure 4.11). This corresponds to the observed larger difference of the I_{50} values of GH-306 for the inhibition of the mutants F614 and F515. The involvement of all spin systems of the benzylpyrimidines in the interaction with LPS supports the assumption that the charged pyrimidine ring of the benzylpyrimidines interacts with the charged phosphate groups. In addition, it can be concluded that this interaction is reinforced by hydrophobic interactions of the substituents at the benzene ring with lipid A, provided they are extended enough to reach down to lipid A.

Part of the increase in antibacterial effect on mutants with membrane defects may also be due to increased permeability. These mutants not only have shorter saccharide chains but at the same time are not able to incorporate proteins in the outer membrane, which leads to domains of glycerophospholipids in the bilayer of the outer membrane [103].

4.3.2 Self-promoted Uptake of Antibacterial Peptides

In Gram-negative bacteria not only the outer leaflet of the plasma membrane but also the outer membrane contains anionic molecules, thus creating a negatively charged surface. This is in contrast to mammalian cell membranes, which possess zwitterionic amphiphiles at the extracellular monolayer of the plasma membrane. This difference is the reason for the selectivity of antimicrobial peptides such as polymyxin, gramicidin, and magainin. Most of these have an α-helical structure and are cationic and amphipathic. Very different structural motifs, such as α-helical peptides and β-sheet peptides, exist, some of which are rich in specific amino acids such as tryptophan or histamine. Others show macrocyclic cysteine knots. Despite these differences, all these peptides can bind to the membrane surface of Gram-negative bacteria and become cytotoxic by disturbance of the bacterial outer and inner membranes.

The mechanism proposed for the transport of the peptide across the outer membrane of Gram-negative bacteria is depicted in Figure 4.12 [113]. The interaction depends on both the structure of the peptide and the composition of the membrane. There is evidence that these peptides directly act upon the lipid membrane by permeabilizing the membrane. No stereospecific interaction with membrane-bound proteins seems to be involved in their mechanism of action. These polycations are able to generate severe disorganization in the outer membrane, which allows them to pass the bilayer by a "self-promoted pathway". It has been shown with fluorescence-labeled lipids that cationic peptides attract smaller clusters of anionic lipid and that magainin-induced leakage depends strongly on the type of anionic lipid [114].

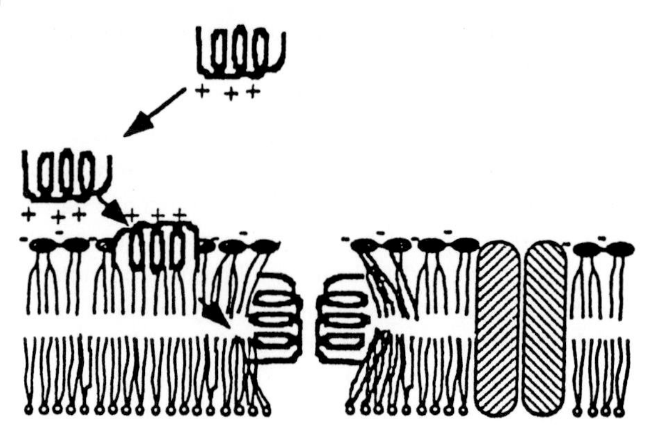

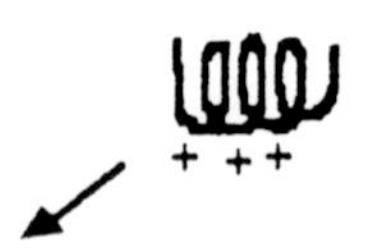

Fig. 4.12 Mechanism proposed for the transport of antimicrobial peptides across the outer membrane of Gram-negative bacteria. The initial recognition involves the negatively charged LPS on the outer leaflet and the cationic peptide. Once in the periplasmic space, the peptide can move into the cytoplasmic membrane. (Reprinted from Fig. 6 of ref. 113 with permission from Elsevier Science.)

Magainin-induced leakage is, for example, much stronger in liposomes of PG than in liposomes of PS [115]. The former is much more abundant in bacterial cells. Beyond that, there are data that suggest that DNA could be the final target. In the context of this book it is of importance to note that the specificity of the antimicrobial peptides for anionic membrane components can be modeled in liposome studies. A detailed review of the mechanism of action of antimicrobial peptides [113] and the structural and charge requirements for activity has recently been published [116].

Studies exploring the molecular mechanism of polymyxin B (PMB)–membrane interactions have been performed on artificial asymmetric planar bilayer membranes utilizing electrical measurements obtained by the patch-clamp technique [117]. Bilayers of different composition were used. The leaflet representing the inner leaflet of the bacterial outer membrane consisted of a mixture of PE, glycerol, and diphosphatidylglycerol in a molar ratio of 81:17:2. The outer leaflet was composed of LPS from deep rough mutants of PMB-sensitive *E. coli* F515 or PMB-resistant strains (*Proteus mirabilis* R45), glycosphingolipid (GSL-1) from *Sphingomonas paucimobilis* IAM 12576, or phospholipids (PG, diphytanoylphosphatidylcholine, DPhyPC). The induction of fluctuations of the current caused by transient membrane lesions was observed in all membrane systems. The authors could demonstrate a good correlation between the diameter of the lesions, the molecular charge, and the surface charge density but not the molecular area (Table 4.32). Only in membrane systems that resembled the lipid matrix of the PMB-sensitive strains were the diameters of the lesions sufficiently large ($d = 2.4\,\text{nm} \pm 8\%$) to allow PMB molecules to permeate. In all other systems they were too small. The results obtained clearly support the importance of self-promoted transport across the membrane for PMB molecules as a prerequisite for their activity. The interaction of antimicrobial peptides with model lipid bilayers and biological membranes has been stressed in a special issue of *Biochimica Biophysica Acta* [118].

Tab. 4.32 Molecular parameters describing the (glyco)lipids constituting the bilayer leaflet at the side of PMB addition of various asymmetric planar bilayers (the second leaflet was always made from the phospholipid mixture) and the average diameters of the induced membrane lesions. (Reprinted from Tab. 1 of ref. 117 with permission from Bertelsmann-Springer.)

(Glyco)lipid	Average molecular mass[a] (Da)	Molecular area[b,c] (nm^2)	Molecular charge[b,d] (e_0)	Charge density[b] (e_0/nm^2)	Diameter of membrane lesions[b,f,g] (nm)
F515 LPS	2237	1.3	–4	–3.1	2.4 ± 0.2
F515 LPS (pH 3)	2237	1.2	–2[e]	–1.7	< 0.7
R595 LPS	2415	1.4	–3.4	–2.4	2.3 ± 0.5
R45 LPS	2465	1.5	–3.0	–2.0	1.0 ± 0.1
GSL-1	758	0.5	–0.5	–1.0	0.7 ± 0.1
PG	750	0.6	–1	–1.7	< 0.5
DPhyPC	846	0.8	0	0	< 0.5

a) Under consideration of chemical heterogeneity as determined by MALDI-TOF mass spectrometry.
b) If not otherwise stated, all data refer to a bathing solution/subphase at pH 7.
c) Taken from monolayer isotherms at 37 °C and a lateral pressure of 30 mN/m^{-1}, SEM ± 5% (an additional error may result from batch-to-batch variations).
d) Calculated according to the number of phosphoryl and carboxyl groups under consideration of substitutions.
e) The state of ionization at pH 3 was derived from monolayer and ζ-potential measurements on the PMB binding stoichiometries.
f) At negative clamp voltages.
g) Calculated from $I = (\pi \cdot \sigma \cdot d^2 \cdot U)/(4 \times l)$ where I is membrane current, σ is specific conductivity of the subphase, d is the diameter of the lesion, U is clamp voltage, and l is membrane thickness.

4.4 Drug Accumulation, Toxicity, and Selectivity

The examples presented underline the influence of drug physicochemical properties and tissue or membrane composition not only on drug distribution profiles, but also on drug localization and orientation in the membrane. This is, in turn, without any doubt, related to drug accumulation, toxicity, selectivity, and effectivity. These properties have to be especially considered in the case of amphiphilic drugs, which interact not only with positively and negatively charged head groups of the phospholipids but also by hydrophobic forces with the hydrocarbon interior of the lipids. This, then, pertains to major groups of drugs currently used. Carboxyl residues are part of many analgesics and anti-inflammatory drugs, and cationic groups as protonated amino groups are present in local anesthetics, antiarrhythmics or β-blockers. As the phosphate groups of phospholipids bear a negative charge, cationic amphiphiles are especially prone to be taken up by phospholipid membranes.

Under *in vivo* conditions, not only the degree of partitioning but also the dissociation constant of the lipid–drug complex is of importance. It is evident that the accumulation of drug molecules in the various cell membranes can have an enormous influence on the pharmacokinetics of drugs and, in consequence, on effectivity, selectivity, and toxicity.

Tab. 4.33 Calculation of equilibrium dissociation constants based on the drug's membrane concentration. (Reprinted from Tab. 3 of ref. 95, with permission from the American Chemical Society)

Drug	$K_{P[mem]}$	K_D', (aqueous conc.)	K_D, (M) (membrane conc.)
Nimodipine[a]	6 300	1.1×10^{-10}	6.9×10^{-7}
Nifedipine[b]	3 000	4.1×10^{-10}	1.2×10^{-6}
Bay K 8644[c]	11 000	2.4×10^{-9}	2.6×10^{-5}
Nisoldipine[b]	13 000	1.4×10^{-10}	1.8×10^{-6}
Amlodipine[d]	19 000	1.2×10^{-9}	2.3×10^{-5}
Iodipine[e]	26 000	3.9×10^{-10}	1.0×10^{-5}

a) Canine cardiac muscle.
b) Rat cardiac muscle.
c) Rabbit cardiac muscle.
d) Rat smooth muscle.
e) Guinea-pig skeletal muscle.

A two-step membrane bilayer pathway has been compared with a one-step aqueous pathway for drug molecules, and diffusion-limited rates of association were calculated for both approaches. In the aqueous model, the drug reaches the receptor by diffusion through the bulk solvent; in the membrane model the drug partitions into the bilayer of the membrane followed by lateral diffusion to the specific binding site (for details see ref. 119). The application of the membrane approach would affect the K_D calculation, i.e. the calculation of the dissociation constant for a drug–receptor complex has to be determined at equilibrium. Normally this is done by Scatchard analysis or a similar approach, in which free and bound fractions are expressed as overall molar concentrations (moles of drug per liter reaction mixture). K_D is obtained by plotting the bound to free drug ratio against free drug. The results for the partitioning of DHPs into membranes have shown that $K_{p,m}$ is $>10^3$ (see Table 4.33) [95], which means that these drugs are distributed in the membrane to such an extent that their concentration in the membrane volume is much higher than in the extramembrane aqueous phase [95].

The diffusional dynamics of rhodamine and rhodamine-labeled nisoldipine in lipid multibilayers made up of sarcolemma was studied. It was found that in fully hydrated bilayers the diffusion constant of membrane-bound drug was 3.8×10^{-8} cm^2/s [120]. This is in agreement with results obtained in other membrane systems and in free volume model calculations. The results suggest that the unbound drug in the process of its specific binding to a receptor protein rapidly diffuses laterally within the membrane. In this case, the use of the drug concentration in the membrane bilayer as the "free" component would be more appropriate for specific binding calculations [121] than the concentration in the aqueous phase. Of course, the fraction of drug unspecifically tightly bound to proteins in the membrane other than the specific receptor has to be accounted for.

In this respect, it is of interest to mention a study by Carvalho *et al.* [122]. These authors studied the partitioning of three calcium antagonists in brain plasma membranes (native synaptic plasma membrane) and compared the determined $K_{p,M}$ values with those obtained from liposomes prepared from lipids extracted from these native membranes. They also compared them with $P_{oct.}$. Two facts became obvious. First, the ranking of the $K_{p,M}$ values was the same for native membranes and the liposomes made from extracted lipids. The absolute values were, however, much larger for $K_{p,M}$ determined in the native membranes, especially when compared with $P_{oct.}$ (Table 4.34). Second, there was a significant increase in $K_{p,m}$ with increase in temperature in native membranes but not in liposomes. This suggests that the presence of membrane proteins positively influenced the uptake of drugs into the native membrane.

Tab. 4.34 Partition coefficients of Ca^{2+} antagonists in synaptic plasma membranes (SPM) and liposomes and octanol-water, at 25 °C. (Reprinted from Tab. 1 of ref. 122 with permission from Elsevier Science)

	Nitrendipine[a]	(–)-Desmethoxy verapamil[a]	Flunarizine[b]	
			Filtration	Centrifugation
Native SPM	334 ± 53	257 ± 36	23×10^3	19×10^3
Liposomes	190 ± 41	118 ± 10	6×10^3	ND
Octanol-water	9.95 ± 0.25	66.6 ± 0.5	–	$10^{5.78}$

a) K_p obtained by the filtration method.
b) K_p obtained by both filtration and centrifugation methods.
ND, Not determined.

A similar difference has been observed for $K_{p,M}$ values of the anticancer drugs tamoxifen and 4-hydroxytamoxifen determined in mitochondria, sarcoplasmic reticulum, and in liposomes made from lipids extracted from the corresponding native membranes [85]. The largest $K_{p,M}$ was observed in mitochondria, followed by sarcoplasmic reticulum and liposomes. The authors introduced for these experiments derivative spectroscopy, a reliable and rapid procedure to estimate drug partitioning into biomembranes. The method used the shift in the absorption spectra of the drug when removed from the aqueous phase to a hydrophobic environment (see Section 3.10).

The true equilibrium constant, K_a, can be calculated from the overall association constant and the membrane-based partition coefficient

$$K_a = K_a' / K_{p,M} \tag{4.44}$$

and correspondingly the K_D for the intra-membrane equilibrium:

$$K_D = K_D' \times K_{p,M} \tag{4.45}$$

The dissociation constants are given in Table 4.33 for the same series of drugs [95].

This means that high overall activity of a compound could be due to high specific affinity for the receptor, a high $K_{p,M}$, or a combination of both. If the high apparent affinity is due to a high $K_{p,M}$, the consequence is a high membrane concentration. A comparison of the K_D and K_D' values of nimodipine and nisoldipine sheds light on the importance of these considerations for drug design. The molar ratio of lipid to DHP receptor in membranes of the canine cardiac sarcolemma has been experimentally determined to be 4.2×10^6 [123], i.e. at a concentration of $\sim 1 \times 10^{-10}$ M in the aqueous phase there should be 10 nimodipine molecules per receptor in the membrane. In contrast, nisoldipine, whose K_D' is similar to that of nimodipine, has a membrane partition coefficient that is about twofold higher. Consequently, the intrinsic affinity of nimodipine is weaker, the intrabilayer concentration required for a response is higher, and the risk of side-effects is greater. However, this discussion is only valid under the assumption that the high-affinity receptor site and the drug are located at the same depth within the membrane. The intrabilayer distribution profile is therefore an important parameter. A detailed consideration of the calculated concentrations in the aqueous and membrane compartment is therefore indicated (Table 4.33). The use of such data in structure–activity relationships analysis may improve the results and facilitate the interpretation.

Not only the amount of drug and its localization in the membrane are important for onset and duration of drug action, but also the equilibrium kinetics, which is related to drug accumulation. To find an explanation for the long duration of action of the β_2-agonist salmeterol Rhodes *et al.* [124] have determined the association–dissociation rate of salmeterol and other drugs in unilamellar and multilamellar liposomes of DOPC containing different amounts of cholesterol. The association rate was found to be rapid and independent of membrane composition. The $K_{p,\,M}$ values determined as a function of time did not vary from 1 min to 1200 min and the rates were independent of whether the liposomes were unilamellar or multilamellar. The association rates were not influenced by the cholesterol content, but it was noticed that cholesterol reduced the $K_{p,\,M}$ values. The partition coefficients, $K_{p,\,M}$,, $P_{oct.}$ and the dissociation rates observed are summarized in Table 4.35. The large $K_{p,\,M}$ value for amiodarone corresponds to its very slow dissociation rate. The slow dissociation of amiodarone from membranes has also been demonstrated by the slow removal of amiodarone from sarcoplasmic reticulum vesicles as a function of number of washes with drug-free buffer. Whereas propranolol and nimodipine are completely or almost completely removed after 20 washes, the amount of amiodarone in the vesicles remains constant after 20 washes (Table 4.36) [125].

Tab. 4.35 Partition coefficients and dissociation $\tau^1/_2$ values. (Reprinted from Tab. 1 of ref. 124 with permission from the American Society for Pharmacology and Experimental Therapeutics)

Drug	$\tau^1/_2$ (min)	K_{pM}	P_{oct}
Nimodipine	< 5	2,700	260
Amlodipine	100	21,800	30
Salmeterol	60	22,500	7,600
Amiodarone	1,000	~1,000,000	350

Tab. 4.36 Comparison of the physical properties of drug-membrane interactions and their pharmacokinetics. (Reprinted from Tab. 1 of ref. 125 with permission from Academic Press)

Drug	Partition coefficient	Washout rate[a] (s)	Bilayer location[b] (Å)	Clinical half-life (h)
Amiodarone	1 000000	~ 0	12	> 1000
Nimodipine	5000	4.8×10^{-3}	6	5.5
Propranolol	1200	1.6×10^{-3}	6	3.9

a) (Half-time for drug removal from the membrane bilayer)$^{-1}$.
b) Measured from the phosphate head group region of the bilayer as calculated from the real space profile structure.

Small-angle X-ray diffraction was used to identify the time-averaged location of amiodarone in a synthetic lipid bilayer. The drug was located about 6 Å from the center of the lipid bilayer (Figure 4.13) [125, 126]. A dielectric constant of $\kappa = 2$, which is similar to that of the bilayer hydrocarbon region, was used to calculate the minimum energy conformation of amiodarone bound to the membrane. The studies were performed below the thermal phase transition and at relatively low hydration of lipid. The calculated conformation differed from that of the crystal structure of amiodarone. Even though the specific steric effects of the lipid acyl chains on the confor-

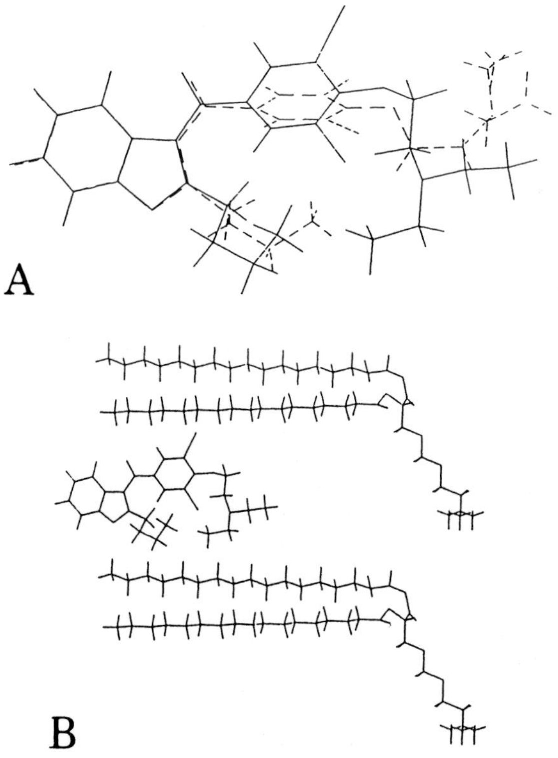

Fig. 4.13 (A) Minimum energy structure of amiodarone (solid bonds) superimposed on the crystal structure (– – –) (B) Amiodarone incorporated into a membrane bilayer, showing the position of the iodine atoms relative to the acyl chains and the size of the molecule relative to the hydrocarbon core region. (Reprinted from Fig. 4 of ref. 126 with permission of the American Chemical Society)

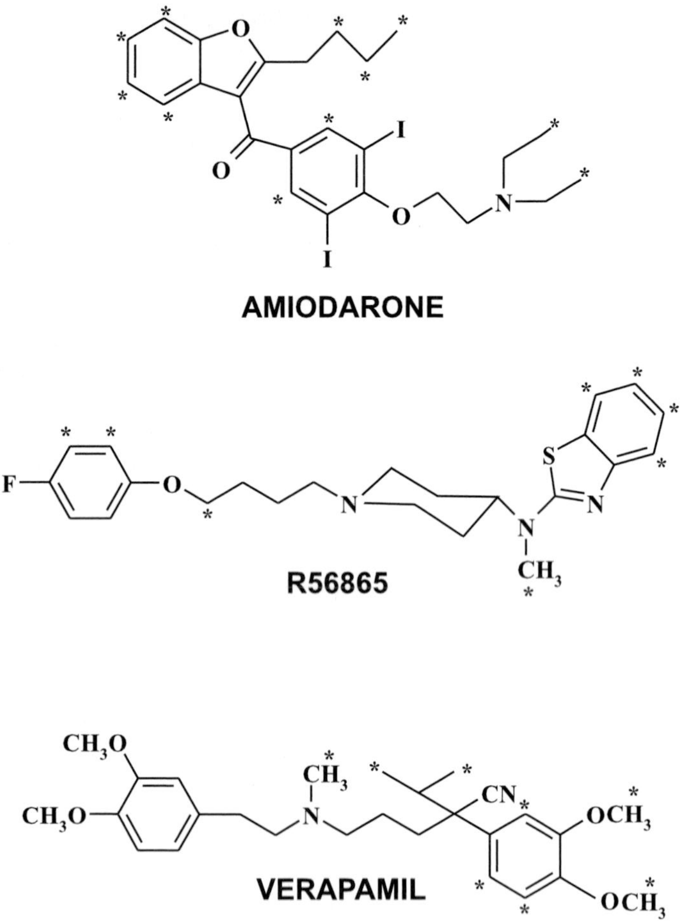

AMIODARONE

R56865

VERAPAMIL

Fig. 4.14 Changes in $1/T_2$ for various spin systems of catamphiphilic drugs (indicated by *) in the presence of lecithin. (Reprinted from Fig. 3 of ref. 83 with permission from Wiley-VCH)

mation of the drug were not considered in this study, it can be inferred that the biologically active conformation of a drug interacting with an receptor within the bilayer may be different from that of the crystal structure [95, 126]. Additional information on the substructures involved in amiodarone–phospholipid interaction was obtained by NMR binding experiments. The change in $1/T_2$ for the various spin systems of the three catamphiphiles amiodarone, R-56865, and verapamil was determined in the presence of lecithin. In the case of verapamil, a clear-cut difference in the involvement of substructures in the interaction with the phospholipid could be derived from the changes in relaxation rate for different spin systems. In contrast, in the case of amiodarone all substructures of the molecule would appear to be involved in the drug–lipid interaction. The changes in $1/T_2$ for various spin systems of the catamphiphiles in the presence of fully hydrated lecithin vesicles are indicated by an asterisk (Figure 4.14) [83]. The results from NMR interaction studies are in agreement with the above discussed X-ray diffraction measurements and the results from the washout experiments showing amiodarone to be deeply buried within the hydrocarbon chains. Partition coefficients for amiodarone have been determined in PC mixed with varying amounts of stearic acid, PE, sphingomyelin, PS, and cholesterol. All lipid systems showed an almost constant partition coefficient of 17 000 (Table 4.37). It is important to note that the partition equilibrium was reached only after about 6 h [127].

Tab. 4.37 Partition coefficient (*P*) of amiodarone between buffer and lipid vesicles of various compositions. (Reprinted from Tab. 1 of ref. 127, with permission from the American Chemical Society)

Lipid species[a]	Lipid composition, (%)	P^a	*n*
Egg PC	100	16,500 ± 900	11
Egg PC/PE	85:15	17,100 ± 750[b]	3
Egg PC/SA	95: 5	16,200 ± 800[b]	3
Egg PC/PS	95: 5	17,000 ± 900[b]	3
Egg PC/PS	85:15	17,500 ± 950[b]	3
Egg PC/Sph	90:10	17,000 ± 1500[b]	3
Egg PC/Ch	90:10	18,000 ± 1600[b]	3
Egg PC/Ch	75:25	17,000 ± 1700[b]	3
PL	–	16,400 ± 1400[b]	4

a) Mean ± SEM.
b) Not significant.
Key: SA, stearic acid; Sph, sphingomyelin; PL, total lipids extracted from the erythrocyte ghost.

Furthermore, simulations of the binding mode of the energy-minimized structures of these three catamphiphiles into a complex of four lipid phosphatidylcholine molecules have been performed (Coats *et al.*, unpublished data), and the findings were in agreement with the discussed X-ray and NMR studies. The highest binding energy is found for amiodarone when buried deeply between the hydrocarbon chain (–83.5 kcal). In contrast, the calculated binding energy for verapamil is significantly larger in the extended conformation, in which the substructure of the molecule, showing no change in $1/T_2$, does not interact with the phosphor groups but is positioned in the water phase (–87.02 kcal). Under conditions in which the verapamil molecule was deeply inserted between the phospholipid hydrocarbon chain molecules the calculated interaction energy was only –74.5 kcal.

The position of amiodarone, deeply buried within the hydrocarbon chain despite its charge, suggests that the sum of the hydrophobic interaction forces exceeds the charge interaction. In contrast to these results, Jendrasiak *et al.* [128] published NMR and X-ray data which indicate that in egg PC liposomes the iodine atoms of amiodarone are located near the hydrocarbon chain–water interface and the tertiary amino group of amiodarone in the head group region of the bilayer.

Later it was shown in detailed studies using UV spectroscopy, light scattering, and fluorescence polarization techniques that the location of the drug depends on many parameters, such as pH, ionic strength, the nature of the membrane model used (liposome, monolayer), the degree of hydration, the drug–lipid ratio, and the proportion of acidic lipids in the membrane [129]. In this respect it is important to note that NMR binding measurements for verapamil in the presence of bovine brain PS show broadening of all resonance signals, even more intensive than in the presence of lecithin (Seydel *et al.*, unpublished results). This means that the strength of interaction and the orientation of drug molecules in the membrane depend on both the structure of the drug and the composition of the membrane.

The extremely extensive incorporation of amiodarone into the bilayer of membranes also finds its expression in the pharmacokinetic parameters, with an extremely large volume of distribution being observed in humans. This exceeds the total body water volume by several fold, indicating a large accumulation in membranes. It also explains the high incidence of side-effects associated with this drug. Many organs show toxic responses to amiodarone, including liver, lung, skin, and nervous system, which in some cases have led to the death of patients. Prolonged treatment can cause pulmonary fibrosis. In rats, after 4 weeks of administration of 50 mg kg/12 h the plasma levels were 2.46 μg/mL and the concentrations in the lung 163 times higher than those in plasma; phospholipidosis-like lesions were observed [130].

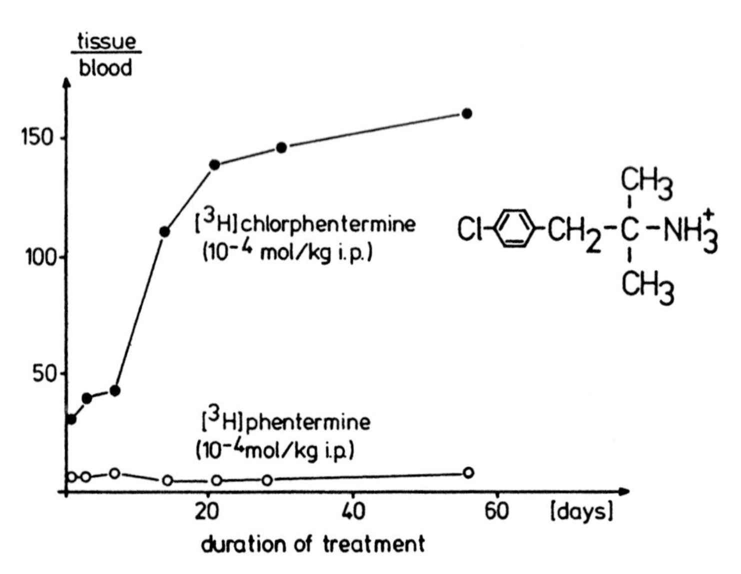

Fig. 4.15 Accumulation of tritiated chlorphentermine and phentermine in the lungs of rats during chronic treatment with the indicated daily dosage. (Reprinted from Fig. 1 of ref. 133 with permission from Taylor & Francis.)

The effect of drug binding to phospholipid membranes on drug toxicity has been studied in detail using the two anorectic drugs phentermine and chlorphentermine [131, 132]. The tritiated compounds were chronically administered to rats by daily injection and the time course of uptake in various tissues monitored. The observed tissue to blood ratios for phentermine and chlorphentermine in the lung are compared in Figure 4.15. The result suggest that chlorphentermine was increasingly accumulated over time, i.e. that the binding reservoir for chlorphentermine increased. This is supported by ultrastructure investigations showing the chlorphentermine-induced formation of lysosomal inclusion bodies, which mainly consisted of phospholipids. Probably the degradation of phospholipids by lysosomal phospholipase was impaired [133]. The impairment of renal function in rats with generalized lipidosis induced by chlorphentermine has also been demonstrated [134].

NMR binding studies on phentermine, chlorphentermine, and other catamphiphiles [135] indicated a very weak interaction with phospholipid vesicles for phentermine but a strong interaction by chlorphentermine and other derivatives studied. The various spin systems were differently involved depending on the structure of the compound (Figure 4.16). The interaction was stronger with phosphatidylethanolamine than with phosphatidylcholine, and no interaction with digalactosyldiglyceride was observed. A decrease in drug–membrane interaction was observed upon addition of cholesterol to the PC liposomes and an increase upon addition of ions (NaCl) to the aqueous phase, indicating the importance of hydrophobic forces in the interaction process between drug and bilayer (Figure 4.17).

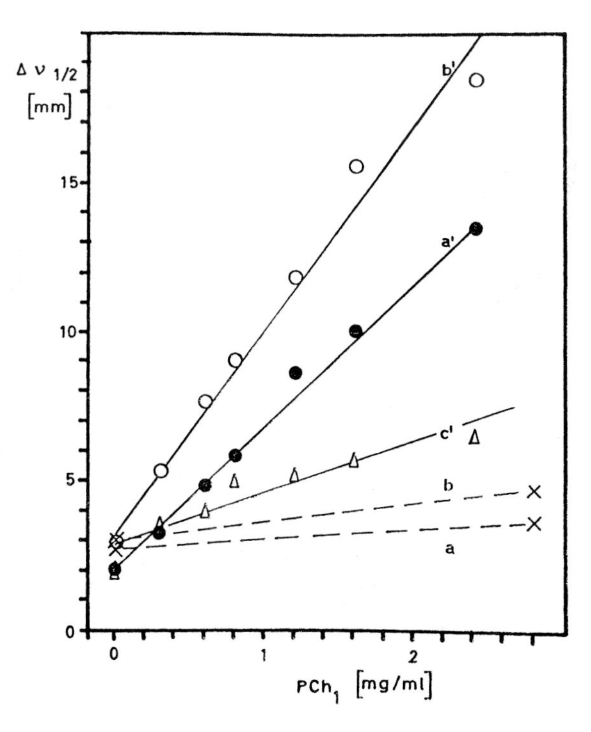

Fig. 4.16 Signal broadening ($\Delta v_{1/2}$, mm) of various spin systems of phentermine (a and b) and chlorphentermine (a′, b′, and c′) as a function of increasing lecithin concentrations (0–2.4 mg/mL). (a, a′ = aromatic protons, b, b′ = methylene protons, c, c′ = methyl protons). For quantitative comparison see the slopes of the regression lines. (Reprinted from Fig. 2 of ref. 135 with permission from Elsevier Science)

The interaction of the monovalent cationic drugs phentermine, chlorphentermine, amitriptyline, and 1-chloroamitriptyline with DPPC bilayer vesicles has also been investigated by optical (fluorescence, 90° light scattering) and calorimetric techniques [136]. In agreement with the other investigations, it can be concluded that the interaction of these catamphiphiles is mainly restricted to the polar head group region of the lipid because no marked change in main transition temperature is observed, in contrast to compounds that penetrate into the interior of the bilayer. Nevertheless, it can be concluded that the difference in strength of interaction observed is governed by differences in the lipophilicity of the compounds but is based on the ionic interaction between the protonated aliphatic chain of the drugs and the negatively charged phosphate groups of the phospholipid. This interaction prevents the drugs' deep penetration into the hydrocarbon chains and prevents a larger perturbation of the liquid crystalline structure.

Another class of catamphiphilic drugs with high toxic potential is the aminoglycosides. Their nephrotoxicity is manifested by lethal injury to the proximal tubular cells by means of the induction of renal cortical lysosomal phospholipidosis. The toxic effect involves three steps: the uptake of the antibiotics into the cells, the intralysosomal lipid storage, and, finally, cell necrosis. The aminoglycosides are highly water

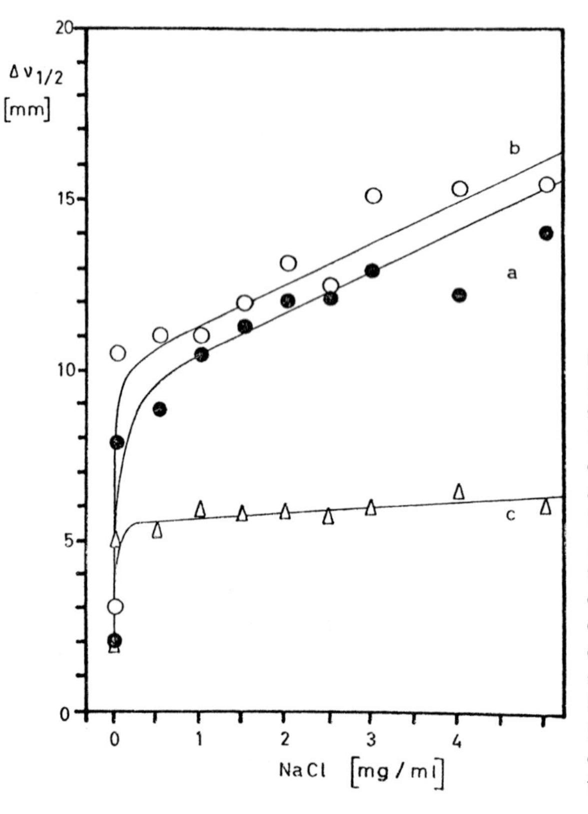

Fig. 4.17 Signal broadening ($\Delta\nu_{1/2}$, mm) of various spin systems of chlorphentermine (a–c) at constant liposome concentration (1.2 mg/mL) as a function of increasing NaCl concentration (0–6 mg/mL). The lowest symbols at zero NaCl concentration represent the control line width of the different proton resonance signals in the absence of both lecithin liposomes and NaCl. (Reprinted from Fig. 4 of ref. 135 with permission from Elsevier Science.)

soluble and can hardly penetrate into cells. They enter the cells of the proximal tube by using a polybase transport system essential for re-uptake of oligopeptides. Aminoglycosides fit onto this transport system and are taken up by receptor-mediated endocytosis. Under *in vitro* conditions, the aminoglycosides bind to negatively charged phospholipids and inhibit phospholipases. In order to evaluate the binding of aminoglycosides to phospholipids, the replacement of Ca^{2+} from PS monolayers and from biomembranes was determined for 13 aminoglycosides [137]. It was found that the affinities of the drugs depend on pH and Ca^{2+} concentration. The ranking in affinity is given in Table 4.38. The interaction between Ca^{2+} and streptomycin was competitive. The interaction for the antibiotics with high affinity showed an unusually steep dose–response curve, which may be caused by positive cooperativity. For the displacement of Ca^{2+} from native biomembranes higher drug concentrations were necessary. The ranking was the same. It can therefore be concluded that aminoglycosides possess high affinities to polar lipids.

Tab. 4.38 Half-maximum concentrations of the aminoglycoside antibiotics investigated to replace Calcium from PS monolayers at a Ca^{2+} concentration of 1.2×10^{-5} M, expressed as ID_{50} ($\times 10^{-7}$ M). (Reprinted from Tab. 1 of ref. 137 with permission from Elsevier Science)

	ID_{50} ($\times 10^{-7}$ M)	ID_{75} / ID_{25}
Streptomycin	60	17
Dihydrostreptomycin	60	30
Amikacin	10	19
Ribostamycin	7	4
Kanamycin	6	5
Paromomycin	6	2
Tobramycin	6	2
Netilmicin	5.5	2
Neomycin	5	2
Framycetin	5	2
Gentamicin	5	2
Dibekacin	4.5	2
Sisomicin	4	2

To characterize the slope of the dose-response curves, the ID_{75}/ID_{25} ratios are given.

To examine the mechanism of intralysosomal lipid storage, binding to phospholipids and the inhibitory effect toward phospholipase A_1 (PLA_1) have been studied separately. The results for binding to PS, the calculated interaction energy, and the percent inhibition of PLA_1 are summarized in Table 4.39 [138]. Derivatives in which the N^1-amino function was replaced by an aminohydroxybutyryl group, BB-K29 or other substituents showed lower inhibitory potency toward PLA_1. In addition, binding and calculated transfer energies of interaction were decreased [139]. It was later

Tab. 4.39 Binding of aminoglycosides to phospholipids, energy of aminoglycoside-phospholipid interaction, and inhibition of phospholipase A_1 activity. (Reprinted from Tab. 1 of ref. 138 with permission from Elsevier Science)

Drug	Aminoglycoside-phospolipid (molar ratio)[a]	Energy of interaction[b]	Phospholipase A_1 inhibition (%)[c]
Streptomycin	0.220:100	–3.5	6
Amikacin	0.682:100	–4.9	41
Gentamicin	1.140:100	–8.4	51
BB-K29	0.232:100	–5.6	22
BB-K89	0.497:100	–4.1	35
BB-K11	0.674:100	–6.5	58
S87351	0.960:100	–5.8	18

a) Aminoglycosides were mixed with liposomes at a molar ratio of 1:100 and were eluted through Sepharose[R] 4B at pH 5.4. The drug-phospholipid ratio in the liposome peak is given.
b) Calculated from the conformational analysis.
c) Inhibition measured at a drug-phospholipid ratio of 1:100.

found that the binding energies vary for different negatively charged phospholipids in the sequence phosphatidylinositol (PI) > PS > phosphatidic acid. The apparent accessibility of the bound aminoglycoside to water varied in an inverse relation with the energy of interaction. Gentamycin adopted an orientation parallel to the hydrophobic–hydrophilic interface. Amikacin, with a lower inhibitory activity than gentamycin, preferred a perpendicular position to the hydrophobic–hydrophilic interface in the three studied liposomes. The authors concluded that the inhibition of PLA_1 depends on the binding to the bilayer, and is modulated by the nature of the phospholipid and the orientation of the drug. In another report it was shown that gentamicin produced a significant increase in PS, PC, and PI as well as total phospholipid in homogenates of a lysosomal fraction of kidney cortex [140]. The induced accumulation of phospholipids was associated with various forms of phospholipase C.

The effect of the aminoglycosides dibekacin, sisomicin, and gentamicin on Ca^{2+} bound to the outer surface of the cardiac plasmalemma on contraction in guinea-pig atria has also been reported. All studied drugs reduced the cellular content of Ca^{2+} by 10–20%; this was accompanied by a decrease in the contractile force of 40–90% [141].

4.4.1 Selectivity

The affinity of certain drugs to tissues with a particular lipid composition, which can lead to unwanted accumulation and toxicity, can, under certain circumstances, be an advantage for the development of drugs with particular selectivity, provided the specific receptor is located in such a membrane and is accessible for the drug.

As an example the inhibition of HMG-CoA reductase (HMGR) will be discussed. HMGR is the target for drugs lowering the plasma total and low-density lipoprotein

(LDL)-cholesterol in hypercholesterolemic patients. It is assumed that the suppression of sterol synthesis outside the liver contributes little to plasma cholesterol lowering. It has further been proposed that the perturbation of non-hepatic isoprenoid metabolism may be partly responsible for adverse side-effects. Some of these drugs that are generally well tolerated, such as lovastatin, are associated with quite a range of adverse reactions (e.g. increased liver enzymes, sleep disturbance, myositis). Attempts have therefore been made to improve the selectivity to liver tissue, thus decreasing the induced adverse reactions. An initial report described the tissue selectivity of pravastatin and lovastatin. Being equally potent in cultured hepatocytes, pravastatin was 100 times less active than lovastatin in skin fibroblasts [142]. Additional *ex vivo* experiments in rats supported this selectivity. Pravastatin inhibited cholesterol biosynthesis only in lipoprotein-producing organs such as liver and intestine. In contrast, lovastatin and mevastatin also inhibited cholesterol biosynthesis in kidney, lung, spleen, prostate, and testis. A newly synthesized HMGR inhibitor, BMY 22089 also showed higher selectivity for hepatic tissue than mevinolin [143]. Later, a whole series of HMGR inhibitors, including lovastatin, pravastatin, fluvastatin, and 12 newly synthesized derivatives, were analyzed for tissue selectivity in liver, spleen, and testis *in vitro*.

The obtained activity values (IC_{50}) were correlated with CLOGP [144]. The results suggested that lipophilicity, which in this set varied between CLOGP 0.04 and 4.82, is an important factor in determining tissue selectivity. It could be concluded that compounds with CLOGP > 2 are more potent in peripheral tissues than in liver and those with CLOGP < 2 possess moderate liver tissue selectivity (tissue–liver ratio > 1). A linear relation is obtained when plotting log tissue–liver ratios against CLOGP.

$$\log \text{spleen/liver} = -0.52\,\text{CLOGP} + 0.93 \quad (4.46)$$

$$n = 14 \quad r^2 = 0.62 \quad s = 0.65 \quad F = 21.6 \quad (P < 0.001)$$

$$\log \text{testis/liver} = -0.65\,\text{CLOGP} + 1.17 \quad (4.47)$$

$$n = 15 \quad r^2 = 0.67 \quad s = 0.75 \quad F = 26.2 \quad (P < 0.001)$$

The potency, IC_{50}, in liver cells was insensitive to variation in lipophilicity in the CLOGP range 0–2.

References

1 Chow, S.-L., Nagwekar, J.B., *J. Pharm. Sci.* **1993**, *82*, 1221–1227.

2 Walter, A., Gutknecht, J., *J. Membr. Biol.* **1984**, *77*, 255–264.

3 Frézard, F., Garnier-Suillerot, A., *Biochim. Biophys. Acta* **1998**, *1389*, 13–22.

4 Xiang, T.-X., Anderson, B.D., *Biophys J.* **1998**, *75*, 2658–2671.

5 Xiang, T.-X., Anderson, B.D., *Biophys. J.* **1997**, *72*, 223–237.

6 Makino, K., Ohshima H., Kondo, T., *Biophys. Chem.* **1990**, *35*, 85–95.

7 Makino, K., Ohshima, H., Kondo, T., *Biophys. Chem.* **1992**, *43*, 89–105.

8 Makino, K., Ohshima, H., Yanagisawa, Y., Kondo, T., *Biophys. Chem.* **1992**, *43*, 21–28.

9 Nook, Th., Doelker, E., Buri, P., *Int. J. Pharmaceutics* **1988**, *43*, 119–129.

10 Wagner, J, G., Sedman, A.J., *J. Pharmacokinet. Biopharm.* **1973**, *1*, 23–50.

11 Higuchi, W.I., Ho, N.F.H., *Int. Pharm. J.* **1988**, *2*, 10–15.

12 Taylor, D.C., Pownall, R., Burke, W., *J. Pharm. Pharmacol.* **1985**, *37*, 280–282.

13 Doluisio, J.T., Tan, G.H., Billups, N.F., Diamond, L., *J. Pharm. Sci.* **1969**, *58*, 1196–1200.

14 Schurgers, N., de Blaey, C.J., *Pharm. Res.* **1984**, *1*, 23–27.

15 Voegele, R.T., Marshall, E.V., Wood, J.M., in *Bioenergetics*, G.C. Brown, Ch.E. Cooper (Eds.), IRL Press, Oxford **1995**, pp. 63–84

16 Balon, K., Riebesehl, B.U., Müller, B.W., *Pharm. Res.* **1999**, *16*, 882–888.

17 Schaper, K.-J., Zhang, H., Raevsky, O.A., *Quant. Struct.–Act. Relat.* **2001**, *20*, 46–54.

18 Raevsky, O.A., in *Computer-Assisted Lead Finding and Optimization*, H. van de Waterbeemd, B. Testa, G. Folkers (Eds.), Verlag Helvetica Chimica Acta, Basel **1997**, pp. 367–378.

19 Sad, E., Katoh, S., Terashima, M., Kawahara, H., Katoh, M., *J. Pharm. Sci.* **1990**, *79*, 232–235.

20 Zhang, H., Raevsky, O.A., Schaper, K.-J., *Jahresbericht Forschungszentrum Borstel*, **1999**, pp. 136–139.

21 Vaes, W.H., Hermens, J.L., *Chem. Res. Toxicol.* **1997**, *10*, 1067–1072.

22 Raevsky, O.A., Schaper, K.-J., *Europ.J. Med. Chem.* **1998**, *33*, 799–809.

23 Raevsky, O.A., Festisov, V.I., Trepalina, E.P., McFarland, J.W., Schaper, K.-J., *Quant. Struct.–Act. Relat.* **2000**, *19*, 366–374.

24 Hillgren, K.M., Kato, A., Borchardt, R.T., *Med. Res. Rev.* **1995**, *15*, 83–109.

25 van de Waterbeemd, H., Camenisch, G., Folkers, G., Raevsky, O.A., *Quant. Struct.–Act. Relat.* **1996**, *15*, 480–490.

26 Abraham, M.H., Chadha, H.S., Mitchell, R.C., *J. Pharm. Sci* **1994**, *83*, 1257–1267.

27 Abraham, M.H., Chadha, H.S., Whiting, G.S., Mitchell, R.C., *J.Pharm. Sci.* **1994**, *83*, 1085–1100

28 Raevsky, O.A., Grigor'ev, V., Kireev, D., Zefirov, N., *Quant. Struct.–Act. Relat.* **1992**, *11*, 49–63.

29 Raevsky, O.A., Schaper, K.-J., Seydel, J.K., *Quant. Struct.–Act. Relat.* **1995**, *14*, 433–436.

30 Biganzoli, E., Cavenaghi, L.A., Rossi, R., Brunat, M.C., Nolli, M.L., *Il Farmaco* **1999**, *54*, 594–599.

31 Norinder, U., Österberg, Th., Artursson, P. *Pharm. Res.* **1997**, *14*, 1786–1791.

32 Artursson, P., Karlsson, J., *Biochem. Biophys. Res. Commun.* **1991**, *175*, 880–885.

33 Lipinski, C.A., Lombardo, F., Dominy, B.W., Feeny, P.J., *Adv. Drug. Deliv. Res.* **1997**, *23*, 3–25.

34 Burton, R., Conradi, A., Higers, A.R., *Adv.Drug Deliv. Res.* **1991**, *7*, 365–386.

35 Karls, M.S., Rush, B.D., Wilkinson, K.F., Vidmar, Th.J., Burton. Ph. S., Ruwart, M.J., *Pharm. Res.* **1991**, *8*, 1477–1481.

36 Conradi, R.A., Hilgers, A.R., Norman, F., Ho, H., Burton, Ph.S., *Pharm. Res.* **1992**, *9*, 435–439.

37 Stenberg P., Luthman, Chr., Artursson, P., *Pharm. Res.* **1999**, *16*, 205–213.

38 Pauletti, G.M., Okumu, F.W., Borchard, R.T., *Pharm. Res.* **1997**, *14*, 164–169.

39 Kansy, M., Senner, F., Gubernator, K., *J. Med. Chem.* **1998**, *41*, 1007 – 1010.

40 Danelian, E., Karlén, A., Karlsson, R., Winiwarter, S., Hansson, A., Löfås, St., Lennernäs, H., Hämäläinen, M.D., *J. Med. Chem.* **2000**, *43*, 2083–2086.

41 Caldwell, G.W., Masucci, J. A., Evangelisto, M., White, R., *J. Chromatogr. A* **1998**, *800*, 161–169.

42 Barbato, F., Capello, B., Miro, A., LaRotonda, M. I., Quaglia, F., *Il Farmaco* **1998**, *53*, 655–661.

43 Barbato, F., La Rotonda, M.I., Quaglia, F., *J. Pharm. Sci.* **1997**, *86*, 225–229.

44 Beigi, F., Gottschalk, I., Lagerquist Häglund Chr., Haneskog, L., Brekhan, E., Zhang, Y., Österberg, Th., Lundahl, P., *Int. J. Pharmaceutics* **1998**, *164*, 129–137.

45 Alifrangis, L.H., Christensen, I.T., Berglund, A., Sandberg, M., Hovgaard, L., Frokjaer, S., *J. Med. Chem.* **2000**, *43*, 103–113.

46 Junginger, H., de Boer, B., Borchardt, G., Bouwstra, J., Breimer, D., Danhof, M., Merkus, F., Verhoef, C. *Progress Report*, Center for Drug Research, University of Leiden, **1998**, 30–31.

47 Saito, H., Noujoh, A., Chiba, Y., Iseki, K., Miyazaki, K., Arita, T., *J. Pharm. Pharmacol.* **1990**, *42*, 308–313.

48 Walter, A., Gutknecht, J. *J. Membr. Biol.* **1986**, *90*, 207–211.

49 Xiang T.-X. *J. Membr. Biol.* **1994**, *140*, 111–122.

50 Ren, S., Das, A., Lien, E.J., *J. Drug Targeting* **1996**, *4*, 103–107.

51 Winiwarter, S., Bonham, N.M., Ax, F., Hallberg, A., Lennernäs, H., Karlen, A., *J. Med. Chem.* **1998**, *41*, 4939–4949.

52 Fagerholm, U., Lennernäs, H., *Eur. J. Pharm. Sci.* **1995**, *3*, 247–253.

53 Palm, K., Sternberg, P., Luthman, K., Artursson, P., *Pharm. Res.* **1997**, *14*, 568–571.

54 Palm K., Luthman, K., Ungell, A.-L., Strandlund, G., Beigi, F., Lundahl, P., Artursson, P., *J. Med. Chem.* **1998**, *41*, 5382–5392.

55 Palm, K., Luthman, K., Ungell, A.-L., Strandlund, G., Artursson P., *J. Pharm. Sci.* **1996**, *85*, 32–39.

56 Rapoport S.I., *NIDA Res. Monogr.* **1992**, *120*, 121–137.

57 Seydel, J.K., Trettin, D., Cordes, H.-P., Wassermann, O., Malyusz M., *J. Med. Chem.* **1980**, *23*, 607–613.

58 Seydel, J.K., Bürger, H., Saxena, A.K., Coleman, M.D., Smith, St.N., Perris, A.D., *Quant. Struct.–Act. Relat.* **1999**, *18*, 43–51.

59 Ohno, K., Pettigrew, K.D., Rapoport, S.I., *Am. J. Physiol.* **1978**, *235*, 299–307.

60 Greig, N.H., Genka, S., Rapoport S.I., *J. Controlled Release*, **1990**, *11*, 61–78.

61 Greig, N.H., Soncrant, T.T., Shetty, H.U., Momma, S., Smith, Q.R., Rapoport, S.I., *Cancer Chemother. Pharmacol.* **1990**, *26*, 263–268.

62 Hansch, C., *Cancer Chemother Rep.* **1972**, *56*, 433–441.

63 Young, R.C., Mitchell, R.C., Brown, Th.H., Gannellin C.R., Griffiths, R., Jones, M., Rana, K.K., Sunders, D., Smith, I.R., Sore, N.E., Wilks, T., *J. Med. Chem.* **1988**, *31*, 656–671.

64 van de Waterbeemd, H., Kansy, M., *Chimia* **1992**, *46*, 299–303.

65 El Tayar, N., Testa, B., Carrupt, P.-A., *J. Phys. Chem.* **1992**, *96*, 1455–1459.

66 Calder, J.A.D., Ganellin, C.R., *Drug Design and Discovery* **1994**, *11*, 259–268.

67 ter Laak, A.M., Tsai, R.S., Donné-Op den Kelder, G.M., Carrupt, P.-A., Testa, B., Timmerman, H., *Eur. J. Pharm. Sci.* **1994**, *2*, 373–384.

68 Abraham, M.H., McGowan, J.C., *Chromatographia* **1987**, *23*, 243–248.

69 Seelig, A., Gottschlich, R., Devan, R.M., *Proc. Natl. Acad. Sci. USA* **1994**, *91*, 68–72.

70 van de Waterbeemd H., Camenisch G., Folkers, G., Raevsky, O.A., *J. Drug Targeting* **1998**, *6*, 151–165.

71 Lombardo, F., Blake, J.M., Curatolo, W.J., *J. Med. Chem.* **1996**, *39*, 4750–4755.

72 Liotard, D.A., Hawkins, G.D., Lynch, G.C., Cramer, C.J., Truhlar, D.G., *J. Comput. Chem.* **1995**, *16*, 422–440.

73 Giesen, D.J., Cramer C.J., Truhlar, D.G., *J. Phys. Chem.* **1995**, *99*, 7137–7146.

74 Mohamadi, F., Richards, N.G.J., Guida, W.C., Liskamp, R., Lipton, M., Cau-

field, C., Chang, G., Hendrickson, T., Still, W.C., *J. Comput. Chem.* **1990**, *11*, 440–467.

75 Shah, M.V., Audus, K.L., Borchard, R.T., *Pharm. Res.* **1989**, *6*, 624–627.

76 Segarra, V., Lopez, M., Ryder, H., Palacios, J.M., *Quant. Struct.–Act. Relat.* **1999**, *18*, 474–481.

77 Crivori, P., Cruciani, G., Carrupt, P.-A., Testa, B., *J. Med. Chem.* **2000**, *43*, 2204–2216.

78 Baláž St., Lukáčová, V., *Quant.Struct.-Act. Relat.* **1999**, *18*, 361–368.

79 Wright, St.E., White, J.C., Huang, L., *Biochim. Biophys. Acta* **1990**, *1021*, 105–113.

80 Smejtek, P., Riker, W.K., Wright, C., Bennett, M.J., *Biochim. Biophys. Acta* **1990**, *1029*, 259–266.

81 Scheufler, E., Vogelsang, R., Wilffert, B., Pegram, B.L., Hunter, J.B., Wermelskirchen, D., Peters, T., *J. Pharmacol. Exper. Ther.* **1990**, *252*, 333–338.

82 Scheufler, E., Peters, T., *Cell. Biol. Intern. Rep.* **1990**, *14*, 381–388.

83 Seydel, J.K., Coats, E.A., Cordes, H.-P., Wiese, M., *Arch. Pharm.* **1994**, *327*, 601–610.

84 Barton, P., Davis, A.M., McCarthy, D.J., Webborn, P.J.H., *J. Pharm Sci.* **1997**, *66*, 1034–1039.

85 Betageri, G.V., Rogers J.A., *Pharm. Res.* **1989**, **6** 399–403.

86 Custodio, J.B.A., Almeida, L.M., Madeira, V.M., *Biochem. Biophys. Res. Commun.* **1991**, *176*, 1079–1085.

87 Welti, R. Mullikin, L.J., Yoshimura, T., Helcamp, Jr., G.M., *Biochemistry* **1984**, *23*, 6086–6091.

88 Jullien, S., Brajtburg, J., Bolard, J., *Anal. Biochem.* **1988**, *172*, 197–202.

89 Rogers, J.A., Davis, S.S., *Biochim. Biophys. Acta* **1980**, *598*, 392–404.

90 Steiner S.H., Moor, M.J., Bickel, M.H., *Drug Metabolism and Distribution* **1991**, *19*, 8–14.

91 Yokogawa K., Nakashima, E., Ishizaki, J., Maeda, H., Nagano, T., Ichimura F., *Pharm. Res.* **1990**, *7*, 691–696.

92 Yata, N., Toyoda, T., Murakami, T., Nishiura, A., Higashi, Y., *Pharm. Res.* **1990**, 7, 1019–1025.

93 Betageri G.V., Rogers J.A., *Int. J. Pharm.* **1988**, *46*, 95–102.

94 Hinderling, P.H., Schmidlin, O., Seydel, J.K., *J. Pharmacokin. Biopharm.* **1984**, *12*, 263–287.

95 Mason R.P., Rhodes, D.G., Herbette, L.G., *J. Med. Chem.* **1991**, *34*, 869–877.

96 Mohr, K., Struve, M., *Biochem. Pharmacol.* **1991**, *41*, 961–965.

97 Mason, P.R., Moring, J., Herbette, L.G., *Nucl. Med. Biol.* **1990**, *17*, 13–33.

98 Oliveira, C.R., Lima, M.C.P., Carvalho, C.A.M., Leyen, J.E., Carvalho, A.P., *Biochem. Pharmacol.* **1989**, *38*, 2113–2120.

99 Miller, G.H., Doukas, P.H., Seydel, J.K., *J. Med. Chem.* **1972**, *15*, 700–706.

100 Bock, L., Miller, G.H., Schaper, K.-J., Seydel, J.K., *J. Med. Chem.* **1974**, *17*, 23–28.

101 Seydel, J.K., Schaper, K.-J., in *Enzyme Inhibitors as Drugs*, M. Sandler (Ed.), Macmillan, London, **1980**, pp. 53–71.

102 Koch, A., Seydel, J.K., Gasco, A., Tironi, C., Fruttero, R., *Quant. Struct.–Act. Relat.* **1993**, *12*, 373–382.

103 Nikaido H., Thanassi, D.G., *Antimicrob. Agents Chemother.* **1993**, *37*, 1393–1399.

104 Vaara, M., Plachy, W.Z., Nikaido, H., *Biochim. Biophys. Acta* **1990**, *1024*, 152–158.

105 Seydel, J.K., Cordes, H-P., Wiese, M., Kansy, M., Kunz, B., in *Trends in Medicinal Chemistry*, Sh. Saral, R. Mechoulam, I. Agranat (Eds.), Blackwell Scientific Publications, Oxford **1992**, pp. 397–401.

106 Schop, H., Wiese, M., Cordes, H.-P., Seydel, J.K., *Eur. J. Med. Chem.* **2000**, *35*, 1–15.

107 Kansy, M., Seydel, J.K., Wiese, M., Haller, R., *Eur. J. Med. Chem.* **1992**, *27*, 237–244.

108 Hachtel, G.R.; Haller, R., Seydel, J.K., *Arzneim.-Forsch.* **1988**, *38*, 1778–1783.

109 Schlecht, S., Westphal, O., *Zbl. Bakter.1.Orig.* **1970**, *213*, 356–381.

110 Nikaido, H., *Biochim. Biophys. Acta* **1976**, *433*, 118–132.

111 Vaara, M., *Antimicrob. Agents Chemother.* **1993**, *37*, 2255–2260.

112 Zähringer, U., Lindner, B., Seydel, U., Rietschel, E.Th., Naoki, H., Unger, F.M., Imoto, M., Kusumoto, S., Shiba, T., *Tetrahedron Lett.* **1985**, *26*, 6321–6324.

113 Epand, M., Vogel, H.J., *Biochim. Biophys. Acta*, **1999**, *1462*, 11–28.

114 Polozov, I.V., Polozova, A.I., Molotkovsky, J.G., Epand, R.M., *Biochim. Biophys. Acta* **1997**, *1328*, 125–139.

115 Matsuzaki, K., Sugishita, K., Ishibe, N., Ueha, M., Nakata, S., Miyajima, K., Epand, R.M., *Biochemistry* **1998**, *37*, 11856–11863.

116 Sitaram, N., Nagaraj, R., *Biochim. Biophys. Acta* **1999**, *1462*, 29–54.

117 Wiese, A., Münstermann, M., Gutsmann, T., Lindner, B., Kawahara, K., Zähringer, U., Seydel, U., *J. Membrane Biol.* **1998**, *162*, 127–138.

118 R.N. McElhaney, E.J. Prenner (Eds.), *Biochim. Biophys. Acta* **1999**, *1462*, 244.

119 Rhodes, D.G., Sarmiento, J.G., Herbette, L.G., *Mol. Pharmacol.* **1985**, *27*, 612–623.

120 Mason, R.P., Chester, D. W., *Biophys. J.* **1989**, *56*, 1193–1201.

121 Herbette, L.G., Rhodes, D.G., Mason, R.P., *Drug Design and Delivery* **1991**, *7*, 75–118.

122 Carvalho, C.M., Oliveira, C.R., Lima, M.P., Leysen, J.E., Carvalho, A.P., *Biochem. Pharmacol.* **1989**, *38*, 2121–2127.

123 Colvin, R.A., Ashavaid, T.F., Herbette, L.G., *Biochim. Biophys. Acta* **1985**, *812*, 601–608.

124 Rhodes, D.G., Newton, R., Butler, R., Herbette, L.G., *Mol. Pharmacol.* **1992**, *42*, 596–602.

125 Herbette, L.G., Trumbore, M., Chester, D.W., Katz, A.M., *J. Mol. Cell Cardiol.* **1988**, *20*, 373–378.

126 Trumbore, M., Chester, D.W., Moring, J., Rhodes, D., Herbette, L.G., *Biophys. J.* **1988**, *54*, 535–543.

127 Chatelain, P., Laruel, R., *J. Pharm. Sci.* **1985**, *74*, 783–784.

128 Jendrasiak, G.L., McItosh, Th.J., Ribeiro, A., Porter R.St., *Biochim. Biophys. Acta* **1989**, *1024*, 19–31.

129 Sautereau, A.-M., Topurnaire, C., Suares, M., Tocanne, J.-F., Paillous, N., *Biochem. Pharmacol.* **1992**, *43*, 2559–2566.

130 Riva, E., Marchi, S., Pesenti, A., Bizzi, A., Cini, M., Veroni, E., Tavbani, E., Boeri, R., Bertani, T., Latini, R., *Biochem. Pharmacol.* **1987**, *36*, 3209–3214.

131 Lüllmann, H., Reil, G., Rossen, E., Seiler, K.U., *Virchows Arch. B. Zellpathol.* **1972**, *11*, 167–181.

132 Lüllmann-Rauch, R., Wassermann, O., *Dtsch. Med. Wschr.* **1973**, *98*, 1616.

133 Lüllmann, H., Mohr, K., in *Metab. Xenobiot.* (Pap. Eur. Meet. ISSX), J.W. Gorrod, J.W. Oelschläger (Eds.), Taylor & Francis, London, **1988**, pp. 13–20.

134 Lüllmann, H., Lüllmann-Rauch, R., Wassermann, O., *Arzneim.-Forsch.* **1981**, *31*, 795–799.

135 Seydel, J.K., Wassermann, O., *Biochem. Pharmacol.* **1976**, *25*, 2357–2364.

136 Seydel, U., Brandenburg, K., Lindner, B., Moll, H., *Thermochim. Acta* **1981**, *49*, 35–48.

137 Lüllmann, H., Vollmer, B., *Biochem. Pharmacol.* **1982**, *31*, 3769–3773.

138 Carlier, M.B., Brasseur, R., Laurent, G., Mingeot-Leclerq, M.P., Ruysschaert, J.M., Tulkens, P.M., *Fd. Chem. Toxic.* **1986**, *24*, 815–816.

139 Carlier, M.B., Brasseur, R., Ruysschaert, J.M., Tulkens, M.P., *Arch. Toxicol.* **1988**, Suppl. *12*, 186–189.

140 Kacew, S., *Toxicol. Appl. Pharmacol.* **1987**, *91*, 469–476.

141 Lüllmann, H., Schwarz, B., *Br. J. Pharmacol.* **1986**, *86*, 799–803.

142 Tsujita, Y., Kuroda, M., Shimada, S., Tanzawa, K., Arai, M., Kaneko, I., Tanaka, M., Watanabe, Y., Fujii, S., *Biochim. Biophys. Acta* **1986**, *877*, 50–60.

143 Balasubramanian,, N., Brown, P. J., Catt, J.D., Han, W.T., Parker, R.A., Sit, S.Y., Wright, J.J., *J. Med. Chem.* **1989**, *32*, 2038–2041.

144 Rosowsky, A., Forsch, R A., Moran, R.G., *J. Med. Chem.* **1991**, *34*, 463–466.

5
Drug-Membrane Interactions and Pharmacodynamics

Joachim K. Seydel

5.1
Drug Efficacy

Having discussed the many ways in which drugs can affect membrane properties and in which membrane properties can influence drug accumulation, orientation, and conformation, it should not be surprising that drug–membrane interactions can influence not only drug pharmacokinetics, toxicity, and selectivity, but also, directly or indirectly, the release of biological (pharmacologic) responses. It is well documented that the composition of the surrounding lipids affects the conformation of membrane-embedded proteins and, vice versa, that proteins inserted into phospholipids affect lipid organization [1–4].

It can be concluded that the integrity of the phospholipid matrix of cell membranes is essential for the cellular function of transmembrane proteins and receptors. Thus, it is logical to expect that incorporation of drugs into the phospholipid membrane can lead to a variety of biological responses. This may be related to:

- Conformational changes during lateral diffusion to the receptor site within the bilayer. In some cases the optimal conformation of the drug may be formed only within the lipid environment.
- A change in cooperativity and/or fluidity, in curvature or in surface charge, or replacement of Ca^{2+} ions by intercalation of the drug with the membrane, provoking an indirect effect on embedded proteins.
- A change in the dynamics of lipids, caused by domain formation of drugs within the membrane close to the phospholipid annulus surrounding the integrated receptor proteins.
- Direct or indirect effects on the defect structures at the phase boundaries of the lateral phase-separated bilayer domains, often the binding locus of enzymes.

It is obvious from the information and examples discussed so far that such effects are most likely evoked by catamphiphiles, which can interact with the membrane surface, i.e. the polar head groups, changing surface charge and the orientation of head groups. Depending on pK_a and lipophilic area, they can also interact with the upper region or with the interior hydrocarbon chains of the bilayer via hydrophobic forces. This is of practical relevance, as a large fraction of drugs used today and belonging to different classes, such as anesthetics, antidepressants, calcium antagonists (β-blocker, antiarrhythmics) and so forth, are cationic drugs (75%), 20% are acids, and less than 5% are non-ionized. Their influence on or responsibility on drug–membrane interaction and thereby their biological effect shall be discussed using a few representative examples.

5.1.1 Effect on Membrane-integrated Enzymes

5.1.1.1 Activation and Inhibition of Protein Kinase C (PKC)

A very detailed investigation was performed to study the role of membrane defects on the regulation of the activity of PKC [5]. The results can be summarized as follows. PKC activation is highly dependent on the fatty acid composition of the various PEs in the liposomes used. If PC/PS mixtures are used, PKC activation activity varies with the fatty acid composition of PS but not that of PC. The results are in agreement with the observation that phospholipids with the lowest bilayer–hexagonal phase transition temperature are most effective in activating PKC. Although PKC is sensitive to the presence of hexagonal phase-forming lipids, its activity is not dependent on differences between gel and lipid crystalline state membranes, nor do membrane effects alone explain the observed effects. This can be deduced from the fact that the introduction of phase boundary defects does not lead to the activation of PKC. Also, cholesterol has no influence, which supports the conclusion that PKC activation is not regulated by the fluidity of the membrane. Instead, activation is sensitive more to the fatty acid composition of the PE fraction of the membrane than to the total membrane fatty acid composition. The true nature of the physicochemical change in the membrane is still to be determined and may reflect a change in the membrane–water surface. It has been suggested that changes in membrane bilayer properties related to the propensity to assume a hexagonal phase are required for activation.

The effect of diacylglycerols (DAGs) on bilayers composed of PC and PS (4:1, mol/mol) was studied by ^{2}H-NMR. DPPC deuterated at the α- and β-position of the choline moiety was used to investigate the surface region of the membrane [6]. The effect of the studied DAGs on the dynamics, measured as change in ^{2}H-NMR spin–lattice relaxation times of the deuterated choline head groups, did not correlate with their effect on PKC activity and seemed not to be a sufficient factor in DAG-induced PKC activation. The directional change in α- and β-quadrupole splitting, however, significantly correlated with the induced PKC activation (Figure 5.1). The event directly responsible is assumed to be not counterdirectional quadrupole splitting but bilayer dehydration, which could increase the tendency of the lipid to form hexagonal phases [7, 8] (see Sections 1.1.3 and 1.1.4 for more details). The conclusion that PKC is activated via membrane interaction is in agreement with previous reports [9].

The opposite effect, the inhibition of PKC, has also been studied. It was concluded that inhibition by mono- and divalent cations is related to the reduction of surface potential of the studied PS/PC liposomes by ion binding according to the Gouy–Chapman theory. Later, the same authors studied the inhibitory effect of the local anesthetics tetracaine and procaine on PKC in liposomes [10]. The local anesthetics significantly reduced the negatively charged surface potential, Ψ, of phospholipid bilayers. The anesthetics were even capable of changing the surface potential to

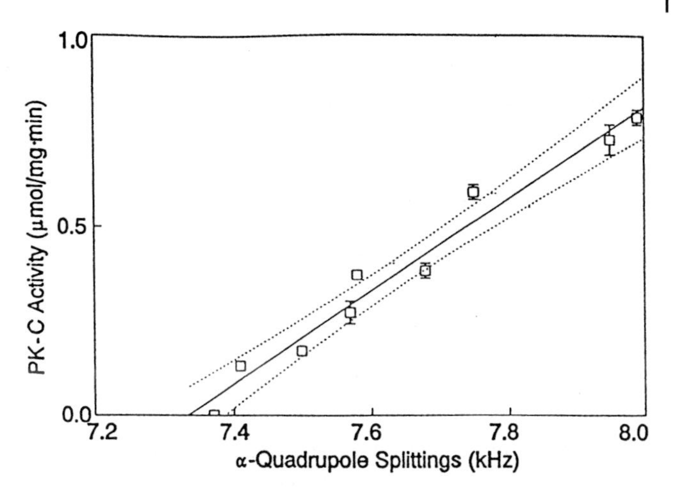

Fig. 5.1 Correlation of PKC activity with α-peak quadrupole splitting in the absence and presence of 15 and 25 mol% DAG ($r = 0.9843$, $P < 0.0001$). Dotted lines indicate 95% confidence interval. (Reprinted from Fig. 5 of ref. 6 with permission from the Rockefeller University Press.)

+10 mV in the presence of positively charged proteins. The data points obtained by the authors by plotting reduction in Ψ – caused by binding of tetracaine and procaine – against percent maximal activity of PKC fit the regression line of the plot of reduction in Ψ by sodium ions against percent maximal PKC activity obtained in a previous study. The result supports the assumption that the inhibition of PKC is related to a reduction in surface potential caused by local anesthetics as well as mono- and divalent cations.

5.1.1.2 Inhibition of Phospholipase A_2 (PLA_2)

Studies of membrane-associated effects of a series of drugs on PLA_2 have been undertaken to investigate the possibility that some proteins and drugs interact with the bilayer at phase boundaries or with defect structures necessary for PLA_2 activity. The direct or indirect effect of drugs on such boundaries or defects could then affect membrane–protein interactions.

This was investigated for a series of drugs including chemotherapeutics and anesthetics with quite different structures. Their effects on PLA_2 inserted into phospholipid bilayers was monitored by means of DSC. The bilayers were composed of phase-separated ternary lipid mixtures of DMPC/palmitoyllecithin (PL) and either hexadecanoic acid (FA_{16}) or hexadecanol [11]. The authors assumed that the defect structures at the boundaries of the lateral phase-separated bilayer domains are the locus of binding for the enzyme. The DSC results on pure DMPC and the ternary mixtures clearly indicated a regioselective interaction of these drugs with the membrane. The more hydrophilic drugs, gentamicin, streptomycin, and cytarabine, did not interact with DMPC but interacted preferably with the DMPC-rich region of the FA_{16}-containing ternary mixture. These drugs moved the thermogram of the acidic ternary mixture phase transition profile to the lower temperature transition, probably involving DMPC-rich phases. In contrast, the more lipophilic drugs, bucaine and dibucaine, showed pronounced effects in the higher-temperature region of the thermogram, probably representing domains relatively rich in PL_{16} and FA_{16}. In conclusion, the results showed that the drugs selectively interacted with different regions of the membrane. In the case of the charged hydrophilic drugs, charge interaction played a significant role. Even if the detailed nature of interaction remains to be evaluated, all studied drugs non-competitively inhibited PLA_2 in the membrane system. A sigmoidal concentration dependence of inhibition, and of action on membrane bilayer and membrane fusion, was observed for cytarabine. This could indicate a cooperative reorganization of the membrane after a certain threshold concentration has been passed. The authors assumed that the observed drug-induced effects were the result of the drug-mediated shift in phase equilibria away from the optimal active phase distribution. As a result, PLA_2 binds with normal affinity to the membrane, but its membrane substrate is not catalytically turned over. The inhibition of PLA_2 has to be seen in the light of the induction of phospholipidosis that is observed on prolonged treatment with some amphiphilic drugs.

Using PLA_2 incorporated into DPPC liposomes, the effect of ambuxol, chloroquine, imipramine, and chlorphentermine on the initial rate of hydrolysis, i.e. activity of PLA_2, has been studied [12]. In the absence of drug, the optimal activity was observed near T_t of DPPC. There was no indication of a lag phase. At lower and higher temperatures, an increase in lag phase duration and a decrease in activity was observed. These two parameters were used to characterize drug effects. Three types of inhibition profiles were observed for the four drugs. Ambroxol and imipramine shifted the temperature–activity profile toward lower temperature without influencing shape and profile. Both inhibited PLA_2 by decreasing T_t of the bilayer. Ambroxol also shifted the lag phase temperature profile to lower temperature. Chloroquine inhibited the activity in a dose-dependent manner and showed a low potency for decreasing T_t. Chlorphentermine, which has been studied most intensively as a lipidosis inducer (see Section 4.3), showed a mixture of the two mechanisms. It induced strong inhibition at T_t and a decrease in inhibitory activity at lower temperatures.

5.1.1.3 Drug–Membrane Interactions and Inhibition of Na^+,K^+-ATPase

The influence of a series of 12 cationic amphiphilic drugs, including anesthetics, antiarrhythmics, and psychotropic agents, has been studied on the binding equilibrium of [^{3}H]-ouabain to membrane suspensions of guinea-pig myocardium. The binding of ^{3}H-labeled ouabain to cardiac Na^+,K^+-ATPase is a method used to characterize the interaction between cardiac glycosides and their receptor.

The drugs inhibited ouabain binding in a concentration-dependent manner and reduced its affinity but did not affect the number of binding sites. The more potent drugs, such as chlorpromazine, propranolol, and dibucaine, diminished the association rate and increased the dissociation rate; the more weakly acting procaine only reduced the association rate of ouabain. This is an important observation. As the authors stated, a competitor would reduce the probability of the agonist hitting an unoccupied receptor but would not influence the dissociation of the agonist–receptor complex, i.e. the studied drugs did not act as competitors. Na^+,K^+-ATPase is a lipoprotein. Therefore, the catamphiphiles can interact with both the receptor protein or the lipid. The interdependence between the protein and its phospholipid annulus has also been investigated for Na^+,K^+-ATPase. The body of knowledge regarding the interaction with lipids as well as the fact that negatively charged lipids such as PS and phosphatidylinosin (PI) are associated with the protein moiety of ATPase and are essential for its function support the notion that lipid–drug interaction is the essential factor. It has been shown that ATPase activity depends on the fluidity of the membrane. For example, the inhibition of Na^+,K^+-ATPase by amiodarone has been reported to depend linearly on the variation induced in the lipid fluidity ($r = 0.95$) [13].

It has been shown that the binding of ouabain is considerably increased by reconstitution of Na^+,K^+-ATPase into PS and the replacement of Ca^{2+} from PS monolayers by catamphiphiles [14]. It can therefore be assumed that the primary action of the drugs involves the lipid portion of the membrane. In agreement with this suggestion, a highly significant correlation was found between the log IC_{50} and log P (r =0.94). The IC_{50} values were not corrected for the protonated form, because most of the compounds were more than 95% ionized at the pH of the experiment, except for 2-aminopyridine, which deviated strongly from the regression (see Table 3.3, Figure 5.2).

An interesting study showed the striking importance of drug–membrane interaction for pharmacological and/or toxicological effects [15]. The authors studied the ef-

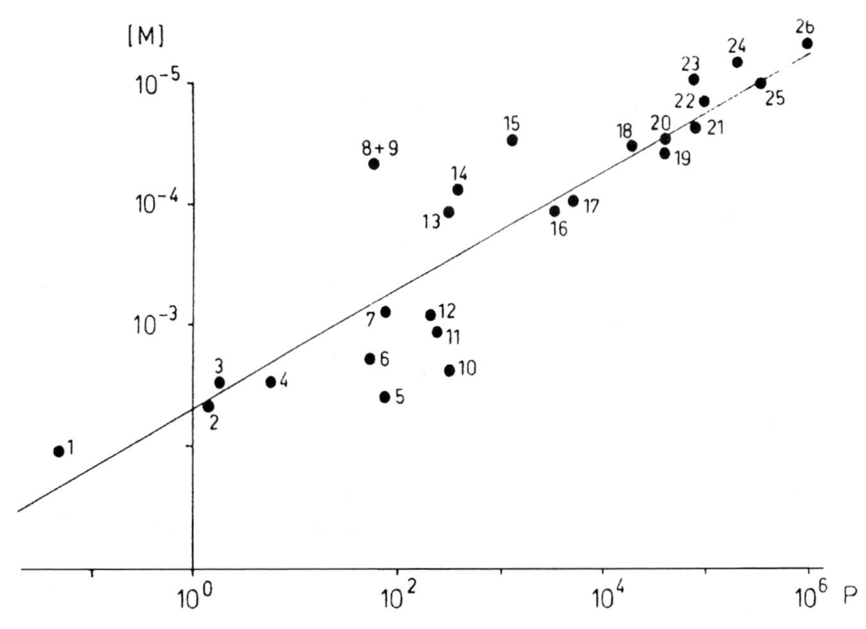

Fig. 5.2 Relationship between the IC_{50} values (ordinate) representing the efficiency of displacing Ca^{2+} from PS monolayers and the partition coefficients (octanol-water) for 26 compounds (abscissa). The regression line is obtained by using the logarithm of both parameters. The numbering corresponds to that given in Table 3.3. (Reprinted from Fig. 6 of ref. 14 with permission from Elsevier Science.)

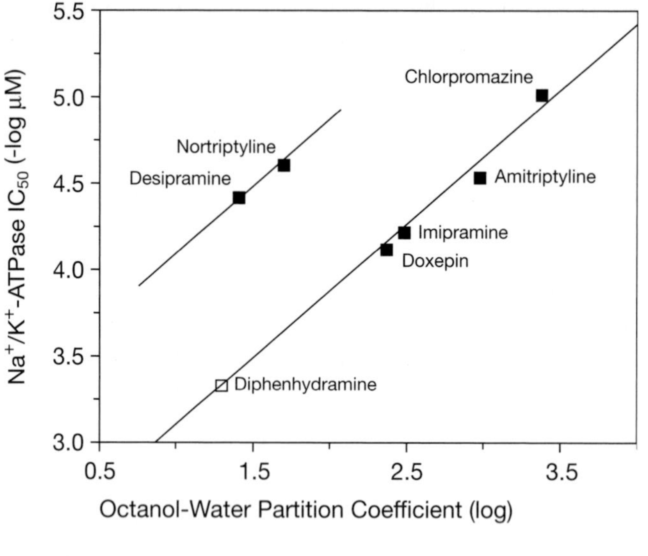

Fig. 5.3 The relationship between drug IC_{50} values and log $D_{oct.}$. (Reprinted from Fig. 2 of ref. 15 with permission from the American Society for Pharmacology and Experimental Therapeutics)

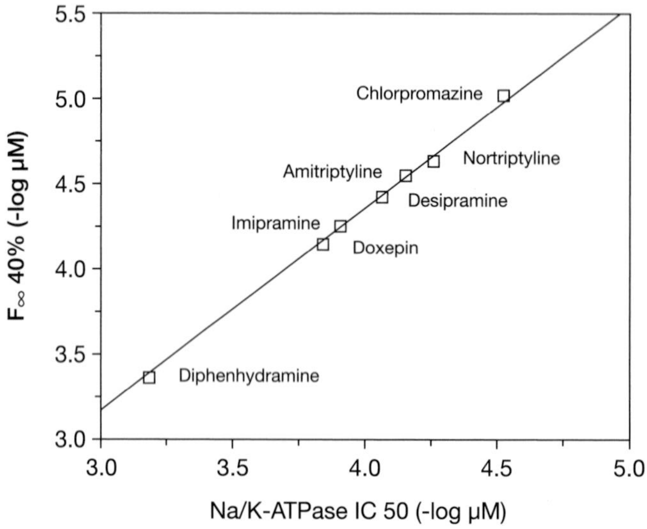

Fig. 5.4 Linear correlation between Na^+,K^+-ATPase IC_{50} and F_∞ 40% values for the tricyclic analogs including nortriptyline and desipramine deviating from the correlation in Fig. 5.3. (Reprinted from Fig. 4 of ref. 15 with permission from The American Society for Pharmacology and Experimental Therapeutics)

fect of seven structurally similar, mainly tricyclic, drugs on the synaptic plasma membrane. These drugs exert similar toxicological side-effects that differ from their pharmacological functions. Rat brain synaptic plasma membrane (SPM) perturbation and changes in Na^+,K^+-ATPase activity were compared with drug lipophilicity. ATPase log IC_{50} decreased with increasing log $D_{oct.}$, except in the case of nortriptyline and desipramine. These two drugs are less lipophilic (Δlog 1.2) than their N-methylated parent drug but were more potent inhibitors (Figure 5.3). The perturbation of SPM was then investigated by the binding of the fluorescence probe ANS to SMP. It was found that the dissociation constant and wavelength maximum of ANS did not change upon drug addition, but an increase in the limiting fluorescence intensity F_∞ of ANS was observed. According to the authors, this indicates that "these cationic drugs bind to the membrane surface, increasing the number but not the polarity of ANS binding sites by canceling charge at anionic phospholipid groups" [15]. The important result is the observed good linear correlation between the drug concentration necessary to increase F_∞ by 40% and the log IC_{50} values for ATPase inhibition, now including the N-demethylated drugs (Figure 5.4). It clearly points to the hydrophobic and electrostatic interaction of the drug molecules with the membrane surface and to a similar mechanism underlying both inhibition and fluorescence change. It is also important to note that the K^+-dependent *p*-nitrophenyl phosphatase is inhibited with the same potency as the Na^+,K^+-ATPase. The authors suggest that electrostatic interactions at the membrane surface between the protonated amino group of the drugs and anionic groups of the enzyme and/or phospholipid near the K^+ binding site are essential for the inhibition. Lipophilicity seems to play more of a modulating role regarding the number and orientation of the interactions. This is supported by the high density of negative surface charge in the lipid annulus of the Na^+ channel, like the Na^+,K^+-ATPase, and could explain the effect of catamphiphiles on other membrane–bound enzymes or receptor proteins [15].

5.1.2 Release of Pharmacological Response

5.1.2.1 Effect of Anesthetics

The molecular mechanism of local anesthetics is a largely unsolved problem. It remains unclear whether the action is predominantly caused by interaction with proteins of the channel or is mainly related to their effects on the membrane matrix and its physical properties. The various hypotheses regarding mechanisms of action of local anesthetics, especially with respect to the involvement of their effects on membrane structure and function, have been studied intensively and frequently and the results have been summarized and reviewed [16]. For this reason, only a few arguments shall be discussed here, principally some of those not discussed in the above review. It has been shown that highly lipophilic drugs can bind directly to the receptor via lateral diffusion through the membrane, and that compounds like chlorpromazine that are non-competitive blockers (NCBs) bind to the binding site at the acetylcholine protein differently from the competitive inhibitors [17]. Photoaffinity labeling indicates that the binding site of NCBs is deep in the pore of the open channel in a transmembrane region [18]. ESR studies have shown that the uncharged form of the local anesthetic first associates with the lipid core of the membrane and binds to the closed channel after diffusion through the membrane whereas the charged form of the drug, like the NCBs, acts on the open channel. The access to the receptor via membrane diffusion is also supported by patch-clamp experiments. Microperfusion of isoflurane into the medium outside of the patch led to an alteration of the channel activity within the patch. This was the case despite the use of a high-resistant membrane patch seal that enclosed the acetylcholine receptor [19]. Because of this and the high lipophilicity of isoflurane it can be suggested that the drug diffused through the membrane to reach the receptor. This idea is supported by several other findings. The penetration of the local anesthetics benzocaine, tetracaine, and procaine into synaptosomal plasma membrane vesicles (SPMVs) was studied using the fluorescent probe technique to evaluate the localization of the drugs in the membrane. It was found that procaine is predominantly distributed at the surface area, whereas tetracaine, in accordance with its high potency, has a greater access to the interior of the lipid bilayer of the SPMVs [20]. As early as 1970 a series of detailed experiments were performed to analyze the effect of local anesthetics on bilayers [21]. In one series of experiments the change in resistance of PC, PS, and phosphatidic acid (PA) membranes upon addition of the drugs on one side of the membrane was determined. The resistance decreased with increasing concentrations of the studied anesthetics. In another series of experiments, mixed membranes composed of the three phospholipids were used and the instability of the membrane due to Ca^{2+} binding on one side of the asymmetric acidic phospholipid membrane was studied in the absence and presence of anesthetics. The presence of the anesthetics stabilized the membrane. The order of potency of the drugs in lowering resistance and inhibiting instability of the asymmetric membrane due to Ca^{2+} membrane binding was in line with the relative concentration necessary for blocking the Na^{+}-channel. This implies that local anesthetics interact with the polar head groups by competing with Ca^{2+} as well as with the hydrocarbon core of the bilayer. The ranking of blocking concentration, relative potency of lowering membrane resistance, and inhibition of instability in asymmetric membranes is the same. Procaine was the least active and nupercaine the most active compound (Table 5.1).

Finally, it is worth mentioning a study in which chromatographic indices on IAMs were determined for 13 local anesthetics [22] (Table 5.2). Regression analysis result-

Tab. 5.1 Relative blocking potency of local anesthetics on natural membranes and various physicochemical effects of local anesthetics on artificial phospholipid membranes. (Reprinted from Tab. 1 of ref. 21 with permission from Elsevier Science)

	Minimum[a] blocking conc. (mM)	Relative blocking potency	Relative[b] blocking potency	Conc. for 5 mV reduction in ζ potential	Relative power of lowering resistance of the membrane[c]	Order of inhibition of instability of phosphatidylserine asymmetric membrane
Procaine	4.6	1	1	5.01	1	Procaine <
Cocaine	2.6	1.8	3.8	0.89	3	Cocaine <
Tetracaine	0.01	460	36	0.10	20	Tetracaine <
Nupercaine	0.005	920	53	0.02	60	Nupercaine

a) Obtained with desheathed frog sciatic nerve at pH 7.0.
b) Obtained with frog sciatic nerve at pH 7.2.
c) pH 7.3, 1 mM local anesthetics.

Tab. 5.2 Half-blocking doses for closed sodium channels, pK_a, and log K_w^{IAM} for a series of local anesthetics. (Adapted from Tab. 4 of ref. 22).

Compound	pD_{50} [a]	pD^i_{50} [b]	log K_w^{IAM}	pK_a
GEA968	–2.94	–2.46	0.38	7.7
Procaine	–1.94	–0.33	0.39	9.0
W36017	–3.18	–2.88	0.49	7.4
Tocainide	–2.82	–2.27	0.53	7.8
Prilocaine	–3.03	–2.48	0.62	7.8
Lidocaine	–2.15	–1.53	0.75	7.9
Mepivacaine	–2.17	–1.76	0.77	7.6
Trimecaine	–1.65	–1.35	1.21	7.4
Bupivacaine	–1.52	–0.74	1.45	8.1
Alprenolol	–1.74	0.46	1.53	9.6
Etidocaine	–0.83	–0.35	1.55	7.7
Tetracaine	0.47	1.60	1.75	8.5
Propranolol	–1.59	0.41	1.81	9.4

a) Half-blocking doses.
b) Half-blocking doses corrected for ionization at pH 7.4 (neutral fraction).

ed in a highly significant correlation with activity data obtained from closed sodium channels, whereas the correlation with log $P_{oct.}$ was less pronounced ($r = 0.766$, $s = 0.91$):

$$pD^i_{50} = 2.033\ (\pm 0.442)\ \log K_W^{IAM} - 3.121\ (\pm 0.505) \tag{5.1}$$

$n = 13$ $r = 0.812$ $s = 0.83$

For the subset of amide linked compounds the following equation was derived:

$$pD^i_{50} = 1.854\ (\pm 0.240)\ \log K_w^{IAM} - 3.355\ (\pm 0.229) \tag{5.2}$$

$n = 9$ $r = 0.946$ $s = 0.254$

However, the outliers in Eq. 5.1 were tetracaine and procaine which showed higher activities than predicted.

$$pD^i_{50} = 2.176\ (\pm 0.234)\ \log K_w^{IAM} - 3.553\ (\pm 0.262) \tag{5.3}$$

$n = 11$ $r = 0.952$ $s = 0.378$

In agreement with other studies assuming a greater contribution of the charged drug forms to membrane interaction, the use of pD_{50} uncorrected for ionization led to poorer regression equation ($r = 0.889$).

An interesting study regarding the possible involvement of drug–membrane interaction on the release of a biological effect was undertaken using the steroids alphaxalone and Δ^{16}-alphaxalone [23]. While alphaxalone shows anesthetic activity and is marketed as Althesin, Δ^{16}-alphaxalone, which differs from alphaxalone only by a double bond in the C-16 position, lacks anesthetic activity. The authors used ^{2}H- and ^{13}C-NMR, DSC, and X-ray diffraction to study the interaction of the two compounds with DMPC vesicles. Solid-state ^{2}H-NMR of the deuterium-labeled steroids at different drug concentrations revealed a poor incorporation of the inactive steroid. The central doublet due to the $COCD_3$ group in the 17-position was used as the measure of the degree of incorporation. The spectra for Δ^{16}-alphaxalone showed lower and nearly equal intensities of the doublet, indicating that at a molar ratio of $\chi = 0.01$ saturation had already occurred. In contrast, the signal of the doublet was much more intense and full incorporation was reached at $\chi = 0.10$ (10% molar ratio) for the active steroid. Again, at the higher molar ratio ($\chi = 0.01$) the active alphaxalone had also a considerably stronger effect on the pretransition temperature, i.e. the two steroids led to different effects on the dynamics and conformational properties of the phospholipid bilayer. The results of X-ray experiments support these findings. The X-ray data indicate a position of the 3-α-hydroxy group in the proximity of the carbonyl groups of the DMPC acyl chain. Comparison of the density profiles of the two steroids again showed a lower degree of incorporation into the bilayer for Δ^{16}-alphaxalone.

Finally, ^{13}C-NMR with cross-polarization and magic angle spinning (^{13}C-CP/MAS) has been used to examine changes in the mesomorphic states of phospholipids at various temperatures in the absence and presence of the two drugs. Three effects were followed:

1) change in peak intensity and line width due to membrane fluidity;
2) change in the chemical shift values of individual carbon nuclei of the phospholipid due to modified phase transition profiles; and
3) the appearance of a specific subset of peaks from the carbon nuclei of the incorporated drugs.

The results can be summarized as follows. Conversion between two mesomorphic states occurred at a rate slower than the NMR time scale. In the presence of alphaxalone this occurred at lower temperatures and indicated a stronger perturbation effect on the lipid chains than for the inactive Δ^{16}-alphaxalone. This is in agreement with the results obtained with the other techniques employed. In the "hydrophobic region" of the spectra, additional peaks of the steroid molecules were observed. At 25.5 °C, the C-18 and C-19 methyl groups appeared at 10.8 ppm and 15.9 ppm, respectively, in the presence of alphaxalone and at 11.9 ppm and 14.7 ppm, respectively in the presence of Δ^{16}-alphaxalone. The C-21 methyl carbon was only found for Δ^{16}-alphaxalone at 26.7 ppm. This indicated a lower mobility for Δ^{16}-alphaxalone. The results at least allow the assumption that, in the case of alphaxalone, drug–membrane interaction is involved in the biological response.

The molecular mechanism of local anesthesia, the location of the local anesthetic dibucaine in model membranes, and the interaction of dibucaine with a Na^+-channel inactivation gate peptide have been studied in detail by ^{2}H- and ^{1}H-NMR spectroscopy [24]. Model membranes consisted of PC, PS, and PE. Dibucaine was deuterated at H_9 and H_1 of the butoxy group and at the 3-position of the quinoline ring. ^{2}H-NMR spectra of the multilamellar dispersions of the lipid mixtures were obtained. In addition, spectra of deuterated palmitic acids incorporated into mixtures containing cholesterol were obtained and the order parameter, S_{CD}, for each carbon

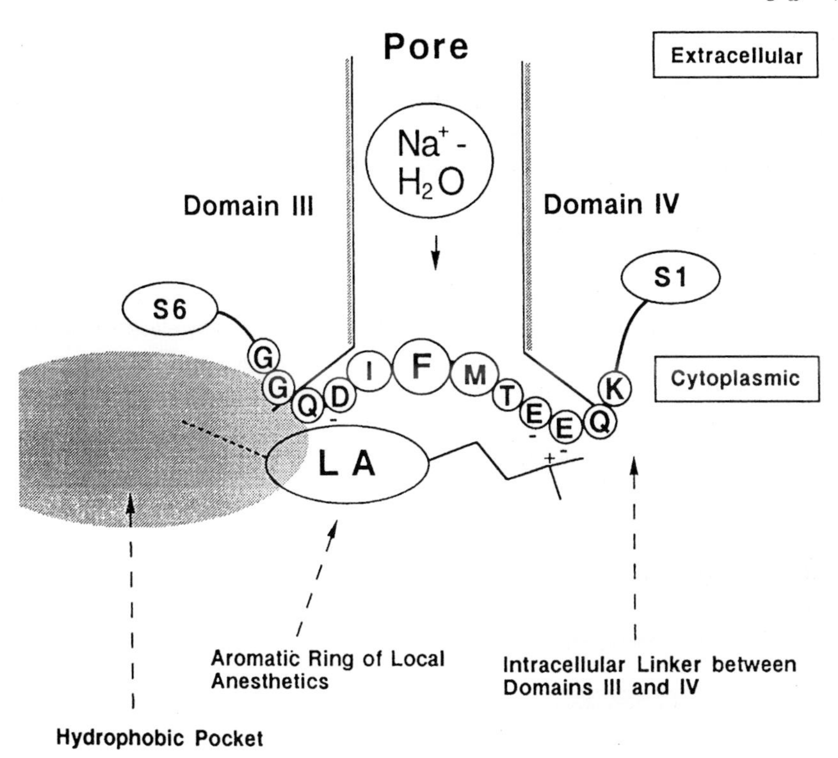

Fig. 5.5 A hinged-lid model for Na^+-channel inactivation and schematic representation of the interaction between a local anesthetic drug and the amino acid residues in the inactivation gate. (Reprinted from Fig. 2 of ref. 24 with permission from the Rockefeller University Press.)

segment was calculated from the observed quadrupole splitting. From these results the authors concluded that the butoxy group of dibucaine penetrates between the acyl chains of lipids in the model membrane and that the quinoline ring is situated at the polar head group region of the lipids, thus allowing for an interaction with a cluster of hydrophobic amino acids (Ile-1488, Phe-1489, Met-1490) within the cellular linker between the domains III and IV of the Na^+-channel protein. These domains function as an inactivating gate. The same amino acid residues were identified to be required for fast Na^+-channel inactivation [25]. This model assumes that a cluster of these hydrophobic amino acids within the linker between the domains III and IV of the protein channel "occludes the intracellular mouth of the activated Na^+-channel and stabilizes the inactivated state, making use of Gly-1484 (or Gly-1485 or both) and Pro-1509 residues on either side of the IFM domain as hinge points" [25] (Figure 5.5). In order to prove that the dibucaine molecule positioned at the surface region of the lipids can interact with the hydrophobic amino acids, the authors synthesized a model peptide, MP-1, that included the hydrophobic amino acids corresponding to the amino acid sequence of the linker part of rat brain type IIA Na^+-channel, and MP-2, in which Phe was replaced by Gln. They measured the ^1H-NMR spectra in phosphate buffer and in PS liposomes. Only in the presence of liposomes were changes in chemical shifts observed (Figure 5.6). It was found that the quinoline ring of dibucaine can interact with the aromatic ring of Phe by stacking of the two rings. This interaction is reinforced in the presence of lipids. Based on these experimental findings, the authors suggested the following molecular mechanism of local anesthetics. "The drug is residing at the polar head group region of the so called boundary lipids in the vicinity of the Na^+-channel pore, binds with the clustered hydrophobic amino acids, especially with the phenylalanine residue, and results in stabilization of the inactivated state thus making it impossible to proceed to the resting state" [25]. Drug binding to phenylalanine could be facilitated by electrostatic interaction of its protonated nitrogen with the negatively charged amino acids on both sides of the IFM domain. This is in accordance with a structure–activity relation derived by Sheldon *et al.* [26], who found an optimal distance between the aromatic ring and the tertiary amine nitrogen for binding to the Na^+ channel.

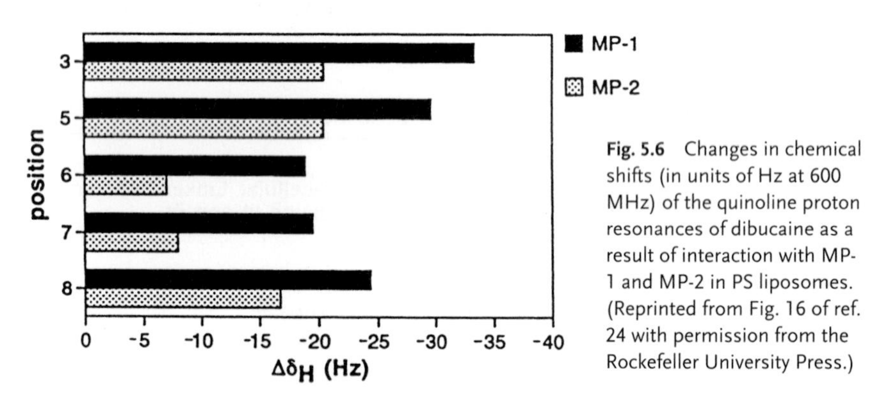

Fig. 5.6 Changes in chemical shifts (in units of Hz at 600 MHz) of the quinoline proton resonances of dibucaine as a result of interaction with MP-1 and MP-2 in PS liposomes. (Reprinted from Fig. 16 of ref. 24 with permission from the Rockefeller University Press.)

5.1.2.2 Negative Chronotropic (Cardiodepressant) Effect

The inhibition of spontaneous beat frequency induced by 21 amphiphilic drugs – six β-blockers, seven antiarrhythmics, and eight catamphiphilic drugs from various groups – was determined. The reduction in contractile frequency indicated a reduced rate of spontaneous heart beat caused by depolarization of the membrane potential in the sinus mode cells [27]. The authors hypothesized that the negative chronotropic effect may be caused by an indirect effect of the drugs on integral membrane proteins, similar to the inhibition of Na^+,K^+-ATPase. The derived correlation between the IC_{50} values and the inhibition of $^{45}Ca^{2+}$ binding to PS monolayers or the reduction in transition temperature of DPPA ($r = 0.85$ and $r = 0.85$ respectively) support this assumption. For some of the compounds, the effect is best described by their capacity to inhibit Ca^{2+} binding; for others the correlation is better when the reduction in T_t is used instead as the independent descriptor. When both parameters were used in combination an even better regression was obtained ($r = 0.94$). The result was interpreted as indicating that the negative chronotropic effect of the studied drugs is determined by both the introduction of charge into the head group region and the disordering effect on the arrangement of the phospholipid molecules. Another explanation could be that the negative chromotropic effect depends only on the disordering effect of intercalated drug molecules but that the extent of this effect is determined by a combination of both the intrinsic capacity of the drug molecule to disarrange the adjacent phospholipid molecules (reflected by T_t) and the number of drug molecules bound to the phospholipid bilayer at a given concentration, i.e. the binding affinity, reflected by the IC_{50} of inhibition of Ca^{2+} binding [27].

5.1.2.3 Anti-inflammatory Effect

Non-steroidal anti-inflammatory drugs (NSAIDs) such as indomethacin, diflunisal, flurbiprofen, and the active metabolite of sulindac, sulindac sulfide, are examples of anionic amphiphiles. Their interaction with liposomes and purple membrane was studied by DSC to determine the thermotropic behavior of these drugs in pure DMPC, DPPC and distearoylphosphatidylcholine (DSPC) and their mixtures. The pH dependence of the drug-induced changes in the thermotropic behavior of DMPC was also investigated [28]. As a sort of control, the inactive prodrug, sulindac, and one of its inactive metabolites were included in the study. To determine the drug localization within the membrane, several membrane reagents with known binding sites in the bilayer were used as reference markers and to monitor the effects released by the NSAIDs under study. Finally, the interaction of the NSAIDs with purple membrane was investigated.

The active drugs, including the active metabolite of sulindac, sulindac sulfide, all showed an intensive broadening of the phase transition signal of the liposomes, but only a very small decrease in phase transition temperature was observed ($< 3\,°C$) (Figure 5.7). Similar thermograms can be induced by hexane or decane, which are known to partition into the inner hydrocarbon region. In contrast, the inactive compounds sulindac and sulindac sulfone showed only a small effect on the phase transition temperature, but no effect on cooperativity, as indicated by the remaining sharp transition peak (Figure 5.7). The results suggested that the active derivatives penetrate deep into the bilayer whereas the inactive compounds interact only with the surface region or carbonyl groups of the bilayer. This finding was validated by the results of NMR and freeze-fraction techniques and by the use of membrane markers, which also induced specific and similar changes in the thermotropic profile of the bilayer. The results showed that the active and inactive compounds interact with different parts of the bilayer and that the charged active drugs can effectively partition into the bilayer. Using equilibrium dialysis the partition coefficient for the active sulindac sulfide in DPPC was determined to be $\log D_{DPPC} = 1.8$; for the prodrug sulindac, $\log D_{DPPC}$ was 0.25. It is interesting to note that similar effects on phase transition of DPPC are exhibited by other anti-inflammatory drugs such as indomethacin, flurbiprofen, and diflunisal [28]. In a recent paper, $\log P_{oct.}$ values were reported for a whole series of drugs with S-atoms in different oxidation states, including sulindac and its metabolites ($\log P_{oct.}^{anion} = 1.8$ for sulindac sulfide and $\log P_{oct.}^{anion} = -0.95$ for sulindac) [29].

The differences in membrane behavior of active and inactive anti-inflammatory drugs became visible in their effects on bacteriorhodopsin incorporated into phos-

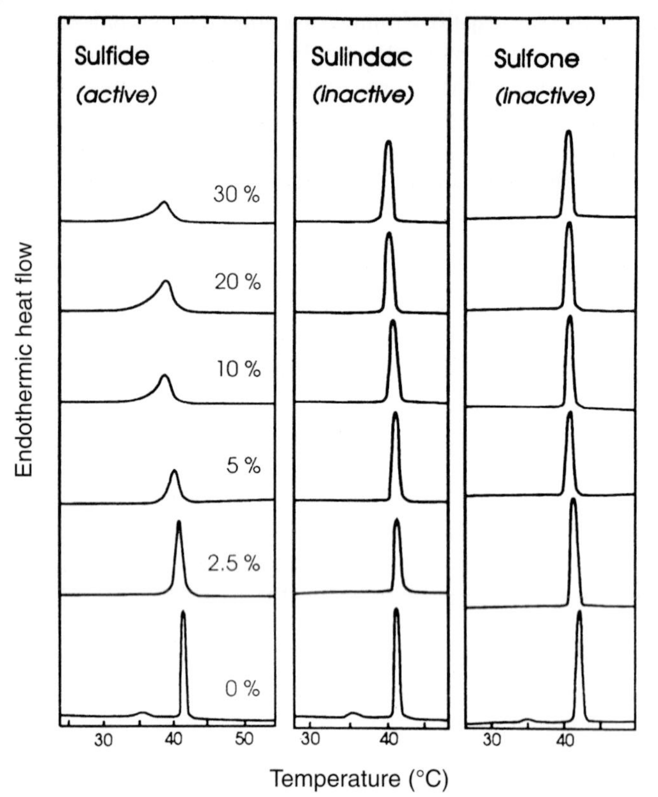

Fig. 5.7 Thermal scans of DPPC multilamellar dispersions with various mole ratios of sulindac sulfide, sulindac, and sulindac sulfone (50 mM phosphate buffer, pH 7.0). (Reprinted from Fig. 6 of ref. 28 with permission from the American Chemical Society.)

pholipid membranes at even smaller concentrations than used in the DSC experiments. Bacteriorhodopsin has a coat of hydrophobic amino acids that forms an interface with the lipid environment, shielding the charged and hydrophilic groups inside. For the active drugs that could partition into the bilayer, the authors found an interaction with the incorporated proteins leading to dissociation of the bacteriorhodopsin lattice even at low concentrations (< 20 µg/mL). This can be considered as an astonishing effect, because the bacteriorhodopsin lattice has been reported to resist even the removal of 80% of its lipids by detergents. Therefore, it is remarkable that the addition of active NSAIDs in small concentrations led to a dissociation of the bacteriorhodopsin lattice whereas the inactive prodrug and metabolite had no effect at all, even at high concentrations. It is also interesting to note that the water solubility at pH 7 of indomethacin, sulindac sulfide, sulindac, and sulindac sulfone was determined to be 0.4, 0.03, 2.7, and 0.5 mg/mL respectively. "Obviously, physical parameters measured in a solution phase related to protein binding and hydrophobicity do not have any simple relation with activities of compounds in a phospholipid membrane" [28].

Similarly, another study on NSAIDs came to the conclusion that the partition coefficients determined on immobilized artificial membranes are more suitable for predicting the effects of NSAIDs than log $P_{oct.}$ [30]. The reason for this was suggested to be the ability of phospholipids to counteract the influence of electrically charged functions of the drugs on lipophilic interactions. This is supported by the observation that log $P_{oct.}$ correlated much better with log K_w^{IAM} than log $D_{oct.7.4}$. As a measure of the biological activity of NSAIDs, the inhibition of two cyclooxygenases has been determined (Table 5.3). The enzyme is considered to be the NSAID locus of action. The following regression equations were derived:

$$pIC_{50} = 1.512\ (\pm 0.273)\ \log K_w^{IAM} - 2.990\ (\pm 0.505) \tag{5.4}$$

$n = 9 \qquad r = 0.902 \qquad s = 0.657$

$$pIC_{50} = 1.14\ (\pm 0.298)\ \log P_{oct.} - 4.583\ (\pm 1.140) \tag{5.5}$$

$n = 9 \qquad r = 0.816 \qquad s = 0.882$

Tab. 5.3 IC_{50} values (μmol/L) of NSAIDs on COX-2 activity in intact cells. (Reprinted from Tab. 4 of ref. 30, with permission from the American Chemical Society)

Compound	Log *P*	Log K_w^{IAM}	IC_{50} [a]
Salicylic acid	2.27	0.05	725
Ibuprofen	3.50	1.12	72.8
Tolmetin	2.79	1.13	27.2
Naproxen	3.18	1.26	5.65
Piroxicam	3.00	1.85	0.604
Flurbiprofen	4.16	2.02	0.102
Indomethacin	4.27	2.39	1.68
Diclofenac	4.40	2.43	1.1
Tolfenamic acid	5.70	2.75	0.0191
Aspirin [b]	1.13	–0.95	278
Sulindac [b]	3.42	1.80	112

a) IC_{50} for inhibition of cyclooxygenase.
b) Compound not considered to derive Eqs. 4 and 5.

Aspirin and sulindac were excluded from the regression. The activity of aspirin was underestimated, probably because of its additional acetylating potential, and sulindac activity was overestimated, which can be explained by its prodrug character.

In contrast to the effect on COX-2, in intact cells no significant correlation was obtained for the effect on COX-1. This might indicate that the two isoforms of cyclooxygenase are located in different subcellular environment. No correlation exists between log K_w^{IAM} and pIC_{50} determined on isolated COX-1 or COX-2 enzyme.

5.1.2.4 Effect of Antiarrhythmics

The mode of action of antiarrhythmics is related to their non-specific interaction with myocardial membranes and it is thought that antiarrhythmics stabilize the membrane [31]. The lipophilicity, Σf, was calculated for 15 antiarrhythmics, and the binding to PC determined by fluorescence spectroscopy [32]. The studied drugs covered a large range of lipophilicity (1.208–6.657). The authors deduced highly significant correlations between the reciprocal logarithm of the drug concentration leading to a 50% increase in the fluorescence signal in the binding experiments and the lipophilicity (Σf) of the drug.

$$\log 1/C = 0.29\ (\pm 0.08)\ \Sigma f + 3.62 \qquad (5.6)$$

$n = 15 \quad r = 0.91 \quad s = 0.20 \quad F = 62.2$

In addition, the strength of the interaction with the PC bilayer correlated well ($n = 8$, $r = 0.97$) with the average daily dose of eight antiarrhythmics for which information about the dose was available.

5.1.2.5 Calcium-Channel Blocking Activity

Retention times determined on IAM columns have been used as descriptors of lipophilicity and polar interactions with PC and have been compared with $\log P_{oct.}$ and $\log D_{oct.}^{7.4}$ for a series of 4-phenyldihydropyridines (DHPs) [33]. Satisfying correlations were found between $\log K_w^{IAM}$ and $\log P_{oct.}$ except for amlodipine, whereas the correlations with $\log D_{oct.7.4,}$ were poor, especially for nicardipine and amlodipine. This again underlines that the interaction of charged DHPs with the IAM phase is determined by lipophilicity and a specific interaction with phospholipids, whereas for neutral drugs similar partition-based mechanisms are involved in both partition systems. The IAM phase can accommodate the charged form of the drugs. Therefore, the $\log K_w^{IAM}$ value for amlodipine is much higher (2.59) than $\log D_{oct.7.4}$ (1.83) (also see Mason *et al.* [34] and ref. 35) and can describe the complex interaction with biomembranes better than $\log P_{oct.}$

For the neutral calcium-channel blockers, nifedipine, nitrendipine, nimodipine, and nisoldipine, receptor binding values were available. They were determined on protein preparations from rat cortical brain [36] and transformed to K_i values [33] – $K_i = IC_{50}/(1 + LC/K_d)$ – where LC is the ligand concentration, K_d the dissociation constant, and IC_{50} the concentration of the drug leading to a 50% inhibition of [^{3}H]-nimodipine specific binding. This means that the K_i value is a measure of the overall effect, including binding to the receptor as well as biomembrane permeation and interaction.

$$\log 1/K_i = 1.581\ (\pm 0.225)\ \log K_w^{IAM} - 3.621\ (\pm 0.511) \qquad (5.7)$$

$n = 4 \quad r = 0.980 \quad s = 0.145$

$$\log 1/K_i = 1.035\ (\pm 0181)\ \log P_{oct.} - 4.244\ (\pm 0.732) \qquad (5.8)$$

$n = 4 \quad r = 0.971 \quad s = 0.176$

The very small data set restricted generalizable conclusions, but according to the authors "the results suggested that membrane solubility may constitute the critical factor that allows the specific binding to the DHP-receptor".

5.1.2.6 α-Adrenoceptor Agonist Activity

Liposome partitioning was used to explain the observed hypotensive (pC25) and hypertensive activity (pC60) of seven and nine α-adrenoceptor agonists respectively [37]. The apparent partition coefficient in octanol–buffer, D, and in four liposome preparations, K_M, were determined and compared (1, DMPC; 2, DMPC/cholesterol/dicetylphosphate, 7:1:2 molar ratio; 3, DMPC/PS, 3.5:1; and 4, DMPC/stearylamine, 3:1) (Table 5.4). The log K_M values for all four systems showed a parabolic relation with log $D_{oct.}$, and in all systems log K_M values were larger than log $D_{oct.}$ In the liposome preparations log K_M values increased, as would be expected, in the following order: positively charged, neutral, negatively charged phospholipids. Regression analysis using various descriptors (log K_M, log $D_{oct.}$, V_w) was performed. The best correlation obtained for the hypotensive effect was with log $K_{M,2}$:

$$pC25 = -2.56\ (\pm 1.13) \log K_{M,2} + 6.33 \tag{5.9}$$

$n = 5 \quad r^2 = 0.94 \quad s = 0.228 \quad Q^2 = 0.84 \quad F = 51.6$

The correlation with log $D_{oct.}$ was poor ($r^2 = 0.429$).
The exceptions were tiamenidine and tetrazoline. In the case of tiamenidine, this is probably because of the heteroaromatic ring. Tetrazoline may deviate because of its

Tab. 5.4 Partition coefficients, log K'_M, of α-adrenoceptor agonists in liposome-buffer systems of different compositions and in the *n*-octanol-buffer (log D') system at 37 °C. (Adapted from Tab. 1 of ref. 37)

Compound	Log K'_M [a]				
	(1)	(2)	(3)	(4)	Log D'
Oxymetazoline	1.94	2.50	2.96	1.16	−0.32
Xylometazoline	1.94	2.40	2.80	1.30	0.40
Cirazoline	1.72	2.23	2.65	1.15	0.53
Tramazoline	1.48	2.17	2.59	0.77	−0.62
Naphazoline	1.34	2.12	2.45	0.70	−0.52
Lofexidine	1.24	1.76	2.20	1.20	0.73
Clonidine	1.15	1.61	2.01	1.17	0.85
Tiamenidine	1.02	1.53	1.94	0.61	−0.17
Tetryzoline	0.95	1.43	1.80	0.55	−0.90

a) The numbers in parentheses represent different liposome compositions: (1) DMPC; (2) DMPC/CHOL/DCP (7:1:2 mol ratio); (3) DMPC/PS (3.5:1 mol-ratio) (the average MW of brain PS was taken as 798 ± 4 on the basis of various fatty acid compositions; (4) DMPC/STA (3:1 mol ratio). The maximum standard deviation (SD) was ± 5%, although in most cases it was < ± 2%.
DCP, dicetylphosphate; STA, stearylamine

steric properties (attachment of the methylene bridge in *ortho*-position of the phenyl ring). Tiaminidine also deviated from the plot log K_M *vs.* log $D_{oct.}$.

The hypertensive effect was best described by a parabolic relation with log $K_{M,4}$:

$$pC60 = 18.80 \log K_{M,4} - 9.77 (\log K_{M,4})^2 - 6.60 \quad (5.10)$$

$n = 9$ $r^2 = 0.85$ $s = 0.875$

The correlation with log $(D_{oct.})^2$ led to $r^2 = 0.158$.

The results again show that liposome partitioning gives superior information and describes biological effects related to drug–membrane interactions more correctly than log $P_{oct.}$, especially in the case of charged drug molecules.

Later, the same pC25 values were correlated with the HPLC retention time of these compounds observed on an IAM column with PPS [38].

$$pC25 = -2.449 (\pm 0.79) \log K_{M,pps} + 3.44 (\pm 0.79) \quad (5.11)$$

$n = 6$ $r^2 = 0.95$ $s = 0.21$ $Q^2 = 0.91$ $F = 81.4$

5.1.2.7 Anticonvulsive Effect

Anticonvulsants are supposed to act on the Na^+-channel that is embedded in phospholipid membranes. In a cooperative study aiming at reducing the number of animal experiments, the interaction of flupirtine (**1**) derivatives with PPC vesicles was

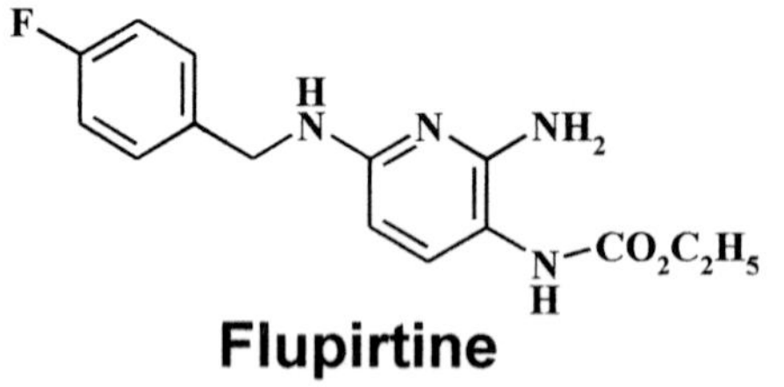

Flupirtine

investigated using NMR techniques [39, 40]. The interaction was quantified as discussed before by the change in relaxation rate, $1/T_2$, as a function of increasing phospholipid (DPPC) concentration. A linear dependence between the signal broadening of the methylene group and the lipid concentration was observed over a large range of lipid concentrations. The slope of such plots was used to characterize the degree of drug–lipid interaction. A highly significant correlation was observed between the degree of interaction and the observed *in vivo* anticonvulsive effect, measured by maximal electro shock (MES) test.

$$\log 1/I_{50(MES)} = 0.772 (\pm 0.12) \log 1/T_{2,slope} + 3.00 (\pm 0.19) \quad (5.12)$$

$n = 13$ $r^2 = 0.94$ $s = 0.086$ $Q^2 = 0.92$ $F = 178$

The observed increase in interaction was paralleled by an increase in the pharmacological response over a large range of activities. The correlation with log $P_{oct.}$ was inferior ($Q^2 = 0.68$). Derivatives bearing a polar substituent at the phenyl ring particularly deviated from the regression, again indicating that the octanol–water partition coefficient does not sufficiently describe the partitioning if other than bulk effects are involved.

A change in lipophilicity at the ethyl carbamate group connected to the pyridine ring led to a decrease in activity. This probably indicates the importance of the distribution of the lipophilic surface on the drug molecule with respect to its cationic center. The "hydrophilic–lipophilic dipole" could influence the orientation of the drug molecule within the membrane. To study this further, the test series was enlarged. As *ortho* substitution appeared to exert an additional effect on activity, the van der Waals volume, $V_{w,\ ortho}$, for the 2,6-positions of the benzyl group was examined as a potential parameter. As found in NMR experiments in dimethylsulfoxide, two conformers predominate, on the basis of observed CH_2-NH vicinal coupling constants: a folded pseudo-*gauche* conformation with a phenyl C-N-pyridyl torsion angle of 30–50° and an extended pseudo-*trans* conformation with the same angle in the 180–210° range. The conformers were related not only to the substitution pattern at the phenyl ring but also to the degree of ionization influenced by the electronic factors present at the pyridine ring. In fully ionized derivatives, both conformers were in equilibrium. To account for possible contributions of changes in the torsion angle to activity or drug–lipid interaction the torsion angles were calculated. Additionally, the van der Waals volume of the *ortho*-substituents at the phenyl ring and of the urethane moiety were considered as descriptors in the regression analysis.

$$\log 1/ED_{50MES} = 0.629\ (\pm 0.224)\ \log \Delta(1/T_2) + 0.040\ (\pm 0.027)\ \text{C-N}_{tors} + 0.035(\pm 0.014)\ V_{ortho} - 5.426\ (\pm 5.557) \quad (5.13)$$

$n = 21 \quad r^2 = 0.885 \quad s = 0.174 \quad Q^2 = 0.825 \quad F = 43.8$

The corresponding equation using $\log k'_{oct.}$ instead of $\log \Delta 1/T_2$ led to an $r^2 = 0.844$ and $Q^2 = 0.781$. It is also worth noting that $V_{urethane}$ can be used to replace the torsion angle.

The largest part of the observed variance in the pharmacological effect can, however, be explained by the change in relaxation rate upon interaction with the phospholipid. This simple method might therefore offer a suitable first screening system for selecting compounds for further animal testing for this class of compounds.

To delineate requirements for selectivity, correlation analysis was conducted for the neurotoxicity data, $\log 1/I_{50}(\text{NT})$, of flupirtine derivatives in a similar way. Unlike the correlation with the MES data, $\log k'_{oct.}$ is now clearly superior to the use of $\log \Delta 1/T_2$ in describing the observed variance in neurotoxic effects.

$$\log 1/I_{50(NT)} = 0.511\ (\pm 0.191)\ \log k'_{oct.} + 0.081\ (\pm 0.0038)\ \text{C-N}_{tors} + 0.042\ (\pm 0.022)\ V_{ortho} - 14.186\ (\pm 8.084) \quad (5.14)$$

$n = 20 \quad r^2 = 0.899 \quad s = 0.233 \quad Q^2 = 0.868 \quad F = 47.8$

Using $\Delta 1/T_2$ instead led to $r^2 = 0.863$, $Q^2 = 0.82$. It seems reasonable that the more unspecific neurotoxic effect of flupirtine analogs can be described by the simple partitioning parameter in octanol.

Further support for the thesis that the observed drug–membrane interaction directly or indirectly affects the "receptor" and does not represent pharmacokinetic influences can be derived from preliminary data of a small set of five derivatives for which some pharmacokinetic parameters were determined in rats [41]. The pharmacokinetic parameters – area under the curve (AUC), elimination rate constant ($k_{el,}$), half-life ($t_{0.5}$), the time of maximal concentration (t_{max}), and maximal concentration (c_{max}) – did not correlate significantly with either log $1/ED_{50(MES)}$, log $\Delta 1/T_2$, or log $K'_{oct.}$. Instead, even for this small set of compounds, log $1/ED_{50(MES)}$ correlated again significantly with both parameters log $\Delta 1/T_2$ and log $K'_{oct.}$ ($r = 0.998$ and 0.973 respectively).

It can be concluded that no significant differences exist between anticonvulsive and neurotoxic effects for this series of derivatives. The variation in the anticonvulsive effect seems to be determined by drug–membrane interactions and not by the pharmacokinetics of these compounds.

5.1.2.8 Antioxidant Effect

The mechanism of the antioxidant effect of the coronary vasodilator dipyridamole and its derivative RA-25 has been investigated [42]. In a previous study, the authors had found a good correlation between the lipophilicity of a series of dipyridamole derivatives and their protective effect. Detailed studies on the inhibition of Fe^{2+}-induced lipid peroxidation in mitochondria were then performed. No significant effect on state IV or II respiration was found at low drug concentrations. This excludes a direct interaction of dipyridamole or RA-25 with the peripheral benzodiazepine receptor. The association constants for dipyridamole and RA-25 in mitochondria were 0.7 and 0.2 mg/mL respectively. Through oxygen consumption studies it could also be shown that the antioxidant effect of the compounds did not involve the initial step of Fe^{2+} oxidation. The concentrations of dipyridamole and RA-25 required to cause a 50% inhibition of iron-induced lipid peroxidation differed by a factor of 100. The data support the assumption that the order of partitioning into the lipid phase of the mitochondrial membrane is decisive for the antioxidant effect of dipyridamole and RA-25 and not a specific binding to membrane proteins.

5.1.2.9 Antineoplastic Activity of Ether Phospholipids

The results of several investigations on the mechanism of action of tumoricidal ether phospholipids support the assumption that biochemical and physicochemical properties of the cell membrane are involved.

The fluidization and increase in permeability of tumor cell membranes has been reported [43, 44], as well as changes in cellular lipid synthesis during ether phospholipid-induced cytolysis [45]. Direct evidence was found by Diomede *et al.* [46] and Principe *et al.* [47] for the importance of the lipid composition of membranes for the sensitivity to antineoplastic ether phospholipids. These authors studied the influence of tumor cell membrane cholesterol content on the sensitivity of leukemic cells [46] and cells derived from three human carcinomas [47] with different rates of cell growth. It was shown that the rate of cell growth had no influence on the antineoplastic activity. Other factors must therefore be responsible for the observed difference in antineoplastic activity. Consequently, lipid composition and cholesterol content were analyzed for the cell lines HT29 (colon carcinoma), A427 (lung carcinoma), and BT20 (breast carcinoma). The three human cell lines showed substantial differences in their membrane lipid composition and a statistically significant difference in their cholesterol content. The following ratios of phospholipid–cholesterol content were found: A427, 6.01; BT20, 2.03; HT29, 2.44. The three ether phospholipids tested were: 1-*O*-octadecyl-2-*O*-methyl-*rac*-glycero-3-phosphocholine (edelfosine), 1-hexadecyl-mercapto-2-methoxymethyl-*rac*-glycero-3-phosphocholine (ilmofosine), and methoxy-3-*N*,*N*-methyl-octadecylamino-2-propyloxyphosphorylcholine (BN52211). The authors found a significant correlation between the membrane cholesterol level and the antineoplastic activity. The ratio phospholipid–cholesterol and the cholesterol content were related to the IC_{50} values of the ether phospholipids, the IC_{50} being highest for BT20 cells with the highest cholesterol content [47].

5.1.2.10 Antimalarial Activity of Chloroquine

Worthy of being cited is a paper that dealt with the ability of chloroquine to stabilize phospholipid membranes against SAG-induced perturbations. This observation is of interest with respect to the antimalarial effect of chloroquine [48]. ^{2}H-NMR spectra of DPPC-d_{62} or DPPS-d_{62} added to lipid extracts from normal human or malaria-infected erythrocytes, malaria parasites or model lipid mixtures were obtained. The spectra of the model lipid mixtures were almost identical to those from the erythrocyte extracts. There also was no difference in the spectra comparing added DPPC-d_{62} or DPPS-d_{62}. However, the spectra obtained from liposomes composed of lipid mixtures extracted from infected erythrocytes or malaria parasites differed from those of liposome preparations derived from normal erythrocytes. Whereas no difference in the order parameters was observed in normal cells and in model lipids, the order parameter in infected and parasite cell extracts was markedly reduced compared with that of normal erythrocytes. The addition of 1-stearoyl-2-*sn*-arachidonoylglycerol (SAG) induced the formation of non-bilayer phases in all lipid systems studied. The addition of chloroquine lead to only a small decrease in the order parameter of the acyl chains of PS but not of PC. In contrast, a strong effect became visible upon addition of chloroquine to the SAG-containing liposomes. The antimalarial agent almost totally reversed the SAG-induced non-bilayer phases. SAG is known to be a potent activator of phospholipase A_2 (PLA_2), and is endogenously formed in erythrocytes. The stabilizing effect of chloroquine on the membrane may be related to the known property of chloroquine to inhibit PLA_2 and therefore may be possibly responsible for its therapeutic and antitoxic mechanisms [48].

5.1.2.11 Conformation of Acetogenin Derivatives in Membranes and the Relation to Cytotoxicity

An interesting paper has been published describing the effect of phospholipid bilayers on the conformation of plant acetogenins known to be very cytotoxic [49]. The authors studied the influence of the alkyl chain length between the two tetrahydrofuran (THF) rings of acetogenins on location and conformation of acetogenin-type compounds in DMPC bilayers. The interaction was followed by DSC and various NMR techniques including NOE, NOESY, and the addition of Mn^{2+} to obtain information on the location of the compounds (see also Chapter 3). Asimicin and its derivatives, parviflorin and longimicin B, were studied. These compounds possess bis-THF rings flanked by hydroxy groups on hydrocarbon chains of various length (C_{15} to C_{24}). They also possess an additional terminal α,β-unsaturated γ-lactone on one of the alkyl chains and a 4-hydroxyl group. The results of ^{1}H-NOEs, ^{1}H-two dimensional NOE and DSC experiments indicated that the THF rings of all three derivatives studied were positioned near the polar interfacial head group region of DMPC. From the ^{1}H-difference NOE spectra it was concluded that the lactone rings of asimicin and parviflorin resided below the glycerol backbone in the membrane. These two derivatives differ only in alkyl chain length by two methylene groups, parviflorin possessing fewer carbon atoms. In contrast, the lactone ring of longimicin B, with an alkyl chain four carbon atoms shorter, is found close to the midplane in the membrane. These findings were supported by the results of broadening measurements by the addition of Mn^{2+} (see Chapter 3). The authors concluded on the basis of these results that, depending on the length of the alkyl chain, these asimicin-type acetogenins can adopt either sickle-shaped or U-shaped conformation in the membrane. The interesting point is the relation between the difference in conformation and the observed cytotoxicity. Parviflorin and asimicin show the same high cytotoxicity, but longimicin B, with the shortest alkyl chain and a U-shape conformation, exhibits no significant cytotoxicity [49].

5.1.2.12 A Membrane-forming and Inflammation-inducing Bacterial Macromolecule

Not only catamphiphilic drug molecules but also amphiphilic biomolecules like endotoxins are known to produce – depending on structure and concentration – biological effects through interaction with the host cell membrane. Endotoxins are integral constituents of the membrane of Gram-negative bacteria and participate in physiological membrane functions of these bacterial cells. They are essential for bacterial viability. In addition, they are important for the interaction with antibacterial drugs (see Section 4.3.1) and with components of the host immune system. At higher concentration, endotoxins are potent toxins in the course of Gram-negative infections. Ultimately, they can lead to septic shock. Endotoxins are composed of three structural units (see Figure 4.10): the O-specific side chain, the core region, and the lipid A component. Lipid A has been found to be the membrane-forming and inflammation-inducing principle of LPS. The structural requirements for endotoxin bioactivity have been analyzed and determined. They were described in a review article [50].

Striking correlations were found between T_t of a specific endotoxin and various biological effects. The most significant correlation with a biological effect was, however, found for the ability of endotoxins to adopt particular supramolecular structures. Lipid A samples with lamellar structure were completely inactive (no monokinase secretion), those with mixed lamellar/cubic structures had a modest activity, and those samples which were in pure non-lamellar form (Q/H_{II}, conical conformation) were highly active [50, 51] (Figure 5.8) [49]. The biological activity of endotoxins is certainly due to the interaction with host cell membranes. Lipid A constitutes the molecular entity that primarily interacts via hydrophobic forces with the host cell membrane. The sugar moiety of LPS has a modulating effect on lipid A bioactivity via its influence on the hydrophobicity of the endotoxin molecule and through this on the critical aggregate concentration (CAC) but also on the fluidity of the acyl chains of lipid A. Detailed studies on T_t of lipid A and the chemical structure of the sugar region were performed [51].

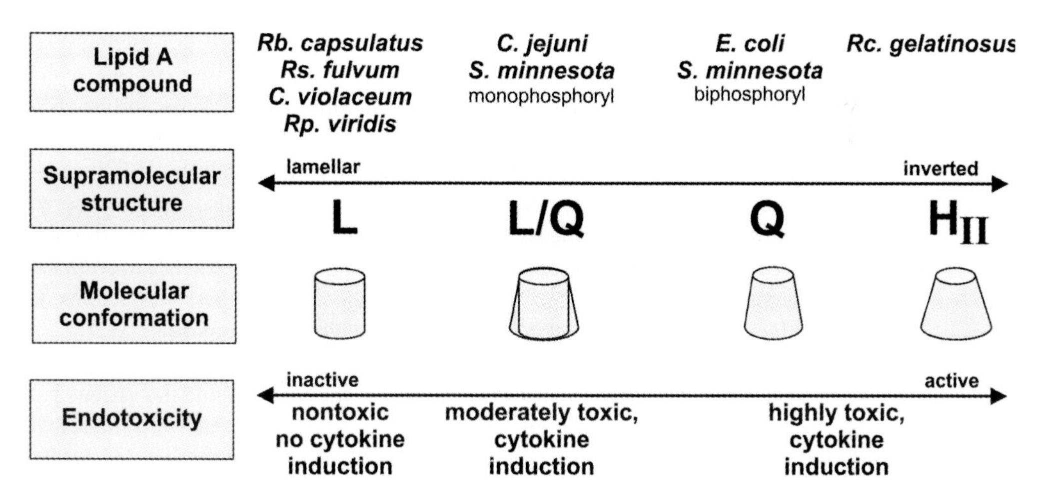

Fig. 5.8 Correlation between supramolecular structure/conformation and biological activity for various forms of lipid A at 37 °C.

In this respect, it is interesting to note that Brade *et al.* [52] found a different correlation between structure and activity for immunogenicity and antigenicity. Activity was low for isolated lipid A, but significantly higher for lipid A incorporated into the lipid matrix of phospholipid liposomes, probably because of the existence of a lamellar phase in the latter case. This can lead to an exposure of critical epitopes necessary for the manifestation of immunological activities.

5.1.2.13 **Drug–Membrane Interactions involved in Alzheimer's Disease**

Metabolite–membrane interaction may be involved in Alzheimer's disease. The disease is characterized by changes in phospholipid metabolism, which lead to perturbations in the level of phosphomonoesters such as L-phosphoserine (L-PS). It is speculated that the change in lipid metabolism may lead to defects in the bilayer structure, which then lead to increased β-amyloid formation. The effect of L-PS on membrane bilayers has therefore been studied by small-angle X-ray diffraction and DSC. Liposomes were composed of lecithin and cholesterol [53]. From the X-ray diffraction data the authors generated a one-dimensional electron density profile of a control DMPC/cholesterol bilayer. After incubation with L-PS, a broad decrease in electron density was observed about 4–12 Å from the lipid center. At the same time, an increase in the width of the phospholipid head groups' electron density occurred and the lipid bilayer width was reduced by 3 Å. This interaction of L-PS was concentration dependent and influenced by the cholesterol content. The result was reproducible when mixtures of egg phospholipid and cholesterol were used instead of DMPC. Millimolar concentration of L-PS decreased the phase transition temperature cooperativity in a concentration-dependent manner. The effect was greater in liposomes containing 10% cholesterol. The authors assumed that these data provide "direct evidence that phosphomonoester levels modulate the biophysical properties of the membrane lipid bilayer which may, in turn, lead to altered structure/function relationships in Alzheimer's disease".

This example concludes the section on drug–membrane interaction and its relation to biological activity. It seems surprising that drugs that differ so much in structure and which belong to various classes can inhibit enzymes catalyzing different reactions and using different substrates. In this respect it should also be remembered that the inhibition of an enzyme such as firefly luciferase embedded in a membrane can be used to estimate and describe the induction of general anesthetics in animals. The observations discussed seem to exclude specific drug–receptor interactions as the only factor or rate-limiting step determining the biological response.

If in the discussed examples the mode of action were determined only by a direct interaction between the drug and a receptor protein then why should the various parameters derived from the observed effects on membranes correlate with the activities but not significantly or less significantly with the octanol–water partition coefficient? Is it not more likely that the intercalation in the membrane and the lateral diffusion to the receptor or the direct interaction of the drug molecules with the membrane lipids can affect the functioning of the membrane-integrated proteins and are responsible for the biological response? The difficulty is that we have an arsenal of techniques to study drug–membrane interactions which allow us to describe and to quantify effects of drugs on model membranes and vice versa, but it is still not possible to relate these effects on a membrane to a definitely defined mechanism of action.

5.2
Drug Resistance

5.2.1
Bacterial Cells

Generally, infectious diseases can effectively be treated with antibiotics. However, the emergence of resistance is changing this situation. Drug resistance is becoming an increasing problem in the therapy of various infectious diseases. Some of possible reasons leading to drug resistance of bacteria are summarized:

- intrinsic resistance
  - structural deficits within the structure of the drug molecule;
  - insufficient properties to reach the receptor through passive diffusion;
- acquired resistance
  - mutation or selection of cells under therapy;
  - change in membrane composition, leading to decreased permeability (permeability barriers);
  - overproduction of a target enzyme;
  - change in the structure of the target enzyme;
  - overproduction of bacterial efflux systems (LMRA);
- protection by intracellular growth
  - drug becomes inactivated within the macrophages;
  - insufficient permeability of the macrophage membrane for the drug;
  - stress situation like for mycobacteria within the macrophages where the synthesis of multilamellar cell membrane components is induced

Owing to the subject of this book, the discussion will focus on possible changes in the permeability of the bacterial membrane. The pharmaceutical industry is for the most part successful in developing drugs that can overcome specific mechanisms of resistance, but less successful in overcoming unspecific mechanisms such as permeability barriers. These may become more and more important. As a quick recap: most Gram-positive bacteria are surrounded by a thick peptidoglycan cell wall that can generally be passed by small molecules such as antibiotics because of its coarse network. Gram-negative bacteria such as *E. coli,* in contrast, are surrounded by a second outer membrane whose outer leaflet is composed of unusual LPS. The absence of unsaturated fatty acids in the lipid probably makes the LPS leaflet even more rigid. A balance between lipophilic and hydrophilic properties of the drug molecule is therefore requested. Even organisms with cell envelopes of relatively high permeability can develop resistance by decreasing pore size or number of channels and/or by changing the composition of the membrane [54].

The role of drug–membrane interactions in overcoming, or at least in reducing, resistance will be discussed using two examples of antibacterials. A series of 5-(substituted)benzyl-2,4-diaminopyrimidines and 4,6-diamino-1,2-dihydro-2,2-dimethyl-1-(3-substituted)phenyl-s-triazines, inhibitors of dihydrofolate reductase (DHFR), were tested against sensitive and resistant *E. coli* cell cultures as well as against *E. coli*-derived DHFR [55] and compared with the effect on sensitive and resistant murine tumor cells. Whereas the triazine were found to inhibit sensitive and resistant *E. coli* strains to the same degree, marked differences were observed for benzylpyrimidines tested against the two strains. The following regression equations were derived for inhibition of *E. coli*-derived DHFR:

$$\log 1/K_i = 1.33\ MR'_{3,5} + 0.94MR'_4 + 5.69 \tag{5.15}$$

$n = 34 \quad r = 0.904 \quad s = 0.281$

where MR represents the molar refractivity of the indicated substituents and is primarily a measure of the volume of the substituent. The prime with MR indicates that MR was truncated (in the 3- and 5-position) at the value of 0.79. This implies that the maximum value of MR at any position is 0.79 and 2.37 for the sum of all three positions (MR for H = 0.1). The best equation for a 50% inhibition of whole cell cultures of sensitive *E. coli* was:

$$\log 1/IC_{50} = 1.15\ (\pm 0.22)\ MR'_{3,4,5} + 0.27\ (\pm 0.22)\ \pi_{3,4,5} - 0.14\ (\pm 0.08)\ \pi^2_{3,4,5} + 3.79\ (\pm 0.31) \tag{5.16}$$

$n = 28 \quad r = 0.916 \quad s = 0.341 \quad F_{1,24} = 14.1 \quad \pi_o = 0.94$

The lack of a hydrophobic term for the isolated enzyme system in Eq. (5.15) and the presence of such a term in Eq. 5.16 for the cell culture was presumed to be due to the interaction with lipophilic components of the cell membrane.

The final equation for the inhibition of resistant cell cultures by the benzylpyrimidines was:

$$\log 1/IC_{50} = 1.39\ (\pm 0.16)\ MR'_{3,4,5} + 0.35\ (\pm 0.08)\ \pi_{3,4,5} + 2.11\ (\pm 0.23) \tag{5.17}$$

$n = 26 \quad r = 0.969 \quad s = 0.238 \quad F_{1,23} = 87.3 \quad (MR'_{3,4,5}/\ \pi_{3,4,5}\ r = 0.04)$

The major difference between the two equations is in the intercept. Of greater interest, though, is the difference in π_o, which describes the optimal lipophilicity. Whereas there was a non-linear (parabolic) dependence of antibacterial activity on lipophilicity in the case of the sensitive strains [$\pi_o = 0.94$ (0.29–1.28)], the activity against the resistant strain linearly increased with lipophilicity at least up to π of 3.2.

The same two patterns of correlation between activity and lipophilicity were found for both triazines and benzylpyrimidines as inhibitors of sensitive and resistant murine tumor cells, and for triazines acting on sensitive and resistant *L. casei* cells. In the sensitive strains, the cut-off point for an increase in activity with increasing lipophilicity of the drugs was observed much earlier. The authors [55] concluded "hence one can design more effective drugs for resistant bacterial or cancer cells by making more lipophilic congeners. In doing so, one can of course, exceed log P_o for the whole animal system".

A later study using a more homologous series of 5-(substituted benzyl)-2, 4-diaminopyrimidines led to a surprising result [39, 56]. All derivatives had a methoxy group in the 3-position of the benzyl ring. Their lipophilicity varied only with the

Tab. 5.5 Inhibitory activity (IC_{50}, uM) of 3-OCH_3, 4-alkoxy-benzylpyrimidines ($x = 1$–6) in cell-free and whole cell systems of *E. coli* ATCC 11775 (TMP sensitive) and *E. coli* RT 500 (TMP resistant) and the lipophilicity descriptor log k'_r. (Reprinted from Tab. 2 ref. 39)

	IC_{50} (uM) cell-free *E. coli* ATCC 11775	RT 500	IC_{50} (uM) whole-cell *E. coli* ATCC 11775	RT 500	Log k'_r
Trimethoprim	0.0018	0.0023	0.97	1147	0.211
GH 01	–	–	1.46	1168	0.706
GH 02	0.02	0.017	4.99	885	1.299
GH 03	0.014	0.013	13.43	738.3	1.849
GH 04	0.015	0.015	17.52	163	2.427
GH 05	0.018	0.018	29.2	55.6	3.001
GH 06	0.022	0.021	31.2	18.6	3.575
Brodimoprim	0.0018	0.0011	0.81	118.4	1.382

length of the alkoxy group in the 4-position, the number of methylene groups being n = 1 to n = 7. The derivatives were tested against DHFR derived from the sensitive *E. coli* ATCC 11775 and resistant *E. coli* RT 500, the latter being an overproducer of DHFR, and against the two corresponding cell cultures. No difference in inhibitory activity was observed against DHFR derived from the sensitive and the resistant strains, indicating no change in the active site of the target enzyme (Table 5.5). The inhibitory activity against sensitive cultures decreased with increasing chain length of the alkoxy group, i.e. with an increase in lipophilicity. In contrast, an increase in inhibitory activity was observed with an increase in lipophilicity in the case of the re-

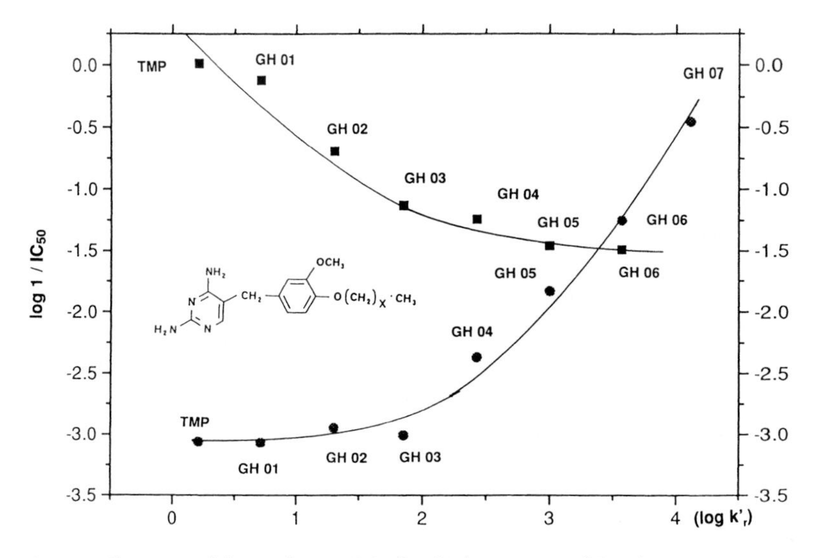

Fig. 5.9 Change in inhibition (log $1/IC_{50}$) of multiplication rate of *E. coli* (■, TPM sensitive; ●, TPM resistant) as a function of lipophilicity (log k_r') of 3-methoxy-4-alkoxypyrimidines. (Reprinted from Fig. 1 of ref. 56.)

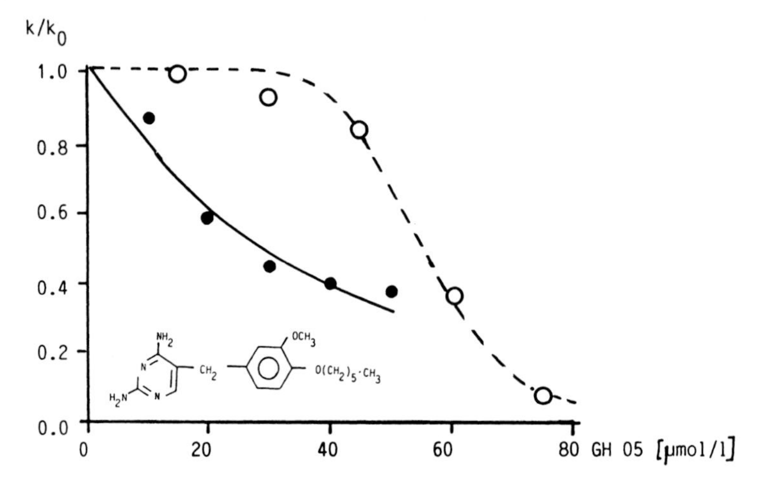

Fig. 5.10 Effect of GH 305 on generation rate, k, of (a) *E. coli* RT 500 (TMP resistant) (– – –), (b) *E. coli* ATCC 11775 (TMP sensitive) (——); k_o, generation rate of control culture. (Reprinted from Fig. 2 of ref. 56.)

sistant cell culture. So far, this was in agreement with the postulate of the data discussed above. The heptoxy derivative had almost the same activity against the resistant and the sensitive strain, $IC_{50} = 0.025\ \mu M$. However, its activity was still about 10 times weaker than the standard, trimethoprim (Figure 5.9). Comparing the dose–response curves for the inhibition of sensitive and resistant strains (this should always be done) a dramatic difference became obvious for derivatives with $(CH_2)_n \gg 4$ (Figure 5.10), indicating a change in the mechanism of action.

Further support for this assumption comes from studies of the activity of combinations of these DHFR inhibitors with sulfonamides, known to act synergistically in combination with DHFR inhibitors. For the combination of trimethoprim and a sulfonamide, a synergism in activity is seen even against resistant cultures. In the same resistant cells no synergism was observed for long-chain derivatives in combination with sulfonamides. A long-chain derivative in combination with trimethoprim was in fact antagonistic, although it was expected to show at least additive effect. In the case of resistant *E. coli* cells, these highly lipophilic benzylpyrimidines are no longer primarily DHFR inhibitors. Instead, they exert their effect by interacting with the membrane components of the resistant cells, probably as detergents, i.e. a cationic soap. The resistant *E. coli* RT 500 used is not only an overproducer of DHFR. Not only that but it has defects in its membrane and is a rough mutant. The changed membrane structure seems to offer suitable interaction sites for these catamphiphiles with membrane components of the resistant strain.

The observed variation in the IC_{50} for the sensitive *E. coli* culture can be described by a parabolic relationship to the lipophilicity (log k') of the benzylpyrimidines:

$$\log 1/IC_{50} = -0.939\ (\pm 0.137) \log k' + 0.120\ (\pm 0.0359\ (\log k')^2 + 0.296\ (\pm 0.109) \quad (5.18)$$

$$n = 7 \qquad r = 0.99 \qquad s = 0.10 \qquad F = 107$$

and for the resistant strain the relation between activity and lipophilicity can be described by an equation to the second power:

$$\log 1/I_{50} = 0.150\ (\pm 0.01)\ \log k'^2 - 3.168\ (\pm 0.065) \qquad (5.19)$$

$n = 7 \qquad r = 0.989 \qquad s = 0.12 \qquad F = 223$

In conclusion, in the case of resistant *E. coli* cultures, the increase in antibacterial activity of the more lipophilic benzylpyrimidines is due not to a favorable influence on the transport through the cell membrane but to their interaction with membrane components leading to membrane destruction and cell death.

For more information, see the recent review on the progress in overcoming drug resistance and on the development of new antibiotics [57].

5.2.2 Reversal of Multidrug Resistance in Tumor Cells

Drug resistance is an increasing problem not only in the chemotherapy of infectious diseases but also in tumor chemotherapy. The so-called multidrug resistance (MDR) phenomenon is often associated with the amplification or overexpression of the *mdr-1* gene, which codes for the expression of a cell-surface P-glycoprotein (P-gp). P-gp acts as an energy-dependent efflux pump, which can transport different cytotoxic agents such as anthracyclines, vinca alkaloids, and other drugs used in the chemotherapy of cancer, out of tumor cells, thus decreasing their intracellular concentration and preventing cytotoxic effects [58]. P-gp belongs to the group of ATP-dependent transmembrane proteins. P-gp and the MDR-associated protein (MRP) are well characterized [59] (Figure 5.11). In some cases, a direct relationship between P-gp expression and degree of resistance has been reported [60].

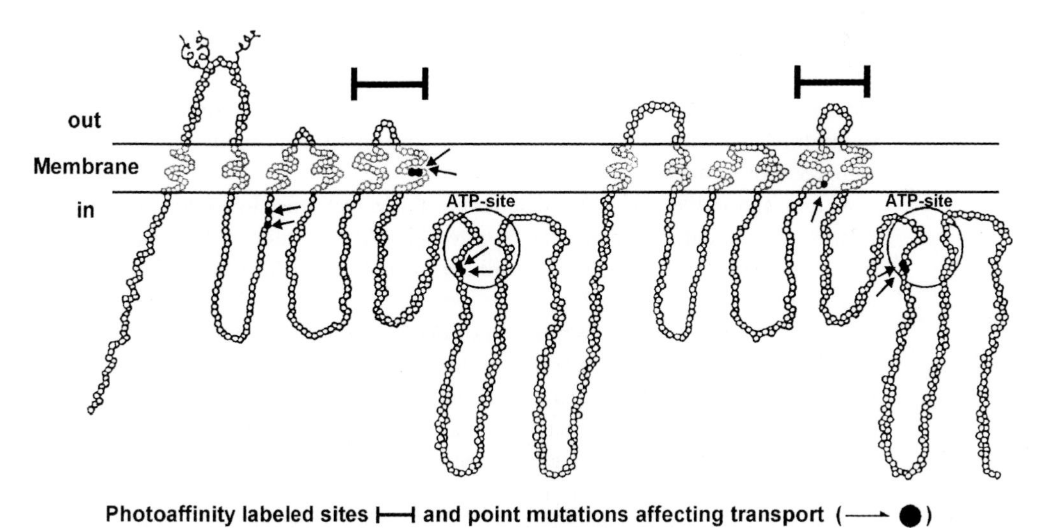

Fig. 5.11 Model of human p-glycoprotein. (Adapted from Fig. 2 of ref. 59.)

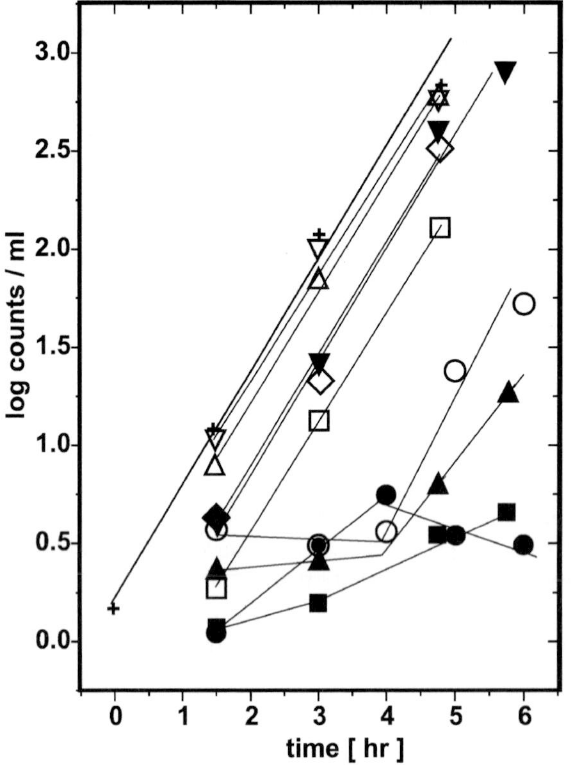

Fig. 5.12 Tetracycline resistance-reversing activity of MDR-reversing drugs against *E. coli*. +, Control; ◇, tetracycline; ▽, verapamil; △, quinacrine; ❑, chlorpromazine; O, trifluoperazine; ▼, tetracycline + verapamil; ▲, tetracycline + quinacrine; ●, tetracycline + chlorpromazine; ■, tetracycline + trifluoperazine. (Reprinted from Fig. 1 of ref. 64.)

During the last decades several drugs and compounds have been identified that to different degrees are able to overcome MDR so that the cells resemble sensitive cells in their chemosensitivity. These drugs mainly include catamphiphilic, membrane-active compounds and belong to various classes of drugs such as calcium channel blockers (verapamil), neuroleptics (flupentixol), anesthetics, antimalarial drugs (quinidine), antiarrhythmics (amiodarone), and many other compounds. Reviews were recently published [61, 157].

As long ago as 1980 the reversal of daunorubicin resistance by its non-cytotoxic derivative *N*-acetyldaunorubicin was reported [62]. Shortly thereafter it was reported that MDR could be reversed by coadministration of the calcium channel blocker verapamil [63].

These reversing drugs show quite dissimilar structures. The only property in common is their catamphiphilic nature. In this respect it is of interest that these drugs with a MDR-reversing activity in tumor cells also modify the resistance of Gram-negative [64] and Gram-positive bacteria [65] as well as *Plasmodia* [66, 67], i.e. of eucaryotic and procaryotic cells. For a series of these MDR-reversing agents it could be demonstrated that the power in reversing the tetracycline resistance in *E. coli* corresponds to that in tumor cells (Figure 5.12) [64]. This is in accordance with the reported similarities of different P-gps [68].

Two facts should be kept in mind for further discussion:

1) Resistance of different cells (tumors, *Plasmodia*, bacteria) can be reversed by the same reversing drug, i.e. P-gp seems to be a quite conserved membrane-embedded protein.
2) The required structural specificity for the reversing drug is not very high, because drugs with a very different structure and conformation can reverse MDR.

5.2.3 Proposed Mechanisms of Action

Currently, several mechanisms of action of MDR-reversing or -modulating drugs are postulated. They are briefly listed and discussed for a better understanding of why drug–membrane interactions as an essential factor in the MDR-reversing process cannot be excluded from the consideration.

The most widely spread concept presumes that the modulating drugs are substrates for P-gp and through competition simply inhibit the efflux of the cytotoxic drugs. This concept relies mainly on the observed ability of MDR-reversing compounds to compete effectively with photolabeled P-gp structures [69–71]. In addition, the presence of the *mdr-1* gene in tumor cells correlates well with their clinical resistance [72]. Several models based on a pump mechanism that accounts for the ability of P-gp to recognize structurally unrelated compounds have been proposed. The major models are the "vacuum cleaner" hypothesis and the "flippase"" hypothesis. The "vacuum cleaner" hypothesis assumes that the multidrug transporter recognizes the drugs by their ability to bind or intercalate into the lipid bilayer [73] and transports them from the inner to the outer leaflet. In the "flippase" model the drugs are transported to the exterior [74]. Another hypothesis claims that P-gp undergoes wide-ranging drug-dependent dynamic reorganizations [75] and that different modifiers may interact with separate or overlapping domains of P-gp [76]. These models for the possible functioning of P-glycoprotein as an efflux pump have been reviewed [77].

However, there have been some contradictory results. It has been reported that some modifiers do not compete with the cytotoxic agent for binding to P-gp even though they act as substrates for it [77]. In some resistant cell lines there was no correlation between an increase in MDR and the P-gp levels [79–81]. The calcium channel blocker SR3357 (**2**) – which is 4–5 times more potent than verapamil – was shown not to compete for the binding site of the labeled [^{3}H]-azidopine on P-gp whereas verapamil did. SR3357 did not bind to P-gp, but to a 65-kDa protein. Inter-

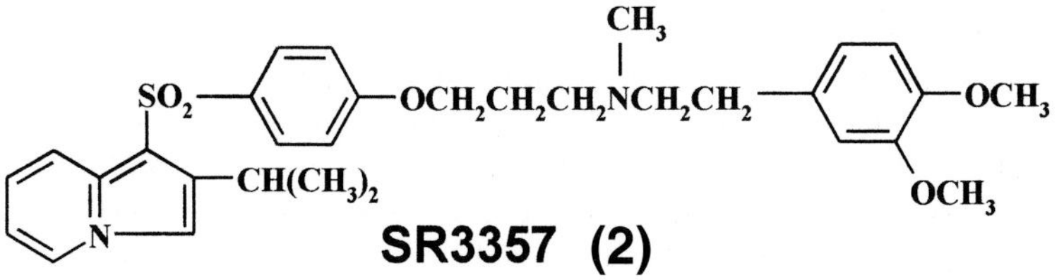

SR3357 (2)

estingly, a concentration of 30 μM SR3357 induced a 72% inhibition in acid lysosomal sphingomyelinase activity, a fivefold increase in sphingosine level, and a 75% inhibition of intracellular PKC activity [82].

This leads to other proposed mechanism of action. The P-gp binding hypothesis of MDR-modifying drugs neglects the effect which these drugs have on the physicochemical properties of the lipid environment of P-gp.

Therefore, in addition to the direct interaction of resistance-modifying drugs with P-gp, alternative mechanisms or combinations of mechanisms are presently being discussed in the scientific community. One of the proposed mechanisms is related to the role of protein kinases, especially PKC, in the activation of P-gp. The calcium-activated PS-dependent PKC has been reported to be mainly responsible for the phosphorylation of P-gp [83]. It has also been demonstrated that PKC inhibitors can modulate *in vitro* P-gp function in tumor cell lines [84–86]. For example, the inhibition of MDR by safingol [(2S,3S)-n-$C_{15}H_{31}$-CH(OH)-CH(NH_2)-CH_2OH] is independent of changes in P-gp substrate activity and correlates with the inhibition of PKC [87]. Membrane defects play an essential role in the regulation of PKC activity [88]. The authors showed that phospholipids with the lowest bilayer–hexagonal phase transition temperature are the most effective in augmenting PKC activity. The enzyme is quite sensitive to the presence of hexagonal phase-forming lipids such as PE. A strong dependence of Ca^{2+}-stimulated PKC activity on PE concentration and also on fatty acid composition was found in LUVs made of palmitoyloleoylphosphatidylserine (POPS) and different PEs as a function of POPS in the membrane (Figure 5.13). Negatively charged lipids are found particularly in the inner leaflet of the plasma membrane. They are also responsible for binding of cytotoxic agents such as doxorubicin and for MDR-reversing drugs.

Thus, direct inhibition of PKC [86] or interaction of catamphiphilic modulators such as flupentixol with PKC substrates like PS could lead to MDR reversal [64]. Staurosporine, a highly effective inhibitor of PKC, binds to one of the protein units of PKC. One of its derivatives has been shown to potentiate the inhibitory activity of vinblastine in P-gp- overexpressing KB-8511 cells [61]. Similar effects have been reported for other PKC inhibitors such as erbastin and neomycin sulfate. In addition, a novel gene for a protein kinase associated with MDR expression was postulated [61]. The predicted location of this kinase is in the cell membrane. A high expression was found in resistant cells in comparison with their sensitive counterparts. In this context it should be mentioned that a series of thioxanthines inhibit PKC and are at the same time active in the reversal of MDR [89]. This activity is stereospecific, with the *trans*-form of various thioxanthines being two- to threefold more active than the *cis*-form in both PKC inhibition and MDR-modulating efficacy. This correlates well with the degree of its interaction with PS (see Section 3.8.7), the assumed target of interaction in the membrane [38, 90].

It has also been demonstrated that PI, whose metabolic turnover generates the important messenger molecules inositol and DAG, is different in adriamycin-sensitive and -resistant cells. The resistant cells show an increased turnover. This could be important for PKC activity, as one of the functions of DAG is the control of PKC activity. Indeed, it has been shown that the activities of PKC in resistant and sensitive cells

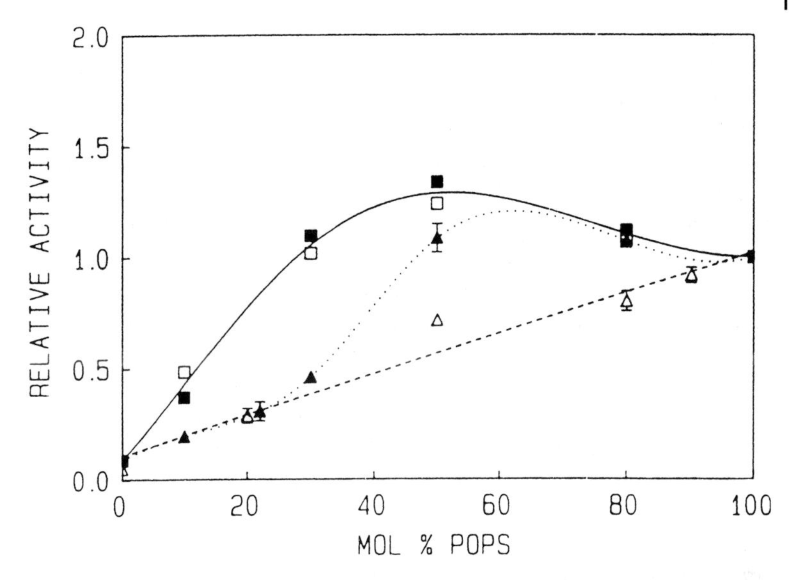

Fig. 5.13 Phosphatidylserine (PS) dependence of Ca^{2+}-stimulated PKC activity as a function of palmitoyl- (Po), oleoylphosphatidylserine POPS concentration in the POPS/PE LUVs. Enzyme activity measured relative to 100% POPS. Total phospholipid concentration in the incubation mixture was 480 µM. The following mixed systems were analyzed: ❑, DOPE/DOPS; ■, DPoPE/DOPS; ▲, POPE/POPS); △, DMPE/POPS. (Reprinted from Fig. 1 of ref. 88 with permission from Academic Press.)

are identical for the membrane form but that its cytosolic activity is increased by 82% in resistant cells. There was no change in the amount of PKC [91].

In parallel with these mechanisms one could speculate that the ability of cationic amphiphiles to interact and change properties of membrane phospholipids is part of the mechanism leading to reversal of MDR. Independent of the obvious lack of structural similarities, all these modifiers are membrane active. They can therefore influence the membrane properties in various ways from changes in membrane fluidity and permeability, to competition for binding sites with cytotoxic drugs such as doxorubicin, to indirect inhibition of P-gp phosphorylation, via inhibition of PKC through interaction with PS and replacement of Ca^{2+} ions, to domain formation and heterogeneity in the membrane close to the membrane-inserted protein. In this way they can lead to disturbances of the efflux pump and finally to MDR reversal. Several reported observations point at a direct or indirect involvement of drug–membrane interactions in the activity of cytotoxic agents and MDR-reversing compounds:

- direct interaction through changes in the conformation of P-gp proteins induced by the MDR modifiers via changes in membrane properties;
- indirect interaction through changes in membrane properties such as fluidity or composition and subsequent changes in drug transport and accumulation.

5.2.4 Change in Composition of Membranes and Influence on P-gp, Cytotoxic Agents, and MDR-Reversing Drugs

5.2.4.1 Comparison of Lipid Composition of Sensitive and Resistant Cells

Changes in membrane composition would affect transport and accumulation of cytotoxic drugs and resistance-modifying agents as well as the functioning of the membrane-embedded proteins. Several studies have been performed to analyze membrane composition and properties with respect to phospholipid content and the ratio of specific phospholipids in sensitive and resistant cell lines. Different techniques were used and significant differences between sensitive and resistant cells were found. For example, changes in fluorescence polarization values and phospholipid content were reported for Friend leukemia cells (FLCs), leading to changes in the ratio of PC/PE and to changes in the electrophoretic mobility (Table 5.6) [92]. Membrane potential differences between sensitive and resistant cell lines were also reported. A fourfold higher potential was measured in sensitive cells compared with resistant FLCs [93]. In a number of carcinoma cell lines a relatively high accumulation of positively charged anthracyclines was found, which correlates with the observed high electronegative potentials found for these cells in the absence of high-level P-gp. In contrast, the resistant cell counterparts showed lower accumulation of positively charged drugs, increased expression of MDR-1, and lower plasma membrane potentials (Table 5.7) [94]. A decreased membrane potential was also found for a series of drug-resistant tumor sublines. Incubation with the resistance-modifying

Tab. 5.6 Phospholipid composition of Friend leukemia cells resistant to different levels to doxorubicin. (Reprinted from Tab. 2 of ref. 92 with permission from Potamitis Press)

	FLC	DOX-RFLC		
		1	2	3
SP H	4.6 ± 0.4	3.0 ± 0.6	3.0 ± 0.1	4.0 ± 0.02
PC	50.9 ± 1.2	65.5 ± 0.9	68.8 ± 0.4	59.4 ± 0.3
PS	3.0 ± 0.3	3.7 ± 0.8	3.5 ± 0.2	4.7 ± 0.04
PI	8.2 ± 0.2	5.5 ± 0.5	4.9 ± 0.4	7.2 ± 0.08
PE	25.2 ± 1.3	16.6 ± 0.6	13.8 ± 0.9	19.0 ± 0.3
DPG	8.1 ± 0.7	5.7 ± 0.4	5.1 ± 0.3	4.9 ± 0.2
PC/PE	2.0	3.9	5.0	3.1
PC/SPH	10.3	21.8	23.1	14.7
PI/PS	2.7	1.5	1.4	1.5

Friend leukemia cells (FLCs) and cell variants resistant to different levels of doxorubicin were grown to the saturation level in drug-free medium. The phospholipid composition was determined in duplicate or triplicate from two separate experiments. The values are expressed as percent of total lipid phosphorus.
FLC; Friend leukemia cells; DOX-RFLC; FLCs resistant to doxorubicin; SPH; sphingomyelin; DPG: diphosphatidylglycerol.

Tab. 5.7 Lower adriamycin levels in normal epithelial (CV-1) than in carcinoma cells (MCF-7) are reversed by verapamil. (Reprinted from Tab. 3 ref. 94 with permission from John Libbey Eurotex)

Drug	Cell type	Intracellular drug ($ng/10^6$ cells)
Adriamycin	CV-1	20
Adriamycin + Verapamil	CV-1	54
Adriamycin	MCF-7	149
Adriamycin + Verapamil	MCF-7	110

Tab. 5.8 Plasma membrane potentials in PDR cell lines[a]. (Reprinted from Tab. 1 of ref. 95 with permission from Bertelsmann-Springer)

Cell lines	Membrane potentials (mean channel numbers)	*P* value
(a) L_0	31.4 ± 2.0	–
(b) L_{100}	19.8 ± 2.0	< 0.001 *vs.* (a)
(c) L_{100} + CsA[a]	30.9 ± 2.9	> 0.1 *vs.* (a); < 0.01 *vs.* (b)
(d) L_{100} + Ver[a]	26.3 ± 1.8	> 0.05 *vs.* (a); < 0.02 *vs.* (b)
(a) H-129	27.5 ± 0.6	–
(b) H-129/DNR	19.3 ± 1.0	< 0.001 *vs.* (a)
(c) H-129/DNR + CsA[a]	25.5 ± 0.8	> 0.05 *vs.* (a); < 0.01 *vs.* (b)
(d) H-129/DNR + Ver[a]	22.9 ± 0.3	< 0.02 *vs.* (a) and (b)

a) Cyclosporin 0.1 μM and verapamil 0.1 μM.
DNR = Daunorubicin

drug verapamil led to a restoration of the membrane potential to the level of the sensitive cell line (Table 5.8) [95]. In plasma membranes of leukemic lymphoblasts resistant to vinblastine, a 50% increase in cholesterol (CHOL) – responsible for the rigidity of membranes – was observed, together with a very significant elevation of the etherlipid content [96]. In doxorubicin-resistant A2780 tumor cells, on the other hand, a decrease in CHOL content as well as in the ratio of free to esterified CHOL compared with the sensitive counterpart was reported [97]. However, if sensitive and resistant cells were cultured in a medium deprived of lipoproteins, the sensitive cells responded with a more striking decrease in CHOL content.

MDR cell lines exhibit several other changes in surface membrane properties. Often, the structural order is increased in resistant cells as analyzed by electron spin resonance (ESR) and fluorescence anisotropy studies [98]. In addition, an increase in intramembranous particles and the rate of fluid-phase endocytosis are reported for resistant cells [99, 100].

The intrinsic plasma membrane fluidity of a series of sarcoma cell lines with various degrees of adriamycin resistance was studied by ESR [91]. The results document a systematic increase in the order parameter, *S*, as the cells become more and more resistant. At the same time, a parallel increase in the sphingomyelin content was ob-

Tab. 5.9 Peak intensity ratios obtained from one-dimensional spectra obtained at day 4 after transplantation of 10^5 cells/mL (mean ± SD). (Reprinted from Tab. 1 of ref. 101 with permission from the American Association for Cancer Research)

Cell line	No. of experiments	CH_2/CH_3	$CHOL/CH_3$	Ino/Ct	Glu/Ct	Eth/Ct
K562wt	5	3.50 ± 0.56	0.66 ± 0.23	0.43 ± 0.08	8.02 ± 1.17	0.30 ± 0.06
K562adr	8	1.63 ± 0.16^{a}	1.07 ± 0.21^{a}	1.04 ± 0.23^{a}	6.14 ± 0.83^{a}	0.30 ± 0.07
K562ads (passages 50-120)	7	$2.25 \pm 0.16^{b,c}$	0.87 ± 0.24	$1.29 \pm 0.21^{b,c}$	6.43 ± 0.49^{c}	0.31 ± 0.05

a) K562wt *vs.* K562adr significantly different.
b) K562adr *vs.* K562ads significantly different.
c) K562wt *vs.* K562ads significantly different.
K562wt, human leukemia cells.
K562adr, human leukemia cells, adriamycin resistant.
K562ads, human leukemia cells, adriamycin resistant cultured in the absence of adriamycin.

served. This implies that the resistant cells possess a more rigid (lower fluidity) plasma membrane.

Lately, NMR spectroscopy has been used to study and compare membrane composition in sensitive and adriamycin- and taxol-resistant K562 leukemia cell lines [101]. Using one-dimensional spectra it was found that both resistant cell lines showed lower fatty acid methylene–methyl ratios. In agreement with previous publications, higher choline–methyl ratios compared with the sensitive strain were also found. Also, a decrease in glutamine content could be detected using 2-D COSY spectra. Very interestingly, the fatty acid signals returned almost to normal after the cells were cultured in the absence of the drug. In addition, verapamil-treated cultures also showed partial recovery of fatty acid signals (Figure 5.14 and Table 5.9). These

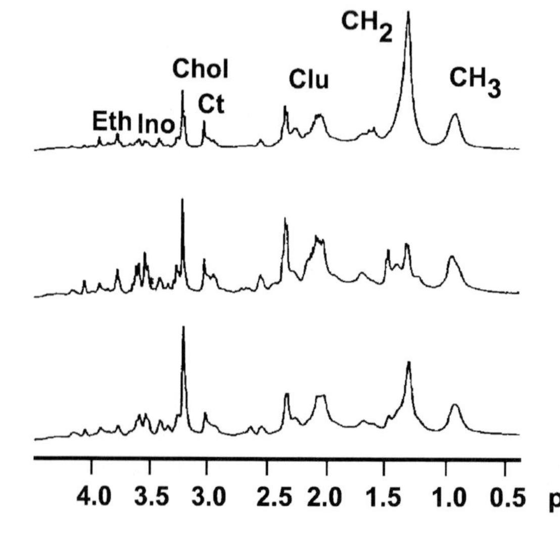

Fig. 5.14 NMR spectroscopy of cellular lipids of sensitive (K562wt), adriamycin-resistant (K562adr), and (K 562 ads) tumor cells (top to bottom). (Reprinted from Fig. 1 of ref. 101 with permission from the American Association for Cancer Research)

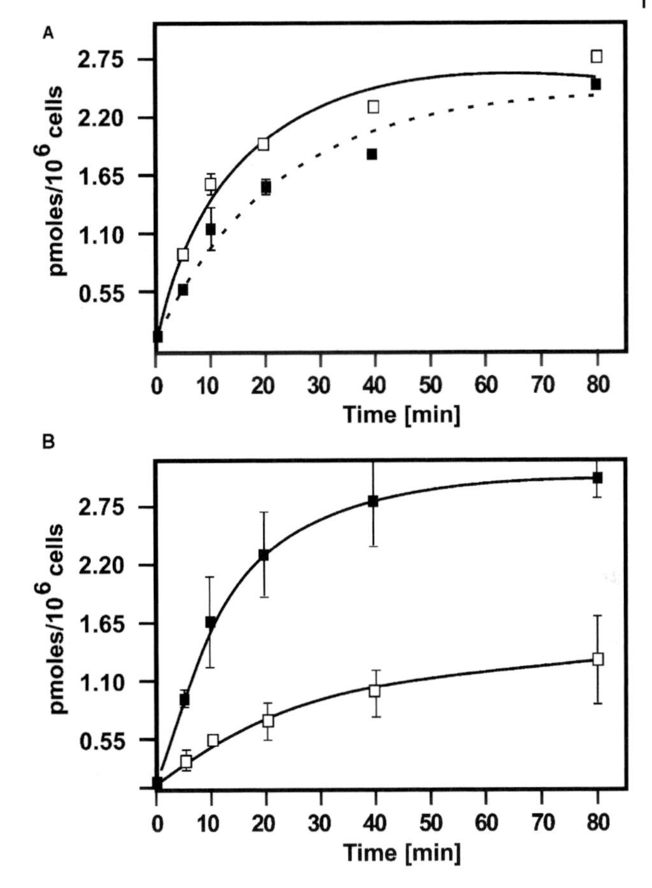

Fig. 5.15 Effect of heptadecanoic acid incorporation on vinblastine uptake. Accumulation of [^{3}H]-vinblastine in (A) AB_1 and (B) CH^RC5 cells grown in normal (□) or $C_{17:0}$-supplemented medium (■). Each point is the mean (± SD) of three independent experiments. (Reprinted from Fig. 1 of ref. 103 with permission from Elsevier Science)

results were paralleled by the observed reversal of the MDR phenotype determined through the measurement of the inhibitory concentrations of adriamycin and vinblastine in the K562adr cells cultured without drug or after short-term exposure to verapamil. Compared with the resistant cells, P-gp, the mRNA expression and DNA amplification of the *mdr* gene were, however, unchanged. This is an important observation as the resistance was reversed without a change in *mdr1* amplification. The findings of the authors support the assumption that lipids play an important role in MDR and MDR modification [101].

To study the possible effect of membrane biophysical properties on drug accumulation, the lipid composition of membranes has deliberately been altered by growing cells in the presence of anionic phospholipids, saturated or unsaturated fatty acids, or other components affecting membrane properties, e.g. cholesterol [102, 103]. The accumulation of drugs in MDR hamster cell lines (CH^RC5) was increased when membrane properties were altered. For this purpose the cells were cultured in a medium supplemented with $C_{17:0}$ fatty acid. The cells showed an extent of vinblastine accumulation similar to that found in the sensitive wild-type counterparts (AB1) (Figure 5.15) [103].

5.2.4.2 Membrane Composition and Functioning of Membrane-embedded Proteins

It has been suggested that the addition of heptadecanoic acid to the growth medium can lead to a change in the composition of the acyl groups of membrane phospholipids. The increase in saturated fatty acids, like the increase in CHOL, is expected to reduce membrane fluidity. On the other hand, it has been shown that the addition of fatty acids, such as linoleic acid, and amphiphiles that increase fluidity can also reverse defects in drug accumulation. These results are in agreement with the assumption that the physicochemical properties of biological membranes are adjusted in such a way that an optimal function of the membrane-embedded proteins is guaranteed. Any deviation leads to a decrease in the activity [104]. Such a dependence of P-gp and P-gp-associated ATPase on lipid composition was indeed described in reconstituted systems for P-gp [102]. The effects of lipids and detergents on ATPase-activity of P-gp have been reported [105]. It was observed that ATPase activity is most stable in CHAPS, retaining 50% of its original activity. At 2–10 μM, Triton X-100 enhanced the activity twofold. P-gp ATPase was sensitive to thermal inactivation but remained fully active in the presence of asolectin, PE (especially with saturated acyl chain), and PS but not PC. In this context, it is interesting to note that PE is often located in the inner leaflet of the plasma membrane of mammalian cells and could therefore easily interact with the ATPase-binding domain of P-gp. This would imply that lipids are essential for membrane integrity and activity.

In contrast, other lipids were able to restore the catalytic activity of P-gp ATPase after delipidation. Unsaturated PC and PS were most effective. The observed ranking in restoration for head groups was PC > PE > PS > PI. A strong preference for unsaturated fluid lipids with low T_t values was seen. These lipids are also preferred by various other ATPases.

The role of CHOL content in P-gp functioning has been studied in P-gp reconstituted in purified lipid. P-gp reconstituted into liposomes was able to bind [^{3}H]-azidopine. Increasing the CHOL content up to 20% (PC/CHOL weight ratio 8:2) led to an increase in the specific binding activity; however, further increase in CHOL content led to a decrease in binding activity. Cholesterol constitutes approximately 20–30% of membrane lipids in normal plasma membranes of human cells. It has been suggested that CHOL content either influences the motional order of phospholipids, thereby decreasing the binding affinity, or is an essential positive effector for P-gp through direct interaction [102]. Interestingly, it has been reported that the membrane content of CHOL affects the activity of Na^+,K^+-ATPase and glucuronosyltransferase, whereas ergosterol shows only minor effects on ATPase activity [106]. Lipid composition (cholesterol, anionic phospholipids) affects not only enzyme activities but also the membrane interaction of those drugs known to reverse MDR [107], so that a dual role of the plasma membrane in MDR cells is indicated.

Drugs reversing MDR are often tested by their ability to reduce photolabeling of P-gp in membrane suspensions. To model this interaction, purified PC liposomes containing no P-gp protein and the fluorescence dye rhodamine 6G (Rh6G) instead of photolabeling agents have been used. The authors examined 16 structurally diverse compounds, some of them known MDR modulators, to determine if their effect on lipid physicochemical properties was correlated with their MDR-reversing activity.

Tab. 5.10 Inhibition of membrane binding of Rh6G (K_i) and reversal of MDR by various drugs. (Adapted from ref. 108)

Compound	Reverse MDR	K_i for Rh6G inhibition (μM)
Verapamil	Intermediate	46.7 ± 10.0
Bepridil	High	27.1 ± 6.0
Clindamycin	None	128.0 ± 10.9
Atropine	None	None
Pirenzepine	None	None
Quinidine	Intermediate	67.1 ± 8.6
Chlorpromazine	High	6.5 ± 1.9
Quinine	Low	103.3 ± 20.7
Cefotaxime	None	None
Colchicine	None	None
Prenylamine	High	3.3 ± 0.6
Amitryptiline	Low	11.3 ± 2.3
Pentazocine	None	88.3 ± 11.9
Lidocaine	None	None
Furosemide	None	None
Promethazine	Intermediate	9.4 ± 1.3

Resistance to Rh6G, characterized by its decreased cellular accumulation, has been shown to be part of the MDR phenotype. It was found that potent modulators inhibit the membrane binding of Rh6G in a concentration-dependent manner. A good correlation of the determined K_i values with the ability to reverse MDR was observed (Table 5.10) [108]. Out of the 16 compounds studied, six showed no ability to modulate accumulation of labeled cytostatics. The effects of these compounds on lipid structure were also negligible and their K_i values for replacement of RhG6 were high. Amitriptyline showed a low modulation of drug accumulation despite its effects on lipids (low K_i for RhG6 replacement). An explanation for this exception could be its low effect on the fluorescence of *N*-phenyl-1-naphthylamine (NPN). This could suggest that amitriptyline, unlike bepridil, prenylamine, promethazine, and chlorpromazine, has no effect on membrane rigidity or does not enhance the hydrophobic environment of NPN. "Alteration in membrane rigidity may be the most critical factor in determining which drug-induced membrane perturbation leads to modulation of MDR" [108]. The results indicate that MDR-modulating drugs can produce strong alterations in membrane properties which affect drug distribution even in the absence of proteins (P-gp). Thus, the limitation to the binding of modulators to P-gp in the discussion of the underlying mechanism seems unrealistic [108].

This argument is further supported by results of a study showing that the potentiation of anticancer drug cytotoxicity by MDR-reversing drugs involves alterations in membrane fluidity, which in turn lead to increases in permeability [109]. The authors could show that the investigated chemosensitizers induced alterations in the bulk membrane fluidity in a dose-dependent manner and in a concentration range

Tab. 5.11 Potentiation by reserpine and verapamil of anticancer drug cytotoxicity in parental AA8 cells and their MDR T19 subline as determined by the clonogenic assay. (Reprinted from Tab. 1 of ref. 109 with permission from Blackwell Science)

Drug	AA8					T19					Resistance (fold)
	LD_{50} (nM)		PI with reserpine	LD_{50} (nM)		LD_{50} (nM)		PI with reserpine	LD_{50} (nM)		
	–	+ Reserpine		+ Verapamil (nM)	PI with verapamil	–	+ Reserpine		+ verapamil (nM)	PI with verapamil	
Doxorubicin	70	2.2	31.8	3.2	21.9	2450	4.1	598	23.4	105	35
Daunorubicin	8.6	1.7	5.1	1.5	5.7	726	3.7	196	10.1	71.9	84.4
Vinblastine	4.7	0.5	9.4	0.7	6.7	46	0.6	77	1.3	35.4	9.8
Taxol	70.5	0.9	78.3	2.3	30.7	1561	1.6	976	17.9	87.2	22.1

Mean LD_{50} values were obtained from 2–4 independent experiments, in which cells were incubated with various concentrations of the cytotoxic drug being tested, in the absence or presence of 5 μM reserpine or 10 μM verapamil. Values were interpolated from cytotoxicity curves generated for each anticancer drug over a wide range of concentrations. PI is defined as the ratio of the LD_{50} value obtained in the absence of modulator and that obtained in its presence. The degree of resistance was calculated by dividing the LD_{50} value obtained with T19 by that found with wild-type AA8 cells.

relevant to their cytotoxic potentiation. Some others caused membrane rigidification. Interestingly, it was found that reserpine potentiated the cytotoxicity of several antitumor drugs in MDR cell lines, thus exceeding the drug resistance by a factor of 2.5–45. Even more surprising, a 5- to 78-fold and 6- to 31-fold potentiation by reserpine and verapamil, respectively, was found in the sensitive parental cell line AA8, possessing no P-gp (Table 5.11). The results strongly indicated the existence of other components of cytotoxicity potentiation independent of MDR. In the same study, P-gp-overexpressing MDR cells and P-gp-deficient cells were used to evaluate the effect of chemosensitizer on taxol influx rates and steady-state accumulation (Table 5.12). The results can be interpreted by assuming that the chemosensitizer – as shown for reserpine and verapamil – can potentiate cytotoxicity in P-gp-containing and P-gp-deficient cells through an increase in the level of cell-associated drug. As these results suggested that MDR-reversing drugs can increase membrane permeability to various lipophilic agents, these authors studied various detergents for their potentiating activity. Non-toxic concentrations of Nonidet P-40 and Triton X-100 led to a 8- to 14-fold potentiation of taxol influx in P-gp-deficient cells and a four- to eightfold potentiation in T19MDR cells. Again, as for the chemosensitizers, these non-ionic detergents potentiate the cytotoxicity of taxol to a higher degree in P-gp-deficient cells than in the resistant counterparts. They can mimic the increased chromophore accumulation and cytotoxicity potentiation found for the MDR-reversing drugs verapamil and reserpine.

Tab. 5.12 Effect of chemosensitizers on taxol influx rates and steady-state levels of cellular taxol accumulation in parental and MDR hamster cells. (Reprinted from Tab. 3 of ref. 109 with permission from Blackwell Science)

Cell line	Modulator	Initial taxol influx rate (pmol 10^{-6}) cells min^{-1}	Taxol accumulation (pmol/10^6 cells)
AA8	None	0.25 (1)	1.22 (1)
	+ Verapamil	1.00 (4)	13.50 (11.1)
	+ Reserpine	0.70 (2.8)	21.00 (17.2)
T19	None	0.13 (1)	0.32 (1)
	+ Verapamil	0.16 (1.2)	3.00 (9.4)
	+ Reserpine	0.35 (2.7)	5.68 (17.8)
Vin^{R1}	None	0.01 (1)	0.15 (1)
	+ Verapamil	0.02 (2)	0.50 (3.3)
	+ Reserpine	0.03 (3)	2.00 (13.3)

Taxol influx rates that were linear for the first 60 s are given as the initial influx rates. Numbers in parentheses describe the relative initial rates and accumulation levels of taxol obtained in the presence of modulators compared with their absence. The taxol accumulation values were obtained after 1 h incubation.

Finally, the effect of different chemosensitizers on the permeability of liposomes loaded with the membrane-impermeable chromophore carboxyfluorescein (CF) was studied as a model for the assumed change in membrane permeability of a number

Tab. 5.13 Chemosensitizer-mediated leakage of CF from charged liposomes. (Reprinted from Tab. 5 of ref. 109 with permission from Blackwell Science)

Compound type	Compound	30% liposome leakage concentration (μM)
Modulator	Bepridil	1.2
	Chlorpromazine	16.5
	Dipyridamole	74
	Prenylamine	1
	Progesterone	143.6
	Quinidine	120
	Quinine	215
	Reserpine	17.8
	Trifluoperazine	4.6
	Verapamil	64.6
Fluidizers	A_2C	0.8
	Benzyl alcohol	8200
Detergents	Chaps	1500
	Digitonin	1.4
	n-Octylglucoside	6625
	Nonidet P-40	83
	Triton X-100	47.5
	Tween 80	1363
Cytotoxic agents	Colchicine	none
	Emetine	none
	Taxol	none

CF-charged liposomes (25 μg/mL) were incubated for 30 min in the absence or presence of modulators or other agents. Following excitation at 490 nm with a Perkin-Elmer MPF-44B spectrofluorometer, the fluorescence emitted from CF that leaked out of permeabilized liposomes was collected at 520 nm. Values are presented as the average of two experiments. None, indicates that no measurable liposome leakage was observed even when using a drug concentration of 200 μM.

of agents. The results are summarized in Table 5.13. They show that reserpine and verapamil produced a strong dose-dependent leakage of CF from the loaded large unilamellar liposomes. The same was observed for the strong membrane fluidizer A_2C. In contrast, the P-gp substrates taxol, colchicine, and emetine did not show an effect. The observed effects of chemosensitizer on the polarization of diphenylhexatriene fluorescence in normal and MDR rodent and human cell lines were in agreement with the assumed fluidization of the membrane. Decreased membrane polarization is regarded as an indicator of increased bulk membrane fluidity. The results of these detailed studies again favor the assumption that MDR-reversing drugs – showing very different structures and varying the degree of lipophilicity – interact with biological membranes rather than with specific proteins. An indirect effect on

P-gp seems more likely. According to the authors. it could proceed according to the following scheme [109]:

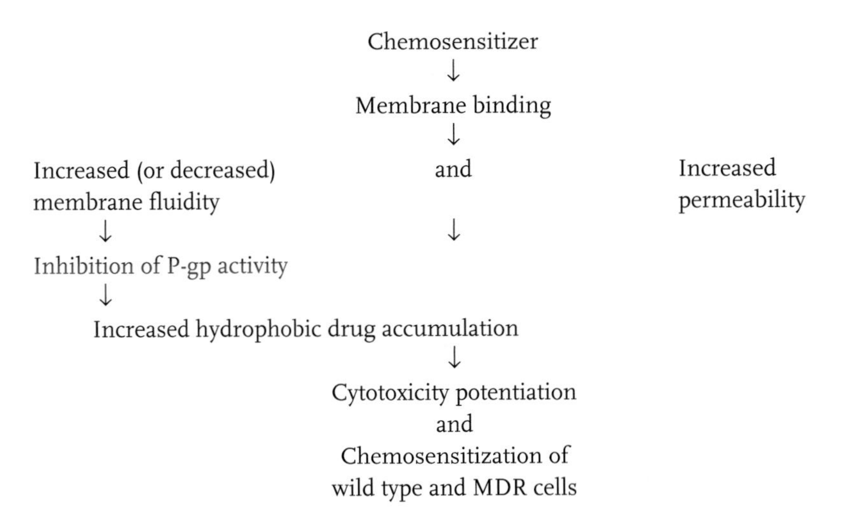

Similarly, the effects of non-ionic detergents on P-gp drug binding and reversal of MDR were investigated [110]. It was found that very low concentrations of the non-ionic detergents Triton X-100 and Nonidet P-40 completely prevented azidopine photolabeling to P-gp and were able to reverse the MDR phenotype. High concentrations of the denaturing agent urea or of the zwitterionic detergent 1-[(3-cholamido-propyl)]dimethylaminol-1-propanesulfate did not inhibit photolabeling of azidopine. Verapamil was less effective in inhibiting azidopine photolabeling of P-gp than Triton X-100. It was, however, more effective in potentiating vinblastine accumulation in resistant cells. The results support the assumption that low non-toxic concentrations of Triton X-100 that do not disrupt the bilayer or extract membrane proteins lead to the reversal of MDR by inhibiting drug binding. In agreement with the information that the photoaffinity drug labeling sites for iodoaryl azidoprazosin of P-gp are within the transmembrane domains 6 and 12, the authors [111] suggest that the sites of P-gp drug binding are related to sequences within the bilayer of the membrane.

5.2.5
Membrane Composition, Drug Binding, and Transport Kinetics

Several studies have been performed to analyze the effect of membrane composition on accumulation and transport of antineoplastic drugs. In particular, the role of anionic lipids in transport and passive diffusion, as well as on the binding of doxorubicin and its effect on the degree of order of lipid acyl chains was investigated. Using ^{2}H-NMR spectroscopy the effect of doxorubicin on mixed bilayers of DOPS, DOPA, dioleoylphosphatidylcholine (DOPC), and DOPE was studied. It was found that doxorubicin does not affect acyl chain order of pure zwitterionic phospholipids but dramatically influences that of anionic lipids [112]. At 25 °C, in bilayers consisting of 67

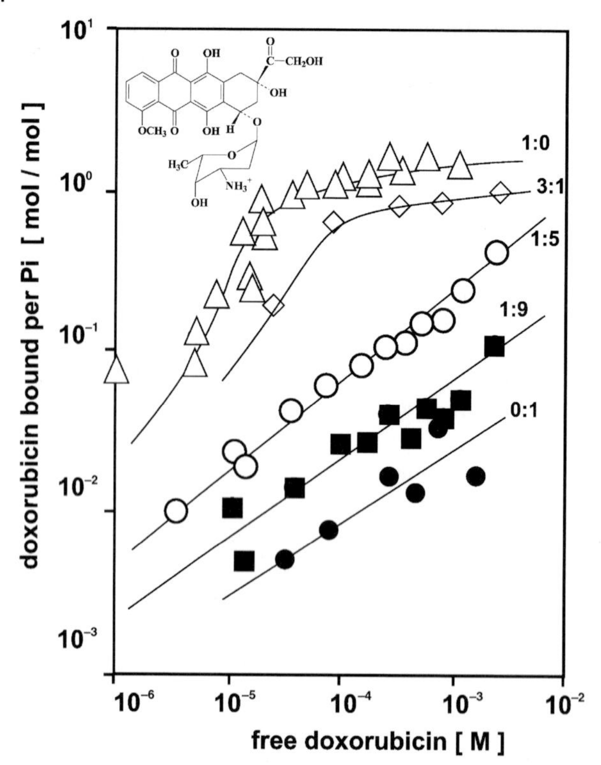

Fig. 5.16 (A) Structure of doxorubicin. (B) Binding of doxorubicin to LUVLET consisting of various DOPG/DOPC mixtures of the ratios indicated (Reprinted from Fig. 1 of ref. 113 with permission from the American Chemical Society.)

mol% DOPE and 33 mol% anionic lipids, lipids adopted the inverted hexagonal H_{II} phase upon addition of doxorubicin. This destabilization occurred only in the presence of anionic lipids and sufficient amounts of DOPE. Under these conditions, a strong binding is observed. Using different ratios of 1-palmitoyl-2-oleoyl-*sn*-glycero-3-phosphoglycerol (DOPG)/DOPC as an example, the preferred interaction of doxorubicin with anionic phospholipids such as DOPG is shown in Figure 5.16 [113]. The preference of doxorubicin to bind to anionic lipids could be demonstrated not only using model membranes, but also with *E. coli* plasma membranes and lipid extracts of these biological membranes (MDR-27). The anionic phospholipid and PE content of wild-type *E. coli* plasma membranes resembles that of the cytoplasmic leaflet of mammalian cells.

Much earlier, the partitioning of [^{14}C]-doxorubicin was determined in subcellular fractions of doxorubicin-sensitive (V-79) and doxorubicin-resistant (LZ) hamster lung fibroblast cell lines. A significant difference was reported (Table 5.14) [114]. At concentrations clinically relevant *in vivo*, doxorubicin preferentially partitioned into the membrane, achieving a 100-fold higher concentration in the membrane fraction than the cytosol fraction. The data in Table 5.14 show the ability of resistant cells to decrease the doxorubicin content in the lipid fraction of the membrane. It is remarkable that the doxorubicin content of the lipid membrane fraction in resistant cells decreased about ninefold, whereas the reduction in whole cells is only fourfold in com-

Tab. 5.14 The partitioning of [^{14}C]-doxorubicin (DOX) in the subcellular fractions of DOX-sensitive (V-79) and DOX-resistant (LZ) Chinese hamster lung fibroblast cell lines. (Reprinted from Tab. 3 of ref. 114, with permission from Elsevier Science)

Fractions	pmol DOX/10^6 cells		Ratio of DOX (V-79/LZ)
	V-79	LZ	
Total particulate	14.1	3.6	3.9
C:M soluble	5.1 ± 0.3	0.6 ± 0.1	8.5
C:M precipitable	7.4 ± 0.7	2.1 ± 0.2	3.5
Aqueous	1.6 ± 0.2	0.9 ± 0.2	1.7
Total cytosolic	5.7 ± 0.3	1.7 ± 0.1	3.6
Total cellular	19.8	5.3	3.7

C:M = chloroform/methanol

parison with sensitive cells. This suggests that lipid-associated doxorubicin is a significant factor in the cytotoxicity of doxorubicin and "that during development of resistance cells acquire mechanisms to preferentially exclude doxorubicin from their membranes" [114]. Such a change may be brought about by increasing the P-gp content of resistant cells. P-gp is a very large 12-transmembrane domain protein and may change membrane properties (fluidity).

The effect of doxorubicin binding to anionic lipids on passive diffusion of doxorubicin across membrane bilayers was studied and its effect on the function and rate of pumping of P-gp was discussed [115]. In accordance with the previous publication, the main extranuclear pool of the related daunorubicin was the intracellular membrane of cancer cells or in liposomes, but not free in the cells [116]. It could clearly be demonstrated that drug binding to the anionic lipids of the inner leaflet significantly lowered the transport rate for doxorubicin. If drug binding is not considered in the calculation, passive efflux rates will be underestimated and thus also will active drug

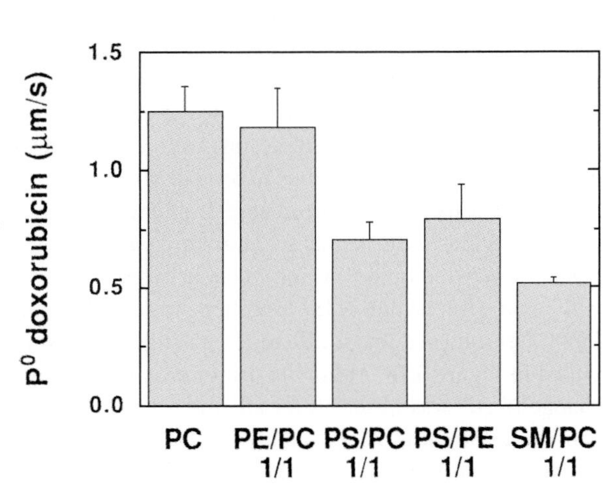

Fig. 5.17 Permeability coefficients, P^o, for doxorubicin transport across model membranes composed of different phospholipids. PC, phosphatidylcholine; PE, phosphatidylethanolamine; PS, phosphatidylserine; SM, sphingomyelin. (Reprinted from Fig. 3 of ref. 115 with permission from the American Chemical Society.)

pumping by P-gp [115]. Depending on the content of anionic lipid, more doxorubicin is bound and the transport rate of doxorubicin will be concomitantly decreased because of decreasing free, transportable drug concentration. In addition, in model membranes a decrease in the intrinsic permeability coefficient is observed, even after correction for drug binding (Figure 5.17). This was unexpected, because the binding to anionic lipids decreases the acyl chain order and should promote diffusion [115]. An explanation could be the insertion of drug molecules between the head groups, which would tighten the interface region, thus overcompensating for the disordering effect on the acyl chain. The insertion of drug molecules would lead to an increase in positively charged drug molecules in the membrane and thus to a decrease in drug transport of unbound drug. The binding of doxorubicin to anionic lipids is determined by electrostatic forces.

Doxorubicin binding to anionic lipids could be reversed by adding chemosensitizers such as verapamil. This is in agreement with the observation that the difference in doxorubicin passive transport in zwitterionic and anionic lipids disappears in the presence of verapamil [115]. The results, which demonstrate that the highest drug (substrate) concentration is to be found at the interphase region of the inner leaflet of the plasma membrane and that cytosolic concentrations are low [115], are important in the discussion of the role of the ATP-dependent P-gp pump, for which the binding sites are assumed to be located or anchored in the inner leaflet of the membrane.

It is not only cytotoxic drugs such as doxorubicin or vinblastine that bind favorably to anionic phospholipids, but also MDR-reversing agents. A series of such drugs, including phenothiazines and structurally related drugs such as verapamil, flunarizine, and lidocaine, were investigated for their ability to interact with phospholipid liposomes [117]. All drugs, except for amiodarone, were known to reverse MDR *in vitro* in resistant tumor cells to varying degrees [118–120]. At the time of these experiments amiodarone was not known to be a reversal compound. Artificial membranes were used to avoid effects that were not membrane related. Phosphatidylcholine (in the form of DPPC), as the major phospholipid in membranes of sensitive and resistant cells and phosphatidylserine (DPPS), as an essential phospholipid of the inner leaflet of bilayers and known to interact specifically with antitumor drugs, were used in the liposome preparations.

DSC and NMR techniques were used to study the type and degree of interaction between drug and bilayer. The results were compared with those from experiments on the ability of the compounds to reverse MDR *in vitro* in resistant tumor cell lines. In the DSC experiments, the change in phase transition, T_t, and enthalpy, ΔH, was recorded. Some of the results are summarized in Table 5.15, together with the MDR-reversing activities and PKC inhibitory activity.

When the interaction of the modifiers was studied on DPPS liposomes rather than DPPC liposomes, a major difference was the observed change in ΔH. This is exemplified in Figure 5.18. At lipid to drug molar ratios ranging from 1:0.01 to 1:0.1, the change in ΔH was observed for all studied compounds with the exception of verapamil and lidocaine. The decrease in ΔH of DPPS, indicating a new type of phospholipid organization, was paralleled by the formation of a new endothermic peak at

Tab. 5.15 Drug-membrane interaction parameters and biological data of the multidrug resistance (*MDR*) modifiers studied. (Reprinted from Tab. 1 of ref. 117 with permission from Bertelsmann-Springer)

Compound	DPPS $\Delta T_{ind.}$ °C	DPPS T_{max} slope	DPPC T_{max} slope	BBPS $1/T_2$ slope	MCF-7/DOX		P388/DOX			PKC inhibition IC_{50} (μM)
					Drug IC_{50} (μM)	MDR ratio	Drug IC_{50} (μM)	Drug + DOX IC_{50} (μM)	MDR ratio	
trans-Flupentixol	31.21 (±0.15)	–47.70 (±1.07)	–243.57 (±7.82)	96.19 (±5.35)	25 (±4)	15.2 (±1.9)	8	0.8	10.0	29 (±3)
cis-Flupentixol	28.83 (±0.08)	–45.97 (±0.82)	–105.42 (±12.77)	88.90 (±5.45)	24 (±4)	4.8 (±0.6)	12	0.8	15.0	71 (±4)
Trifluoperazine	28.52 (±0.06)	–32.74 (±1.03)	–29.18 (±1.60)	38.97 (±6.40)	19 (±3)	3.4 (±0.4)	8	0.8	10.0	100 (±30)
Chlorpromazine	23.67 (±0.05)	–28.64 (±1.47)	–26.73 (±0.54)	36.42 (±2.42)	8 (±1)	1.6 (±0.3)	20	8.0	2.5	50 (±5)
Triflupromazine	23.24 (±0.05)	–24.94 (±1.18)	–24.75 (±0.27)	38.52 (±2.56)	16 (±3)	2.0 (±0.3)	20	8.0	2.5	170 (±30)
Imipramine	20.53 (±0.09)	–27.98 (±5.23)	–16.97 (±0.14)	NM	19 (±5)	2.5 (±0.9)	60	20.0	3.0	520 (±10)
Quinacrine	13.51 (±0.06)	–24.12 (±2.37)	–16.35 (±0.65)	NM	3 (±1)	1.3 (±0.1)	NA	NA	NA	850 (±110)
Verapamil	19.38 (±0.05	–24.01 (±1.66)	–16.26 (±0.41)	42.81 (±0.82)	NA	NA	> 100	2.0	> 50	NA
Flunarizine	21.58 (±0.03)	–12.77 (±0.96)	–23.43 (±1.17)	NM	NA	NA	19	3.4	5.6	NA
Lidocaine	NM	–11.10 (±1.82)	–21.29 (±0.82)	1.89 (±0.13)	NA	NA	NA	NA	0	NA
Amiodarone	NM	–36.24 (±2.79)	NM	81.94 (±3.88)	NA	NA	NA	NA	NA	NA

ΔT_{ind}, T_{max} slopes and $1/T_2$ slope are characteristic parameters of the interaction between the drugs and the phospholipids dipalmitoylglycerophosphoserine (*DPPS*), dipalmitoylglycerophosphocholine (*DPPC*), and bovine brain phosphatidylserine (*BBPS*) respectively. *IC_{50}*, the concentration that produces 50% inhibition of either the cell growth or protein kinase C (*PKC*) activity; *Drug IC_{50}*, IC_{50} of the drug alone; *Drug + DOX IC_{50}*, IC_{50} of the drug in presence of doxorubicin. The MDR ratio in MCF-7/DOX is IC_{50} of doxorubicin alone divided by IC_{50} of doxorubicin in presence of the drug; MDR ratio in P338/DOX is IC_{50} of the drug alone divided by IC_{50} of the drug in presence of doxorubicin; the MDR ratio for lidocaine is given as 0 to indicate that the drug possesses no MDR-reversing activity in P338/DOX. *NM*, parameters not measured; *NA*, data not available.

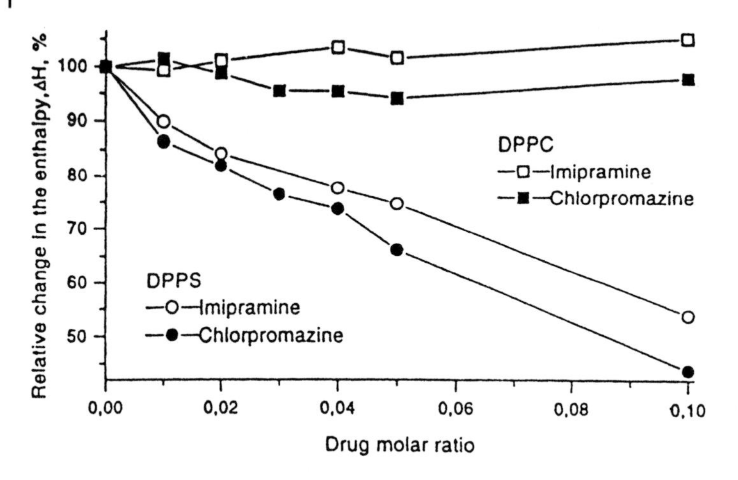

Fig. 5.18 Effect of increasing molecular ratios of chlorpromazine and imipramine on the enthalpy, ΔH, of the main gel to liquid crystalline phase transition of dipalmitoylglycerophosphoserine (DPPS) and dipalmitoylglycerophosphocholine (DPPC). The relative change in ΔH at a given drug molar ratio is shown in relation to ΔH of the control (no drug, 100%). (Reprinted from Fig. 1 of ref. 117 with permission from Bertelsmann-Springer.)

a temperature below that of the main phase transition. This drug-induced peak, T_{ind}, is characteristic for a particular reversing agent, the lowest position being at 21.82 °C in the case of *trans*-flupentixol and the highest at 39.52 °C in the case of quinacrine.

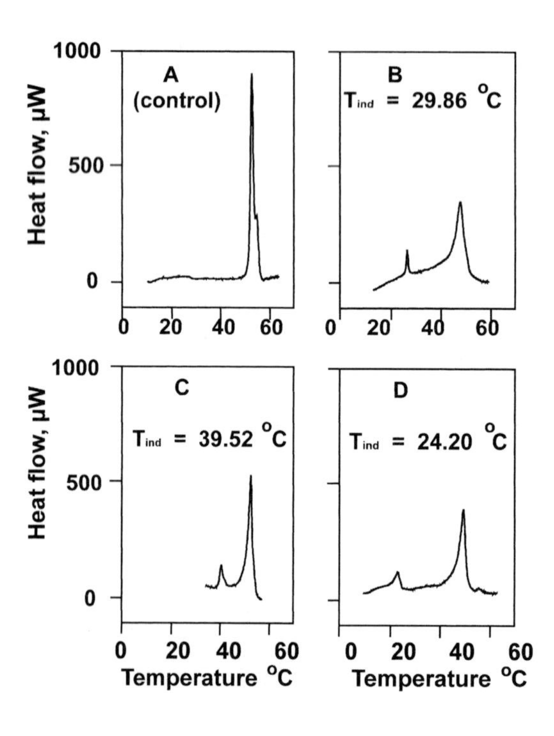

Fig. 5.19 Tracings of DSC thermograms of DPPS alone and in presence of MDR modifiers at a lipid-drug ratio of 1:0.05 (drug concentration of the ionized form). The small peaks on the left of the main phase transition peaks represent the drug-induced peaks, $T_{ind.}$. A: control; B: trifluperazine; C: quinacrine; D: cis-flupentoxol. (Reprinted from Fig. 3 of ref. 117 with permission from Bertelsmann-Springer.)

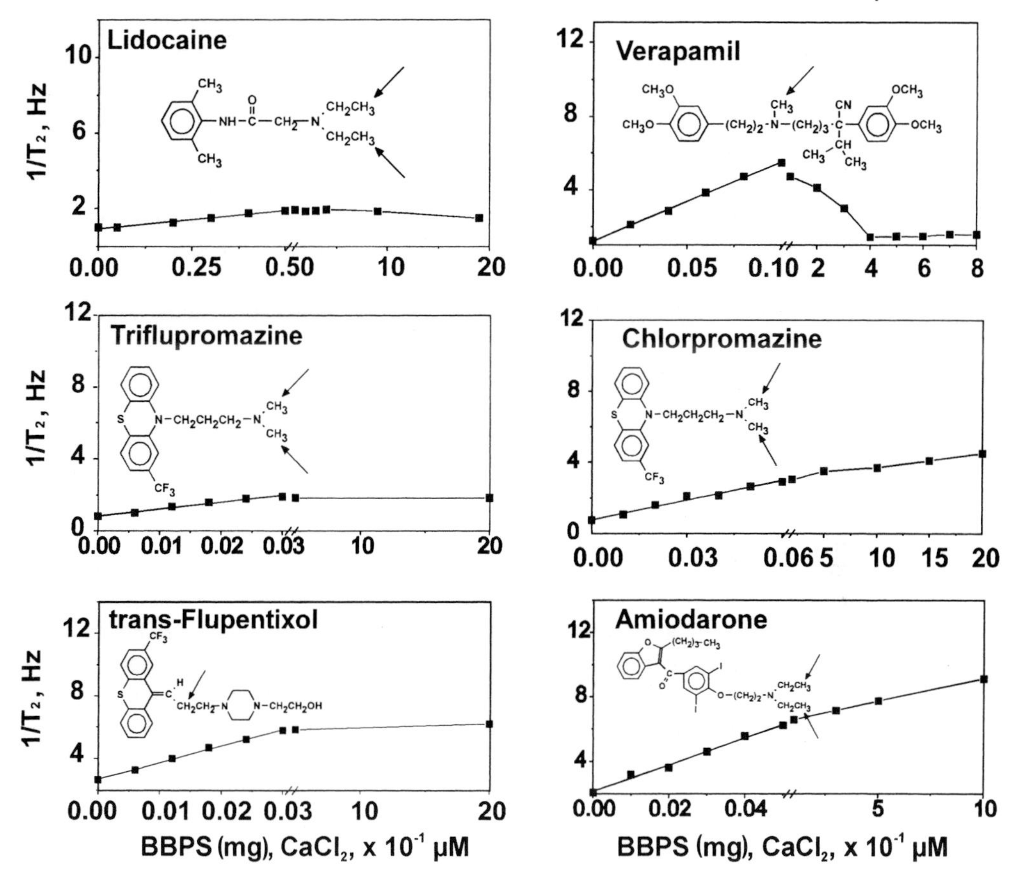

Fig. 5.20 Changes in drug proton relaxation rate ($1/T_2$) as a function of increasing concentration of bovine brain phosphatidylserine (BBPS) alone and after addition of $CaCl_2$. The spin systems measured are indicated by arrows on the drug structures. (Reprinted from Fig. 6 of ref. 117 with permission from Bertelsmann-Springer.)

T_{ind} remained almost constant for a particular drug (Figure 5.19), independent of the lipid to drug molar ratio; only the heights of the peak increased with increasing lipid to drug ratio. For *cis*-flupentixol, such a new peak at almost the same temperature (24.2 °C) was also observed when interacting with DPPC (24.4 °C). The T_t values for DPPC and DPPS in control experiments were 42.35 °C (SD ± 0.03 °C) and 53.05 °C (SD ± 0.07 °C) respectively [117].

Additionally, NMR measurements were performed and the changes in $1/T_2$ relaxation rates recorded as a function of increasing bovine brain phosphatidylserine (BBPS) at constant drug concentrations. The change in $1/T_2$ is related to a decrease in rotational freedom (for details, see Chapter 3). In the NMR measurements *cis*- and *trans*-flupentixol showed the strongest interaction, lidocaine the weakest. This is in agreement with thermotropic experiments (Table 5.15). Broadening was dependent on liposome concentration in a linear manner, and the slope was taken as a measure of the degree of interaction [39]. Figure 5.20 shows such a plot [117]. Upon addition of Ca^{2+} ions, the binding of drugs to the anionic lipid could be reversed only in the case of those drugs with a weaker degree of interaction. In the case of amiodarone an increase in the interaction was in fact observed upon addition of the ions. This effect could also be achieved by the addition of NaCl, indicating a strong involvement of hydrophobic interaction forces for these highly lipophilic molecules. This is in agreement with the observation that, in the case of amiodarone, the spin systems of all protons are involved in the interaction with liposomes to the same degree. In contrast, in verapamil only part of the structure loses rotational freedom. X-ray studies on amiodarone have shown it to be deeply buried in the hydrocarbon chain region [121].

^{31}P-NMR and DSC have also been used to study the effect of membrane-active, positively charged antitumor drugs, and it was found that the two studied anthracyclines, adriamycin and 4-*epi*-adriamycin, induced structural phase separation in DPPE–cardiolipin liposomes (2:1), indicated by the formation of a hexagonal H_{II} phase. In contrast, in the presence of ethidium bromide and 2-*N*-methylellipticinium, the phospholipid remained in an organized bilayer configuration. It was found that the anthracyclines are preferentially localized in the interface and interact electrostatically with the negatively charged phosphates of cardiolipin [122]. Recently, the interaction of the MDR-reversing agent trifluoperazine with various lipid preparations was studied by DSC and fluorescence spectroscopy [123]. Domain formation was observed to be induced by trifluoperazine in zwitterionic PC vesicles but not in charged phosphatidylglycerol bilayers. The authors speculated that defects produced at the domain boundaries could lead to an increase in membrane fluidity and permeability, thus explaining at least part of the mode of action of the MDR-reversing activity of trifluoperazine [123].

These results are in agreement with the aforementioned findings and support the assumption that drug–membrane interactions leading to severe changes in membrane structure (domain formation and heterogeneity) can affect the functioning of embedded proteins such as Na^+, k^+-ATPase, P-gp, and PKC.

MDR-reversing drugs bind to phospholipids, preferably anionic phospholipids, as do the cytotoxic drugs, but they also compete with the cytotoxic drugs for binding sites in the bilayer. This competition has been observed not only for verapamil and doxorubicin, but also for doxorubicin and flupentixol. Figure 5.21 represents a plot of the increasing interaction of doxorubicin (increase in $1/T_2$ for the indicated aromatic protons with increasing BBPS concentrations in NMR experiments [124]. The results of NMR measurements obtained at 4 mM doxorubicin and increasing BBPS concentrations are shown. The broadening of the proton resonance signals are linearly dependent on the lipid concentration within the range studied. The slope of the proton signal of the CH_3 group is about 1.5-fold higher than that of the aromatic proton. NMR experiments performed at lower doxorubicin concentrations (1 and 2 mM) resulted in the same ratios of the slopes. However, at the same drug concentration, the slope obtained for the doxorubicin CH_3-proton signal (40.8) was lower than that of the aliphatic protons of *trans*-flupentixol and verapamil (96.2 and 42.8 respective-

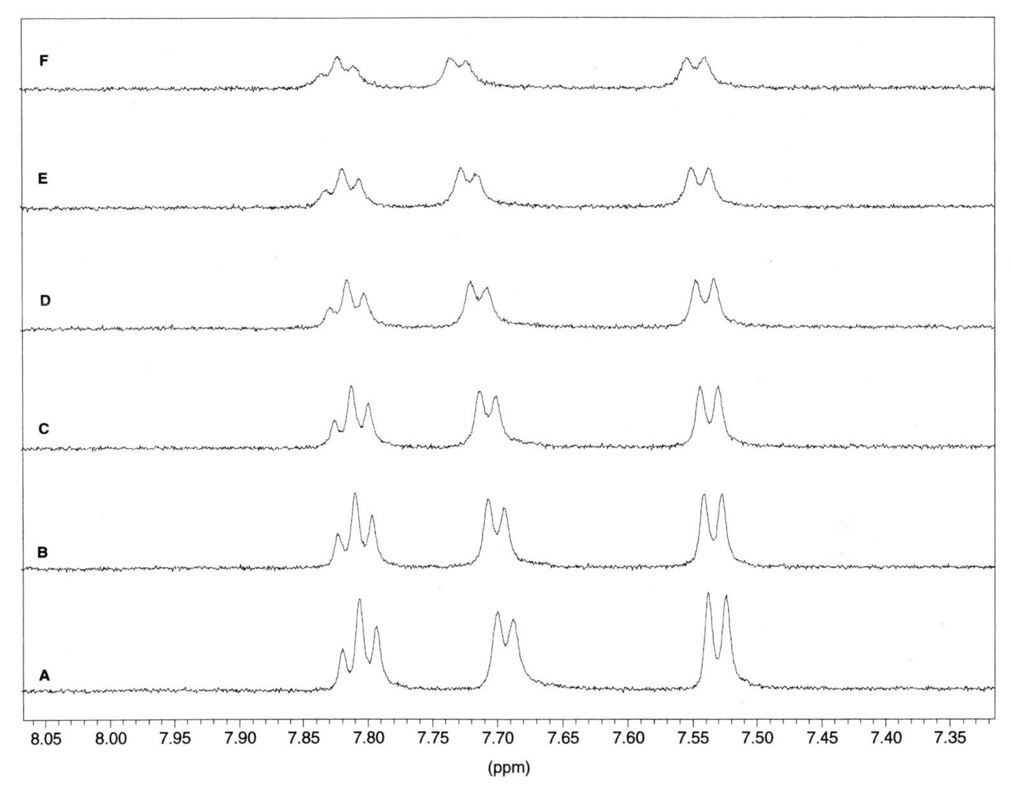

Fig. 5.21 NMR spectra of doxorubicin in the absence (bottom trace) and presence of increasing concentrations of bovine brain phosphatidylserine (BBPS) [124].

ly), but still the direct comparison between doxorubicin and modifiers is not fully correct as the spin systems used are not identical [117]. According to the slopes, doxorubicin shows similarities with drugs that exhibit a strong interaction with negatively charged PS [122, 124]. Interestingly, the addition of *trans*-flupentixol led to a complete reversal of the broadening of doxorubicin resonance signals, i.e. doxorubicin bound to the phospholipid was displaced by *trans*-flupentixol. It was also observed that the liposomes precipitated after some days, and doxorubicin resonance signals could no longer be observed in the supernatant. The precipitate was insoluble in DMSO, but upon addition of *trans*-flupentixol the complex resolved, showing the normal NMR spectrum of the unbound doxorubicin [124].

Another observation that underlines the importance of drug–membrane interactions in tumor therapy is the increase in antineoplastic activity against resistant cell lines with increasing lipophilicity of the cytotoxic drugs at constant degree of ionization (pK_a ~8.4). Independently of the anthracycline used to induce resistance, cross-

Tab. 5.16 Intracellular accumulation of idarubicin (IDA) and dausorubicin (DAU) after 1-hr exposure to the same extracellular drug concentration in the presence or absence of 10 μM verapamil. (Reprinted from Tab. 2 of ref. 125, with permission from Elsevier Science)

Drug	(ng/mL)[a] ext	LoVo			LoVo-IDA-1			LoVo-DOX-1		
		–VER[b]	+VER[c]	Increase[d]	– VER	+VER	Increase	– VER	+VER	Increase
Idarubicin	300	106 ± 19	110 ± 10	1.0	26 ± 5 (4.1)e	87 ± 9	3.3	37 ± 5 (2.9)	86 ± 6	2.3
Idarubicin	1000	302 ± 17	341 ± 25	1.1	72 ± 4 (4.4)	272 ± 47	3.8	123 ± 22 (2.8)	292 ± 51	2.4
Idarubicin	3000	1164 ± 168	1326 ± 112	1.1	304 ± 24 (3.8)	969 ± 131	3.2	398 ± 38 (3.1)	887 ± 134	2.2
Dausorubicin	300	55 ± 8	69 ± 11	1.3	14 ± 2 (3.7)	38 ± 7	2.7	20 ± 4 (2.8)	38 ± 3	1.9
Dausorubicin	1000	177 ± 21	192 ± 23	1.1	48 ± 10 (3.8)	156 ± 20	3.3	76 ± 13 (2.7)	152 ± 12	2.0
Dausorubicin	3000	531 ± 70	611 ± 104	1.1	98 ± 20 (5.4)	340 ± 43	3.5	173 ± 28 (3.3)	377 ± 44	2.2

a) Extracellular drug concentration.
b) Intracellular drug concentration expressed as ng/10^6 cells in the absence or presence of 10 μM VER (c).
d) Ratio of b to c.
e) The relative decrease in idarubicin and dausorubicin uptake in resistant cell as determined by the ratio between drug uptake in sensitive and resistant cells is given in brackets.
LoVo, lovostatin.

Tab. 5.17 Extracellular and intracellular idarubicin (IDA) and daunorubicin (DAU) concentrations inhibiting cell growth by 50% (IC_{50}-int) in LoVo-sensitive and resistant cell lines. (Reprinted from Tab. 3 of ref. 125 with permission from Elsevier Science)

	$IC_{50\,ext}$ (ng/mL)[a]		RI_{ext}[b]		$IC_{50\,int}$ (ng/10^6 cells)		RI_{int}[c]	
Cell line	**IDA**	**DAU**	**IDA**	**DAU**	**IDA**	**DAU**	**IDA**	**DAU**
LoVo	10.4 ± 1.3	47.8 ± 3.5	1	1	5.5 ± 0.9	11.6 ± 2.4	1.0	1.0
LoVo-IDA-1	239.2 ± 25.7	5216.7 ± 1056	23	109	19.9 ± 1.3	161.2 ± 22.7	3.6	13.9
LoVo-IDA-2	217.4 ± 17.8	5788.3 ± 521	21	121	17.4 ± 1.5	169.0 ± 31.7	3.2	14.6
LoVo-DOX-1	106.1 ± 20.8	1709.4 ± 398	10	36	12.2 ± 2.7	74.2 ± 15.1	2.2	6.4
LoVo-DOX-2	116.5 ± 11.0	1514.7 ± 203	11	32	12.6 ± 1.6	71.3 ± 8.4	2.3	6.2

a) Cell exposure was 1 hr. Data obtained from at least three independent experiments.
b) IC_{50ext} resistant cells/IC_{50}ext sensitive cells.
c) IC_{50int} resistant cells/IC_{50}int sensitive cells.

resistance to the more lipophilic idarubicin was lower than for daunorubicin (for partition coefficients see Table 5.21). The resistance was paralleled by an increase in P-gp. On the other hand, the intracellular content of idarubicin was greater than that of daunorubicin in sensitive as well as in resistant colon carcinoma cells. The uptakes were reduced in resistant cells. The relative decrease in idarubicin uptake – the ratio between sensitive and resistant cells – was similar to the observed decrease in daunorubicin uptake (Table 5.16) [125]. However, the idarubicin efflux rate was slower than that of daunorubicin in both the sensitive and resistant cells. In consequence, the ratio between the percentage of idarubicin retained in resistant and sensitive cells at specific times of the efflux kinetics was similar to that observed for DAU. No difference between idarubicin and daunorubicin in terms of inhibition of P-gp photolabeling was detected in azidopine binding experiments. This excludes changes in the affinity of idarubicin and daunorubicin to P-gp because of differences in their lipophilicity. Together with the other findings, this is a clear indication that other effects must be involved in the better circumvention of MDR by idarubicin. The effect of verapamil on MDR reversal was also similar for both drugs. All resistant cell lines tolerated higher intracellular levels of both anthracyclines compared to the sensitive cells (Table 5.17). The potency of idarubicin seemed to be greater, possibly because of its higher lipophilicity. The results demonstrate that reduction in drug incorporation cannot explain resistance, because MDR cells need much higher intracellular concentrations of both idarubicin and daunorubicin than sensitive cells for the same inhibitory effect. Distribution studies, however, show a different pattern for the two anthracyclines. In resistant cell lines a shift of daunorubicin from nucleus (N) to the cytoplasm (C) is observed [125] (Table 5.18), whereas no shift in the N/C ratio was seen for idarubicin when comparing sensitive and resistant cells. Idarubicin was preferably distributed in the cytoplasm of sensitive and resistant strains. The authors assumed that the reduced resistance of MDR cells to idarubicin is due to the difference in the N/C distribution ratio of these drugs.

Similarly, the circumvention of MDR was investigated as a function of anthracyline lipophilicity and charge [126]. Adriamycin and five derivatives were studied in

Tab. 5.18 Subcellular idarubicin and daunorubicin distribution in LoVo-sensitive and MDR cell lines. (Reprinted from Tab. 5 of ref. 125 with permission from Elsevier Science)

Cell line	N/C ratio[a]			
	IDA		DAU	
	–VER	+VER	–VER	+VER
LoVo	0.19 ± 0.1	0.20 ± 0.10	3.40 ± 2.7	3.7 ± 1.9
LoVo-IDA-1	0.16 ± 0.1	0.19 ± 0.04	0.31 ± 0.2	1.3 ± 0.5
LoVo-DOX-1	0.20 ± 0.1	0.17 ± 0.12	0.41 ± 0.1	1.1 ± 0.3

a) N/C ratio was determined at the $IC_{50\ ext}$ value in resistant LoVo-IDA-1 and LoVo-DOX-1 cells. In sensitive cells, or when verapamil (VER) was used, 1000 ng/mL of IDA or DAU was used, as at lower concentrations background noise was too high.

sensitive and resistant cell lines. The results are summarized in Table 5.19. Compared with the sensitive cell line, the resistance of a highly resistant strain, resistance index 4534, was almost totally overcome by highly lipophilic anthracycline derivatives that do not possess a positively charged nitrogen atom. The charge seems to become less important for MDR reversing activity with increasing lipophilicity of the modifier. It was found that these lipophilic anthracycline derivatives are predominantly localized in the cytoplasm and mitochondria rather than the nucleus, thus overcoming P-gp- MDR. Circumvention of P-gp-related MDR in comparison with doxorubicin and idarubicin was also reported for the novel, uncharged anthracycline,

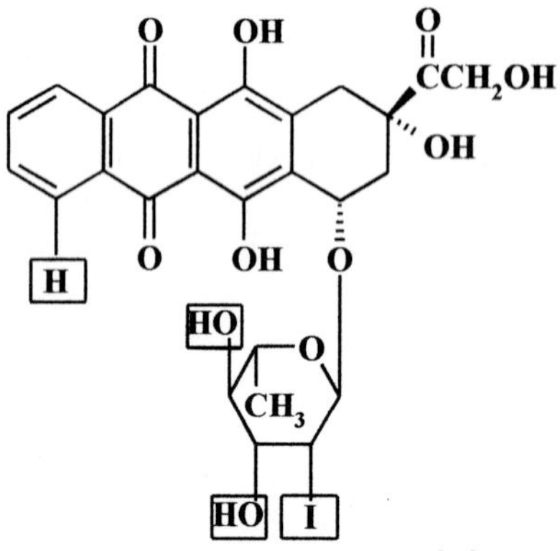

Anamycin **(3)**

anamycin (**3**). It is less potent in sensitive cells than doxorubicin and idarubicin. Its levels in the cell were not affected by verapamil in either sensitive or resistant strains (Table 5.20) [127].

Differences in the distribution patterns of doxorubicin, daunorubicin, idarubicin, and idarubicinol were also observed by CD measurements in LUVs of PC–phospha-

Tab. 5.19 Effect of charge and lipophilicity of anthracyclines on their accumulation in cardiac muscle and non muscle (fibroblast) cells. (Reprinted from Tab. 1 of ref. 126 with permission from the American Chemical Society)

Positiv charged

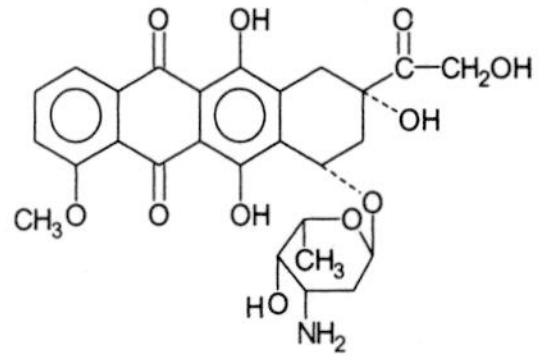

Neutral counterpart

Adriamycin (ADR)

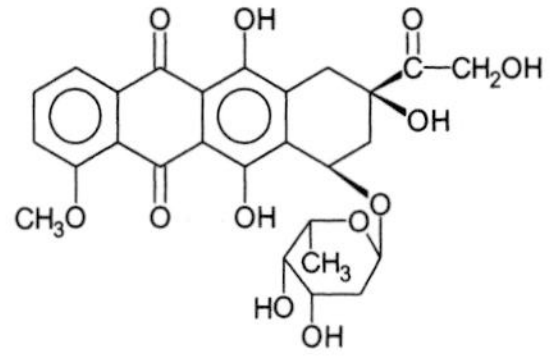

AD OH (WP 159)

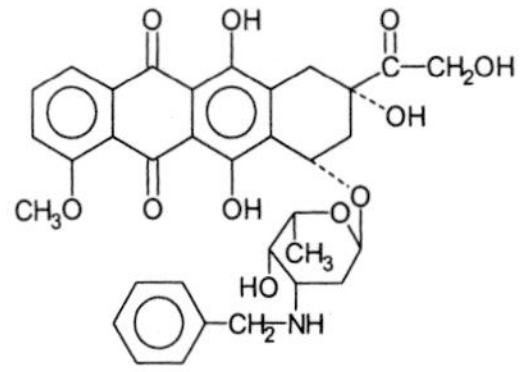

AD 288

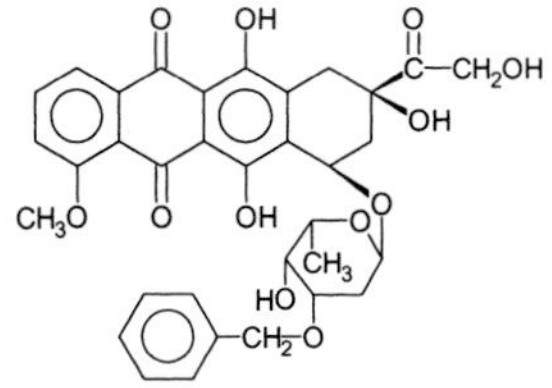

WP 546

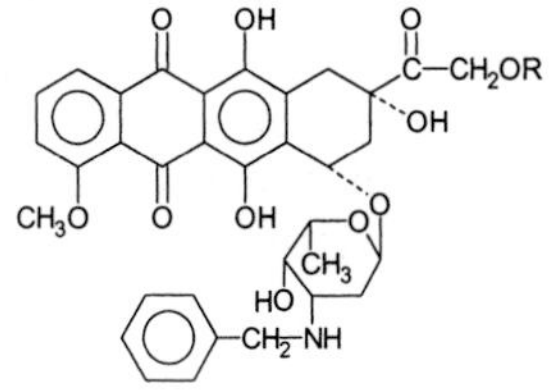

AD 198, R = $COCH_2CH_2CH_2CH_3$

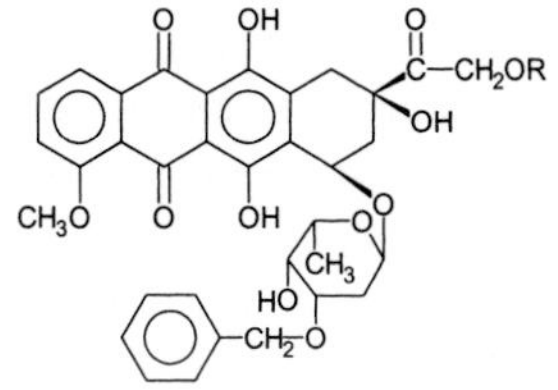

WP 549, R = $COCH_2CH_2CH_2CH_3$

Drug	Localization[a]	Charge	Log D^b	CM/FB[c]	(ID_{50}, [M])		Resistance index[d]
					MDR– (FLC)	MDR+ (ARN15)	
ADR	Nucleus	+	0.08	+++/+	6.78×10^{-9}	3.06×10^{-5}	4513
WP 159	Nucleus	0	1.40	+++/++	1.58×10^{-8}	6.29×10^{-6}	411
AD 288	Nucleus and some mitochondria	+	> 1.99	+++/++	2.53×10^{-8}	8.42×10^{-7}	33
WP 546	cyto/diffuse and mito/ nuc < AD 288	0	> 1.99	++/++	6.51×10^{-7}	2.90×10^{-6}	4
AD 198	Mito/cyto part and diffuse metachromatic	+	> 1.99	++/++	2.0×10^{-8}	2.10×10^{-7}	11
WP 549	Diffuse (mito) < AD 198 more diffuse straining in fibroblasts	0	> 1.99	++/+++	5.0×10^{-7}	1.56×10^{-6}	3

a) Mito, mitochondria; cyto, cytoplasm; nuc, nucleus; part, particulate.
b) Partition coefficient pH 2,4.
c) Relative fluorescence brightness in cardiac muscle cells (CM) *vs.* cardiac fibroblasts (FB).
d) Ratio of doses yielding 50% growth inhibition in MDR+ (adriamycin-induced)/MDR– (parental) Friend leukemic cells.

Tab. 5.20 *In vitro* cytotoxicity of doxorubicin (DOX), idarubicin (IDA), and annanycin (ANN) with and without verapamil (VER) in sensitive HL-60S and resistant HL-60/DOX cells. (Reproduced from Tab. 3 of ref. 127 with permission from the American Society of Hematology)

	ID_{50} (μg/mL)		
Drugs	**HL-60S cells**	**HL-60/DOX Cells**	**RI**
ANN	0.300 ± 0.17	0.80 ± 0.04	2.6
ANN + VER	0.290 ± 0.18	0.77 ± 0.06	2.6
IDA	0.004 ± 0.003	0.16 ± 0.16	40
IDA + VER	0.004 ± 0.002	0.08 ± 0.02	20
DOX	0.078 ± 0.003	9.13 ± 0.55	117.5
DOX + VER	0.057 ± 0.006	1.46 ± 0.68	25.6

RI, resistance index.

tidic acid (PA)–cholesterol mixtures at various ratios. The hydrophilic doxorubicin was electrostatically bound to LUVs, whereas the dihydroanthrachinone remained in the watery phase. Daunorubicin showed additional hydrophobic interactions, whereas the interaction of the more lipophilic idarubicin involved a complex of two or three idarubicin molecules in a right-handed conformation with one cholesterol molecule and also PA molecules. The interaction of the highly lipophilic idarubicinol with the liposome vesicles again differed from that of idarubicin (in contrast to doxorubicin and daunorubicin, which cannot recognize cholesterol-rich membranes), but was also influenced by the presence of CHOL [128].

Because of the observed discrepancies between drug accumulation at steady state in drug-sensitive and highly resistant cell lines and the degree of resistance as well as the observed distribution patterns of MDR-reversing drugs and cytotoxic drugs, kinetic studies on the uptake and efflux of drugs have been undertaken. Probably the first scientist to consider this important aspect was Tapiero [129]. He observed differences in uptake and efflux kinetics of doxorubicin compared with THP-doxorubicin (pirarubicin) and found that the amount of THP-doxorubicin remaining in resistant cells was lower than that of doxorubicin. Detailed kinetic experiments were done and reported by Garnier-Suillerot and coworkers [130–132] and others [109, 133, 134].

The influx (k_+) and efflux (k_a) coefficients, Michaelis–Menten constants (K_m), and resistance factors (RF) were determined for a series of anthracyclines (**4–13**) with different pK_a values and octanol–water partition coefficients, D, at pH 7.4 [130, 135]. The results are summarized in Table 5.21. Increases in D favorably influenced k_+ and unfavorably affected k_a and $1/K_m$; k_+ became faster, and k_a increased, as did $1/K_m$. This implies that, in the case of the anthracyclines studied, lipophilicity affects uptake kinetics more than the kinetics of P-gp-mediated efflux. Their RF value therefore decreased with increasing lipophilicity. However, possible differences in the distribution pattern of these molecules as well as the influence of membrane binding on their uptake rates were not considered.

Drug	R_4	R_9	R_{10}	R'_3	R'_4	R''_4
(4) Doxorubicin	OCH_3	$COCH_2OH$	H	NH_2	OH	H
(5) Daunorubicin	OCH_3	$COCH_3$	H	NH_2	OH	H
(6) F-idarubicin	H	$COCH_3$	H	NH_2	F	H
(7) Idarubicin	H	$COCH_3$	H	NH_2	OH	H
(8) THP-Doxorubicin (Pirarubicin)	OCH_3	$COCH_2OH$	H	NH_2	x*	H
(9) Br-Daunorubicin	OCH_3	$COCH_3$	H	H	Br	H
(10) Carminomycin	OH	$COCH_3$	H	NH_2	OH	H
(11) Aclacinomycin	OH	C_2H_5	CO_2CH_3	$N(CH_3)_2$	x**	H
(12) Hydroxyrubicin	OCH_3	$COCH_2OH$	H	OH	OH	H
(13) Iodo-doxorubicin	OCH_3	$COCH_2OH$	H	NH_2	I	H

Anthracyclines (4) to (13) (also see Table 5-21)

Tab. 5.21 Kinetic parameters for the uptake and release of anthracyclines by MDR cells. (Reprinted from Tab. 1 of ref. 130 with permission from Bentham Science Publishers)

	Drugs	pK_a	D	k_+ ($\times 10^{-10}$ s^{-1})	k_a ($\times 10^{-10}$ s^{-1})	K_m (μM)	RF
1	Doxorubicin	8.4	0.5 1.2	0.2	2.9	> 2	30
2	Daunorubicin	8.4	1.5 7.8	1.5	6.8	2.2 1.5	20
3	F-idarubicin			20	10	0.9	11
4	Idarubicin	8.4	7.6 13.1	40	8.4	0.9	3.3
5	THP-doxorubicin	7.7		40	41	0.5	7
6	Bromodaunorubicin	6.5		70	38	0.3	3
7	Carminomycin	7.5	76	60			6
8	Aclacinomycin	7.0	> 200	80			3
9	Hydroxyrubicin	–		2	4	2	2.5
10	4'-Iodoxorubicin	6.4	31.4				

D, partition coefficient between 1-octanol and buffer at pH 7 or 7.4.; k_+, mean influx coefficient, k_a: mean active efflux coefficient; K_m, Michaelis Menten constant; RF, resistance factor.

The importance of uptake rates in relation to the RF was considered in a later paper [131] for four anthracyclines with identical amino sugar moiety (doxorubicin, daunorubicin, idarubicin, and 8(*S*)-fluoroidarubicin). The kinetics of uptake showed large differences among these anthracyclines. The rates of uptake for daunorubicin, 8(*S*)-fluoroidarubicin, and idarubicin were, respectively, 7,5-, 100- and 200-fold higher than for doxorubicin. The uptake rates increased with increasing lipophilicity. On the other hand, all four anthracyclines were extruded at a similar rate, the rates for idarubicin and 8(*S*)-fluoroidarubicin (the most lipophilic compounds) being only threefold higher than for doxorubicin. The K_m values for the four anthracyclines were also similar. Therefore, the faster uptake could be the decisive factor. The uptake rate and binding to membrane components were not differentiated in this paper.

To exclude the complications inherent in whole-cell studies, artificial membranes were also used to investigate the role of passive transbilayer drug movement in MDR and its modulation [133]. Drug movement can be followed by fluorescence technique in the case of drugs that alter fluorescence when transported from the aqueous phase into the membrane bilayer. This technique was used to determine the transbilayer movement through unilamellar vesicles of the P-gp substrate rhodamine 123. The quenching of fluorescence by this dye occurred in two steps. The first step, leading to 50% total fluorescence decrease, was too fast to be recorded. The second step in fluorescence decrease was slow and followed a first-order process with a lifetime of about 3 min. However, this technique could not be applied to the MDR modulators quinidine and quinine because their rate of movement was too fast for this technique (subseconds). Thus, the transbilayer movement rate was estimated from their equilibrium rates across artificial multilamellar vesicles consisting of phosphatidylcholine (11 μmol) and cardiolipin (2 μmol). The rates of five modifiers and five MDR-type cytotoxic drugs were determined. The equilibration rates of the five MDR modulators were much faster [133] than for the five representative MDR-type drugs, which were similar to the equilibration rate of rhodamine 123. The equilibration rate of the slowest modulator, verapamil, was fast compared with the equilibration rate of the most rapid MDR-type drug, taxol.

In addition, the octanol–buffer partition coefficients and the binding affinity were determined (Table 5.22) for the investigated compounds and compared with their equilibration rates. It is very important to note that no correlation between these parameters was found. The fast influx rates of modifiers in comparison with the slow rates of cytotoxic agents such as doxorubicin and the maximal turnover rate of about 1500/min of P-gp could lead to a situation that P-gp can handle cytotoxic drugs alone. This does not apply, however, in the presence of fast-moving modifiers that can re-enter the cell faster than the maximal capacity of the efflux pump. According to the authors, the result could also suggest another mechanism of action. The slow movement across lipid membranes shown for cytotoxic agents such as doxorubicin "raises the possibility that the composition of the cell-plasma membrane can modulate the transbilayer movement rate of the cytotoxic drugs across the membrane and consequently affect 'competition' between the active efflux of drugs and passive diffusion". A non-P-gp-mediated mechanism of MDR that considers the observed differences in membrane composition and properties which lead to a reduction in passive influx of the cytotoxic drugs could therefore be proposed. It would be reversible by modifiers shown to be able to increase the fluidity of membranes [109].

Tab. 5.22 Solubility and binding properties of drugs and modulators. (Reprinted from Tab. 1 of ref. 133 with permission from the American Society for Biochemistry and Molecular Biology)

	Concentration μM	Partition coefficient (octanol/ aqueous phase)	Binding coefficient (lipid/ aqueous phase)
Daunomycin	200	7.4	260
Doxorubicin	200	1.1	155
Mitoxantrone	200	7.2	450
Vinblastine	0.05	45.0	240
Taxol	0.02	12.0	480
Rhodamine 123	5	6.9	112
Quinidine	200	9.9	270
Quinine	200	9.1	250
Progesterone	0.02	360	925
Trifluoperazine	200	27	245
Verapamil	0.02	8.3	996
Verapamil	200	8.3	550

Binding coefficient, the concentration ratio of the drug in the lipid phase relative to the aqueous phase, was determined by equilibrium dialysis. The binding coefficient is the calculated concentration ratio in the lipid bilayer and the aqueous medium.

Bilayer composition and temperature are important factors affecting the permeability of lipid bilayers to anthracyclines. The uptake into LUVs as a function of these factors was investigated using doxorubicin, daunorubicin, and pirarubicin as an example. DNA encapsulated inside the LUV was used as a driving force (drug binding to DNA), and the decay of the anthracycline fluorescence in the presence of DNA-containing liposomes was used to describe the diffusion through the bilayer. Two steps are involved in the process of crossing the membrane. The first step involves the interaction with the polar head groups and is fast compared with the kinetics of translocation through the hydrocarbon region. It depends on the type and degree of interaction with the polar head groups. The uptake of the anthracyclines occurs through passive diffusion of the neutral form [136]. Three types of LUVs were prepared:

1) PC/PA/CHOL at molar ratios of 75:5:20, buffer at pH 6;
2) PC/PA/CHOL 60:20:20 at pH 6;
3) PC/PA/CHOL 60:20:20 at pH 7.4; and
4) PC/PA/CHOL 52:3:45 at pH 6.

As expected, the permeability coefficient, P^o, decreased with increasing negative charge of the phospholipid in the bilayer for the three anthracyclines. This can be explained by a decrease in the amount of the neutral form of the drugs within the polar head group region, which decreases with increasing charge of the phospholipid. P^o was also negatively affected by the increase in CHOL in the bilayer because of the consequent increase in rigidity of the membrane. Whereas the activation energy, E_a, for the passive diffusion of the neutral form remained within the range 100 ± 16 kJ/mol for daunorubicin and pirarubicin, E_a decreased for the most hydrophilic compound, doxorubicin, to 57 kJ/mol with increasing amounts of PA in the lipid mixture. Surprisingly, on increasing the CHOL content E_a increased to 168 kJ/mol for the most lipophilic pirarubicin but remained around 100 kJ/mol for doxorubicin and daunorubicin.

These aspects of MDR as well as the presented data underline the statement that permeability properties of compounds – especially their amphiphilic nature – cannot sufficiently be described through their partition coefficient in the octanol–buffer system because of special interactions with the phospholipids constituting the membrane.

Summarizing the data on MDR, it can be stated that the affinity (K_m) to P-gp of the various modifiers does not depend on a specific structure. MDR modulators generally possess very similar K_m values. In contrast, strong effects of structure modifications are shown for both modifiers and cytotoxic agents in regard to their various interactions with membranes. Structure or property changes of the drugs lead to changes in their uptake, binding, distribution, localization in the membrane, and competition for binding sites as a function of membrane composition. The structure and properties of drugs affect different membrane properties: fluidization, hexagonal (H_{II}) phase formation, domain formation, and so forth. These changes have been shown to influence the functioning of the embedded proteins. Although the rate-limiting step in the complex mechanism of reversing agents is not yet known, and there might be differences in different resistant patterns and tumor cell lines, there is no doubt that the reported drug–membrane interactions play an essential role in drug resistance and reversal of resistance. This even more true, as differences in cell membranes for sensitive and resistant cells have been observed.

5.2.6
SARs and QSARs for Cytotoxic Agents and MDR Modifiers

Several SARs and QSARs have been derived from data on antineoplastic activity as well as for MDR-reversing activity. The Hansch approach, Free–Wilson, and neural network analysis have been applied. The importance of lipophilicity, molar refractivity (MR), and charge for the description of activity is common to all derived relations.

Biedler and Reihm [137] were the first to describe the MDR phenomenon as a function of drug property. They observed that the exposure of several lines of Chinese hamster cells to increasing concentrations of actinomycin D led to cross-resistance to a variety of drugs. They suggested that cross-resistance was correlated with the MW of these drugs. In another interesting report it was for the first time suggested that a correlation exists between the degree of partitioning of these drugs into the hydrophobic phase and cross-resistance generated in the resistant cells, pointing to the plasma membrane as the site of alteration [138]. In 1990, Selassie *et al.* [139] analyzed the data of Biedler and Rheim [137] (Table 5.23) by multiple regression analysis and arrived at the following final equation:

Tab. 5.23 Cross-resistance of Chinese hamster cells resistant to actinomycin D. (Reprinted from Tab. 4 of ref. 139 with permission from the American Chemical Society)

	Log CR				
Antineoplastic agent	**Observed**	**Predicted**[a]	**Deviation**	**Log MW**[b]	**Log** ***D***[c]
Mithramycin	2.83	2.62	0.21	3.04	-0.25
Vincristine	2.28	2.50	–0.22	2.97	2.57
Puromycin	1.92	1.42	0.50	2.67	0.86
Daunomycin	1.46	1.59	–0.13	2.72	0.66
Demecolsine	1.26	1.07	0.19	2.57	1.37
Mitomycin C	0.49	0.58	–0.09	2.52	–0.38
Proflavine	0.46	0.75	–0.29	2.49	1.10
Novobiocin[a]	0.28	1.95	–1.67	2.80	1.58
Bromodeoxyuridine	0.08	0.49	–0.41	2.49	–0.29
Nitroquinoline-*N*-oxide	0.04	–0.06	0.10	2.11	–0.89
Amethopterin	0.04	0.01	0.03	2.66	–2.52
6-Mercaptopurine	–0.30	–0.42	0.12	2.23	0.01
Hydrocortisone[d]	–0.40	1.53	–1.93	2.69	1.20
Nitrogen mustard[e]	–0.40	–0.50	–0.17	2.28	–2.00[f]
Actinomycin D	1.89	1.72	0.17	2.52	–0.38
Vinblastine	2.38	2.12	0.26	2.96	2.64

a) Predicted from Eq. 5.21.
b) (log MW *vs.* Log *P*, $r^2 = 0.21$).
c) Partition coefficient in octanol-phosphate buffer, pH 7.4.
d) not included in the derivation of Eq. 21.
e) Partition coefficient not measurable.
f) Estimated.

$$\log CR = 3.85 (\pm 0.91) \log MW + 0.26 (\pm 0.15) \log D - 0.09 (\pm 0.06) (\log P)^2 - 9.00 (\pm 2.3) \quad (5.21)$$

$n = 13 \quad r = 0.947 \quad s = 0.30 \quad \log P_o = 1.44 \quad F_{1,9} = 10.3$

where CR = IC_{50R}/IC_{50S} is the cross-resistance to actinomycin D, being the molar concentrations of the drug inducing a 50% inhibition of growth in resistant and sensitive cells respectively. Similarly, the authors analyzed cross-resistance to 29 heterogeneous antineoplastic drugs in a methotrexate-resistant leukemia cell line:

$$\log CR = 7.44 (\pm 2.10) \log MW - 14.97 (\pm 3.94) \log (\beta 10^{\log MW} + 1) - 0.13 (\pm 0.06) \log D - 13.13 (\pm 4.22) \quad (5.22)$$

$n = 29 \quad r = 0.87 \quad s = 0.39 \quad F_{1,24} = 14.68 \quad \log \beta = -2.60 \log MW_o = 2.60$

The example indicates that cross-resistance is increased for hydrophilic drugs of moderate size. The authors suggest that "current chemotherapeutic regimens may be improved by treating resistant cells with antineoplastic agents displaying physicochemical characteristics opposite to that of the original inducing agent." Both equations show the importance of molecular size (expressed as MW) in gaining excess to the cytosolic compartment of resistant cells.

The dependence on charge and lipophilicity of eight rhodamine derivatives for differential accumulation, cytotoxicity, and sensitivity to modulators in sensitive and resistant cell lines has been reported [140]. The physicochemical properties that were shared by MDR-modulating compounds in resistant leukemic cells were investigated [141]. It was found that the lipophilicity at physiological pH, charge, and MR and perhaps structural similarity were important for the modulating activity of a heterogeneous series of drugs. The correlation coefficient with log P was $r = 0.79$ for 12 derivatives when the minimum effective concentration that enhanced vinca alkaloid cytotoxicity in CEM/VBL100 cells was plotted against log P.

In a study of flavonoid-related modulators of MDR, the important role of lipophilicity was again observed. However, exceptions led to the assumption that additional parameters determined the modulating activity [142]. The relation between substructure and MDR-reversing activity for a homologous series of verapamil analogs was reported [143].

The sodium channel blocker propafenone (**24**) was identified as a modulator of P-gp MDR, and large series of analogs were synthesized, tested for MDR-reversing activity, and QSAR analysis was performed [144]. Using calculated log P values according to Ghose *et al.* [145] highly significant regression equations could be derived ($r = 0.99$). It was, however, found that the reduction of the carbonyl group as well as the conversion to a methyl ether led to remarkable decreases in activity not accounted for by the change in lipophilicity. In addition, the relative position of the acyl and propylamine side chains proved to be essential, i.e. the distance between the carbonyl group and the nitrogen atom. Therefore, log P described the variation in MDR-reversing activity only within strongly homologous series. Later, the same research group extended their investigation to acylpyrazolones [146] and again found a significant correlation between log $1/IC_{50}$ and log P:

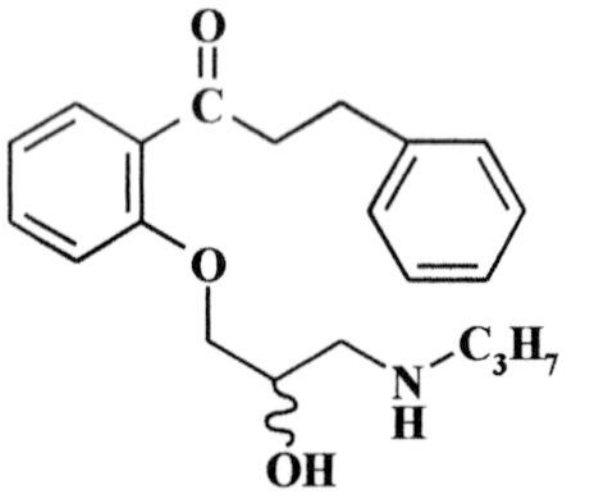

Propafenone (24)

$$\log 1/IC_{50} = 0.69\ (\pm\ 0.24) \log P - 2.72\ (\pm\ 0.60) \qquad (5.23)$$

$n = 15 \quad r = 0.86 \quad s = 0.39 \quad F = 38.54 \quad Q^2 = 0.68$

To account for possible steric effects of the phenyl analogs and for the possible importance of hydrogen-bond acceptor strength of the carbonyl oxygen ,the STERIMOL parameter L and the charge, Ch, at the acyl carbonyl oxygen atoms calculated by the Gasteiger–Hückel algorithm were added:

$$\log 1/IC_{50} = 0.82\ (\pm 0.11)\ \log P - 50.24\ (\pm\ 24.12)\ Ch - 0.32(\pm\ 0.10)\ L - 21.52\ (\pm 9.56) \quad (5.24)$$

$n = 15 \quad r = 0.98 \quad s = 0.17 \quad F = 87.17 \quad Q^2 = 0.92$

When a series of benzofuran analogs of propafenone-type modulators of MDR was compared with propafenone derivatives, a highly significant correlation with log P was found for both series. Two regression lines were obtained with parallel slopes but different intercepts. To explain the difference in intrinsic activity, the interaction with artificial membranes (lecithin liposomes) was studied by NMR and DSC for selected derivatives of both series. NMR interaction measurements showed that propafenone and benzofuran analogs bearing a hydroxyl group interact in the vicinity of this substructure. Dehydroxybenzofuran, on the other hand, showed strong interactions over the whole (phenylethyl)benzofuran region. This difference in the interaction pattern of various benzofuran derivatives had no influence on the MDR-reversing activity. It could be speculated that the differences in MDR-reversing activity for the two series are due “to a decrease in flexibility of the benzofurans compared to propafenones or to the lack of an appropriate hydrogen-bond receptor” [147].

For a series of dihydrobenzopyrans and tetrahydroquinolines, molar refractivity was superior to log P in describing the MDR-reversing activity [148]. Although it has been shown that some MDR-reversing compounds, such as steroids or non-ionic detergents, and also highly lipophilic antineoplastic drugs such as anthracycline analogs, do not possess a charged nitrogen atom, it has recently been shown that the MDR-reversing activity of 12 propafenone-related amines, amides, and anilines with a wide range of pK_a values, and including esters, can be explained by their calculated hydrogen acceptor strength, ΣC_a, using the software package HYBOTPLUS. In contrast, no correlation was found with pK_a (Table 5.24) [149]:

$$\log 1/IC_{50} = 0.58\ (\pm 0.07)\ \Sigma C_a - 3.55\ (\pm 0.44) \quad (5.25)$$

$n = 12 \quad r = 0.93 \quad F = 65.9 \quad Q^2 = 0.82$

A large combined data set of 48 propafenones was then analyzed by both Free–Wilson analysis and a combined Hansch/Free-Wilson approach using an artificial neural network (ANN). With this approach it was possible, in contrast to conventional MLR analysis, to correctly predict the MDR-reversing activity of 34 compounds of the data set after the ANN was trained by only 14 compounds. Best results were obtained using those descriptors showing the highest statistical significance in MLR analysis [150].

Phenothiazines and thioxanthenes are powerful MDR-reversing drugs known, and their reversal activity has been determined in several resistant cell lines. The influence of molecular alterations on their ability to reverse doxorubicin resistance in a breast carcinoma cell line, MCF-7/DOX, was analyzed by Ford *et al.* [119, 120]. The following substructures were found to be essential for MDR-reversing activity: hydrophobic tricyclic ring, positively charged tertiary amine, incorporation of the amino moiety into a cyclic structure for phenothiazines, and halogenated tricyclic rings, piperazinyl amino side groups, and *trans*-isomerism for thioxanthenes. An al-

Tab. 5.24 Chemical structure, physicochemical parameters, and P-gp-inhibitory activity of compounds GP05 to GP570. (Reprinted from Tab. 1 of ref. 149 with permission from the American Society of Pharmacology and Experimental Therapeutics)

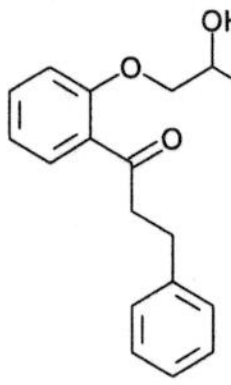

General structure of compounds GP05 to GP388

Compound	R1	Log *P*	pK_a	C_a	EC_{50} Rhodamine 123	EC_{50} Daunomycin	
						Observed	Predicted
GP05		3.67	8.44	5.90	1.45	0.65	1.49
GP29	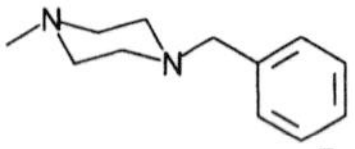	4.43	7.45	6.70	0.62	0.38	0.49
GP31	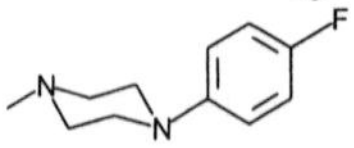	4.93	7.33	7.24	0.26	0.14	0.26
GP62	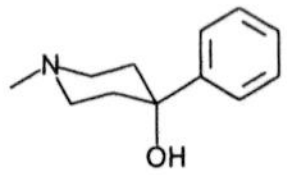	3.98	7.56	7.45	0.11	0.06	0.23
GP240	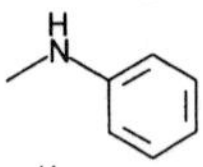	4.23	3.25	4.72	17.61	6.47	6.71
GP339	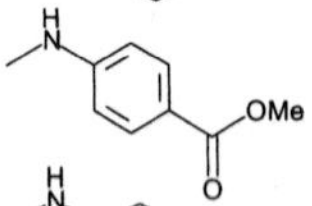	4.05	1.70	5.87	4.57	1.54	1.43
GP358	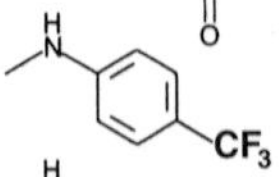	5.15	1.73	4.75	17.87	8.49	6.05
GP359	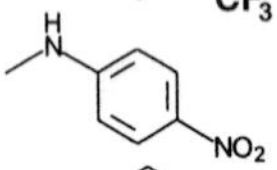	4.26	0.23	5.21	5.89	2.57	3.62
GP360	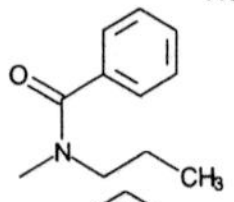	4.60	–1.46	6.48	2.57	1.65	0.58
GP366	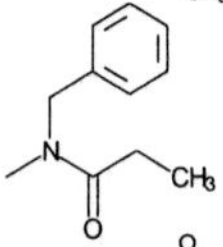	4.60	NA	6.55	2.92	2.04	0.51
GP388	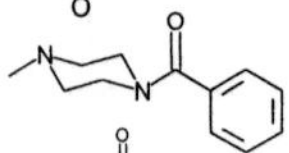	3.87	6.67	8.07	0.36	0.13	0.06
GP570	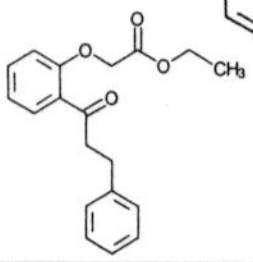	3.35	NA	3.50	44.45	30.80	36.25

NA, not applicable.

most similar conclusion was derived from an analysis of 232 phenothiazines and related drugs in P388/DOX-resistant murine leukemia cells [118, 151]. Later, the Free–Wilson analysis was used to analyze separately a set of MDR-reversing thioxanthenes (Table 5.25) and phenothiazines (Table 5.26) using classical MLR and genetic algorithms (GA) for feature selection [152, 153]. In the case of the thioxanthenes, the side chain length, L1, between the ring system and tertiary nitrogen, the type of the tertiary nitrogen substituent, NALK, and, interestingly, the stereoisomerism, were found by both MLR and GA to be significant in explaining the MDR-reversing activity:

Tab. 5.25 MDR-reversing activity and indicator variables of thioxanthenes. (Reprinted from Tab. 1 of ref. 152 with permission from Wiley-VCH).

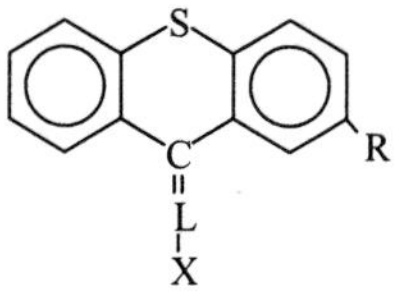

R→ CL: -Cl　　CF3: $-CF_3$

L→ L1: $=CH-CH_2-$　　L2: $=CH-CH_2-CH_2-$

X→ NALK: $-N(CH_3)_2$ or $-N(C_2H_5)_2$　　NPRZ: $-N$(piperazine)$N-(CH_2)_2-OH$　　NPRD: $-N$(piperidine)

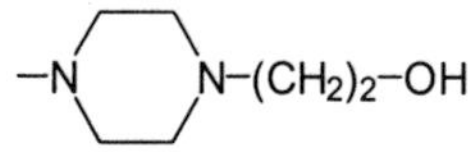

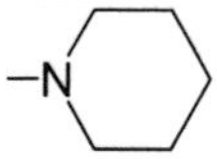

Compounds	Log MDR[a]	R[b]		L		X			CIS	TRANS
		CL	CF3	1	2	NALK	NPRZ	NPRD		
cis-762	0.041	1	0	1	0	1	0	0	1	0
trans-762	0.114	1	0	1	0	1	0	0	0	1
cis-chlorprothixene	0.301	1	0	0	1	1	0	0	1	0
trans-Chlorprothixene	0.845	1	0	0	1	1	0	0	0	1
cis-768	0.602	1	0	0	1	1	0	0	1	0
trans-768	0.857	1	0	0	1	1	0	0	0	1
cis-796	0.255	0	1	0	1	1	0	0	1	0
trans-796	0.447	0	1	0	1	1	0	0	0	1
cis-753	0.826	1	0	0	1	0	0	1	1	0
trans-753	1.176	1	0	0	1	0	0	1	0	1
Racemic 751	0.255	1	0	0	1	0	1	0	0.5	0.5
trans-7006	0.204	1	0	1	0	0	1	0	0	1
cis-Clopenthixol	0.415	1	0	0	1	0	1	0	1	0
trans-Clopenthixol	1.176	1	0	0	1	0	1	0	0	1
cis-Flupentixol	0.681	0	1	0	1	0	1	0	1	0
trans-Flupentixol	1.182	0	1	0	1	0	1	0	0	1
789	0.987	1	0	0	1	0	1	0	0	0

a) Log MDR = log (MDR fold reversal) is the MDR-reversing activity of the compounds.
b) CL, CF3, L1, L2, NALK, NPRZ, NPRD, CIS and TRANS are the indicator variables.

Tab. 5.26 MDR-reversing activity and indicator variables of phenothiazines and related drugs. (Reprinted from Tab. 2 of ref. 152 with permission from Wiley-VCH)

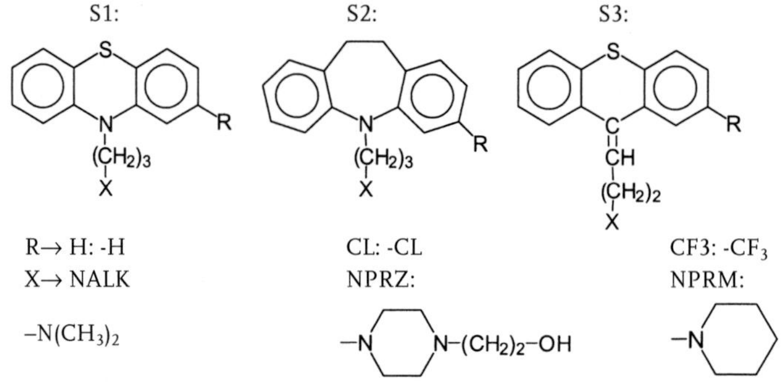

S1: S2: S3:

R→ H: -H CL: -CL CF3: $-CF_3$

X→ NALK NPRZ: NPRM:

$-N(CH_3)_2$ $-N$ (piperazine) $N-(CH_2)_2-OH$ $-N$ (piperidine)

Compound	Log MDR[a]	RS[b]			R			X			CIS	TRANS
		S1	S2	S3	H	Cl	CF3	NALK	NPRZ	NPRM		
Promazine	0.079	1	0	0	1	0	0	1	0	0	0	0
Chlorpromazine	0.204	1	0	0	0	1	0	1	0	0	0	0
Triflupromazine	0.301	1	0	0	0	0	1	1	0	0	0	0
Perfenazine	0.301	1	0	0	0	1	0	0	1	0	0	0
Fluphenazine	0.431	1	0	0	0	0	1	0	1	0	0	0
Prochlorperazine	0.415	1	0	0	0	1	0	0	0	1	0	0
Trifluoperazine	0.531	1	0	0	0	0	1	0	0	1	0	0
Imipramine	0.398	0	1	0	1	0	0	1	0	0	0	0
Chlorimipramine	0.301	0	1	0	0	1	0	1	0	0	0	0
cis-Chlorprothixene	0.301	0	0	1	0	1	0	1	0	0	1	0
trans-Chlorprothixene	0.845	0	0	1	0	1	0	1	0	0	0	1
cis-796	0.255	0	0	1	0	0	1	1	0	0	1	0
trans-796	0.447	0	0	1	0	0	1	1	0	0	0	1
cis-Clopentixol	0.415	0	0	1	0	1	0	0	1	0	1	0
trans-Clopentixol	1.176	0	0	1	0	1	0	0	1	0	0	1
cis-Flupentixol	0.681	0	0	1	0	0	1	0	1	0	1	0
trans-Flupentixol	1.182	0	0	1	0	0	1	0	1	0	0	1

a) Log MDR = Log (MDR fold reversal) is the MDR-reversing activity of the compounds.
b) S1, S2, S3, H, CL, CF3, NALK, NPRZ, NPRM, CIS and TRANS are the indicator variables.

$$\log MDR = -0.617\,(\pm 0.127)\,L1 - 0.278\,(\pm 0.099)\,NALK - 0.344\,(\pm 0.099)\,CIS + 1.037\,(\pm 0.081) \quad (5.26)$$

$n = 16$ $r = 0.896$ $s = 0.194$ $F = 16.28$ $Q^2 = 0.621$

For the set of 17 phenothiazines and related structures, four features were found to be of significance in explaining the observed variation in MDR-reversing activity: the ring system type, S1, S2 (phenothiazine, imipramine), the side chain type, NALK, and the *cis*-isomerism:

$$\log MDR = -0.719\,(\pm 0.065)\,S1 - 0.545\,(\pm 0.093)\,S2 - 0.259\,(\pm 0.057)\,NALK - 0.611\,(\pm 0.073)\,CIS + 1.054\,(\pm 0.057) \quad (5.27)$$

$n = 16 \quad r = 0.968 \quad s = 0.095 \quad F = 41.5 \quad Q^2 = 0.883$

The thioxanthene *trans*-flupentixol, substituted with CF_3 in position 2, a piperazine moiety with a four-bond distance from the ring system, and *trans*-isomerism, was found to be the most potent compound. Molecular modeling has been performed assuming that, in agreement with the observed two- to threefold stronger interaction with phospholipid bilayers in NMR binding studies, the observed two- to threefold difference in MDR-reversing activity compared with the *cis*-form could be due to different preferable conformations in the lipid environment. The optimized conformations were compared with those derived from ^{1}H-NMR analysis in lipid environment, some of them corresponding significantly. Finally, electrostatic and lipophilic fields of *cis*- and *trans*-isomers were compared to determine if molecular properties can be related to the observed activity difference. The results showed that the *cis*- and *trans*-forms may adopt different (mirror-image) forms when entering the phospholipid bilayer in terms of the orientation of their ring system. It can be suggested that, with regard to MDR activity, the form adopted by the *trans*-isomer is a better fit to the "MDR-reversal receptor" than the form adopted by the *cis*-isomer [152].

Finally, correlations between the parameters involved in the interaction of chemosensitizers with phospholipid bilayers, derived from DSC and NMR measurements, and their potency in reversing MDR should be discussed. The latter data were taken from the literature. A problem in handling these data is the fact that different definitions for the potentiating effect were used. In the case of MCF-7DOX cell lines, the criterion used was MDR ratio or fold reversal, defined by dividing the IC_{50} of doxorubicin alone by the IC_{50} of doxorubicin plus chemosensitizing drug at drug concentrations $\leq IC_{10}$ [119, 120]. The ratio can be considered as the increase in apparent potency of doxorubicin. In contrast, the definition used in P388/DOX cell lines was the IC_{50} value of the modifying drug alone divided by the IC_{50} of the drug plus doxorubicin at a subinhibitory concentration of 0.02 μM [119, 151]. This ratio can be considered as an increase in the apparent cytotoxicity of the drug in the presence of doxorubicin. The following drug–membrane interaction parameters derived from DSC measurement were used in the regression analysis (Table 5.27): $\Delta T_{ind.}$, and the difference observed for T_t for the phospholipid (PS) in the absence and presence of drug, $T_{t,\,slope}$ is calculated from the linear change in T_t as a function of lipid–drug ratio for DPPC and DPPS respectively. In spite of the fact that the number of compounds is small, the biological data complex, and the range of activity small, the results of the regression analysis show a significant correlation between strength of drug–membrane interaction and MDR-reversing activity (Table 5.27) [64, 117]. It is interesting to note that again a similar dependence of MDR-reversing activity and PKC inhibition on drug–membrane interactions was found. This supports the assumption that an indirect effect rather than a direct is involved, i.e. modifiers indirectly can affect protein function via changes in membrane properties which would affect proteins rather than a direct binding to these proteins, which possess different

Tab. 5.27 Correlation coefficients between drug-membrane interaction parameters and biological data of the seven modifiers studied. (*cis-* and *trans-*flupentixol, chlorpromazine, trifluoperazine, triflupromazine, imipramine, and quinacrine). Chlorpromazine was omitted from regression corresponding to the MCF-7/DOX cell line. Quinacrine data were not available for the P388/DOX cell line. (Reprinted from Tab. 2 of ref. 117, with permission from Bertelsmann-Springer)

	ΔT_{ind}	T_t slope DPPS	T_t slope DPPC	Log(1/MDR_{ratio})	Log (1/$IC_{50\ Drug}$)	Log (1/$IC_{50\ PKC}$)
MCF-7/DOX tumor cell line						
$\Delta T_{ind.}$	1.0000					
T_t slope DPPS	–0.8085	1.0000				
T_t slope DPPC	–0.6693	0.8756	1.0000			
Log(1/MDR_{ratio})	–0.8638	0.9312	0.8562	1.0000		
Log(1/$IC_{50\ Drug}$)	–0.8456	0.6479	0.4942	0.8153	1.0000	
Log(1/$IC_{50\ PKC}$)	0.8826	–0.7131	–0.6669	–0.6533	–0.5598	1.0000
P388/DOX tumor cell line						
ΔT_{ind}	1.0000					
T_t slope DPPS	0.8594	1.0000				
T_t slope DPPC	0.7744	0.8729	1.0000			
Log(1/MDR_{ratio})	0.8815	0.8642	0.5963	1.0000		
Log(1/$IC_{50\ Drug}$)	0.9307	–0.6278	–0.5686	–0.7319	1.0000	
Log(1/$IC_{50\ Drug+DOX}$)	0.9727	–0.8061	–0.6263	–0.9356	0.9253	1.0000

functions and structures. However, a conclusion can still not be derived from the present structure–activity relationships.

Finally, two recent review articles on the relationships between P-gp-mediated MDR and the lipid phase of the cell membrane should be mentioned [154, 155] as they support the presented arguments. The authors of the first review [154] emphasized "that the lipid phase of the membrane cannot be overlooked while investigating the MDR-phenotype. The properties that make a compound able to interact with P-gp as a substrate or as modulator, make it also able to interact with a lipid bilayer. Increasing evidence shows that lipids have many different functions, such as signal transduction or modulation of peripheral or integral proteins" [154, 155]. The second, very recent, review examined the possible key role of lipids in MDR [156]. In this review, the various factors that could be involved in MDR are discussed. For example, the links between the different forms of MDR and lipid metabolism, the effect of lipids on drug influx, the role of cholesterol content, the influence of DAG production on P-gp-mediated transport, the role of exocytic factors in MDR, the fusogenic properties of lipid composition, and the contribution of changes in acidic phospholipid. The authors conclude that "the involvement of an exogenic component in MDR is fairly strong" and that "the importance of lipids could be major" [156]. However, a generalization about the role of lipids in MDR is not yet possible, because of the limited studies performed and the diversity of MDR types [156].

References

1 Heimburg, T., Hildebrandt, P., Marsh, D., *Biochemistry* **1991**, *30*, 9084–9089.
2 Heimburg T., Biltonen, R., *Biochemistry* **1994**, *33*, 9477–9488.
3 Watts, A., *Biochim. Biophys. Acta* **1998**, *1376*, 297–318.
4 Glaubitz, Cl., Gröbner, G., Watts, A., *Biochim.Biophys. Acta* **2000**, *1463*, 151–161.
5 Senistra, G., Epand, R.M., *Arch. Biochem. Biophys.* **1993**, *300*, 378–383.
6 Goldberg, E.M., Lester, D.S., Borchardt, D.B., Zidovetzki, R., *Biophys. J.* **1995**, *69*, 965–973.
7 Yeagle, P.L., Sen, A., *Biochemistry* **1986**, *25*, 7518–7522.
8 Ulrich, A.S., Watts, A., *Biophys. J.* **1994**, *66*, 1441–1449.
9 Epand, R.M., Lester, D.S., *Trends Pharmacol. Sci.* **1990**, *11*, 317–320.
10 Trudell, J.R., Costa, A.K., Hubbell, W.L., *Ann. N.Y. Acad. Sci.* **1991**, *625*, 743–747.
11 Koehler, K.A., Hines, J., Mansour, E.G., Rustum, Y.M., Jahagidar, D.V., Jain, M.K., *Biochem. Pharmacol.* **1985**, *34*, 4025–4031.
12 Gärtner, R., *Biochem. Pharmacol.* **1987**, *7*, 1063–1067.
13 Chatelain, P., Laruel, R., Gillard, M., *Biochim. Biophys. Acta* **1985**, *129*, 148–154.
14 Lüllmann, H., Plösch, H., Ziegler, A., *Biochem. Pharmacol.* **1980**, *29*, 2969–2974.
15 Carfagna, M., Muhoberac, B.B., *Mol. Pharmacol.* **1993**, *44*, 129–141.
16 Aloin, R.C., Curtain, C.C., Gordon, L.M. (Eds.), *Drug and Anesthetic Effects on Membrane Structure and Function*, Wiley-Liss, New York **1991**.
17 Peper, K., Bradley, R.J., Dreyer, F., *Physiol. Rev.* **1982**, *62*, 1271–1340.
18 Heidman, T., Changeux, J.-P., *Proc. Natl. Acad. Sci. USA* **1984**, *81*, 1897–1901.
19 Brett, R.S., Dilger, J.P., Yland, K.F., *Anesthesiology* **1988**, *69*, 161–170.
20 Yun, I., Kim. I., Yum, S.M., Baek, S.Y., Kang, J.S., *Asia Pacific J. Pharmacol.* **1990**, *5*, 19–26.
21 Ohki, S., *Biochim. Biophys. Acta* **1970**, *219*, 18–27.
22 Barbato, F., Rotonda, M.I., Quaglia, F., *Pharm. Res.* **1997**, *14*, 1699–1705.
23 Mavromoustakos, T., Daliani, I., Matsoukas, J., in *Bioactive Peptides in Drug Discovery and Design: Medical Aspects*, J. Matsoukas, T. Mavromoustakos (Eds.), IOS Press Amsterdam **1999**, pp. 13–24.
24 Kuroda, Y, Ogawa, M., Nasu, H., Terashima, M., Kasahara, M., Kiyama, Y., Wakita, M., Fujiwara, Y., Fujii, N., Nakagawa, T., *Biophys. J.* **1996**, *71*, 1191–1207.
25 Patton, D.E., West, J.W., Catterall, W.A., Goldin, L.A., *Proc. Natl. Acad. Sci. USA* **1992**, *89*, 10905–10909.
26 Sheldon, R.S., Hill, R., Taouis, M., Wilson, L.M., *Mol. Pharmacol.* **1991**, *39*, 609–614.
27 Lüllmann, H., Mohr, K., in *Metab. Xenobiot. (Pap. Eur. Meet. ISSX)*, J.W. Gorrod, J.W. Oelschläger (Eds.), Taylor & Francis, London **1988**, pp. 13–20.
28 Hwang, S.-B., Shen, T.Y., *J. Med. Chem.* **1981**, *24*, 1202–1221.
29 Caron, G., Gaillard, P., Carrupt, P.A., Testa, B., *Helv. Chim. Acta* **1997**, *80*, 449–462.
30 Barbato, F., La Rotonda, M.I., Quaglia, F., *J. Pharm. Sci.* **1997**, *86*, 225–229.
31 Gourley, D.H.R., in: *Interaction of Drugs with Cells*, Charles C. Thomas (Ed.), Springfield, IL **1971**, Chapter 5, pp. 79.
32 Voigt, W., Mannhold, R., Limberg, J., Blaschke, G., *J. Pharm. Sci.* **1988**, *77*, 1018–1020.
33 Barbato, F., LaRotonda, I.M., Quaglia, F., *Eur. J. Med. Chem.* **1996**, *31*, 311–318.
34 Mason, R.P., Rhodes, D.G., Herbette, L.G., *J. Med. Chem.* **1991**, *34*, 869–877.
35 Herbette, L.G., Vecchiarelli, M., Satarni, A., Leonardi, A., *Blood Pressure* **1998**, *Vol. 2*, 10–17.
36 Bellemann, P., in *Pharmacochemistry Library*, Vol. 9, W.T. Nauta, R.F. Rekker (Eds.), Vol. Ed. A.F. Harms, Elsevier Science, Amsterdam **1986**, pp. 23–46.
37 Choi, Y.W., Rogers, J.A., *Pharm. Res.* **1990**, *7*, 508–512.

38 Seydel, J.K., Coats, E.A., Cordes, H.-P., Wiese, M., *Arch. Pharm.* **1994**, *327*, 601–610.
39 Seydel, J.K., Albores Velasco, M., Coats, E.A., Cordes, H.-P., Kunz, B., Wiese, M., *Quant. Struct.–Act. Relat.* **1992**, *11*, 205–210.
40 Seydel, J.K., Schaper, K.-J., Coats, E.A., Cordes, H.-P., Emig, P., Engel, J., Kutscher, B., Polymeropoulos, E.E., *J. Med. Chem.* **1994**, *37*, 3016–3022.
41 Clement, B., Karnatz, A., Seydel, J.K. (unpublished results).
42 Nepomuceno, M.F., de Oliveira Mamede, M.E., Vaz de Macedo, D., Alves, A.A., Pereira-da-Silva, L., Tabak, M., *Biochim. Biophys. Acta* **1999**, *1418*, 285–294.
43 Noseda, A., Godwin, P.L., Modest, E.J., *Biochim. Biophys. Acta* **1988**, *945*, 92–100.
44 Van Blitterswijk, W.J., Hilkmann, H., Storme, G.A., *Lipids* **1987**, *22*, 820–823.
45 Hermann, D.B.J., *J. Natl. Cancer Inst.* **1985**, *75*, 423–430.
46 Diomede, L., Bizzi, A., Magistrelli, A., Modest, E.J., *Int. J. Cancer* **1991**, *49*, 1–5.
47 Principe, P., Diomede, L., Sidoti, C., Salmona, M., Broquet, C., Braquet, P., *Int. J. Oncol.* **1991**, *1*, 713–719.
48 Zidowetzki, R., Sherman, I.W., Cardenas, M., Borchard, D.B., *Biochem. Pharmacol.* **1993**, *45*, 183–189.
49 Shimada, H., Grutzner, J.B., Kozlowski, J.F., McLaughlin, J.L., *Biochemistry* **1998**, *37*, 854-866.
50 Seydel, U., Ulmer, A.J., Uhlig, St., Rietschel, E. Th., in: *Membrane Structure in Disease and Drug Therapy*, G. Zimmer (Ed.), Marcel Dekker, New York **2000**, pp. 217–241.
51 Brandenburg, K., Mayer, H., Koch, M.H.J., Weckesser, J., Rietschel, E.Th., Seydel, U., *Eur. J. Biochem.* **1993**, *218*, 555–563.
52 Brade, L., Brandenburg, K., Kuhn, H.-M., Kusumoto, S., Macher, I., Rietschel, E.Th., *Infect. Immun.* **1987**, *55*, 2636–2644.
53 Mason, R.P., Trumbore, M.W., Pettegrew J. W., *Ann. N.Y. Acad. Sci.* **1996**, *777*, 368–373.
54 Nikaido, H., *Antimicrob. Agents Chemother.* **1989**, *33*, 1831–1839.
55 Coats, E.A., Genther, C.S., Dias Celassie, C., Strong, C.D., Hansch, C., *J. Med. Chem.* **1985**, *28*, 1910–1916.
56 Seydel, J.K., Wiese, M., Cordes, H.-P., Chi, H.L., Schaper, K.-J., Coats, E.A., Kunz, B., Engel, J., Kutscher, B., Emig, H., in *QSAR: Rational Approaches to the Design of Bioactive Compounds*, C. Silipo, A. Vittoria (Eds.), Elsevier, Amsterdam **1991**, pp. 367–376.
57 Rosen, B.P., Mobashery, S. (Eds.), *Resolving the Antibiotic Paradox*. Kluwer Academic/Plenum Press, New Jork, **1998.** (*Advances in Experimental Medicine Biology*, Vol. 456,)
58 Bradley, G., Juranka, P.F., Ling, V., *Biochim. Biophys. Acta* **1988**, *948*, 87–128.
59 Gottesmann, M.M., Pastan, I. *Annu. Rev. Biochem.* **1993**, *62*, 385–427.
60 De Vita, V.T., Jr., in *Drug Resistance: Mechanisms and Reversal*, E. Mihich (Ed.), John Libbey CIC, New York **1990**, pp. 7–27.
61 Ecker, G., Chiba, P., *Exp. Opin. Ther. Patents* **1997**, *7*, 589–599.
62 Scovsgard, T., *Cancer Res.* **1980**, *40*, 1077–1083.
63 Tsuruo, T., Iida, H., Tsukagoshiu, S., Sakurai, Y., *Cancer Res.* **1981**, *41*, 1967–1972.
64 Seydel, J.K., Coats, E.A., Pajeva, I.K., Wiese M., in: *Bioactive Compound Design, Possibilities for Industrial Use.* M.G.Ford, R. Greenwood, G.T. Brooks, R. Franke (Eds.), Bios Scientific Publishers, Ltd., Oxford, **1996**, pp. 137–147
65 Renau, Th.E., Lèger, R., Flamme, E.M., Sangalang, J., She, M.W., Yen, R., Gannon, C.L., Griffith, D., Chamberland, S., Lomovskaya, O., Hecker, S.J., Lee, V.J., Ohta, T., Nakayama, K., *J. Med. Chem.* **(1999)** *42*, 4928–4931
66 Martin, S.K., Oduola, A.M.J., Milhous, W.K., *Science* **1987**, *235*, 899–901.
67 Tanabe, K., Kato, M., Izumo, A., Hagiwara, A., Doi, S., *Experimental Parasitology* **1990**, 70, 419–426.
68 van Veen, H.W., Venema, K., Bolhuis, H., Oussenko, I., Kok, J., Poolman, B., Driessen, A.J., Konings, W.N., *Proc. Natl Acad. Sci. USA* **1996**, *93*, 10668–10672.

69 Safa, R.A., Glover, C.J., Meyers, M.B., Biedler, J.L., Felsted, R.L., *J. Biol. Chem.* **1986**, *261*, 6137–6140.

70 Beck, W.T., Qian. X., *Biochem. Pharmacol.* **1992**, *43*, 89–93.

71 Safa, A.R., in: *Cancer Cells*, S. Gupta, T. Tsuruo Eds., Wiley & Sohns, Ltd., **1996**, pp.231–249

72 Holzmayer, T.A., Hilsenbeck, S., Von Hoff, D.D., Roninson, I.B., *JNC,* **1992**, *84*, 1486–1491.

73 Raviv, Y., Pollard, H.B., Bruggeman, E.P., Pastan, I., Gottesman, M.M., *J. Biol. Chem.* **1990**, *265*, 3975–3980.

74 Higgins, C.F., *Cell* 79, (**1994**) 393–395.

75 Pawagi, A.B., Wang, J., Silverman, M., Reithmeier, R.A., Deber, C.M., *J. Mol. Biol.* **1994**, *235*, 554–564.

76 Safa, A.R., Agresti, M., Bryk, D., Tamai, I., *Biochemistry* **1994**, *33*, 256–265.

77 Sharom, F.J., *J. Membr. Biol.* **1997**, *160*, 161–175.

78 Garnier-Suillerot, A., Borrel, M.N., Pereira, E., Fiallo M., *Anti-Cancer Drugs* **1994**, *5* (Suppl.1) 30.

79 Huet, S., Schott, B., Robert, J., *Br. J. Cancer* **1992**, *65*, 538–544.

80 Laredo, J., Huynh, A., Mullert, C., Jaffrezou, J.-P., Bailly, J.-D., Cassar, G., Laurent, G., Demur, C., *Blood* **1994**, *1*, 229–237.

81 Sognier, M.A., Altenberg, G.A., Eberle, R., Tucker, M., Belli, J.A., *Anti-Cancer Drugs* **1994**, *5* (Suppl. 1), 7.

82 Jaffrezou, J.-P., Herbert, J.-M., Levade, Th., Gau, M.-N., Chatelain, P., Laurent, G., *J. Biol. Chem.* **1991**, *266*, 19858–19864.

83 Blobe, G.C., Sachs, C.W., Khan, W.A., Fabbro, D., Stabel, S., Wetsel, W.C., Obeid, L.M., Fine, R.L., Hannun, Y.A., *J. Biol. Chem.* **1993**, *268*, 658–664.

84 Sato, W., Jusa, K., Natio, M., Tsuruo, T., *Biochem. Biophys. Res. Commun.* **1990**, *173*, 1252–1257.

85 Chambers, T., Zheng, B., Kuo, J.K., *Mol. Pharmacol.* **1992**, *41*, 1008–1015.

86 Bates, S.E., Lee, J.S., Dickstein, B., Spolyar, M., Fojo, A.T., *Biochemistry* **1993**, *32*, 9156–9164.

87 Sachs, Cl.W., Safa, A.R., Harrison, St.D., Fine, R.L., *J. Biol. Chem.* **1995**, *44*, 26639–26648.

88 Senisterra, G., Epand, R.M., *Arch. Biochem. Biophys.* **1993**, *300*, 378–383.

89 Hait, W.N., Aftab, D.T., *Biochem. Pharmacol.* **1992**, *43*, 103–107.

90 Aftab, D.T. Hait, W.N., *Anal. Biochem.* **1990**, *187*, 84–88.

91 Tritton, T.R., Posada, J., in *Anticancer Drugs*, Vol. 191, H. Tapiero, J. Robert, T.J. Lampidis (Eds.), INSERM/John Libbey Eurotext, London, **1989**, Colloques INSERM Vol. 191, pp. 289–300.

92 Tapiero, H., Mishal, Z., Wioland, M., Silber, A., Fourcade, A., Zwinfelstein G., *Anticancer Rev.* **1986**, *6*, 649–652.

93 Hasmann, M., Valet, G.K., Tapiero, H., Trevorrow. K., Lampidis, T., *Biochem. Pharmacol.* **1989**, *38*, 305–312.

94 Lampidis, T.J., Savaraj, N., Valet, G.K., Trevorrow, K., Fourcade, A., Tapiero, H., in *Anticancer Drugs*. Tapiero, T. J. Lampidis (Eds.), INSERM/John Libbey Eurotext, London, **1989**, Colloques INSERM Vol. 191, pp. 29–38

95 Vayuvegula, B., Slater, L., Meador, J., Gupta, S., *Cancer Chemother.* **1988**, *22*, 163–168.

96 May, G.L., Wright, L.C., Dyne, M., Mackinnon, W.B., Fox, R.M., Mountford, C.E., *Int. J. Cancer* **1988**, *42*, 728–733.

97 Mazzoni, A., Trave, F., *Oncology Res.* **1993**, *5*, 75–82.

98 Callaghan, R., Van Gorkom, L.C.M., Epand R.M., *Br. J. Cancer* **1992**, *66*, 781–786.

99 Arsenault, L.A., Ling, V., Karzner, N., *Biochim. Biophys. Acta* **1988**, *938*, 315–321.

100 Sehestedt, M., Skovsgaard, T., Van Deurs, B., Winther-Nielsen, H., *Br. J. Cancer* **1988**, *56*, 747–751.

101 Le Moyec, L., Tatoud, R., Degeorges, A., Calabresse, C., Bauza, G., Eugene, M., Calvo, F., *Cancer Res.* **1996**, *56*, 3461–3467.

102 Saeki, T., Shimabuku, A.M., Ueda, K., Komano, T., *Biochim. Biophys. Acta* **1992**, *1107*, 105–110.

103 Callaghan, R., Stafford, A., Epand, R.M., *Biochim. Biophys. Acta* **1993**, *1175*, 277–282.

104 Hui, S.W., Sen, A., *Proc. Natl. Acad. Sci. USA* **1989**, *86*, 5825–5829.

105 Doige, C.A., Yu, X., Sharom, F., *Biochim. Biophys. Acta* **1993**, *1146*, 65–72.

106 Rotenberg, M., Zakim, D., *J. Biol. Chem.* **1991**, *265*, 18753–18756.

107 Escriba, P.V., Ferrer-Montiel, A.V., Ferragut, J.A., Gonzales-Ros, J.M., *Biochemistry* **1990**, *29*, 7275–7282.

108 Wadkins, R.M., Houghton, P.J., *Biochim. Biophys. Acta* **1993**, *1153*, 225–236.

109 Drori, St., Eytan. G.D., Assaraf, Y., *Eur. J. Biochem.* **1995**, *228*, 1022–1029.

110 Zordan-Nuodo, T., Ling, V., Liu, Z., Georges, E., *Cancer Res.* **1993**, *53*, 5994–6000.

111 Greenberg, L.M., *J. Biol. Chem.* **1993**, *268*, 11417–11425.

112 de Wolf, F.A., Nicolay, K., de Kruijff, B., *Biochemistry* **1992**, *31*, 9252–9262.

113 de Wolf, F.A., Staffhorst, R.W.H.M., Smits, H.-P., Onwezen, M.F., de Kruijff, B., *Biochemistry* **1993**, *32*, 6688–6695.

114 Awasthi, S., Sharma, R., Awasthi, Y.C., Belli, J.A., Frenkel, E.P., *Cancer Lett.* **1978**, *63*, 109–116.

115 Speelmans, G., Staffhorst, R.W.H.M., de Kruijff, B., De Wolf, F.A., *Biochemistry* **1994**, *33*, 13761–13768.

116 Rutherford, A.V., Willigham, M.C., *J. Histochem. Cytochem.* **1993**, *41*, 1573–1577.

117 Pajeva, I.K., Wiese, M., Cordes, H.-P., Seydel, J.K., *J. Cancer Clin. Oncol.* **1996**, *122*, 27–40.

118 Ramu, A., Ramu, N., *Cancer. Chemother. Pharmacol.* **1992**, *30*, 165–173.

119 Ford, J.M., Bruggeman, E.P., Pastan, I., Gottesman M., Hait, W.N., *Cancer Res.* **1990**, *50*, 1748–1756.

120 Ford, J.M., Prozialeck, W.C., Hait, W.N., *Mol. Pharmacol.* **1989**, *35*, 105–115.

121 Herbette, L.G., Trumbore, M., Chester, D.W., Katz, A.M., *J. Mol. Cell Cardiol.* **1988**, *20*, 373–378.

122 Nicolay, K., Sauterau, A.-M., Tocanne, J.-F., Brasseur, R., Huart, P., Ruysschaert, J.-M., de Kruijff, B., *Biochim. Biophys. Acta* **1988**, *940*, 197–208.

123 Hendrich, A.B., Wesolowska, O., Michalak, K., *Biochim. Biophys. Acta* **2001**, *1510*, 414–425.

124 Pajeva, I.K., Todorov, D.K., Seydel, J.K., in *Non antibiotics: a New Class of Unrecognized Antimicrobics*, Vol. 2, A.N. Chakrabarty, J. Molnar, S.G. Dastidar, N. Motohashi (Eds.), **2000**, in press.

125 Toffoli, G., Simone, F., Gigante, M., Boiocchi. M., *Biochem. Pharmacol.* **1994**, *48*, 1871–1881.

126 Lampidis T.J., Kolonias, D., Podona, T., Safa, A.R., Lothstein. L., Savraj, N., Tapiero, H., Pribe, W., *Biochemistry* **1997**, *36*, 2679–2685.

127 Consoli, U., Pribe, W., Ling, Y.-H., Mahadevia, R., Griffin, M., Zhao, S., Perez-Soler, R., Andreef, M., *Blood* **1996**, *88*, 633–644.

128 Gallois, L., Fiallo, M., Garnier-Suillerot A., *Biochim. Biophys. Acta* **1998**, *1370*, 31–40.

129 Tapiero, H., Munck, J.-N., Fourcade, A., *Drugs Exptl-Clin. Res.* **1986**, *12*, 257–263.

130 Garnier-Suillerot, A., *Current Pharmaceutical Design* **1995**, *1*, 69–82.

131 Mankhetkorn, S., Dubru, F., Hesschenbrouk, J., Fiallo, M., Garnier-Suillerot, A., *Mol. Pharmacol.* **1996**, *49*, 532–539.

132 Frezard, F., Garnier-Suiollerot, A., *Biochim. Biophys. Acta* **1998**, *1389*, 13–22.

133 Eytan, G.D., Regev, R., Oren, G., Assaraf, Y.G., *J. Biol. Chem.* **1996**, *271*, 12897–12902.

134 Stein, W., Cardarelli, C., Pastan, I., Gottesman, M., *Mol. Pharmacol.* **1994**, *45*, 763–772.

135 Marbeuf-gueye, C., Ettori, D., Priebe, W., Kozlowski, H., Garnier-Suillerot, A., *Biochim. Biophys. Acta* **1999**, *1450*, 374–384.

136 Harrigan, P.R., Wong, K.F., Redelmeier, T.E., Wheeler, J.J., Cullis, P.R., *Biochim. Biophys. Acta* **1993**, *1149*, 329–338.

137 Biedler, J.L., Reihm, H., *Cancer Res.* **1970**, *30*, 1174–1184.

138 Bech-Hansen, N.T., Till, J.E., Ling, V.J. *Cell Physiol.* **1976**, *88*, 23–31.

139 Selassie, C.D., Hansch, C., Khwaga, T.A., *J. Med. Chem.* **1990**, *33*, 1914–1919.

140 Lampidis, T.J., Castello, C., Del Giglio, A., Pressman, B.C., Viallet, P., Trevorrow, K.W., Valet, G., Tapiero, H.,

Savaraj, N., *Biochem. Pharmacol.* **1989**, *38*, 4267–4271.

141 Zamora, J.M., Pearce, H.L., Beck, W., *Mol. Pharmacol.* **1988**, *33*, 454–462.

142 Ferté, J., Kühnel, J.-M., Chapuis, G., Rolland, Y., Lewin, G., Schwaller, M.A., *J. Med. Chem.* **1999**, *42*, 478–489.

143 Toffoli, G., Simone, F., Corona, G., Raschack, M., Cappelletto, B., Gigante, M., Boiocchi, M., *Biochem. Pharmacol.* **1995**, *50*, 1245–1255.

144 Chiba, P., Ecker, G., Schmid, D., Drach, J., Tell, B., Goldenberg, S., Gekeler, V., *Mol. Pharmacol.* **1996**, *49*, 1122–1130.

145 Ghose, A.K., Pritchett, A., Crippen, G.M., *J. Comput. Chem.* **1988**, *9*, 80–90.

146 Chiba, P., Holer, W., Landau, M., Bechmann, G., Lorenz. K., Plagens, B., Hitzler, M., Richter, E., Ecker, G., *J. Med. Chem.* **1998**, **41**, 4001–4011.

147 Ecker, G., Chiba, P., Hitzler, M., Schmid, D., Visser, Kl., Cordes, H.-P., Csöllei, J., Seydel, J.K., Schaper, K.-J., *J. Med. Chem.* **1996**, **39**, 4767–4774.

148 Hiesböck, R., Wolf, Ch., Richer, E., Hitzler, M., Chiba, P., Kratzel, M., Ecker, G., *J. Med.Chem.* **1999**, **42**, 1921–1926.

149 Ecker, G., Huber, M., Schmid, D., Chiba, P., *Mol. Pharmacol.* **1999**, *56*, 791–796.

150 Tmej, C., Chiba, P., Schaper, K.-J., Ecker, G., Fleischhacker, W., in *Drug Resistance in Leukemia and Lymphoma, III* Kaspers, G.J.L., (Ed.), KluwerAcademic/PlenumPress, New York **1999**, Ad. Expr. Med Biol. Vol. 457, pp. 95–105.

151 Ramu, A., in *Resistance to Antineoplastic Drugs*, D. Kessel (Ed.), CRC Press, Bocca Raton, FL **1989**, pp. 63–80.

152 Pajeva, I.K., Wiese, M., *Quant. Struct.–Act. Rel.* **1997**, *16*, 1–10.

153 Wiese, M., Pajeva, I.K., *Pharmazie* **1997**, *52*, 679–685.

154 Ferté, F., *Eur. J. Biochem.* **2000**, *267*, 277–294.

155 Divecha, N., Irvine, R.F., *Cell* **1995**, *82*, 693–696.

156 Pallarés-Trujillo, J., Lopéz-Soriano, F.J., Argilés, J.M., *Int. J. Oncol.* **2000**, *16*, 783–798.

157 Wiese, M., Pajeva, I., *Curr. Med. Chem.* **2001**, *8*, 685–713.

6
Computer Simulation of Phospholipids and Drug–Phospholipid Interactions

Michael Wiese

The computer simulation of molecules and their properties has a short history compared with experimental methods, and only in recent decades have computer simulations become a widely used tool, particularly in studies of proteins and nucleic acids. Biomembranes have been investigated less than other classes of biomolecules for several reasons. Structures derived from X-ray crystallography are usually the starting point for computational studies. The physiologically relevant phases of membranes are highly variable and difficult to study experimentally. Thus, only limited atomic-level structural data are available from X-ray and neutron diffraction compared with proteins and nucleic acids. In addition, the fluidity of membranes is important as it determines to a great extent the properties and functions of membranes. Fluidity results from the flexibility of lipids, which can adopt numerous low-energy conformations. Thus, reliable computer simulation of phospholipids must include the dynamic nature of lipids, making static models unrealistic. For this reason, simulations of lipids and drug–lipid interactions have in the past attracted far less attention than the molecular modeling of small molecules and proteins and their interaction. More recently, however, as increasing computational power has become available, the simulation of lipids has become a rapidly growing field, as evidenced by the large increase in related publications.

This chapter will not review all of the published studies, but instead will focus on examples of computer simulations of phospholipid membrane systems ranging from simple models through descriptions of lipid and water in full atomic detail to complex membranes containing small solutes, lipids, and proteins. The chapter is aimed at medicinal chemists who are interested in drug–phospholipid interactions. Before discussing the results of different simulations, the currently applied methodologies will briefly be described.

6.1
Modeling Strategies for Studying Phospholipids and Drug–Phospholipid Interactions

The first computational investigations of phospholipids were undertaken using purely statistical methods to describe overall properties. In the 1980s the simulation of molecules and their behavior using an explicit atomic description was extended to large systems such as proteins and lipids. To simulate the static and dynamic behavior of a molecule or an ensemble of molecules, a force field is required to describe the potential energy of the system as function of the atomic coordinates and to take into account covalent and non-bonded interactions. Among the available methods for studying proteins and lipids and their interaction with other molecules, empirical force fields based on "molecular mechanics" were exclusively used because of their computational speed. Thus, in the beginning, mostly molecular mechanics methods were applied to study the interactions of small molecules with phospholipids, resulting in static models of conformations and interactions. Simulations of the dynamic and/or time-averaged molecular properties of phospholipids were hampered mainly by the limited computational power available. However, shortly afterward came the first simulations that provided a valuable insight into the macroscopic properties of membranes at a molecular level.

In the study of the dynamic properties of phospholipids, two main levels of detail are used: mean field and all-atom representation of the system. Likewise, there are two main simulation techniques: Monte Carlo (MC) and molecular dynamics (MD) simulations.

6.1.1 Types of Representation of the Simulated System

6.1.1.1 Mean Field Simulations

At this level, the phospholipid and surrounding water are not included explicitly but are described by an empirical potential that characterizes partitioning of hydrophilic and hydrophobic parts of a molecule into the phospholipid. In the simplest way, the phospholipid bilayer is represented by a lipophilic slab oriented in the x, y-plane and the potential energy function of the force field is supplemented with a hydrophobic interaction term (Eq. 6.1):

$$E = E_{\text{bond}} + E_{\text{non-bonded}} + E_{\text{hydrophobic}} \tag{6.1}$$

This type of simulation has been often used to study peptide–phospholipid interactions. In these simulations, the hydrophobic term has usually been derived from hydrophobicity scales of amino acid side chains, but more detailed descriptions based on the transfer energy from water to a hydrophobic environment and the accessible molecular surface have also been developed [1, 2]. The hydrophobic contribution then takes the following form:

$$E_{\text{hydrophobic}} = -a \sum h_i f(z_i) \tag{6.2}$$

where a is a scaling term to weight the hydrophobic energy term, and can also be included in h_i or $f(z_i)$ and h_i is the hydrophobic contribution of an amino acid side chain or a group of atoms. h_i is positive for hydrophobic groups and negative for hydrophilic ones and depends on the hydrophobicity scale employed, which is therefore a crucial point in mean field simulations. Most of the currently applied scales assume a single transfer value that can be assigned to each amino acid.

In an attempt to find a better way of considering the interfacial region, molecular dynamics simulations of different helices were performed in phospholipid environment. From the analysis of the amino acid–lipid interaction energies, an extension to hydrophobicity scales for proteins in an bilayer was derived [3]. The function $f(z_i)$ characterizes the hydrophobicity of the environment in the z-direction. In the simplest case, it is a step function, but more often a type of sigmoidal function is used. The range of $f(z_i)$ determines what kind of interactions are considered by the $E_{hydrophobic}$ term. If it runs from zero to 1, attractive and repulsive contributions are considered for one of the phases only. A schematic representation of the mean field model is shown in Figure 6.1. The molecule whose behavior is to be simulated can be treated as a flexible or rigid body. The system is further evaluated by Monte Carlo or molecular dynamics simulation.

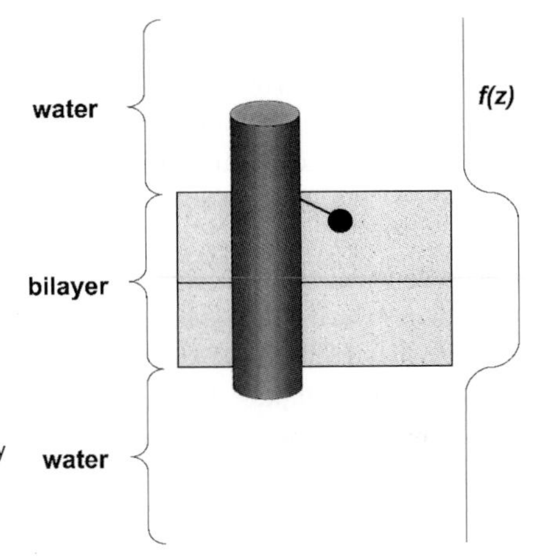

Fig. 6.1 Mean field model of the interaction of a side chain with a lipid bilayer. The hydrophobic contribution of the side chain is multiplied by a function $f(z_i)$, which is shown schematically on the right. The overall interaction energy is obtained as sum over all side chains according to Eq. 6.2.

To refine the mean field representation further, another term that considers the special properties of the head group region was recently introduced. An additional dipole potential was included in Eq. 6.1 that accounted for change in dielectric constant across the water–lipid interface from high (in bulk water) to low (in the lipid interior) values [4].

6.1.1.2 All-atom Simulations

In this case, the phospholipids and the surrounding water molecules are treated explicitly. Thus, this type of simulation provides a much more detailed view at the atomic level. However, all-atom simulations face some difficulties. First of all, the choice of force field is of crucial importance. Although several force fields have been developed for simulation of small molecules or proteins and nucleic acids, no special force fields have been designed for modeling of phospholipids. For this reason, computer simulations use mostly force fields developed for proteins and nucleic acids. Of these, the AMBER [5] and CHARMM [6, 7] force fields are very popular and have been used by several research groups. The GROMOS [8] force field and the consistent valence force field (CVFF), employed in the Discover program [9], are also applied in many investigations. For calculation of non-bonded interactions, the "optimized parameters for liquid systems" (OPLS) parameter set for intermolecular interactions [10] has often been used in conjunction with the above-mentioned force fields. It soon became apparent that those force fields could reproduce rather good experimental data related to geometry, order, and short time dynamics. However, some adaptations of charges and/or other force field parameters were necessary in some cases in order to obtain more consistent results. In an early simulation, Egberts *et al.* [11] noticed that the original charges of the GROMOS force field led to a phospholipid bilayer in the gel state, despite the fact that the simulation temperature was well above the gel to liquid crystalline phase transition temperature of the studied phospholipid. The authors scaled down the charges by a factor of 2 and obtained a system that behaved in the expected manner. In the early literature, many modifications of original force fields that deal with such empirical corrections are reported. However, several attempts have also been undertaken to parametrization force fields systematically in order to better reproduce experimental results for lipids. This involves adjustment of parameters for torsion angles and the Lennard–Jones potentials and use of high-level *ab initio* calculations to derive charges. In the early 1990s, Stouch and coworkers [12–14] published a series of papers on development of parameter sets including charges for lipid molecules to be used with the CVFF of the Discover program. New parameters for the simulation of phospholipids have also been developed for the CHARMM force field. In 1996, a set of potential functions for saturated phospholipids that led to a better agreement with experimental data was published [15]. It was further extended to include unsaturated hydrocarbon chains [16] and recently further refined [17]. In 1998, a new force field was developed based on the AMBER force field for the simulation of biological systems, including phospholipids. It yields an improved correspondence between the experimental and simulated polar head group conformations [18]. Owing to the refinements of the force field parameters that have been undertaken, today simulations of phospholipids yield systems whose properties are generally in good agreement with available experimental data.

Another aspect of all-atom simulations is the level of detail of the atomic representation, e.g. the use of united atoms, in which the hydrogens of methylene and methyl groups are merged into the central carbon and not explicitly considered, compared to a true all-atom representation that includes all hydrogens explicitly. A united atom representation is very attractive, as the number of non-bonded interactions is greatly reduced, thus speeding up the calculation considerably. However, it has been shown that with united atoms the calculated diffusion coefficients in the membrane are far too high [19], making this representation unsuitable for such studies. As computational power has increased, simulations with explicit hydrogens have become more and more common. However, there is still interest in united atom representations for complex systems and long time scales to save computer time. This is reflected by the ongoing research aimed at improving united atom force field parameters. For the GROMOS force field, the Lennard–Jones parameters of the united atoms were adapted in such a way as to give perfect agreement with the experimental density in a molecular dynamics simulation under constant-pressure conditions [20]. And recently, a united atom force field for phospholipid membranes was developed based on the AMBER force field. The torsional parameters were redefined to reproduce geometry and energy profiles obtained from *ab initio* calculations of model compounds. In a simulation under constant pressure conditions the developed force field parameters gave an excellent agreement with the experimental results [21].

6.1.2 Monte Carlo Simulations

Monte Carlo simulations are based on statistical thermodynamics. According to this theory, a system can be fully described by a weighted sum of all possible configurations. As it is impossible to sample the whole configuration space, even for the smallest system of interest, approximations must be used. The importance sampling scheme developed by Metropolis *et al.* [22] provides such an approximation that converges to the exact solution at infinite number of sampling steps. The starting point is a set of atomic coordinates representing the initial configuration of the system. The energy of the system is then calculated according to the employed force field. Usually, the intermolecular interactions are truncated at some distance, using either a smoothing function or a simple cut-off. By applying periodic boundary conditions, an infinite system can be simulated. A random perturbation is applied to the system, called a Monte Carlo move. This can be a movement of an atom, change in a torsion angle, rotation of a molecule, or any other move. The energies before (H_{old}) and after (H_{new}) the perturbation are compared (Eq. 6.3).

$$\Delta H = H_{new} - H_{old} \tag{6.3}$$

If $\Delta H \leq 0$, the energy is lower than before and the move is accepted. If ΔH 0, the probability of acceptance, P, is calculated according to the Boltzmann distribution (Eq. 6.4).

$$P = e^{\frac{-\Delta H}{kT}} \tag{6.4}$$

where k is the Boltzmann constant and T the temperature of the system. The probability, P, is compared with a random number, z, generated in the range between zero and 1. If $P > z$ the move is accepted, otherwise it is rejected.

The possibility of non-physical moves, such as translation and rotation of whole peptides and lipids to sample system configurations, makes Monte Carlo simulations very effective in systems with mean field approximation of the surrounding environment. In systems with an atomic-level description of the environment, most of the random moves result in higher energy owing to the density of neighboring particles that lead to a high probability of unfavorable steric interactions. Thus the acceptance ratio P/z can become very small, making the algorithm computationally inefficient.

Static properties of the system are obtained by averaging structural or any other properties of interest over the sampled configurations. The dynamics of the system can be simulated by association of the MC steps with an artificial time parameter. Naturally, the Monte Carlo steps do not correspond to a real time scale and the dynamics do not represent physical phenomena.

6.1.3 Molecular Dynamics Simulations

Molecular dynamics (MD) simulations are the most powerful methods of obtaining detailed information on the dynamics of phospholipids and the interactions of molecules with phospholipid membranes. The basic principle of MD is to follow the development of a system in time. Again, a starting configuration is chosen, then the system is heated to the desired temperature and simulated for some period of time to allow equilibration. Finally, the productive phase starts from which dynamic and average properties of the system are derived. MD simulations are based on Newton's law of motion. The position of an atom is obtained by solving this equation for all atoms as a function of the underlying force field. As this cannot be done analytically for a multibody system, numerical integration methods have to be employed. Owing to the fast vibrational motion of the atoms, the integration time step must be small to avoid errors in the integration. Usually, a time step of 1 fs (10^{-15} s) is used. To allow an increase in the time step, the SHAKE algorithm [23] can be used to constrain the bond lengths of the hydrogen atoms during the simulation. Through this constraining, the step size of the integration can be increased to 2 fs. Because of the short step size that can be used in the integration of the equation of motion, the times accessible to such simulations remain short. Whereas about 10 years ago the simulation times were in the range of a few hundred picoseconds, they have now increased to approximately 10 ns. Even with these "long" time simulations, many of the dynamic motions of lipid bilayers cannot be studied because of their much longer time scale. Nevertheless, MD simulations have an enormous potential to provide insights into the structure and function of lipids.

Several experimentally available macroscopic features have been used to validate the results of MD simulations. Among them are:

- phospholipid surface area;
- electron density distribution across the bilayer normal;
- relationship of *trans*/*gauche*/kink conformations in the aliphatic chain as determined by FTIR;
- orientation of the individual methylene fragments of the aliphatic chain, described by the order parameter as obtained from ^{2}H-NMR measurements.

The order parameters are mostly used to judge the correspondence of the obtained results to the experiment. Three different kinds of order parameters are used in the literature to characterize the conformational state of the lipid alkyl chains. The first is S^{CD}, which can be obtained directly from the quadrupolar splitting of selectively deuterated phospholipids by NMR. S^{CD} is related to the observed quadrupolar splitting by Eq. 6.5:

$$\Delta\nu = 3/4 \times \text{constant} \times \left| S^{CD} \right| \qquad (6.5)$$

where "constant" denotes the static deuteron quadrupolar coupling constant (see chapter 3.8) for which different values have been determined and used in calculation of S^{CD}. In the pioneering work of Seelig and Seelig [24] a value of 170 kHz was used for hydrocarbon chains, while a slightly smaller value of 167 kHz was determined in

Table 6.1 $|S^{CD}|$ order parameters as a function of temperature and position in the *sn*-1 chain of DMPC. The values were calculated from the quadrupolar splittings taken from ref. 100, using a deuteron quadrupolar coupling constant of 167 kHz according to ref. 101

Carbon	Temperature (K)								
	293	**298**	**303**	**308**	**313**	**318**	**323**	**328**	**333**
Pure DMPC									
2-6		0.242	0.217	0.201	0.187	0.180	0.173	0.165	0.159
7		0.241	0.208	0.188	0.173	0.166	0.155	0.145	0.141
8		0.241	0.201	0.180	0.168	0.159	0.147	0.137	0.129
9		0.221	0.182	0.161	0.149	0.141	0.130	0.119	0.114
10		0.209	0.167	0.148	0.133	0.125	0.116	0.107	0.100
11		0.182	0.148	0.129	0.116	0.110	0.102	0.094	0.088
12		0.157	0.124	0.110	0.099	0.093	0.087	0.079	0.075
13		0.117	0.092	0.080	0.073	0.068	0.063	0.059	0.056
14		0.034	0.027	0.025	0.021	0.020	0.018	0.017	0.015
In the presence of 30 mol% cholesterol									
2-8	0.453	0.438	0.417	0.402	0.384	0.369	0.357		
9	0.429	0.413	0.398	0.382	0.368	0.351	0.339		
10	0.413	0.395	0.377	0.357	0.339	0.328	0.316		
11	0.379	0.359	0.339	0.321	0.299	0.285	0.271		
12	0.323	0.305	0.285	0.267	0.253	0.239	0.229		
13	0.248	0.232	0.212	0.198	0.187	0.177	0.169		
14	0.066	0.063	0.058	0.054	0.052	0.049	0.046		

ref. 25, and a value of 180 kHz was applied in ref. 84 for estimation of the agreement between simulation and experiment.

In simulations, S^{CD} is calculated according to Eq. 6.6:

$$S^{CD} = \tfrac{1}{2} < 3\cos^2\theta - 1 > \qquad (6.6)$$

where θ is the angle between the CD bond vector and the bilayer normal and the brackets denote averaging over time and over all the lipids. The values of S^{CD} are assumed to be negative, based on experimental and theoretical reasoning. They can range from –0.5 for a fully ordered chain in all-*trans* conformation to zero for a chain undergoing isotropic rotation. Typical values of S^{CD} for lipid bilayers in the fluid phase are about –0.2 at the top of the fatty acid chain and close to zero in the terminal methyl groups. Values of S^{CD} for the *sn*-2 chains of dilaureoylphosphatidylcholine (DLPC), DMPC, and DPPC at different temperatures are tabulated in ref. 26. In Tables 6.1 and 6.2 the S^{CD} values for the *sn*-1 *and sn*-2 chains of DMPC are summarized as collected from the literature. Table 6.3 presents experimental order parameters of DPPC.

The second order parameter is termed "average alkyl chain order", S_{mol}. It is calculated in the same way as S^{CD}, with the difference that the angle θ is taken between the long molecular axis and the bilayer normal. It is related to S^{CD} by Eq. 6.7:

Tab. 6.2 $|S^{CD}|$ order parameters as a function of temperature and position in the *sn*-2 chain of DMPC

Carbon	Temperature (K)										
	296[a]	298[b]	303[a]	308[a]	313[a]	318[a]	323[a]	328[a]	333[a]	298[c, d]	298[b, d]
2R	0.096	0.088	0.094	0.094	0.094	0.093	0.093	0.093	0.093	0.130	0.127
2S	0.162	0.158	0.147	0.141	0.137	0.133	0.130	0.128	0.126	0.258	
3	0.236	0.232	0.221	0.212	0.204	0.198	0.193	0.190	0.189	0.374	0.365
4	0.248	0.239	0.232	0.220	0.212	0.204	0.196	0.190	0.185	0.433	0.389
5	0.250		0.232	0.220	0.212	0.204	0.196	0.190	0.185	0.433	
6	0.252	0.251	0.234	0.222	0.212	0.204	0.196	0.190	0.185	0.433	0.420
7	0.250		0.230	0.216	0.206	0.198	0.192	0.185	0.179	0.433	
8	0.244	0.230	0.224	0.210	0.199	0.190	0.181	0.173	0.165	0.433	0.419
9	0.230		0.208	0.198	0.184	0.174	0.168	0.160	0.154	0.433	
10	0.213	0.210	0.190	0.176	0.164	0.156	0.146	0.140	0.134	0.433	0.402
11	0.194	0.204	0.177	0.161	0.154	0.148	0.132	0.126	0.122	0.398	0.384
12	0.170	0.187	0.150	0.139	0.129	0.121	0.114	0.107	0.101	0.354	0.354
13	0.141	0.157	0.121	0.111	0.102	0.096	0.089	0.086	0.078	0.294	0.289
14	0.037		0.031	0.027	0.025	0.023	0.019	0.018	0.018	0.078	

a) Data from ref. 26.
b) Data from ref. 103.
c) Data from ref. 102.
d) DMPC + 30 mol % cholesterol.

$$S_{mol} = -2 \times S^{CD} \tag{6.7}$$

For a chain parallel to the bilayer, normally $S_{mol} = 1$ and decreases to zero for a fully unordered, isotropic chain. For a chain oriented parallel to the bilayer the value of S_{mol} is –0.5.

The third order parameter, S^{CC}, corresponds for a carbon n to the direction of the C_{n-1}–C_n bond and is calculated from experimentally available S^{CD} values. In contrast to S^{CD}, it shows a marked odd–even dependence in the liquid crystalline state. From S^{CC} it is also possible to calculate the average length of the alkyl chain [26].

Several factors can be of crucial importance for the outcome of a MD simulation: starting structure, equilibrium time, boundary conditions, long-range electrostatics and non-bonded cut-off, and the kind of system that is simulated.

6.1.3.1 Starting Structure, and Equilibrium Time

Conformations based on the few available X-ray structures of phospholipids in the crystalline state, generated from energy minimization or from conformational libraries of phospholipid alkyl chains, have been used as starting structures for the simulation. It has been argued that the latter is preferable in order to save time necessary for equilibration when starting from all-*trans* conformations of the alkyl chains observed in X-ray structures. To build up the configuration of the system, e.g. the lateral positions of the phospholipids, either crystal structure data or programs

Tab. 6.3 $|S^{CD}|$ order parameters as a function of temperature and position in the *sn*-1 and *sn*-2 chain of DPPC

	Temperature (K)								
	314[a] *sn*-2	**314[c]**	**317[a] *sn*-2**	**323[b] *sn*-1**	**323[a] *sn*-2**	**323[c]**	**330[a] *sn*-2**	**330[c]**	**338[a] *sn*-2**
Carbon									
2R[d]	0.100	0.113	0.099		0.097	0.096	0.100	0.097	0.096
2S[d]	0.153	0.185	0.144		0.142	0.153	0.133	0.136	0.128
2		0.240		0.282		0.217		0.209	
3	0.218	0.218	0.212	0.266	0.203	0.209/0.185	0.193	0.191/0.177	0.184
4	0.241	0.242	0.230	0.289	0.217	0.217	0.198	0.208	0.187
5	0.241	0.236	0.230	0.284	0.217	0.209	0.198	0.193	0.187
6	0.241		0.230	0.286	0.217		0.198		0.187
7	0.241		0.230	0.274	0.217		0.198		0.187
8	0.241		0.229	0.269	0.217		0.193		0.187
9	0.232	0.237/0.221	0.216	0.256	0.205	0.193	0.186	0.172	0.178
10	0.227	0.221/0.216	0.212	0.249	0.190	0.183	0.174	0.162	0.160
11	0.217		0.201	0.236	0.179		0.162		0.146
12	0.200	0.178	0.181	0.226	0.162	0.146/0.136	0.145	0.123	0.128
13	0.185		0.166	0.207	0.147		0.133		0.118
14	0.155	0.158/0.142	0.137	0.190	0.122	0.119/0.106	0.107	0.097	0.096
15	0.123	0.119/0.098	0.110	0.143	0.098	0.089/0.077	0.087	0.077/0.067	0.076
16	0.033		0.029		0.026		0.023		0.020

a) Data from ref. 26.
b) Data from ref. 104.
c) Splittings were not assigned to individual alkyl chains. S^{CD} values were calculated from splittings reported in ref. 24, using a deuteron quadrupolar coupling constant of 167 kHz according to ref. 101.
d) *pro*R and *pro*S positions according to ref. 105.

for placement of adjacent lipids at energetically feasible positions and orientations have been applied. The system is heated to the desired temperature for several picoseconds and then equilibrated until some system variables no longer change. Factors that are most used to judge the status of the simulation include energy, temperature, and surface area per lipid. Although in almost all studies, the system is termed "equilibrated" after about 100–200 ps, a recent investigation shows that the equilibration of the lipid orientation about its long axis has a relaxation time of about 1.5 ns. The authors concluded that about three relaxation times are necessary (leading to a required simulation time of 4.5 ns) for the full equilibration of reorientation around the molecular long axis [27]. As this is far longer than earlier simulation times, these systems are better termed "nearly equilibrated" rather than "equilibrated". In the same study, a united atom representation was found to be inadequate as it never yielded correct values for both the order parameter and the surface area per lipid simultaneously.

6.1.3.2 Boundary Conditions

Though the systems that are simulated by MD consist of several thousands to tens of thousands atoms, they are still very small on the macroscopic scale. Thus, the simulated system is surrounded by a vacuum. This leads to the so-called "boundary effects", i.e. deviations of the behavior of molecules at the border of the simulated system from those in the center. In the worst case, the system will evaporate just like a water drop in a vacuum. To tackle this problem, two strategies are mainly used: some restraint of the movement of the outermost atoms (stochastic boundary conditions) or periodic boundary conditions. Restraint can be achieved by surrounding the system with repulsive walls or by restraining the outermost atoms to their starting position by a harmonic force. The stochastic boundary conditions do not simulate the effectively infinite bilayer system, and it has been argued that they may inhibit chain tilting of the boundary lipids, which would affect tilting of the rest of the phospholipids.

Periodic boundary conditions are in some way "natural" for the membrane, as they emulate an effectively infinite system. In this case, copies of the simulated system are put aside in each direction and the interaction of a molecule with its surrounding is calculated from both the cell and the images surrounding the simulated system. To prevent spurious contributions, non-bonded cut-offs that are shorter than half the size of the simulated box must be used. The size and geometry of the simulated cell can be set or made variable depending on the type of system that is simulated (see below). Although computationally more expensive, in recent years almost all simulations have been performed using periodic boundary conditions as the effects of restraints on the rest of the system are difficult to analyze in simulations with stochastic boundary conditions.

6.1.3.3 Long-range Electrostatics and Non-bonded Cut-off

To speed up the calculation, non-bonded interactions between two atoms or groups are usually ignored if the distance between them is larger than a predefined maximum, the cut-off value. For van der Waals interactions, a cut-off distance of 8 Å has been shown to be sufficient, owing to the fast decline of the Lennard–Jones potential with increasing distance. However, in the case of electrostatic interactions, such a short cut-off distance can lead to artifacts in the simulation because of their long-range nature. In principle, electrostatic interactions can be fully considered with infinite cut-off using the Ewald summation. As this technique is very costly from a computational point of view, it is rarely used for large systems like phospholipids. Only since the development of the fast particle mesh Ewald summation algorithm [28, 29], have electrostatic interactions been treated throughout the whole system in several simulations. However, nowadays most MD simulations use a cut-off for electrostatics that is set to larger distances to minimize possible artifacts due to the truncation of long-range electrostatic interactions. The perturbing effect of cut-offs that are too short has been known for a long time. For example, in several simulations of pure water, perturbations in the orientational structure were observed compared with the results obtained when the electrostatics were fully considered via the Ewald summation technique. In addition, the radial structure of water around a chloride ion was found to differ between simulations using a cut-off in the range of 8–10 Å in contrast to infinite cut-offs. To circumvent the truncation of interactions by a cut-off, switching functions have been employed that smooth electrostatic interactions within a distance range to zero. But their use has also been questioned [30].

The effect of the cut-off distance on the behavior of water in a phospholipid–water system has been studied in detail with an extensive series of MD simulations [31]. The system consisted of a dimyristoylphosphatidylcholine (DMPC) monolayer of 18 DMPC molecules covered by a water layer of approximately 60 Å thickness. An all-atom representation with flexible water molecules was used. Two-dimensional periodic boundary conditions were employed in the plane of the monolayer, while the water molecules furthest away from the lipid layer were restrained by a repulsive wall. Two cut-offs were used: a "long" one for interactions involving the head groups of the phospholipids and a "short" one for all other interactions (water–water, water–hydrocarbon, hydrocarbon–hydrocarbon). In the initial simulation, 14 Å was chosen for the "long" cut-off, and 10 Å for the "short" one. Although these values are considerably higher than 8 Å, which has been often used, a strong dependence of the water behavior on the distance from the phospholipids was observed. The average number of hydrogen bonds per water molecule was less than 3 close to the head group region and increased to more than 3.5 near the repulsive wall (Figure 6.2). In addition, the average movement of the water molecules showed an unexpected behavior. Near to the phospholipid it was 20–25% larger than in bulk water (Figure 6.3). Increasing the "long" cut-off led to a more uniform behavior of the water molecules across the water layer and the movement close to the head groups became less than in bulk water. Only at cut-off values of 24 Å and larger did the behavior of the water molecules converge. A noticeable, but much smaller, effect on the water structure was observed if the "small" cut-off was increased from 10 to 14 Å. The structure of the water close to the head group of the phospholipid was also influenced by the cut-off. With increasing cut-off, the hydration shells of the phosphate and trimethylammonium groups were more sharply defined, and for the phosphate group a third hydration shell became evident when the large cut-off was used [32]. The authors

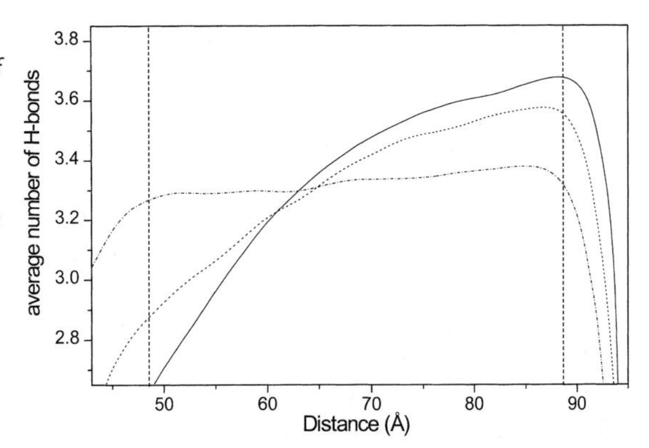

Fig. 6.2 Effect of different "long" cut-offs on the average number of H-bonds formed by water molecules as a function of distance from the bottom of the simulated systems: solid line, 14 Å; dashed line, 18 Å; dash dot line, 30 Å. The lipids reside in the region below the dotted line at 47.5 Å and the dotted line at 89 Å represents the cut-off of the boundary wall force. (Adapted from Fig. 6 of ref. 31 with permission from the American Institute of Physics).

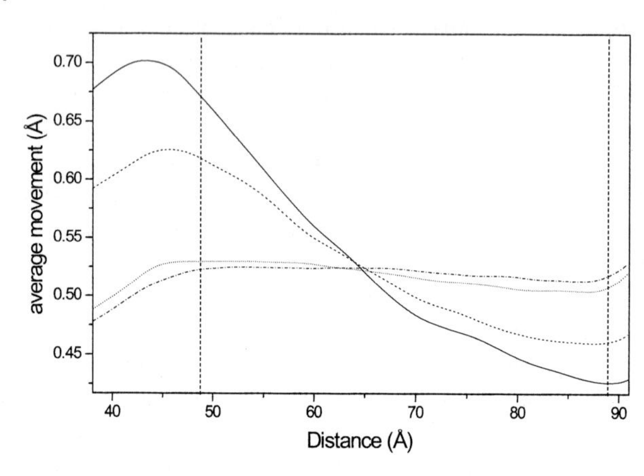

Fig. 6.3 Effect of different "long" cut-offs on average movement of water molecules as a function of distance from the bottom of the simulated systems: solid line, 14 Å; dashed line, 18 Å; dotted line, 26 Å; dash dot line, 30 Å. The lipids reside in the region below the dotted line at 47.5 Å, and the dotted line at 89 Å represents the cut-off of the boundary wall force. (Adapted from Fig. 8 of ref. 31 with permission from the American Institute pf Physics).

concluded that, to avoid artifacts, the inclusion of long-range electrostatic interactions is necessary for simulations of systems, such as phospholipids, in which the electrostatics make an important contribution.

The influence of a cut-off relative to the full treatment of electrostatic interactions by Ewald summation on various water parameters has been investigated by Feller *et al.* [33]. These authors performed simulations of pure water and water–DPPC bilayers and also compared the effect of different truncation methods. In the simulations with Ewald summation, the water polarization profiles were in excellent agreement with experimental values from determinations of the hydration force, while they were significantly higher when a cut-off was employed. In addition, the calculated electrostatic potential profile across the bilayer was in much better agreement with experimental values in case of infinite cut-off. However, the values of surface tension and diffusion coefficient of pure water deviated from experiment in the simulations with Ewald summation, pointing out the necessity to reparameterize the water model for use with Ewald summation.

6.1.3.4 Kind of Simulation System

Another parameter that can have a great influence on the results obtained is the type of the simulation performed. Generally, simulations are carried out at constant particle number (*N*). The volume (*V*) and energy (*E*) of the simulated system can be held constant, leading to a so-called NVE, or microcanonical, ensemble. When the volume and temperature are held constant, this yields a canonical or NVT ensemble. In both cases, the size of the simulated system is chosen in such a way as to represent the desired state of the phospholipid, mostly the liquid crystalline L_α phase. The surface per lipid and the thickness of the bilayer are set based on experimental values and remain unchanged during the simulation. Therefore, the system is not able to adjust its size and thickness.

To achieve the desired flexibility of the system a constant pressure is applied in one or more directions, allowing the size of the simulated system to change. Three basic classes have been used: constant normal pressure along the bilayer normal with fixed surface area (NPAT), constant pressure in all directions (NPT), and constant pressure along the bilayer normal with constant surface tension in direction of the bilayer (NPγT). The first simulation protocol provides some flexibility but allows only the thickness of the bilayer to be adjusted, whereas the other two schemes allow the simulation cell to adapt fully during the simulation. Simulations of DPPC using NPAT conditions have been performed for four different surface areas per phospholipid (59.3, 62.9, 65.5, and 68.1 $Å^2$) to study the influence of the surface area on the behavior of the simulation [34]. By comparing the available structural information on hydrated DPPC in the fluid L_α phase with values calculated from the simulation, the authors found that with a surface area of 62.9 $Å^2$ the experimental deuterium order parameters and electron density profiles could be most closely reproduced. With a surface area of 65.5 $Å^2$ the agreement with the experiment was slightly worse, while the lowest and highest surface areas resulted in strong deviations from the experiment. Very recently, the same strong influence of the chosen surface area on the calculated order parameter of the alkyl chains has also been reported in a series of simulations involving a phospholipid–detergent mixture [35]. These investigations impressively show the importance of a correctly chosen surface area for the performance of MD simulations under NPAT and NVT conditions. A thorough discussion of the thermodynamics of interfacial systems such as bilayers and the problems involved in constant pressure simulations is given in ref. 36.

The result of simulations employing NPT and NPγT conditions depends strongly on the balance of attractive and repulsive forces that result from the force field used to yield realistic areas and densities. In a number of studies with *ab initio*-derived charges, the simulated system behaved as if it was in a gel state, with much smaller surface area of the phospholipids and higher order parameters than experimentally determined, despite the fact that the temperature was well above the phase transition temperature. The expected fluid crystalline state could only be achieved by scaling down the charges of the phospholipids [11, 37]. In a simulation that used semiempirical charges, the fluid crystalline state was observed without modification of the charges [38].

The explicit inclusion of the surface tension (NPγT) leads to anisotropic pressures along and perpendicular to the membrane. In such a simulation, the liquid crystalline state emerged spontaneously from the gel-state X-ray structure that served as starting point [39]. Although the thickness of the membrane was in good agreement with experimental data, the surface area per phospholipid was approximately 57 $Å^2$, corresponding to the low boundary of the experimental values estimated for DPPC (58–71 $Å^2$). Nevertheless, many experimental measures of the structure of fluid membranes have been found to be in good agreement with values calculated from the simulation. Based on these encouraging results, this simulation protocol has been claimed to be superior.

The superiority of NPγT over NPAT simulations has also been claimed in a more recent 120-ps simulation of monolayers of different phospholipids, based again on the agreement of calculated order parameters with experimental data [40]. However, it was also observed that the surface area of the phospholipids did not deviate much from its starting value. To investigate this further, another simulation was performed in which the surface tension deviated significantly from the experimentally measured pressure. Despite this difference, no significant change in the surface area was found, implying that this type of simulation was sensitive to the starting conditions. This result can be understood by taking into account the short simulation time of only 120 ps and results from other longer simulations described below, showing that the adaptation of the surface area needed considerable time.

The dependence of the calculated surface pressure on other system variables has been investigated too. In simulations applying NPAT conditions, it was demonstrated that the calculated surface tension was somewhat insensitive to changes in the surface area of up to 8 $Å^2$ [35]. And Feller and Pastor [41] found that in MD simulations with fixed surface area the calculated surface tension was dependent on the number of phospholipids. The surface tension increased with decreasing system size with values of 33.5, 39.2, and 57.2 dyn/cm for systems of 72, 32, and 18 lipids respectively.

The effect of the applied surface tension on several properties, such as surface area and molecular order parameters, was investigated by the same authors in a series of longer simulations [42]. The applied surface tension was found to have a strong influence on calculated surface areas per lipid and molecular order parameters, with a surface tension in the range 35–45 dyn/cm yielding values in closest agreement with the experiment. Notable is the large variation in surface area in two repeated simulations performed at surface tensions of 35 and 50 dyn/cm, and the small difference when going from $\gamma = 40$ to 45 dyn/cm (Table 6.4). The surface areas tended to change from the starting value of 62.9 $Å^2$ over the whole simulation time of 1 ns, suggesting that equilibrium was probably not reached even after 1 ns. From the drift in surface areas, the authors estimated that simulation times of about 10 ns would be necessary to reach equilibrium. These investigations explain the insensitivity of surface area to surface tension reported from the short 120-ps simulation described above [40]. No significant differences were observed comparing NPAT with the corresponding NPγT simulation for molecular order parameters, lateral diffusion, and other parameters.

Tab. 6.4 Comparison of mean surface areas per DPPC molecule as function of applied surface tension. Data taken from ref. 42

γ (dyn/cm)	Surface area ($Å^2$)
0	54.3
35	58.9
35	61.9
40	66.4
45	64.7
50	75.8
50	69.0
55	66.6

Thus, the preferred ensemble for use in MD simulations is still open and controversial. A detailed discussion on the advantages and disadvantages of the different simulation ensembles is given in refs. 43 and 44.

6.2 Computer Simulations with Phospholipids

Various dynamic processes have been investigated using computer simulations of phospholipids. These include the dynamics of the alkyl chain movement of the phospholipid, the structure of water at the interface, diffusion of small molecules, interactions of phospholipids with water, drugs, peptides, and proteins, and the effect of unsaturation or the presence of cholesterol on the phospholipid conformation.

6.2.1 Distribution of Solutes

The distribution of solutes in the interface between water and membranes is of great theoretical and practical interest. Several investigations have been undertaken to explore this phenomenon on a theoretical basis. The models employed differed largely in the level of detail at which the molecular structure was explicitly considered. In the simplest approach, the interface was treated by a mean field with an instantaneous change from bulk water to bulk lipid phase. A more general set of mean field approaches approximated the lipid–water interface by a gradual change from high to low dielectric medium and/or hydrophobic interaction forces. Sanders and Schwonek [45] developed an empirical energy function to describe solute interactions in the water–phospholipid interface that yielded good correspondence between calculated and experimental values for several investigated properties. In this model. the dielectric constant varied from 78.5 in water to 2 in the membrane in a sigmoidal fashion, as shown in Figure 6.4 and given by Eq. 6.8 for inside the lipid ($z < 0$) and in Eq. 6.9 for the watery phase ($z \geq 0$):

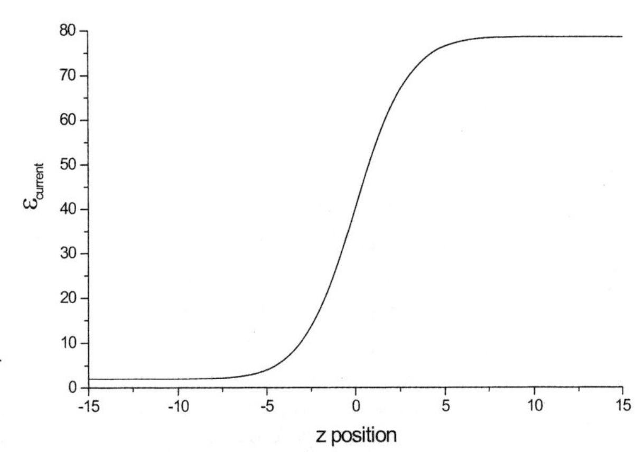

Fig. 6.4 Local dielectric constant, $\varepsilon_{current}$, calculated according to Eqs. 6.8 and 6.9 [45].

$$\varepsilon = 76.5 \times e^{-\left(\frac{-z-z_0}{A}\right)^2} + 2 \quad (6.8)$$

and

$$\varepsilon = 80.5 - \left(76.5 \times e^{-\left(\frac{-z-z_0}{A}\right)^2} + 2\right) \quad (6.9)$$

Here z_0 was set to 3.85, so that the midpoint of the change from low to high dielectric was at $z = 0$, and A was set to 4.63, based on results from neutron diffraction studies.

Hydrophobic interactions were considered by atomic fragment values based on those of Viswanadhan *et al.* [46] and refitted for hydrocarbons to better reproduce log *P* values of hydrocarbons. Additionally, hydrogen bond donor and acceptor parameters taken from ref. 47 were applied. The fragment values, together with the hydrogen bond parameters, are listed in Table 6.5. Empirical correction factors for the number of carbon atoms were introduced to reproduce differences in partitioning in the systems octanol–water and hydrocarbon–water, leading to Eq. 6.10 for the interaction energy of an atom in the interface. It should be noted that the interaction energy was calculated only for values of $\varepsilon_{current}$ in the range 2–10, corresponding to hydrocarbon and octanol respectively.

$$G_{int} = \frac{-2.3\,RTP_0}{2} + \frac{\varepsilon_{current} - 10}{8} \times 2.3RT(3.4\alpha + 1.96\beta + 0.091\,N_c + 0.31\,N_{CX} + 0.045\,N_{Ar}) \quad (6.10)$$

where P_o is the atomic fragment value, $\varepsilon_{current}$ is the distance-dependent dielectric constant (Figure 6.4), α is the hydrogen bond acceptor constant, and β is the hydrogen bond donor constant. N_C denotes the number of C_{sp3} carbons to which no heteroatom is attached, N_{CX} is the number of C_{sp3} carbons to which a heteroatom is attached, and N_{Ar} is the number of aromatic carbons.

Partition energies for the solutes were obtained from the difference of G_{int} in water and organic solvent according to Eq. 6.11. Where $\Sigma\, G_{H_2O}$ and $\Sigma\, G_{nonpolar\ solvent}$ are the interfacial energies from Eq. 6.10 for the solute placed in water ($\varepsilon = 78.5$) and in the apolar phase (whith $\varepsilon = 10$ for octanol and $\varepsilon = 2$ for hydrocarbon).

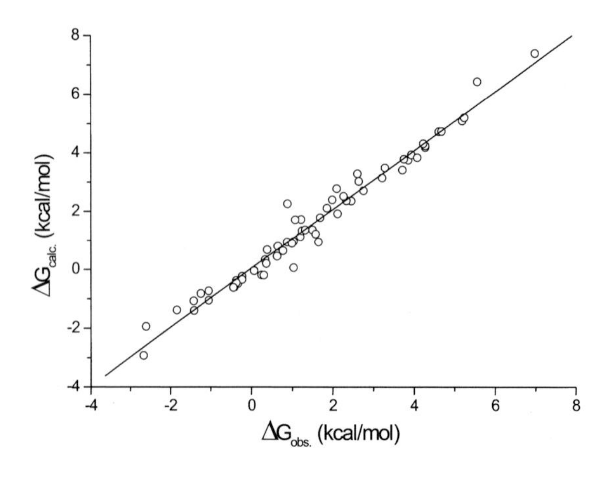

Fig. 6.5 Experimental *vs.* calculated partitioning energies into octanol using Eq. 6.10 for 65 test solutes. Data taken from ref. 45.

Tab. 6.5 Atomic fragment values and hydrogen bond donor and acceptor values used in the simulation of drug-octanol and drug-membrane partitioning. Values are taken from ref. 45

Atom type[a, b]	Atomic hydrophobicity	α	β
$\underline{C}H_3R$	–0.6771	0	0
R-$\underline{CH_3}$[c]	0.8690	0	0
$\underline{C}H_2R_2$	–0.4873	0	0
R-$\underline{CH_2}$-R[c]	0.5800	0	0
$\underline{C}HR_3$	–0.3633	0	0
$R_3\underline{CH}$[c]	0.1810	0	0
$\underline{C}R_4$	–0.1366	0	0
$\underline{C}H_3X$	–1.0824	0	0
$\underline{C}H_2RX$	–0.8370	0	0
$\underline{C}H_2X_2$	–0.6015	0	0
$\underline{C}HR_2X$	–0.5210	0	0
$\underline{C}HRX_2$	–0.4042	0	0
$\underline{C}HX_3$	0.3651	0	0
$\underline{C}R_3X$	–0.5399	0	0
$\underline{C}R_2X_2$	0.4011	0	0
$\underline{C}RX_3$	0.2263	0	0
$\underline{C}X_4$	0.8282	0	0
=$\underline{C}H_2$	–0.1053	0	0
=$\underline{C}HR$	–0.0681	0	0
=$\underline{C}R_2$	–0.2287	0	0
=$\underline{C}HX$	–0.3665	0	0
=$\underline{C}RX$	–0.9188	0	0
=$\underline{C}X_2$	–0.0082	0	0
Ar R-$\underline{C}$H-R	0.0068	0	–0.017
Ar R-$\underline{C}$R-R	0.1600	0	–0.017
Ar R-$\underline{C}$X-R	–0.1033	0	–0.017
Ar R-$\underline{C}$H-X	0.0598	0	–0.017
Ar R-$\underline{C}$R-X	0.1290	0	–0.017
Ar R-$\underline{C}$X-X	0.1652	0	-0.017
Ar X-$\underline{C}$H-X	0.2975	0	–0.017
Ar X-$\underline{C}$R-X	0.9421	0	–0.017
Ar X-$\underline{C}$X-X	0.2074	0	–0.017
R-($\underline{C}$=X)-R	0.0956	0	0
Ar-($\underline{C}$=X)-R	–0.1116	0	0
R-($\underline{C}$=X)-X	0.0709	0	0
$\underline{H}$-C^0_{sp3}	0.4418	0	0
$\underline{H}$-C^1_{sp3}	0.3343	0	0
$\underline{H}$-C^2_{sp3}	0.3161	0	0
$\underline{H}$-C^3_{sp3}	–0.1488	0	0
$\underline{H}$-C^0_{sp2}	0.3343	0	0
$\underline{H}$-C^1_{sp2}	0.3161	0	0
$\underline{H}$-C^2_{sp2}	–0.1488	0	0
$\underline{H}$-OR	–0.6200	–0.31	0
$\underline{H}$-NHR	–0.3260	–0.025	0
$\underline{H}$-NR_2	–0.3260	–0.025	0
$\underline{H}$-S	–0.3260	–0.025	0
(C=O)N-$\underline{H}$	–0.3260	–0.25	0
(C=O)O-$\underline{H}$	–0.3260	–0.55	0
Ar-O$\underline{H}$	–0.3260	–0.61	0
Ar-N$\underline{H}$	–0.3260	–0.13	0
$\underline{H}$-αC	0.2099	0	0
$\underline{H}$-C^0_{sp3}-CX	0.3695	0	0
$\underline{H}$-C^0_{sp3}-CX2	0.2697	0	0
$\underline{H}$-C^0_{sp3}-CX3	0.3647	0	0
R-$\underline{O}$-H	–0.0804	0	–0.51
X-C_{sp3}-C-$\underline{OH}$	–0.5680	–0.32	–0.48
(C=O)-$\underline{O}$-H	0.4860	0	–0.15
Ar-$\underline{O}$-H	0.4860	0	–0.23
R-(C=$\underline{O}$)-R	–0.3514	0	–0.48
R-(C=$\underline{O}$)-O	–0.3514	0	–0.30
R-(C=$\underline{O}$)-N	–0.3514	0	–0.45
R-$\underline{O}$-R	0.1720	0	–0.47
R-(C=O)-$\underline{O}$-R	0.2712	0	–0.15
R-$\underline{N}H_2$	0.1187	0	–0.68
R2-$\underline{N}$H	0.2805	0	–0.70
R3-$\underline{N}$	0.3954	0	–0.68
Ar-$\underline{N}H_2$	0.3132	0	–0.50
Ar-$\underline{N}$HR	0.4238	0	–0.50
Ar-$\underline{N}R_2$	0.8678	0	–0.50
R(C=O)-$\underline{N}$	–0.0528	0	–0.33
Ar $\underline{N}$	–0.1106	0	–0.42

a) Atoms for which parameters apply are underlined. Where more than one atom is underlined, the parameters apply to the entire group.

b) R, any group linked through carbon; X, hetero atom; Ar, aromatic. For C^n_{sp3} and C^n_{sp2}, n is the oxidation number, defined as the number of bonds to hetero atoms with double bonds counting twice.

c) Aliphatic group parameters apply to carbon atoms separated by at least three bonds from the nearest hetero atom, or two or more bonds from the nearest carbonyl, aromatic, or alkenyl carbon.

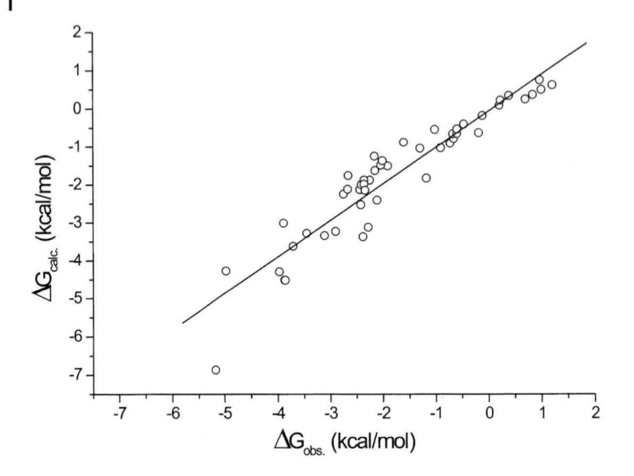

Fig. 6.6 Experimental *vs.* calculated partitioning energies into hexane using Eq. 6.10 for 49 test solutes. Data taken from ref. 45.

$$\Delta G_{\text{calc.}} = \sum G_{H_2O} - \sum G_{\text{nonpolar solvent}} \tag{6.11}$$

For a number of solutes, the partition coefficients in octanol could be well reproduced by Eq. 6.11 (Figure 6.5), as could the differences in the partitioning between octanol–water and hydrocarbon–water systems. A relatively large difference of 1.7 kcal/mol between experimental and calculated values was observed only for glycerol, which has a low partition coefficient in octanol and a far lower one in hexane (Figure 6.6). Experimental partitioning energies for various solutes between water and phosphatidylcholine bilayer were also well reproduced within the experimental error.

To further validate their approach, the authors performed a 4.8-ns MD simulation of β-hexylglucopyranoside within the simulated interface. The conformational preferences obtained were in fairly good agreement with those derived from NMR experiments. Although the ^{13}C dipolar coupling constants could not be reproduced quantitatively, their signs agreed with the experiment, showing the power of the mean field approximation.

Xiang and Anderson [48] proposed a statistical mechanical theory that relates distribution properties of solutes within the interface to the size, shape, and orientation of the solute and the structure of the interface. In this model, the lateral pressure as obtained from a MD simulation and solute–solvent interaction parameters were used to calculate distributions in the interface.

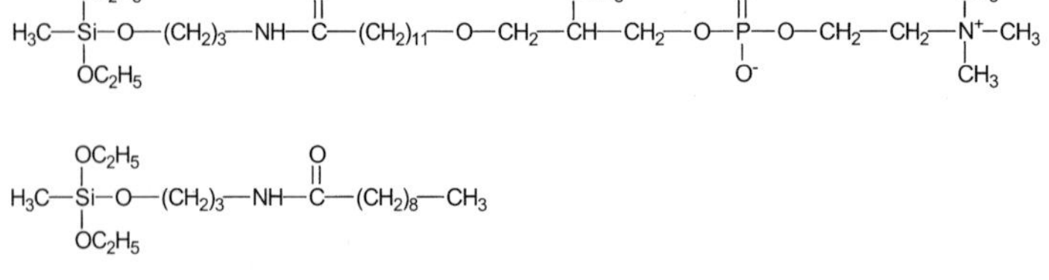

$H_3C{-}Si(OC_2H_5)_2{-}O{-}(CH_2)_3{-}NH{-}C(=O){-}(CH_2)_{11}{-}O{-}CH_2{-}CH(OCH_3){-}CH_2{-}O{-}P(=O)(O^-){-}O{-}CH_2{-}CH_2{-}N^+(CH_3)_3$

$H_3C{-}Si(OC_2H_5)_2{-}O{-}(CH_2)_3{-}NH{-}C(=O){-}(CH_2)_8{-}CH_3$

Fig. 6.7 Chemical structures of PC and decanoate ligands used in the simulation of an IAM.PC surface.

The behavior of IAMs, which are used in partitioning studies (see Chapter 2), has been investigated by MD simulation [49]. The simulated system was similar to a previously synthesized IAM surface, but differed from the commercially available IAM.PC in some respects. The covalently attached phospholipid was an ether analog of PC with only one long alkyl chain and a methyl group in the *sn*-2 position. The system consisted of 36 of these ether lipids and seven decanoyl moieties, all attached to silica propylamine (Figure 6.7). The decanoate was included in a ratio that corresponded to that experimentally observed after end capping of the IAM surface. For the placement of the lipids, an energy minimization procedure was applied. The ether lipids were represented by point charges of +5 and the decanoates by +1, which were initially randomly placed on a square of 54 × 54 Å. The system of repulsive point charges was then energy minimized to give the final positions at which the silica methyl groups were fixed. In this way, a maximal distribution of the alkyl chains was achieved.

During a 250-ps simulation, the structure of the water and the order of the alkyl chains were investigated. The distribution of the functional groups in the head group region was found to be very similar to that obtained from a simulation of a fluid POPC membrane. In addition, the distribution of water in the interfacial region was the same. In both cases, water penetrated into the membrane up to the C-1 carbon of the alkyl chains. As expected from the covalent attachment, the order parameters of the lipid ether alkyl chain were higher than in a fluid membrane. However, the order of the decanoate ligands was much lower, showing that the anchoring of the chains was not responsible for this effect. A detailed analysis of the *trans/gauche* ratio showed that the decanoates behaved like unconstrained alkyl chains, whereas for the PC ligands an increase in *trans* conformations was evident, pointing to a restriction of their movement. The similarities found between simulation and experimental findings show that they are useful tools to explose the behavior of IAMS theoretically.

6.2.2
Mechanism of Diffusion through Phospholipid Membranes

The experimentally observed permeation range of small molecules through phospholipid membranes is very large and varies by several orders of magnitude, ranging roughly from 10^{1} to 10^{-6} cm/s. Permeability depends on both the solubility in the membrane and the diffusion across it. Experimentally, it has been shown that the size dependence differs for very small solutes (molecular weight < 50) and larger solutes with molecular weights in the range of 50–300. In the case of larger solutes, a size dependency according to the Stokes–Einstein equation was observed if the permeabilities were corrected for hydrophobicity, but for very small solutes a much steeper dependence was found. Thus, different mechanisms of diffusion were proposed: a "hopping" mechanism for very small solutes and "normal" diffusion for larger solutes. Several MD simulations have been performed in order to obtain a better understanding of the mechanisms of diffusion of solutes of varying size.

It should be noted in advance that the time required for a penetrating molecule to diffuse through the membrane is much longer than can be simulated. Therefore, in all simulations special techniques have been used to estimate the diffusion rate.

The first atomic level molecular dynamics simulation was reported in 1993 [50]. The diffusion of benzene through a bilayer membrane composed of DMPC was investigated. The system consisted of 36 DMPC molecules arranged in a bilayer and hydrating water. In three simulations a single benzene molecule was inserted at different positions and in a fourth four benzene molecules were inserted into the membrane to increase the sampling statistics. The temperature was 320 K, well above the phase transition temperature of DMPC yielding a fluid crystalline state, and simulation times were in the range of 500–1000 ps. The properties of these systems were compared with an unperturbed membrane simulated under the same conditions. While no change in membrane thickness was observed, slight changes in other parameters were observed as a function of benzene concentration. The order parameters of the carbons near the carbonyl group decreased, but no consistent effect of benzene on this parameter was found further away. Controversially, the number of *gauche* conformations was slightly reduced and the conversion of *trans-* to *gauche* conformations was lower, pointing to a rigidifying of the membrane. Diffusion coefficients were calculated from the mean-squared displacement of the benzene molecules and were found to vary according to the position within the membrane, being higher in the center. It was found that the benzene molecules showed no uniform movement; instead, jumps occasionally occurred, leading to movements over 5–8 Å in a very short time of 5 ps. The benzene molecules near the bilayer center exhibited larger and more frequent jumps. A detailed analysis of the trajectories showed that the jumps were mediated by concerted movements of the alkyl chains of the phospholipids, which opened a way between two voids, allowing the benzenes to make jumps between the two voids. As a consequence the average diffusion rate in the center of the bilayer was about 3 times larger than in the head group region. An additional simulation of benzene in a hydrocarbon consisting of unoriented tetradecane yielded diffusion coefficients intermediate between those near the head group and those near the center of the bilayer (Table 6.6).

Although some benzene jumps were also observed in the hydrocarbon, they were smaller than those observed for benzene molecules near the center of the bilayer,

Tab. 6.6 Comparison of diffusion of benzene molecules in DMPC as function of location and in tetradecane. Data taken from ref. 50

	Diffusion rate (10^{-6} cm^2/s)	**Most common location**
Benzene in phospholipid simulation no.		
2	2.7	Moves between different regions
4 (molecule A)	2.1	Moves between different regions
3	1.4	Resides near carbonyl or head group region
4 (molecule B)	1.3	Resides near carbonyl or head group region
1	4.0	Resides in center of bilayer
4 (molecule C)	3.8	Resides in center of bilayer
4 (molecule D)	4.6	Resides in center of bilayer
Benzene in tetradecane	1.9	

and no movements greater than 5 Å occurred. However, on average, they were larger than those found for benzenes near the head group region. The results obtained show that the diffusion of non-polar molecules such as benzene within a membrane is not uniform and differs from that in bulk hydrocarbons.

The simulation with four inserted benzene molecules was extended to study the influence of temperature on the position and movement of the benzene molecules [51]. At different temperatures, the benzene molecules appeared to favor different regions of the bilayer. At the lowest temperature studied (310 K) the benzenes moved to the center of the bilayer, while at 340 K they resided mostly near the head group region. To study the size dependence of diffusion of non-polar molecules, comparative simulations with the smaller methane and the larger adamantane molecules were performed [52]. As in the case of benzene, the diffusion rate of methane was found to be larger in the center of the bilayer than in the head group region, again caused by jumps between different voids in the bilayer. This effect was even more pronounced in case of the smaller methane, leading to a four- to sixfold higher diffusion in the bilayer center. However, for adamantane, no differences were observed, the diffusion rate being the same regardless of the position within the bilayer. This could be related to the fact that adamantane did not exhibit the jumps observed in the case of benzene and methane.

Thus, these results show again that the mechanism of diffusion through a lipid bilayer seems to be complex and depends on several factors.

To investigate the importance of hydrophobicity, the diffusion of water, ammonia, and oxygen was studied by molecular dynamics simulations [53]. These three molecules are of similar size and differ largely in their hydrophobicity, ranging from the hydrophilic water over ammonia to the fully hydrophobic oxygen. The simulation system consisted of 64 DPPC molecules arranged in a bilayer and hydrated with water. As the spontaneous entry of hydrophilic solutes such as water into membrane is too slow, constraint MD simulations were performed, in which the solutes were inserted into the lipid membrane at many equidistant positions across the bilayer, constraining their position with respect to the bilayer normal. Additionally, a potential of mean force was derived by constraining the solutes to different regions within the membrane and determining the force needed to keep the constraint. The applied method allowed the simultaneous calculation of the local diffusion constant and free energy profile. For the hydrophilic water and ammonia, the calculated excess free energy profiles across the membrane increased smoothly from bulk water to the interior of the bilayer, being highest in the region of ordered alkyl chains. In contrast, a favorable interaction was observed for the lipophilic oxygen (Figure 6.8). Like the studies described above, a much higher diffusion rate in the center of the bilayer was calculated. The lowest diffusion rate for all three solutes occurred in the head group region. From the excess free energies and the diffusion rates it could be calculated that the local resistance to permeation for water and ammonia is highest in the region of ordered alkyl chains close to the head group region. This study shows clearly the large hydrophobicity dependence of the permeation process through a membrane. While for hydrophilic solutes the rate-limiting step is the crossing of the region of ordered alkyl chains, which are densely packed, for a hydrophobic solute like oxygen the rate-limiting step is the diffusion through the water layer.

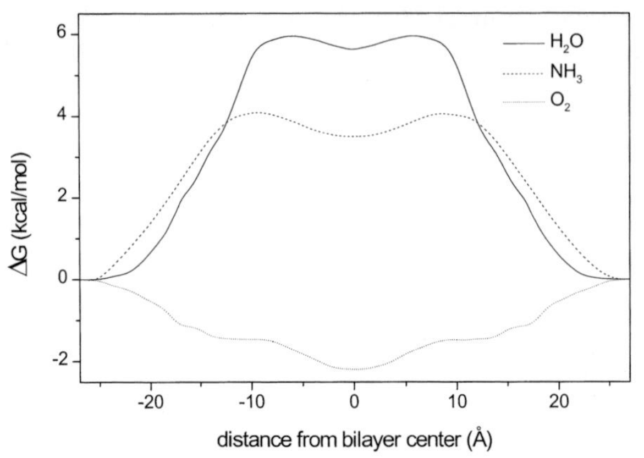

Fig. 6.8 Calculated excess free energy profiles of water, ammonia, and oxygen across the DPPC bilayer. (Adapted from Fig. 2 of ref. 53 with permission from the American Chemical Society).

The influence of size and shape on the diffusion of hydrophobic solutes was estimated by simulations involving artificial Lennard–Jones particles those intermolecular interaction parameters were based on those for ammonia or oxygen, respectively. The results on the size dependence of diffusion confirmed that the membrane interior differs strongly from a bulk hydrocarbon. In the center of the bilayer, the excess free energy for hydrophobic Lennard–Jones particles remained low irrespective of the size of the particles. This can be explained by the large fraction of accessible volume in that region.

Using Lennard–Jones particles based on oxygen atoms the influence of shape on diffusion was investigated. By rigidly connecting two, three or five particles more and more aspherical ones were constructed and their excess free energy in the membrane calculated. With increasing elongation, generally a stabilization within the membrane was observed. However, the decrease in excess free energy differed across the membrane, being most pronounced in the region of ordered alkyl chains. For the most elongated particle, consisting of five oxygens, this region became the favored one. This stabilization was a result of the more favorable interactions of an elongated molecule in an ordered region compared with a disordered one. It could

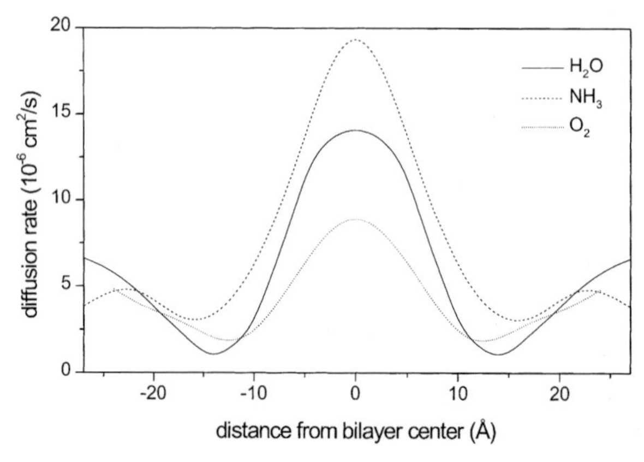

Fig. 6.9 Calculated diffusion rate profiles of water, ammonia. and oxygen across the DPPC bilayer. (Adapted from Fig. 2 of ref. 53 with permission from the American Chemical Society).

account for the experimentally observed more favorable interaction between flat molecules and membranes compared with bulk hydrocarbons.

Using a similar approach, the proton transport across a lipid membrane by water pores has been studied [54]. A water pore was induced in the membrane, and its conformational behavior and life time investigated. From the simulation it became obvious that the formation of a water pore spanning the whole membrane was a rare event and that such pores were short lived, lasting, on average, only a few picoseconds. The calculated transport rates through the pores could be considered to be in agreement with the experimentally observed high transport rate only if a proton is transported through the membrane every time a pore is formed. The authors concluded that this requires that the entry of the proton is not rate limiting, which is a reasonable conclusion only if high local proton concentrations near the pore are assumed. This could be the case if local concentration fluctuations at the membrane exist and a pore is most likely formed at a region of local high proton concentration. Obviously, the transport of protons through membranes is a process that is not fully understood at the molecular level.

The passing of dimethylsulfoxide (DMSO) through a model bilayer composed of glycerol-mono-octadecanoate has also been investigated by non-equilibrium constraint molecular dynamics [55]. In this study, the DMSO molecule was pulled by constraints through the membrane, allowing equilibration at defined positions along the bilayer normal for a short time. Using an united atom approach and rigid DMSO, no influence of the presence of DMSO on the structure of the interior of the bilayer was observed. In the simulation, the DMSO molecule passing the water interface introduced a hole in the membrane, allowing a water molecule to co-penetrate into the membrane. The short time of the simulation (300 ps) meant that the calculation of the mean force acting on DMSO was not possible because of convergence problems. This exemplifies the problems in performing simulations of events that need considerable time, such as the diffusion of a molecule through a membrane.

The influence of DMSO on membrane structure was investigated by Smondyrev and Berkowitz [56] by comparison of a DPPC/water system with a system where the water was replaced by DMSO. No influence of DMSO on the structure and ordering of the hydrocarbon part of the membrane was found, but changes in the head group region were observed. The phosphorus–nitrogen (P–N) vector distribution was shifted to lower angles. In water, the P–N vector pointed toward water at an average angle of 9°, whereas in presence of DMSO it pointed more to the interior of the membrane, with an average angle of –4°. Most noticeably, the electrostatic potential across the membrane changed its sign and became smaller in magnitude. From the results of the simulation the authors concluded that DMSO reduced the repulsive hydration force between bilayers by several mechanisms.

6.2.3 Small Molecules and their Interaction with Phospholipids

Relatively few studies have been performed to investigate the interaction of small molecules with phospholipids by computer simulation. In addition to studies of the mechanism of diffusion, small molecules that have been studied involved mainly general anesthetics and calcium channel blockers of the dihydropyridine type.

6.2.3.1 Anesthetics

There is a long history of controversy in the literature regarding the mode of action of general anesthetics. Experimental results derived from model systems of lipids alone or lipid–cholesterol are somewhat controversial. To mention just a few, using Raman spectroscopy it was found that, at clinical concentrations, halothane had no influence on the hydrocarbon chain conformations, and it was concluded that the interaction between halothane and the lipid bilayer occurs in the head group region [57]. This idea was also supported by ^{19}F-NMR studies. The chemical shifts of halothane in a lipid suspension were similar to those in water and differed from those in hydrocarbons. In contrast, from ^{2}H-NMR experiments, it was concluded that halothane is situated in the hydrocarbon region of the membrane (see also chapter 3.3).

The influence of foreign molecules including general anesthetics on the phase behavior of phospholipids has been studied in some detail by MC simulation and statistical thermodynamics of simplified models [65, 66]. From these investigations it was concluded that the gel to fluid transition of phospholipids is strongly influenced by the presence of anesthetics, and that anesthetics have a high affinity to the interfaces of coexisting gel and fluid lipid domains, leading to locally high concentrations.

On a more detailed level, only a few studies have been undertaken to investigate the interaction of general anesthetics with phospholipids by MD simulation. One study investigated the influence of trichloroethene on the structure and dynamics of phospholipid bilayers [58]. A single trichloroethene molecule was studied in a bilayer consisting of 24 DOPC molecules. Initially, the trichloroethene was introduced into the middle of the hydrocarbon chains of one leaflet of the bilayer. The authors observed an increase in the ratio of the effective areas of hydrocarbon chains and polar head groups that suggested a tendency toward the hexagonal phase transition. A small increase in *gauche* conformations of the lipid hydrocarbon chains and a significant increase in mobility of the lipids was also found. In this study, the simulation time was only 150 ps, and, in light of newer results on the time scales necessary to reach equilibrium, one should be careful when considering the correctness of the obtained results.

A far longer simulation time of 1.5 ns was used in a later study of halothane with 64 DPPC molecules [59]. Four halothane molecules were initially placed into the glycerol region, two on each side of the bilayer, to increase the sampling statistics. Three of the four molecules remained in the interface region, while the fourth moved toward the center of the bilayer during the simulation. This movement was similar to that observed for benzene [51], resembling a "rattling in a cage" with occasional jumps to a new position.

No influence of halothane on the lipid hydrocarbon chain conformations was found although the overall geometry changed slightly as a consequence of a small lateral expansion, accompanied by a small contraction of the bilayer thickness. Very

Fig. 6.10 Formula of the hypothetical nifedipine analog used in the MD simulation of Alper and Stouch [60].

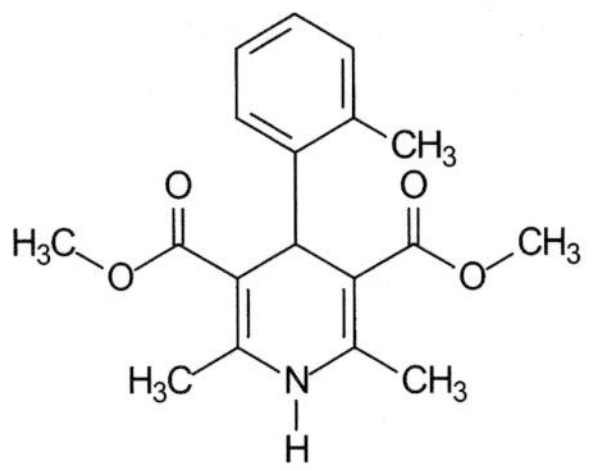

interesting and surprising results were obtained regarding the hydration of halothane. Two of the three halothane molecules that resided near the glycerol region showed hydration characteristics very close to those in bulk water despite their location, nicely explaining the experimental results of ^{19}F-NMR studies.

6.2.3.2 **Dihydropyridines**

Alper and Stouch [60] studied the behavior of a nifedipine analog in which the 2-nitro substituent on the phenyl ring of nifedipine was replaced by a methyl group. Apparently, the dihydropyridine ring was replaced by a pyridine ring, leading to a hypothetical chemical structure with a C_{sp3} carbon in position 4 having only three bonds (Figure 6.10). Unfortunately, nothing was reported about the geometric properties of the (dihydro)pyridine moiety used in the simulations.

The phospholipid bilayer corresponded to the one used in the studies of benzene diffusion described above. Two simulations of 2 ns duration each were performed, differing in the initial placement of the nifedipine analog. In one simulation, the solute was placed into the center of the bilayer and in the other it was placed close to the hydrocarbon–water interface. The influence of the solute on the membrane and the diffusion and conformational and orientational changes of the solute were investigated. Almost no perturbation of the membrane due to the presence of the nifedipine analog occurred, and only small changes in the order parameters were found relative to those from the simulation of a neat bilayer. In contrast to the studies of benzene, the rate of diffusion did not vary with location in the bilayer and was in both cases 1.3×10^{-6} cm^2/s, similar to that of benzene when located in the head group region (see Table 6.5). Also, no sudden jumps between voids took place in the case of the nifedipine analog, as the transient voids in the center of the bilayer were too small to accommodate such a large molecule.

The rate of rotation and reorientation of the molecule was dependent on its location, being much slower when residing in the interface region compared with the bilayer center. When situated in the head group region, the nifedipine analog showed extensive H-bonding between the NH group of the didydropyridine ring and the carbonyl oxygens of the phospholipids. The preferred orientation of the solute showed a significant tilt of its long axis relative to the bilayer normal that enabled formation of additional H-bonds between one of its ester carbonyl groups and water molecules.

Another difference between the two simulations was related to conformational transitions from *trans* to *cis* orientation of the ester groups of the nifedipine analog.

These conformational changes were highly correlated in the case of the solute situated in the head group region, and when one changed from *trans* to *cis* the other changed in the opposite direction at the same time. No correlations were observed for the nifedipine analog situated at the bilayer center. The significance of this observation is unclear as only a few conformational transitions were observed.

In another simulation, the effect of nifedipine and lacidipine on the structure of a phospholipid bilayer was investigated [61]. In this study, the dihydropyridines were inserted into the middle of the bilayer and remained there during the 600 ps of MD simulation. No global effects of the dihydropyridines on the phospholipid structure were observed, only local effects in the vicinity of the drugs. Some local ordering was observed due to the presence of nifedipine, as judged by the decrease in the isomerization rate between *trans* and *gauche* conformations of the alkyl chains, while lacidipine did not significantly affect this ratio. It should be noted, however, that this investigation contained some methodological flaws that make the judgment of the validity of the presented results difficult. For the phospholipids, an united atom model was used, whereas the dihydropyridines were treated with explicit hydrogen atoms. The charges of the phospholipids were derived from *ab initio* calculations and those of the dihydropyridines were derived by semiempirical methods. The system was simulated under NVE conditions without any boundary conditions.

6.2.4 Effect of Cholesterol on Membrane Structure

Cholesterol is an essential constituent of mammalian membranes and is required for their normal functioning. While a large number of experimental studies have been undertaken to investigate the effect of cholesterol on the physical properties of phospholipid membranes, our understanding of changes in these properties at a molecular level is still incomplete. Computer simulations can help in this respect, but in relation to the importance of cholesterol relatively few have been undertaken. This can be explained by the fact that the influence of cholesterol is highly dependent on its molar ratio compared with the phospholipids, also on the type of phospholipid used, and new phases appear in presence of cholesterol, thus making the simulation of such process rather difficult.

Early simulations of cholesterol–phospholipid mixtures in the 1990s used the Monte Carlo technique and concentrated on the effect of the presence of cholesterol on the ordering of the alkyl chains of the phospholipids [62, 63]. Though based on a relatively simple model of the cholesterol–phospholipid interaction, with phospholipids represented only by the alkyl chains fixed in the plane, the simulations led to results that were in qualitative agreement with later studies using more elaborate simulation protocols. An increase in the order parameter was observed that was most pronounced for phospholipids that were neighbors of more than one cholesterol molecule. Owing to the limitations of the simulation model, the absolute values of the order parameters were about twice as large as experimental values. The presence of cholesterol also significantly decreased the number of *gauche* conformations in the alkyl chains but did not lead to all-*trans* conformations. Using a similar system of alkyl chains and molecular dynamics simulation with constant area and volume, the same findings basically emerged, but the calculated order parameters for the pure phospholipid were in much better agreement with experimental data from deuterium NMR [64].

Jørgensen's group [65–68] has studied the influence of cholesterol and general anesthetics on membrane heterogeneity, the formation of lipid domains, and phase transition temperature by MC simulations of large systems containing up to several thousand lipids. Their model was based on a simple intermolecular interaction model of the main transition of a lipid bilayer and statistical thermodynamics. In these investigations, several experimental results, such as changes in the specific heat of the lipid molecules during the phase transition, and smearing out with increasing concentrations of general anesthetics, could be reproduced.

Robinson *et al.* [69] studied the influence of cholesterol on molecular ordering of phospholipids by MD simulation. They used a more detailed description of the phospholipids including the head groups and charges. The simulated system contained two cholesterol molecules and 18 DMPC molecules in each leaflet of the bilayer (10 mol% cholesterol) and was simulated after equilibrium for 400 ps employing NVT conditions. They observed an increase in the fraction of *trans* conformations of the lipid alkyl chains with a decrease in kinks. Also, the dynamic and conformation of the flexible cholesterol side chains was characterized and it was found that they had a smaller tilt angle than the lipid chains with respect to the bilayer normal.

More recently, MD simulations under constant pressure involving full representation of phospholipids and cholesterol have been carried out. They dealt either with DMPC– [70, 71] or DPPC–cholesterol mixtures [72, 73]. The results of these four simulations were generally in agreement but differed in some aspects.

In the simulation of Tu *et al.* [72] with 12.5 mol% cholesterol, the starting configuration was obtained from a simulation of a pure DPPC bilayer composed of 64 DPPC molecules. Four DPPC molecules per leaflet were replaced by cholesterol to generate the DPPC–cholesterol system. The system was simulated for 1.4 ns; the area per phosholipid decreased during the first 700 ps of the simulation but then remained stable for the rest of the simulation. As pointed out by the authors, the decrease in the surface area was due to the replacement of the phospholipids by the thinner cholesterol and not to an ordering effect of cholesterol. Accordingly, little effect of cholesterol on the order of the alkyl chains was observed in this simulation, in contradiction to the results of a cholesterol–DPPC simulation performed by Smondyrev and Berkowitz that is described below.

Gabdouline *et al.* [70] performed a MD simulation of DMPC with 50 mol% cholesterol. Three short simulations in the range of 100–300 ps were performed at 200, 300 and 325 K. During the simulation, the area per DMPC–cholesterol heterodimer increased with increasing temperature, reaching approximately 87 and 91 $Å^2$ for temperatures of 300 and 325 K respectively. A rather small value of 50 $Å^2$ was obtained for the area of the phospholipid, accompanied by a very low fraction of *gauche* conformations compared with the experimental findings. These results are contradictory to those obtained from longer simulations and should be regarded with caution. Possible explanations for the observed discrepancy could be the start from a mini-

PL	PL	CH	CH
CH	CH	PL	PL
PL	PL	CH	CH
CH	CH	PL	PL
PL	PL	CH	CH
CH	CH	PL	PL
PL	PL	CH	CH
CH	CH	PL	PL

CH	CH	CH	CH
PL	PL	PL	PL
PL	PL	PL	PL
CH	CH	CH	CH
CH	CH	CH	CH
PL	PL	PL	PL
PL	PL	PL	PL
CH	CH	CH	CH

PL	CS	PL	CS
CS	PL	CS	PL
PL	CS	PL	CS
CS	PL	CS	PL
PL	CS	PL	CS
CS	PL	CS	PL
PL	CS	PL	CS
CS	PL	CS	PL

Fig. 6.11 Schematic representation of initial structures of the DPPC/cholesterol and DPPC/cholesterol sulfate membranes. PL: phopholipid, CH: cholesterol, CS: cholesterol sulfate. Left: system A, middle: system B [73], right: DPPC/cholesterol sulfate [75]

mized crystal structure and the short simulation time, which did not allow for equilibration of the system.

Smondyrev and Berkowitz [73] performed a rather long simulation (2 ns) of a system with a low cholesterol content (ratio of 1:8 or 11 mol%) and of two systems that contained cholesterol at 1:1 molar ratio and differed in the arrangement of the cholesterol molecules (Figure 6.11). In the simulation with low cholesterol content, the area per lipid was about 62 Å^2, close to that found in a simulation of pure DPPC for the first approximately 800 ps. However, after this time it started to decrease and reached a value of 58.3 Å^2 after an additional 500 ps, where it remained for the last 750 ps of the simulation. In the simulation with high cholesterol content, the area per DPPC–cholesterol heterodimer surface area showed also a decrease over the simulation time that was approximately exponential (Table 6.7). In system A (Figure 6.11), with a more uniform distribution of cholesterol, the surface area tended to decrease over the full simulation time of 2 ns, whereas for system B a stable and somewhat lower value was obtained after 1 ns. Although the authors concluded that system A, with more uniformly distributed cholesterol, might be trapped in a metastable state, this system was used in a later publication for comparison of the effects caused by cholesterol sulfate (see below).

As found previously, the order parameter of the alkyl chains of DPPC increased in the presence of cholesterol. This increase was found to be most pronounced in the region from carbon 4 to carbon 11 and corresponded to the size and location of the rigid

Tab. 6.7 Summary of changes in bilayer geometry caused by the presence of different cholesterol concentrations. Data for DPPC are taken from ref. 73 and those for DMPC from ref. 71

Cholesterol (mol%)	DPPC		DMPC
	Area	Lamellar spacing	Area
0	61.6	59.0	60.2
11	58.3	63.0	–
22	–	–	58.4
50	44.7	77.9	–

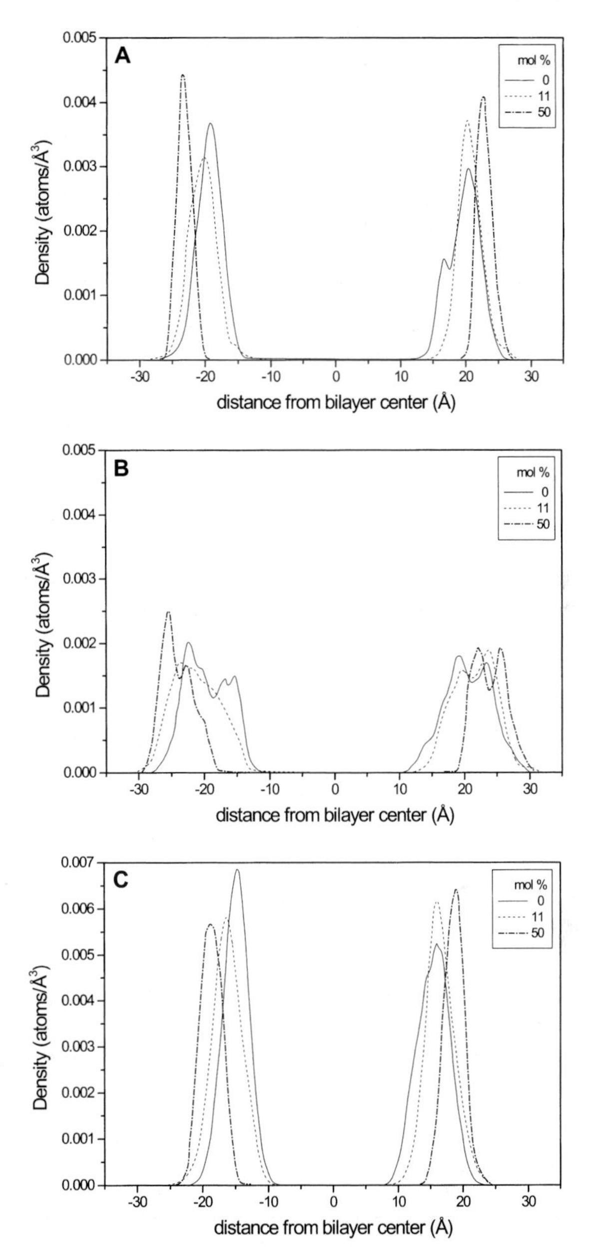

Fig. 6.12 Change in distribution of phosphorus (a), nitrogen (b), and carbonyl oxygen (c) positions along the bilayer normal in pure DPPC and in membranes with 11 and 50 mol% cholesterol. (Adapted from Fig. 9 of ref. 73 with permission from the Biophysical Society).

cholesterol ring system, being located below the carbonyl atoms of the phospholipids. The order parameter, averaged over all carbon atoms in both chains, was also in good agreement with experimental results from NMR measurements. For the system with 11 mol% cholesterol the increase in the order parameter was from 0.34 to 0.38. For the systems with 50 mol% cholesterol, a further increase to 0.72 was found, in close correspondence with the experimental results, which showed an increase of a factor of about 2 for DMPC membranes. The presence of cholesterol also influenced the distribution of the polar head group atoms across the membrane (Figure 6.12). The presence of cholesterol led to a sharpening of the distributions of the phosphorus, nitrogens, and carbonyl oxygens. And with increasing cholesterol concentration the membrane became thicker, as evidenced by the shift of the distributions to values further away from the bilayer center.

The presence of cholesterol also led to a change in the orientation of the P–N vector of the head groups [73]. Whereas in pure DPPC the P–N vectors were largely parallel to the membrane, with a small tendency to point outward into the water layer, in the case of the system with low cholesterol content two kinds of lipid molecules in the bilayer were found. For those lipids in close contact with cholesterol a slightly larger average area per lipid was observed and their P–N vectors mostly pointed toward the bilayer interior. In contrast, the lipids without any contact with cholesterol had P–N vectors with larger angles pointing higher above the membrane. In the systems with high cholesterol content the latter population dominated (Figure 6.13). Cholesterol was found to be located below the carbonyl groups of the phospholipids within the bilayer in such a way that its hydroxy group could form hydrogen bonds with the carbonyl atoms of the phospholipid ester groups and with water present in the membrane interface.

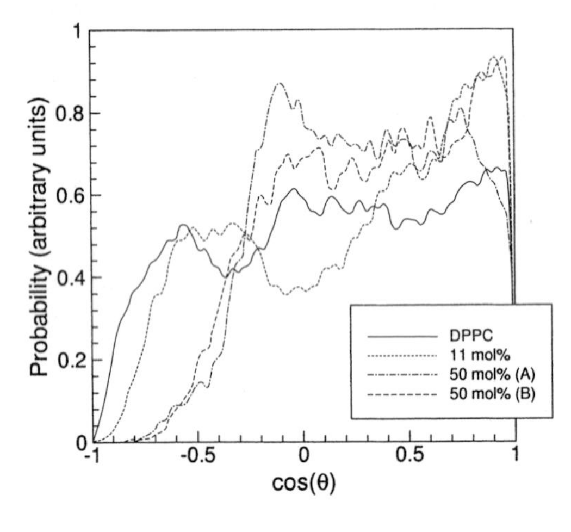

Fig. 6.13 Distribution of the angle between P–N vector and bilayer normal in pure DPPC membrane (solid line) and in membranes containing cholesterol: 11 mol% (dotted line), 50 mol% structure A (dash dot line), and structure B (dashed line). When cosine is positive, the P–N vector points into the water layer.
(Adapted from Fig. 7 of ref. 73 with permission from the Biophysical Society).

Table 6.8 shows the average number of hydrogen bonds formed between the phospholipid and water molecules. With an increase in cholesterol content, a slight increase in the total number of H-bonds between DPPC and water can be seen, while the number of H-bond bridges decreased.

From an even longer 5-ns simulation of 22 mol% cholesterol in DMPC performed by Pasenkiewicz-Gierula *et al.* [71], a qualitatively similar picture emerged. Here again the surface area and number of *gauche* conformations reached a plateau after a few hundred picoseconds, and remained there for more than 1 ns. After that time, a

Tab. 6.8 Average number of hydrogen bonds per DPPC or cholesterol oxygen, formed with water molecules reported from several simulations. Data for DPPC and DPPC-cholesterol are taken from ref. 73 and data for DMPC-cholesterol sulfate are from ref. 75. Data for DMPC-cholesterol are taken from ref. 71 and data for pure DMPC are taken from ref. 74 and 106.

Oxygens	Pure DPPC	DPPC-cholesterol		DPPC-cholesterol sulfate	Pure DMPC	DMPC cholesterol
		11 mol%	50 mol%[a]	50 mol%		22 mol%
Ester phosphate O	0.75	0.81	0.81/0.89	0.74	(0.3) 0.43[b]	0.47[c]
Non-ester phosphate O	3.37	3.27	3.36/3.43	2.83	(3.5) 3.7	3.9
Carbonyl O	1.92	1.99	1.98/1.96	1.28	(0.7) 0.80	0.92
Total H-bonds	6.04	6.07	6.15/6.31	4.85	(5.3) 5.4	5.7
Bonds in bridges	2.23	2.24	1.67/1.64		(1.7)	
O_{chol}		0.72	0.89/0.97			1.1

a) Values for system A and B respectively.
b) Values from the same simulation after different times. Values in parenthesis are from a 1.6-ns simulation and averages over the last 500 ps [106], while the others are from a continuation up to 5.1 ns and averaged over the last 2000 ps [71].
c) Averages of the last 2000 ps from a 5 ns simulation.

further decrease in both parameters occurred, ending at about 2.3 ns total simulation time (Table 6.7). These results point to the need for very long simulations to reach full equilibrium.

The trends in the number of H-bonds between water and cholesterol were similar, but the absolute numbers of H-bonds formed between water and the oxygens of the polar head groups differed significantly (Table 6.8). A more detailed comparison of the different types of H-bonds reported in both simulations is unfortunately not possible, because of the different ways in which they were reported. In the simulation of Smondyrev and Berkowitz [73] the H-bonds between water and the different oxygens were reported, but the authors did not differentiate between single H-bonds and bridges. In the investigation of Pasenkiewicz-Gierula *et al.* [71], these were reported separately but with less differentiation with respect to the oxygens involved in H-bonding.

Recently, Smondyrev and Berkowitz also investigated the influence of cholesterol sulfate on membrane ordering [75]. Like their simulation of DPPC–cholesterol at 50 mol% the system was composed of 16 DPPC and 16 cholesterol sulfate molecules per leaflet, but the initial arrangement of DPPC and cholesterol sulfate differed from the configurations used in the case of cholesterol (Figure 6.11). Cholesterol sulfate led to less ordering of the alkyl chains, a much larger surface area per heterodimer, and a large decrease in the average number of hydrogen bonds per DPPC molecule formed with water, especially for the non-ester phosphate and the carbonyl oxygens, as compared with cholesterol (Table 6.8).

6.2.5
Interactions of peptides with phospholipids

The interactions of several peptides with phospholipids have been studied by computer simulation. Emphasis has been given to several aspects of protein–phospholipid interactions, including the way of association and orientational preference of peptides in contact with a bilayer, the effect of phospholipids on the preference and stability of helical conformations, and the effect of the inserted peptide on the structure and dynamics of the phospholipids. These investigations have been extended to bundles of helices and even whole pore-forming proteins. In particular, the simulation of ion channels and of peptides with antimicrobial action has attracted a great deal of attention in theoretical studies.

6.2.5.1 Mean Field Simulations

Early studies performed by Milik and Skolnick [76] investigated the way of insertion of peptide chains into lipid bilayers. These authors used a mean field approach to represent the water and lipid combined with Monte Carlo simulations. In these in-

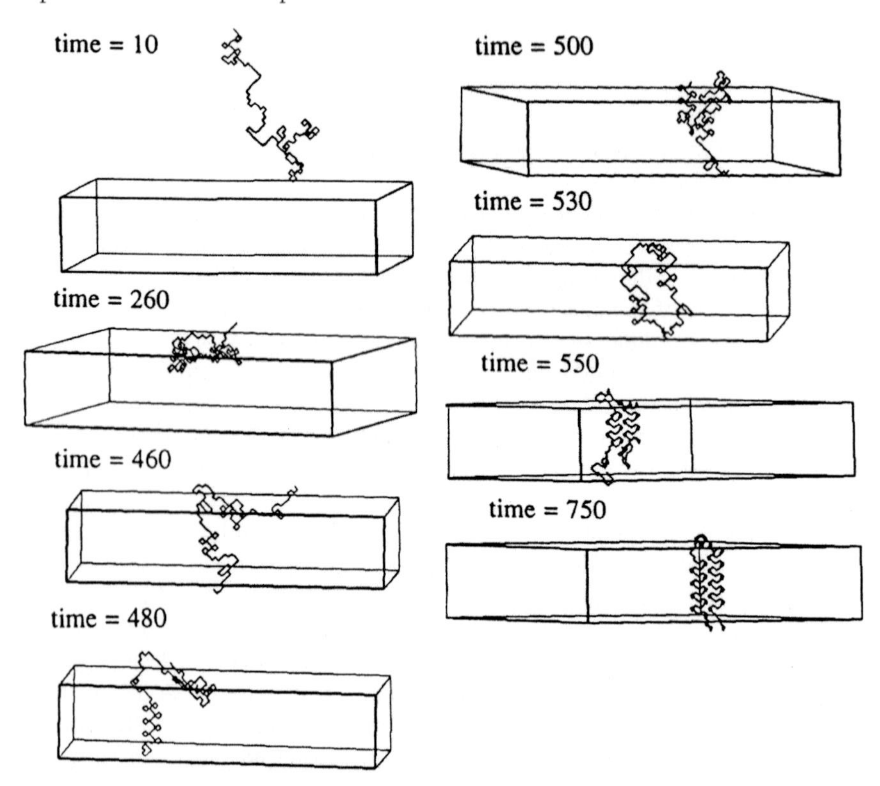

Fig. 6.14 Representative snapshots of the insertion process of the model hairpin with a hydrophilic turn region. (Reprinted from Fig. 3 of ref. 76 with permission from the author).

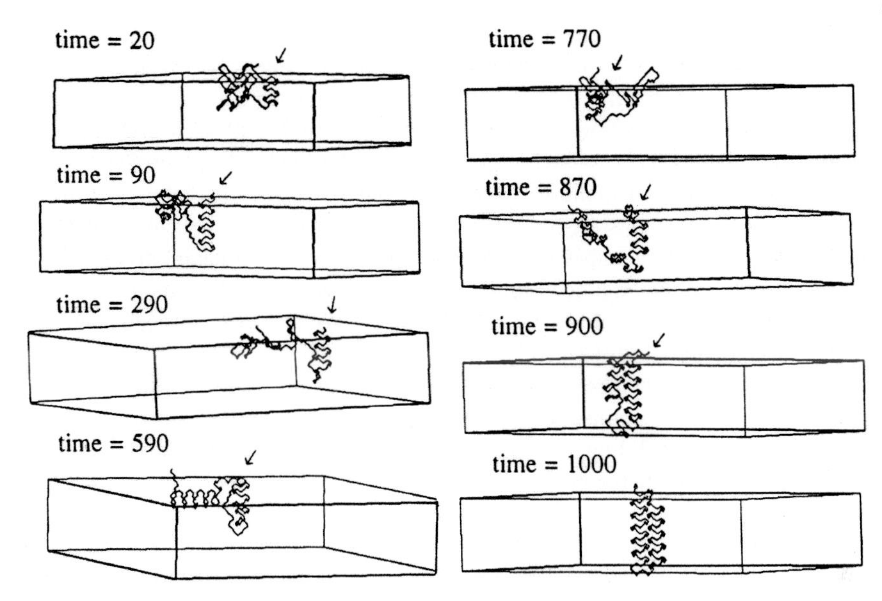

Fig. 6.15 Representative snapshots of the insertion process of the model hairpin with hydrophilic ends. The arrow points to the N-terminus of the chain. (Reprinted from Fig. 4 of ref. 76 with permission from the author).

vestigations, the peptide was represented as a tetrahedral lattice chain. Each amino acid consisted of only three united atoms representing the amide, the Cα carbon, and the side chain. Because of the choice of the diamond lattice configuration, only three torsion angles were possible (180° and ± 60°), but great emphasis was placed on the representation of intermolecular forces such as torsional, Lennard–Jones, and hydrogen bond contributions. The physicochemical properties of the amino acids were grouped into six types ranging from very hydrophilic and charged, with preference for the water phase, through neutral with no preference for either phase, to very hydrophobic. In this system, the insertion of two peptide sequences into the membrane was studied. The first constructed sequence had two slightly hydrophobic fragments connected by a hydrophilic linker and neutral ends. For this sequence, with a hydrophilic middle part, adsorption on the lipid took place and one fragment inserted spontaneously into the membrane, followed by the formation of a hairpin; thereafter the second fragment inserted into the membrane, leading to two antiparallel helices crossing the membrane and connected by a hairpin. Figure 6.14 shows some representative snapshots of the insertion process.

The second sequence studied had one very lipophilic fragment, as found in leader sequences of real proteins, a neutral linker, and a second slightly lipophilic fragment identical to that of the first sequence. However, both ends were highly hydrophilic. Again adsorption on the lipid occurred, but in this case a helix in the opposite direction was formed, pulling the neutral linker through the membrane (Figure 6.15). Finally, two antiparallel helices were formed that were inserted in the lipid in the opposite direction to the first sequence.

This approach was later extended to off-lattice models and a more detailed description of the transfer energy of the different amino acid residues [77]. Magainin, melittin, and several other amphipathic peptides were simulated. In these simulations, differences in the interaction of the peptides with the lipid phase were observed. For example, magainin only showed adsorption onto the lipid and no crossing of the lipid occurred, whereas melittin crossed the lipid and formed a stable transmembrane helix. These results are in full agreement with later studies reported by other research groups presented below, involving more elaborate simulation protocols and representations of the peptides and the lipid. These examples show the potential of computer simulations even when some simplifications have to be made to make the system computationally tractable.

The association and orientational preferences of peptides with regard to a lipid bilayer have also been studied. In particular, peptides with antimicrobial properties have been investigated, to gain insight in their mode of action at a molecular level. Two possible mechanisms are discussed that could lead to observed antimicrobial activity. The first is a carpet-like mechanism, whereby peptide molecules lie on the membrane surface until a critical concentration is reached. At this stage a detergent-like effect is assumed, with peptide and membrane components forming aggregates that leave the membrane, causing its disruption. The other possible mechanism involves formation of a pore. In this case, a peptide bundle spans the lipid bilayer, leading to a pore accompanied by increased ion and water permeability, finally leading to cell death.

Because of the long time scales involved, it is currently not possible to simulate the process of peptide insertion at full atomic level description. As a consequence, in MD simulations a configuration is given, either parallel to the bilayer or inserted into it, and the evolution of the system is followed. Another approach is to perform simulations of orientational preference with the mean field approximation of lipids and water, retaining an atomic level description of the peptide. Several such simulations have been performed to determine the most likely mechanism of action for a given antimicrobial peptide. A recent review on simulations performed with antimicrobial peptides is given in ref. 78.

A combined approach was applied in the study of nisin interaction with model membranes [2]. Using a mean field representation of the lipid bilayer and water, a MC simulation was performed to estimate preferred orientations of nisin at the interface. Simulations were performed keeping the peptide rigid in a conformation based on NMR data and additionally with flexible torsion angles. In all cases, similar orientations were obtained, with the shorter N-terminal part of the L-shaped peptide being inserted in the lipophilic part and the longer part residing in the interface region. To study the experimentally observed influence of phospholipids on nisin activity, a static approach was used. Different kinds of phospholipids were systematically assembled around the peptide by rigid body translation and rotation, choosing the position of best interaction between the peptide and the phospholipid, until it was fully surrounded by the lipids. In all cases the intermolecular interaction energies between the lipids were lowered by the presence of nisin, but only in case of PG did the assembled lipid adopt an arrangement leading to interface curvature.

Although this was an essentially static model based on optimized interactions between the peptide and the phospholipids treated as rigid bodies, the results agreed well with experimental data, showing that increasing PG content in the membrane was paralleled with a nisin-induced leakage.

6.2.5.2 All-atom Simulations

In some studies the effect of the phospholipid on the conformational behavior of peptides adjacent to, or inserted in, the membrane has been investigated. Using a model peptide of 32 alanine residues Shen *et al.* [79] investigated its conformational behavior in a membrane composed of 32 DMPC molecules. The peptide was inserted into an equilibrated bilayer in an ideal α-helical conformation, the system was equilibrated again, and the simulation was run for 1.6 ns under constant-volume conditions. The part of the poly-Ala peptide that was in the core of the membrane showed a stable helical conformation, while in the head group region the helix became less stable and fluctuated more, and those parts exposed to water possessed the greatest flexibility. During the simulation the tilt angle of the helical part increased from 10° after equilibration to an average of 32° over the simulation, allowing more of the hydrophobic alanine residues to interact with the hydrophobic core of the bilayer. Considerable movement of the whole peptide was also observed, especially in the direction of the membrane normal, leading to a non-symmetric placement of the polypeptide chain.

The authors reported that the hydrophobic peptide had little influence on the lipid structure, as lipid lateral diffusion rates, lipid conformations, and head group orientations were identical to a neat bilayer. The distance distribution of the phospholipid atoms surrounding the peptide was rather broad, pointing to the absence of special interactions between the peptide and the surrounding phospholipids.

A different influence on membrane order and dynamics was reported for an amphiphilic peptide in two investigations employing both ^{2}H-NMR measurements and molecular dynamics simulations [80]. Two model peptides were studied having the sequence

Ac-Lys-Lys-Gly-$[Leu]_n$-Lys-Lys-Ala-NH_2

with $n = 16$ and $n = 24$ to investigate the influence of hydrophobic mismatch on the structure and dynamics of the surrounding phospholipids. The simulations were performed by inserting the peptide into a rather small bilayer consisting of only 12 phospholipids, six in each leaflet, and run for 1 ns after equilibration of the initial system using NVE conditions. Based on the calculated order parameters of the alkyl chain carbons, a substantial ordering of the phospholipid alkyl chains was reported in MD simulations when the helical length of the peptide was greater than the lipid bilayer thickness. A comparison of the experimental ^{2}H-NMR order parameters reported in this study with those determined previously [26] showed some increase in the range of carbons 3–10 of the acyl chains, partly supporting the increased ordering found in the simulation. However, there was some discrepancy between the order parameters obtained from the simulation at 325 K and those determined by

^{2}H-NMR at a slightly higher temperature of 333 K as the calculated order parameters were somewhat higher than the experimental ones.

When interpreting the reported results, it must be kept in mind that the simulated system was very small, with only six lipid molecules surrounding the peptides in each leaflet; thus, all lipid molecules were in a direct contact with the peptide. Additionally, the ratio of peptide to lipid molecules was rather high, and thus these studies should be mainly considered as investigations of the influence of the lipid surrounding on the conformational preference of the peptides.

Melittin found in bee venom is another example of a peptide that has been investigated by computer simulation. It is a small amphipathic peptide of 26 amino acid residues with pronounced membrane-lytic properties, which are thought to be associated with an increase in membrane permeability. From neutron scattering and NMR experiments with specifically deuterated side chains, different binding modes of melittin depending on its protonation state have been proposed. While the peptide with an unprotonated N-terminus bound parallel to the membrane, upon protonation of the N-terminus binding occurred in a transbilayer way. However, the mechanism of lysis caused by melittin is still not completly clear, and different mechanisms are discussed.

Based on these experimental results, MD simulations of melittin were performed with both an unprotonated and a protonated N-terminus to study the molecular details of the lysis caused by melittin [81]. In the simulation, melittin with an unprotonated N-terminus was inserted into one leaflet of the membrane composed of DMPC and its influence on membrane structure was studied. To avoid possible problems with constant-pressure algorithms due to the anisotropy of the system, the simulation was performed under constant volume and energy conditions. Melittin caused a reduction in the order of the alkyl chains in the "upper" leaflet to which it was bound. The presence of the peptide led to a local curvature of the membrane with an increase in water penetration into the hydrocarbon core. After 600 ps of simulation the N-terminus was protonated and the simulation was continued for another 500 ps. The number of water molecules near the N-terminus in the hydrocarbon chain region began to increase and reached 15–25 water molecules forming a continuous cluster of water after only about 100 ps.

Later, Bachar and Becker [82] simulated the interaction of melittin with DPPC using constant pressure and area (NPAT) conditions. In this simulation, melittin with a protonated N-terminus was inserted into the membrane in a transbilayer orientation. The vertical position of the peptide was chosen in accordance with experimental data showing that the amino acid tryptophan-19 was positioned near the acyl groups of the alkyl chains. In this transbilayer orientation melittin did not span the whole bilayer, but only two-thirds, with the protonated N-terminus lying in the hydrophobic region of the "lower" leaflet of the membrane. During the simulation, the orientation of the peptide changed from perpendicular to the membrane surface at the beginning and it adopted a tilt angle of approximately 20°. In this simulation, melittin lowered the order of the alkyl chains in the "lower" leaflet, which was not spanned by it. This result seemed contradictory and was explained by the authors assuming that alkyl chains in close contact with a loosely bound peptide become less ordered, but otherwise the effect of the peptide on the ordering is negligible. Unfortunately, nothing was reported by the authors regarding changes in water penetration due to the presence of melittin.

6.2.6
Simulations of Pore-forming Peptides and of the Diffusion of Ions Through Ion Channels

The simulation of ion channels and other pore-forming peptides and proteins at atomic detail is nowadays also possible. With the increase in computational power, these complex systems have attracted much more interest, and several simulations have been reported. Very often, only the transmembrane segments of the channel-forming proteins are included in the simulation to reduce the size and complexity of the system. The simulated systems range from synthetic model ion channels to a bacterial porine protein.

Probably the first pore-forming peptide to be studied by computer simulation was gramicidin A. In 1994 a MD simulation of the gramicidin channel in a phospholipid bilayer was published [83], followed by a more detailed simulation in 1996 [84]. Some ordering of the lipid hydrocarbon chains in direct contact with gramicidin A was detected, but not uniformly, owing to the irregular surface shape of the protein. Decomposition of the interaction energies showed differences depending on the type of amino acid residue exposed to the lipid environment. While for leucine residues mainly van der Waals interactions were observed, for tryptophan residues both van der Waals and electrostatic interactions yielded nearly equal energetically favorable interactions.

Another even more intensively studied example is alamethicin (Alm), an α-helical channel-forming peptide that forms voltage-activated ion channels in lipid bilayers. It consists of 20 amino acids and exists as two major variants:

Ac-Aib-Pro-Aib-Ala-Aib-Ala-Gln-Aib-Val-Aib-Gly-Leu-Aib-Pro^{14}-Val-Aib-Aib-Glu^{18}-Gln-Phol

and a second form in which Glu^{18} is replaced by Gln. As seen from the sequence it contains many α-aminoisobutyric acid (Aib) residues, leading to a preferred helical conformation. Close to the center is a proline at position 14 that introduces a kink in the helix as observed in the X-ray structure. At the C-terminal it contains a phenylalaninol residue (Phol), and therefore no carboxyl group. Alamethicin has been investigated both for its membrane action and as a model for ion channels. It has been shown experimentally that the propensity to form channels is dependent on the transbilayer potential. The channel activation was suggested to correspond to a voltage-induced switch of Alm helices from a surface-associated to a bilayer-inserted orientation. The channels were assumed to be formed by parallel bundles of 5–8 transmembrane Alm helices.

Several simulations have been undertaken to study this interesting peptide in detail. In an early simulation, a mean field approximation was used representing the bilayer by a hydrophobic potential [85]. The starting structure was a surface-bound helix oriented in parallel to the surface of the membrane. After a few time steps, the N-terminus spontaneously moved and inserted into the bilayer, adopting a transmembrane orientation. The angle of the kink in the helix introduced by the proline decreased on insertion and in this way helped to span the bilayer. Because the simulation did not take into account the viscosity of the environment, the fast movement and switching was probably an artifact of the mean field approximation.

A number of MD simulations involving the atomic detail of lipid and water have been performed, with the aim of exploring the different stages of channel formation. The stability of the Alm helix both in water and bound to the surface has been studied and compared. The first system consisted of one negatively charged Alm plus a sodium ion in a box of water [87]. A second simulation used a fully equilibrated bilayer of 128 POPC molecules [86]. Alm was placed close to the interface and the helix was oriented to be parallel to the bilayer surface. The system was hydrated and a sodium ion added. MD simulations under NPT conditions were performed for 2 ns. In pure water, large conformational changes about the Gly-Leu-Aib-Pro hinge and in the C-terminal part were observed showing the flexibility of the peptide in water. In the second half of the simulation, at approximately 1.5 ns, the peptide folded back on itself, adopting a more globular conformation that remained for the rest of the simulation. This illustrates that larger conformational changes require long simulation times and will probably not be observed in short simulations even if the final conformation is energetically preferred and more stable. In contrast to the simulation in water, the loosely surface-bound Alm showed far less conformational changes and the helical conformation remained stable over the whole simulation in the presence of POPC. Although the duration of simulation was definitely too short to allow major movement of Alm from a loosely bound to a more tightly bound or even inserted orientation, it showed that the conformation of a membrane-active peptide can be stabilized by a lipid bilayer, even when its hydrophobic part does not penetrate into the membrane.

The conformational behavior of Alm inserted in a POPC bilayer has been investigated using the same system as described above [87]. To insert the Alm molecule, a hole was introduced by restraints and one POPC molecule was removed. The Alm was placed into the bilayer, the system was hydrated, and one sodium ion added. A second simulation was performed in which no sodium ion was added and Glu^{18} was protonated instead. During the simulation period of 1000 ps, the Alm molecule remained in its α-helical conformation, showing only small fluctuations in the root mean square distance (RMSD) of the $C\alpha$ atoms underlining its conformational stabilization by the lipid bilayer.

Even a pore composed of six Alm molecules in a lipid bilayer has been simulated recently [88]. As the ionization state of the Glu residues within the pore was questionable, two simulations were performed and compared: one with negatively charged Glu^{18} residues and a second in which these residues were protonated and uncharged. In the case of the negatively charged Glu residues the system was made neutral by adding six sodium ions. In both simulations, mainly the C-terminal of the helices underwent conformational fluctuations. These were more pronounced, as observed for a single Alm helix in POPC, but far less than in water [87]. The helices showed different variations in their conformational behavior, leading to a loss of the exact sixfold symmetry. The pore formed by the neutral Glu residues was found to be more stable, in agreement with pK_a calculations suggesting that either none, or one Glu-residue is likely to be ionized. In case of the pore formed by the negatively charged Glu residues a tendency to expand during the simulation was observed, leading to a considerable disruption of the packing of the C-terminal end, in order to allow to move the negatively charged Glu side chains away from each other.

The altered dynamics of water molecules within the pore clearly emerged from the simulations. A very interesting result is that water molecules within the pore were found to be highly oriented by the electrostatic field created by the parallel dipoles of the α-helices. As found already in earlier *in vacuo* simulations [89] the orientation of the water molecules led to strong interactions and significantly contributed to the stability of the pore. For the pore formed by Alm with neutral Glu-residues the calculated single-channel conductance was found to be in good agreement with experimental values, providing evidence that such simulations can be of great help in understanding the water and ion transport through pores at a microscopic level.

The structure of pores formed by two synthetic leucine-serine ion channels has been investigated by Randa *et al.* [90]. The systems consisted of tetrameric bundles of peptides with the sequence:

Leu-Ser-Leu-Leu-Leu-Ser-Leu-Leu-Ser-Leu-Leu-Leu-Ser-Leu-Leu-Ser-Leu-Leu-Leu-Ser-Leu-NH_2

known as LS2, and hexameric bundles of

Leu-Ser-Ser-Leu-Leu-Ser-Leu-Leu-Ser-Ser-Leu-Leu-Ser-Leu-Leu-Ser-Ser-Leu-Leu-Ser-Leu-NH_2

known as LS3. These ion channels were shown to possess different ion selectivities.

The helix bundles were embedded in a pre-equilibrated lipid bilayer of about 100 POPC molecules. After equilibration the systems were simulated for 4 ns under NPT conditions.

During the simulation a column of water molecules was formed in each of the pores. In both pores, the water molecules showed reduced movement but with differences in their diffusion coefficient. In the pore formed by the hexameric LS3 the diffusion coefficient was greatly reduced to about 10% of bulk water, but the water remained mobile enough to allow diffusion of monovalent cations. A conductance of 64 pS was calculated for the simulated LS3 pore from its average radius, the diffusion coefficient of water, and the resistivity of 0.5 M KCl, in excellent agreement with experimental values of 70 pS.

In contrast to the LS3 pore, the water molecules were "frozen" in the tetrameric LS2 pore, with diffusion coefficients of zero. They were found to be aligned antiparallel to the helix dipole caused by the orientation of the hydroxyl groups of the serine residues. This enabled formation of a water wire network important for the transport of protons by the proton wire or Grotthüs mechanism.

The behavior of the influenza A M2 proton-selective channel has been investigated by MD simulation using tetrameric bundles of the transmembrane domains [91].

The simulations were performed under conditions corresponding to the presumed closed form of the channel. During the simulation the channel underwent breathing motions, leading to alternating structures of a nearly quadratically symmetric tetramer and an elongated configuration with two short and two long distances between adjacent helix bundles, which could be termed a dimer of dimers. The water in the pore was either trapped in a pocket or formed a water channel that was broken at the C-terminal end of the pore. Similar to the results obtained for the tetrameric LS2 channel, the movement of the water molecules inside the pore was greatly restricted. Though the different simulations were run for 2 ns, it was noted that the results were sensitive to the exact starting structure.

Recently, simulations of a bacterial potassium channel KcsA have been reported by two research groups [92, 93]. The X-ray structure of the core region of the protein served as starting point. A series of five simulations of the KcsA channel in a POPC bilayer were performed, that differed in the number and position of potassium ions included, ranging from zero to three ions [92]. Owing to the applied cut-off of 17 Å for electrostatic interactions in the simulation with three ions not all of them could "see" each other. As mentioned by the authors, one must retain a degree of caution in interpretation of the obtained results. The systems were simulated for 1 ns, about an order shorter than the time necessary for the diffusion of an ion through a pore. Also, in this case, breathing motions of the protein were observed and in one of the simulations a potassium ion left the channel.

In all five simulations, no significant differences in overall RMSD were observed, suggesting that the presence or absence of potassium ions had no effect on the overall conformation of the protein. However, a distinct influence of potassium ions on the conformation of the part of the helix that acts as a selectivity filter was found. The simultaneous presence of two potassium ions stabilized the conformation observed in the X-ray structure. The potassium ions and water molecules within the pore showed a concerted movement of a water–K^+–water–K^+ column on a time scale of several hundred picoseconds.

In the other study, the channel was simulated in a DPPC bilayer and potassium and chloride ions were added corresponding to an ionic strength of 150 mM [93]. Two potassium ions were inserted into the selectivity filter region of the channel in four configurations and a third one in the central cavity of the channel. The first configuration corresponded to that observed in the X-ray structure, while the others differed in the presence of additional water molecules between the two potassium ions. Two of the alternative configurations were found to be unstable during the equilibrium and were not further simulated, while the third one was stable. Each of the two systems was simulated at 315 K, the phase transition temperature of DPPC, and additionally at 330 K. Electrostatic interactions were taken into account by the particle mesh Ewald summation technique. A transition of a potassium ion in the channel accompanied by a concerted movement of water–K^+ was also observed in this simulation. The alternative configuration, with two water molecules separating the two potassium ions, converted into the X-ray configuration after some hundreds of picoseconds. From the analysis of the trajectories of the different investigated configurations, the authors concluded that the two most stable states have two and three K^+ ions separated by single water molecules in the selectivity filter of the channel.

Models of the nicotinic acetylcholine receptor consisting of the five M2 helices that form the pore have been investigated in a series of MD simulations [94]. The behavior of single helices in water and in lipid was compared, and the α-helical structure was again found to be stabilized by the membrane. Two simulations of bundles of five M2 helices forming a pore and differing in the ionization state of the ionizable amino acid residues were performed. When these residues were included in their ionized states, the bundle expanded over the 2 ns of the simulation, while it remained stable if calculated pK_a values were used to determine the ionization states of ionizable residues. In these simulations structural fluctuations were also observed on a several hundred picosecond time scale that transiently closed the pore.

The behavior of such a large system as a pore formed by a bacterial porine (*E. coli* OmpF) has been simulated in a lipid bilayer of palmitoyloleoylphosphatidylethanolamine (POPE) [95]. Despite the use of united atoms, the final system of the trimeric porin embedded into 318 POPE molecules and solvated with water consisted of more than 65 000 atoms in total. During the 1 ns of the MD simulation the trimeric structure remained stable, with almost all flexibility in the loops and turns outside the β-strands. The movement and orientation of the water molecules was investigated in detail. As found in case of the pore formed by the hexameric LS3 helix bundle [90], the diffusion of the water was decreased to about 10% of that of bulk water. Some ordering of the water molecules was evident from the average water dipole moments, which showed a strong dependence on the vertical position within the porine.

6.2.7 Non-equilibrium Molecular Dynamics Simulations

Partial lipid extraction is considered as biologically important in several processes involving membranes, for example the fusion of membranes or enzymatic reactions with participation of phospholipids. To simulate this process, non-equilibrium or steered MD simulations can be used. This type of simulation was employed for the first time by Grubmüller *et al.* [96] to study the rupture of the binding in the streptavidin–biotin complex. Inspired by this pioneering work, similar studies have been undertaken to characterize the extraction of a phospholipid from the membrane.

Stepaniants *et al.* [97] investigated the extraction of a dilauroylphosphatidylethanolamine (DLPE) molecule from a DLPE monolayer in three systems. In the first case, the DLPE was removed from the membrane into the water phase, while in the other two simulations it was pulled from the membrane into the active site of phospholipase A_2 (PLA_2) forming either a loose or a tight complex with the DLPE monolayer. The authors found that the forces required to displace the lipid into the binding pocket of PLA_2 were similar for both complexes and larger than those required to pull the lipid into the water phase. This is not in agreement with the current hypothesis about the interaction of PLA_2 with phospholipids, which assumes a destabilizing effect of the enzyme [98]. A possible explanation for this rather unexpected result might be the steric hindrance experienced by the lipid on its way into the active site, and the formation of H-bond networks as a result of the presence of hydrophobic residues surrounding the entrance to the binding pocket. Because of the relatively short simulation time of 500 ps for the displacement, the lipid had to be displaced with a relatively high velocity of 14 Å/ns to be at least partially extracted from the membrane. Indeed, this could be the reason for the unexpected result as was shown in several longer simulation using different displacement speeds.

In the simulations performed by Marrink *et al.* [99] the force required to displace two DPPC molecules into the water phase from both sides of a DPPC bilayer simultaneously was recorded as a function of the displacement speed. The positions of the extracted DPPC molecules were chosen in such a way that the distance between them was greater than the applied cut-off so that no direct interactions could occur. Going from very fast displacement to the slowest movement speeds of 2 Å/ns resulted in an exponential decrease in the force necessary for displacement. The simulations were run for times allowing full transfer of the DPPC from the lipid to the water phase (40 Å), except for the slowest one, in which case even after 4 ns the DPPC molecules were only partially displaced into the water phase. Another observation of interest is the conformational change that the lipids underwent upon extraction from the membrane. When moving out of the membrane the number of *gauche* conformations of the alkyl chains decreased from approximately four per lipid tail to 2.5–2.7, depending on the speed of movement. This decrease occurred at low extraction speed until a displacement of about 20 Å, at which point the lipid was about to leave the membrane completely. After that the lipids began to adopt a more globular, folded conformation, as evidenced by the increase in the number of *gauche* conformations. At high extraction speed, the number of *gauche* conformations remained low. These results show the large influence of displacement speed on the results of the simulation and emphasize the importance of using conditions as close as possible to biological and/or experimental ones. However, in such cases as just described, this is still limited by the current computational resources.

6.3 Concluding Remarks

Computer simulations of lipids have progressed from simple to atomic level representations during the past 15–20 years. During this period, impressive improvements in simulation time and the complexity of the simulated systems have been achieved. Comparisons between simulations and experimental results have shown that the calculations are generally in good agreement with measurements and can provide additional insight into details that are unavailable experimentally at the molecular level. Simulations of similar systems performed by several research groups using different MD programs and settings have also led to concordant results, suggesting that such simulations can model experimental systems. Simulations have progressed to the point at which valuable insight into the molecular details of membranes and their interactions with drugs and proteins has been achieved. Non-equilibrium molecular dynamics simulation allow the investigation of events that are too rare or take place on a longer time scale. Despite the improvements, the simulation of lipids and lipid–protein systems remains a challenging area. One obvious need is to increase the length of the simulated time scale by at least one order of magnitude to allow the investigation of slower processes and to provide better statistics on processes already investigated by computer simulation. Another requirement is a significant increase in the size of the simulated system, in order to study phenomena present only over larger distances. With continuing increases in computer speed it might be possible to address these points in the near future and to perform more routinely simulations in which the influence of parameters determining the structure and dynamic properties of lipids can be varied and compared with experimental results in a systematic way.

References

1 Lins, L., Brasseur, R., FASEB J. **1995**, *9*, 535–540.
2 Lins, L., Ducarme, P., Breukink, E., Brasseur, R., *Biochim. Biophys. Acta* **1999**, *1420*, 111–120.
3 Woolf, T.B., *Biophys. J.* **1998**, *74*, 115–131.
4 La Rocca, P., Shai, Y., Samson, M.S.P., *Biophys. Chem.* **1999**, *76*, 145–159.
5 Weiner, S.J., Kollman, P.A., Case, D.A., Singh, U.C., Ghio, C., Alagona, G., Profeta, S., Weiner, P., *J. Am. Chem. Soc.* **1984**, *106*, 765–784.
6 Brooks, B.R., Bruccoleri, R.E., Olafson, B.D., States, D.J., Swaminathan, S., Karplus, M., *J. Comput. Chem.* **1983**, *4*, 187–217.
7 MacKerrel, A.D.Jr., Brooks, B., Brooks, C.L. III., Nilsson, L., Roux, B., Won, Y., Karplus, M., in *Encyclopedia of Computational Chemistry*, Vol. I, P.v.R. Schleyer, N.L.A.T. Clark, J. Gasteiger, P.A. Kollman, H.F. III Schaefer, P.R.S. Schreiner (Eds.), John Wiley & Sons, Chichester **1998**, pp. 271–277.
8 van Gunsteren, W.F., Berendsen, H.J.C., *Groningen Molecular Simulation (GROMOS) Library manual*, Biomos, Nijenborgh 4, 9747 AG Groningen, The Netherlands **1987.**
9 Molecular Simulations, San Diego, CA.
10 Jørgensen, W., Tirado-Rives, J., *J. Am. Chem. Soc.* **1988**, *110*, 1657–1666.
11 Egberts, E., Marrink, S.-J., Berendsen, H.J.C., *Eur. Biophys. J.* **1994**, *22*, 423–436.
12 Stouch, T.R., Ward, K.B., Altieri, A., Hagler, A.T., *J. Comp. Chem.* **1991**, *12*, 1033–1046.
13 Williams, D.E., Stouch, T.R., *J. Comp. Chem.* **1993**, *14*, 1066–1076.
14 Liang, C., Ewig, C.S., Stouch, T.R., Hagler, A.T., *J. Am. Chem. Soc.* **1993**, *115*, 1537–1545.
15 Schlenkrich, M., Brickmann, J., MacKerrel Jr., A.D., Karplus, M., in *Biological Membranes: A Molecular Perspective from Computation and Experiment*, K.M. Merz, B. Roux (Eds.), Birkhäuser, Boston **1996**, pp. 31–81.
16 Feller, S.E., Yin, D., Pastor, R.W., MacKerell Jr., A.D., *Biophys. J.* **1997**, *73*, 2269–2279.
17 Feller, S.E., MacKerrel Jr., A.D., *J. Phys. Chem.* **2000**, *104*, 7510–7515.
18 Vergoten, G. *Biospectroscopy* **1998**, *4*, (Suppl.) S41–S46.
19 Müller-Plathe, F., Rogers, B.C., van Gunsteren, W.F., *Chem. Phys. Lett.* **1992**, *199*, 237–243.
20 Berger, O., Edholm, O., Jähnig, F., *Biophys. J.* **1997**, *72*, 2002–2013.
21 Smondyrev, A.M., Berkowitz, M.L., *J. Comp. Chem.* **1999**, *20*, 531–545.
22 *Applications of the Monte Carlo Method in Statistical Physics*, K. Binder (Ed.), Springer Verlag, Heidelberg **1984.**
23 Ryckaert, J.-P:, Ciccoti, G., Berendsen, H.J.C., *J. Comput. Phys.* **1977**, *23*, 327–341.
24 Seelig, A., Seelig, J., *Biochemistry* **1974**, *13*, 4839–4845.
25 Burnett, L.J., Müller, B.H., *J. Chem. Phys.* **1971**, *55*, 5829–5831.
26 Douliez, J.-P., Leonard, A., Dufourc, E.J., *Biophys. J.* **1995**, *68*, 1727–1739.
27 Takaoka, Y., Pasenkiewicz-Gierula, M., Miyagawa, H., Kitamura, K., Tamura, Y., Kusumi, A., *Biophys. J.* **2000**, *79*, 3118–3138.
28 Darden, T., York, D., Pedersen, L., *J. Chem. Phys.* **1993**, *98*, 10089–10092.
29 Essman, U., Perera, L., Berkowitz, M.L., Darden, T., Lee, H., Pedersen, L.G. J. Chem. Phys. **1995**, *103*, 8577–8593.
30 Tasaki, K., McDonald, S., Brady, J.W., *J. Comput. Chem.* **1993**, *14*, 278–284.
31 Alper, H.E., Bassolino, D., Stouch, T.R., *J. Chem. Phys.* **1993**, *98*, 9798–9807.
32 Alper, H.E., Bassolino-Klimas, D., Stouch, T.R., *J. Chem. Phys.* **1993**, *99*, 5547–5559.
33 Feller, S.E., Pastor, R.W., Rojuckarin, A., Bogusz, S., Brooks, B.R., *J. Phys. Chem.* **1996**, *42*, 17011–17020.
34 Feller, S.E., Venable, R.M., Pastor, R.W., *Langmuir* **1997**, *13*, 6555–6561.
35 Schneider, M.J., Feller, S.E., *J. Phys. Chem.* **2001**, *105*, 1331–1337.

36 Zhang, Y., Feller, S.E., Brooks, B.R., Pastor, R.W., *J. Chem. Phys.* **1995**, *103*, 10252–10266.

37 Marrink, S.-J., Berkowitz, M., Berendsen, H.J.C., *Langmuir* **1993**, *9*, 3122–3131.

38 Huang, P., Perez, J.J., Loew, G.H., *J. Biomol. Struct. Dyn.* **1994**, *11*, 927–956.

39 Chiu, S.-W., Clark, M., Balaji, V., Subramaniam, S., Scott, H.L., Jakobsson, E., *Biophys. J.*, **1995**, *69*, 1230–1245.

40 Mauk, A.W., Chaikof, E.L., Ludovice, P.J., *Langmuir* **1998**, *14*, 5255–5266.

41 Feller, S.E., Pastor, R.W., *Biophys. J.* **1996**, *71*, 1350–1355.

42 Feller, S.E., Pastor, R.W., *J. Chem. Phys.* **1999**, *111*, 1281–1287.

43 Tieleman, D.P., Marrink, S.J., Berendsen, H.J.C., *Biochim. Biophys. Acta* **1997**, *1331*, 235–270.

44 Tobias, D., Tu, J.K., Klein, M.L., *Curr. Opin. Coll. Int. Sci.* **1997**, *2*, 15–26.

45 Sanders, C.R.II., Schwonek, J.P., *Biophys. J.* **1993**, *65*, 1207–1218.

46 Viswanadhan, V.N., Ghose, A.K., Revankar, G.R., Robins, R.K., *J. Chem. Inf. Comput. Sci.* **1989**, *29*, 163–172.

47 Kamlet, M.J., Doherty, R.M., Abraham, M.H., Marcus, Y., Taft, R.W., *J. Phys. Chem.* **1988**, *92*, 5244–5255.

48 Xiang, T.-X., Anderson, B.D., *Biophys. J.* **1994**, *66*, 561–573.

49 Sheng, Q., Schulten, K., Pidgeon, C., *J. Phys. Chem.* **1995**, *99*, 11018–11027.

50 Bassolino-Klimas, D., Alper, H.E., Stouch, T.R., *Biochemistry* **1993**, *32*, 12624–12637.

51 Bassolino-Klimas, D., Alper, H.E., Stouch, T.R., *J. Am. Chem. Soc.* **1995**, *117*, 4118–4129.

52 Bassolino, D., Alper, H., Stouch, T.R., *Drug Design Discov.* **1996**, *13*, 135–141.

53 Marrink, S.J., Berendsen, H.J.C., *J. Phys. Chem.* **1996**, *100*, 16729–16738.

54 Marrink, S.J., Jähnig, F., Berendsen, H.J.C., *Biophys. J.* **1996**, *71*, 632–647.

55 Paci, E., Marchi, M., *Mol. Simul.* **1994**, *14*, 1–10.

56 Smondyrev, A.M., Berkowitz, M.L., *Biophys. J.* **1999**, *76*, 2472–2478.

57 Craig, N.C., Bryant, G.J., Levin, I.W., *Biochemistry* **1987**, *26*, 2449–2458.

58 Huang, P., Bertaccini, E., Loew, G.H., *J. Biomol. Struct. Dyn.* **1995**, *12*, 725–754.

59 Tu, K., Tarek, M., Klein, L., Scharf, D., *Biophys. J.* **1998**, *75*, 2123–2134.

60 Alper, H.E., Stouch, T.R., *J. Phys. Chem.* **1995**, *99*, 5724–5731.

61 Aiello, M., Moran, O., Pisciotta, M., Gambale, F., *Eur. Biophys. J.* **1998**, *27*, 211–218.

62 Scott, H.L., Kalaskar, S., *Biochemistry*, **1989**, *28*, 3687–3692.

63 Scott, H.L., *Biophys. J.* **1991**, *59*, 445–455.

64 Edholm, O., Nyberg, A.M., *Biophys. J.* **1992**, *63*, 1081–1089.

65 Jørgensen, K., Ipsen, J.H., Mouritsen, O.G., Benett, D., Zuckermann, M.J., *Biochim. Biophys. Acta* **1991**, *1062*, 227–238.

66 Jørgensen, K., Ipsen, J.H., Mouritsen, O.G., Zuckermann, M.J., *Chem. Phys. Lipids* **1993**, *65*, 205–216.

67 Mouritsen, O.G., Jørgensen, K., *Chem. Phys. Lipids* **1994**, *73*, 3–25.

68 Mouritsen, O.G., Dammann, B., Fogedby, H.C., Ipsen, J.H., Jeppesen, C., Jørgensen, K., Risbo, J., Sabra, M.C., Sperotto, M.M., Zuckermann, M.J., *Biophys. Chem.* **1995**, *55*, 55–68.

69 Robinson, A.J., Richards, W.G., Thomas, P.J., Hann, M.M., *Biophys. J.* **1995**, *68*, 164–170.

70 Gabdouline, R.R., Vanderkooi, G., Zheng, C., *J. Phys. Chem.* **1996**, *96*, 15942–15946.

71 Pasenkiewicz-Gierula, M., Rog, T., Kitamura, K., Kusumi, A., *Biophys. J.* **2000**, *78*, 1376–1389.

72 Tu, K., Klein, M.L., Tobias, D.J., *Biophys. J.* **1998**, *75*, 2147–2156.

72 Smondyrev, A.M., Berkowitz, M.L., *Biophys. J.* **1999**, *77*, 2075–2089.

74 Pasenkiewicz-Gierula, M., Takaoka, Y., Miyagawa, H., Kitamura, K., Kusumi, A., *J. Phys. Chem.* **1997**, *101*, 3677–3691.

75 Smondyrev, A.M., Berkowitz, M.L., *Biophys. J.* **2000**, *78*, 1672–1680.

76 Milik, M., Skolnick, J., *Proc. Natl. Acad. Sci. USA* **1992**, *89*, 9391–9395.

77 Milik, M., Skolnick, J., *Proteins: Struct. Funct. Genet.* **1993**, *15*, 10–25.

78 La Rocca, P., Biggin, P.C., Tieleman, D.P., Sansom, M.S.P., *Biochim. Biophys. Acta* **1999**, *1462*, 185–200.
79 Shen, L., Bassolino, D., Stouch, T., *Biophys. J.* **1997**, *73*, 3–20.
80 Belohorcova, K., Qian, J., Davis, J.H., *Biophys. J.* **2000**, *79*, 3201–3216.
81 Berneche, S., Nina, M., Roux, B., *Biophys. J.* **1998**, *75*, 1603–1618.
82 Bachar, M., Becker, O.M., *Biophys. J.* **2000**, *78*, 1359–1375.
83 Woolf, T.B., Roux, B., *Proc. Natl. Acad. Sci. USA* **1994**, *91*, 11631–11635.
84 Woolf, T.B., Roux, B., *Proteins: Struct. Funct. Genet.* **1996**, *24*, 92–114.
85 Biggin, P., Breed, J., Son, H.S., Sansom, M.S.P., *Biophys. J.* **1997**, *72*, 627–636.
86 Tieleman, D.P., Berendsen, H.J.C., Sansom, M.S.P. *Biophys. J.* **1999**, *76*, 3186–3191.
87 Tieleman, D.P., Sansom, M.S.P., Berendsen, H.J.C., *Biophys. J.* **1999**, *76*, 40–49.
88 Tieleman, D.P., Berendsen, H.J.C., Sansom, M.S.P., *Biophys. J.* **1999**, *76*, 1757–1769.
89 Breed, J., Sankararamakrishnan, R., Kerr, I.D., Sansom, M.S.P., *Biophys. J.* **1996**, *70*, 1643–1661.
90 Randa, H.S., Forrest, L.R., Voth, G.A., Sansom, M.S.P., *Biophys. J.* **1999**, *77*, 2400–2410.
91 Forrest, L.R., Kukol, A., Arkin, I.T., Tieleman, D.P., Sansom, M.S.P. *Biophys. J.* **2000**, *78*, 55–69.
92 Shrivastava, I.H., Sansom, M.S.P., *Biophys. J.* **2000**, *78*, 557–570.
93 Berneche, S., Roux, B., *Biophys. J.* **2000**, *78*, 2900–2917.
94 Law, R.J., Forrest, L.R., Ranatunga, K.M., La Rocca, P., Tieleman, D.P., Sansom, M.S.P., *Proteins: Struct. Funct. Genet.* **2000**, *39*, 47–55.
95 Tieleman, D.P., Berendsen, H.J.C., *Biophys. J.* **1998**, *74*, 2786–2801.
96 Grubmüller, H., Heymann, B., Tavan, P., *Science* **1996**, *271*, 997–999.
97 Stepaniants, S., Izrailev, S., Schulten, K., *J. Mol. Model.* **1997**, *3*, 473–475.
98 Zhou, F., Schulten, K., *Proteins: Struct. Funct. Genet.* **1996**, *25*, 12–27.
99 Marrink, S.-J., Berger, O., Tieleman, P., Jähnig, F., *Biophys. J.* **1998**, *74*, 931–943.
100 Douliez, J.P., Leonard, A., Dufourc, E.J., *J. Phys. Chem.* **1996**, *100*, 18450–18457.
101 Burnett, L.J., Müller, B.H., *J. Chem. Phys.* **1971**, *55*, 5829–5831.
102 Dufourc, E.J., Smith, I.C.P., Jarrel, H.C., *Biochim. Biophys. Acta* **1984**, *776*, 317–329.
103 Urbina, J.A., Pekerar, S., Le, H., Patterson, J., Montez, B., Oldfield, E., *Biochim. Biophys. Acta*, **1995**, *1238*, 163–176.
104 Douliez, J.P., Ferrarini, A., Dufourc, E.J., *J. Chem. Phys.* **1998**, *109*, 2513–2518.
105 Engel, A.K., Cowburn, D., *FEBS Lett.* **1981**, *126*, 169–171.
106 Pasenkiewicz-Gierula, M., Takaoka, Y., Miyagawa, H., Kitamura, K., Kusumi, A., *Biophys. J.* **1997**, *76*, 1228–1240.

Index

b

d

e